AF545009

INTERACTIONS OF MICROORGANISMS WITH RADIONUCLIDES

RADIOACTIVITY IN THE ENVIRONMENT

A companion series to the Journal of Environmental Radioactivity

Series Editor
M.S. Baxter
Ampfield House
Clachan Seil
Argyll, Scotland, UK

Volume 1: Plutonium in the Environment – Edited Proceedings of the Second Invited International Symposium (A. Kudo, Editor)

Volume 2: Interactions of Microorganisms with Radionuclides (M.J. Keith-Roach and F.R. Livens, Editors)

INTERACTIONS OF MICROORGANISMS WITH RADIONUCLIDES

Editors

M. J. Keith-Roach

Risø National Laboratory, Roskilde, Denmark

F. R. Livens

University of Manchester, Manchester, UK

ELSEVIER

2002

AMSTERDAM – BOSTON – LONDON – NEW YORK – OXFORD – PARIS
SAN DIEGO – SAN FRANCISCO – SINGAPORE – SYDNEY – TOKYO

ELSEVIER SCIENCE Ltd
The Boulevard, Langford Lane
Kidlington, Oxford OX5 1GB, UK

© 2002 Elsevier Science Ltd. All rights reserved.

This work is protected under copyright by Elsevier Science, and the following terms and conditions apply to its use:

Photocopying
Single photocopies of single chapters may be made for personal use as allowed by national copyright laws. Permission of the Publisher and payment of a fee is required for all other photocopying, including multiple or systematic copying, copying for advertising or promotional purposes, resale, and all forms of document delivery. Special rates are available for educational institutions that wish to make photocopies for non-profit educational classroom use.

Permissions may be sought directly from Elsevier Science Global Rights Department, PO Box 800, Oxford OX5 1DX, UK; phone: (+44) 1865 843830, fax: (+44) 1865 853333, e-mail: permissions@elsevier.co.uk. You may also contact Global Rights directly through Elsevier's home page (http://www.elsevier.com), by selecting 'Obtaining Permissions'.

In the USA, users may clear permissions and make payments through the Copyright Clearance Center, Inc., 222 Rosewood Drive, Danvers, MA 01923, USA; phone: (978) 7508400, fax: (978) 7504744, and in the UK through the Copyright Licensing Agency Rapid Clearance Service (CLARCS), 90 Tottenham Court Road, London W1P 0LP, UK; phone: (+44) 171 631 5555; fax: (+44) 171 631 5500. Other countries may have a local reprographic rights agency for payments.

Derivative Works
Tables of contents may be reproduced for internal circulation, but permission of Elsevier Science is required for external resale or distribution of such material.
Permission of the Publisher is required for all other derivative works, including compilations and translations.

Electronic Storage or Usage
Permission of the Publisher is required to store or use electronically any material contained in this work, including any chapter or part of a chapter.

Except as outlined above, no part of this work may be reproduced, stored in a retrieval system or transmitted in any form or by any means, electronic, mechanical, photocopying, recording or otherwise, without prior written permission of the Publisher.
Address permissions requests to: Elsevier Science Global Rights Department, at the mail, fax and e-mail addresses noted above.

Notice
No responsibility is assumed by the Publisher for any injury and/or damage to persons or property as a matter of products liability, negligence or otherwise, or from any use or operation of any methods, products, instructions or ideas contained in the material herein. Because of rapid advances in the medical sciences, in particular, independent verification of diagnoses and drug dosages should be made.

ISBN: 0-08-043708-7
First edition 2002

British Library Cataloguing in Publication Data
A catalog record from the British Library of Congress has been applied for.

Library of Congress Cataloging in Publication Data
A catalog record from the Library of Congress has been applied for.

∞ The paper used in this publication meets the requirements of ANSI/NISO Z39.48-1992 (Permanence of Paper).

Printed in The Netherlands.

CONTENTS

Preface

Many environmental processes are influenced, if not controlled, by microbial action and it is becoming increasingly important to develop an understanding of microbial rôles in geochemistry. Focussing on environmental radioactivity, it may initially be questioned whether understanding these interactions will really be of any practical or predictive benefit? Are microbial effects on radionuclides really that important? This book goes some way to answering these questions, bringing together a representation of the state of the art research investigating these interactions and the extent to which they affect or can be used to control radioactive elements. The basic principles and fundamental mechanisms by which microbes and radionuclides interact are outlined, the methodology described, potential microbial influences on waste repositories examined, direct and indirect effects on transport both on local and global scales considered and potential technological applications identified. Here, we briefly consider these topics and explain why this emerging field will be of increasing importance in the future.

To start with the overall effect of a microbial population on its environment has to be considered, and the perturbations induced by changes in redox potential, the forms and abundance of organic matter and complexing ligands, and the chemistry of stable elements, such as iron and manganese. Such indirect changes to these will inevitably affect contaminant radionuclides and, perhaps of more specific relevance, direct interactions involving particular radionuclides have been identified, including, for example, the exploitation of radionuclides as terminal electron acceptors in respiration, sorption to cell walls or incorporation into biominerals or sequestration of the nuclide internally within a cell. The relative importance of these mechanisms depends on the physicochemical environment, and on the properties of the radionuclides and microbial population. In terms of waste disposal, microbial interactions with the containment materials are also very important, in that leaching and transport becomes much more likely as the physical barriers are eroded. Moreover, in heavily polluted environments such as uranium mine wastes, where the microbes which have adapted to living there may affect uranium in a very different way to those living in systems containing only trace concentrations of uranium. The abundance and diversity of microorganisms and their ability to adapt to different conditions and survive are therefore all extremely important in assessing the extent or nature of the induced changes. Obtaining a very detailed understanding of the biochemistry of microbial interactions with radionuclides may also lead to new technologies which reduce the activity released into the environment, and are potentially simpler, cheaper and more effective that chemical technologies

Taking the example of uranium, which is discussed in detail in this book, specific metabolic interactions have been characterised at the molecular level and biochemical mechanisms of uranium resistance have been identified. The effects of different microbial interactions in complex environmental systems and global-scale cycling processes have also been identified. This has aided research that has identified ways of potentially manipulating microbes in situ to provide novel, safe remediation techniques for uranium-contaminated land. Gaining a true understanding of these interactions and their relevance is therefore extremely important, not just as academic curiosities, but in order to handle and predict the consequences of radioactive waste and contamination.

Since the interactions of microorganisms, radionuclides and the environment can be viewed on many different levels, we have tried to assemble a complete picture of current knowledge relating to the environmentally relevant interactions of microbes with radionuclides. These range from removal of radionuclides at source to understanding their fate and behaviour in the environment, or from understanding interactions in simplified, model systems in fine detail to measuring net effects in complex, heterogeneous systems. Microbial activity and diversity is considered in different systems, including sediments, uranium waste piles and potential waste repositories, as this is fundamental to achieving a predictive understanding in this area. This research area clearly depends heavily on a range of disciplines and illustrates the strength of multidisciplinary research, which has produced a wealth of information. We hope that this book helps bridge gaps between workers in this area, and stimulates the interest of others with relevant knowledge and skills.

Miranda J. Keith-Roach
Risø National Laboratory
DENMARK

Francis R. Livens
University of Manchester
UK

INTERACTIONS OF MICROORGANISMS WITH RADIONUCLIDES
Miranda J. Keith-Roach and Francis R. Livens (Editors)
© 2002 Elsevier Science Ltd. All rights reserved

Chapter 1

Natural microbial communities

Clare H. Robinson, Kenneth D. Bruce

Division of Life Sciences, King's College, University of London, Franklin-Wilkins Building, 150 Stamford Street, London SE1 9NN, UK

1. Introduction

Microbes, or microorganisms, are a large and diverse group of microscopic organisms that exist as single cells or cell clusters; this group also includes viruses that are microscopic but not cellular (Madigan et al., 2000). Thus, the concept of the microorganism is mainly morphological and has limited phylogenetic significance (Zavarzin, 1995). One fundamental classification within microbes, into either eukaryotes or prokaryotes, depends on whether they possess or lack a nuclear membrane. In this chapter, the prokaryotic and eukaryotic microorganisms present in natural environments are detailed, and their roles in ecological processes are examined.

Basic descriptions

True fungi belong to the Kingdom Fungi in the Eukaryota, and are unicellular or filamentous, usually consisting of multicellular hyphae which are collectively called the mycelium. They can reproduce sexually or asexually, and have roles in ecosystems as decomposers (saprotrophs), mutualists in mycorrhizas and lichens, or parasites of animals and plants. A mycorrhiza is a mutualistic symbiosis between a fungus and a plant root. Three widespread types are recognised: vesicular-arbuscular (VAM), ecto- and ericoid mycorrhizas. Excellent reviews of the classification and morphology of these mycorrhizal types are provided by Smith & Read (1997). The unifying function of mycorrhizas is that external hyphae supply soil-derived nutrients to the plant. Detailed roles of mycorrhizal fungi in temperate biomes, in agriculture, horticulture and managed environments are provided by Smith & Read (1997). Mycorrhizas are potentially very important in remediation of contaminated land (e.g. Meharg & Cairney, 2000).

For the purposes of this chapter, it is worth noting two fungal groupings, the basidiomycetes and the microfungi. Basidiomycetes are members of the phylum Basidiomycota, the diagnostic character of which is the presence of a typically macroscopic fruit-body, the basidium, bearing sexually-produced basidiospores. Enzymes produced by basidiomycete mycelium present in soil, plant litter or wood are often capable of decomposing lignocellulose. Microfungi have microscopic fruit-bodies, and this artificial grouping of convenience includes fungi in soils and plant litter which produce large numbers of spores asexually. Enzymes from microfungal mycelium are usually unable to degrade lignin.

Eukaryotic microorganisms also include Algae, a large and diverse group of chlorophyll-containing species which carry out aerobic photosynthesis (Madigan et al., 2000). Protozoa are also eukaryotic microorganisms, which are unicellular and lack cell walls.

Prokaryotic microorganisms include both the Bacteria and Archaea. Bacteria have been studied much more intensively than Archaea. As a consequence, much more is known about their growth requirements, metabolic diversity, genetics and ecology. Bacteria typically are unicellular organisms that divide by binary fission to form two separate cells. Bacterial growth can be very rapid, e.g. under optimised growth conditions, *Escherichia coli* can double in cell numbers approximately every 20 minutes. However, not all Bacteria are growing actively at every time point with, for example, certain species of Bacteria present in environments as resting stages or spores. The metabolic diversity of Bacteria allows the exploitation of many different sources of nutrients in the environment. This metabolic diversity is reflected by the crucial roles that Bacteria play at many stages of a wide range of biogeochemical cycles (Atlas and Bartha, 1998). The importance of Bacteria to biotechnological developments is becoming clear (Rondon et al., 1999). For example, attempts have been made to harness these natural processes to remediate heavy metal-impacted waste material (Gadd, 2000; see Chapter 6, this volume). Although Bacteria are small (around 1–5 μm long), they are abundant in terrestrial environments (Killham, 1994).

Archaea are considered to be as distinct from Bacteria as they are from the Eukarya. However, certain features typical in Bacteria are also found in Archaea, e.g. binary fission (Horn et al., 1999). Archaea were however once considered to inhabit only 'extreme' environments, including those that were highly anoxic or had a high temperature, high salinity, or lay at the extremes of the pH scale. Nevertheless, recent molecular ecological studies have demonstrated their widespread presence in aquatic and temperate terrestrial environments including those of agricultural importance (Aravalli et al., 1998; De Long, 1998). Studies have also shown that the Archaea can be divided into Crenarchaeota and Euryarchaeota (Woese et al., 1990). Much remains to be understood about the environmental role of these organisms.

Viruses, described as non-living microscopic infectious agents, are also present in terrestrial systems. A wide variety of viruses can be found in soil environments. These viruses are able to infect many different life forms including animals, plants and microbes. The organisms that an individual virus can infect constitute its host range. Given their medical and economic importance, the most studied viruses are those associated with animal and plant disease. Human enteric viruses, for example, often enter environments through the discharge of domestic waste (Straub et al., 1993). In addition, however, natural environments also contain viral particles that infect prokaryotes, frequently referred to as phage. These phages may be important in controlling bacterial numbers (Ashelford et al., 1999). Archaeal viruses are present in both the Crenarchaeota and Euryarchaeota Kingdoms (Zillig et al., 1996). The application of viruses to biotechnological problems in terrestrial systems and concerns over the fate of both the virus and introduced DNA (England et al., 1998) are all topics of current interest.

For the remainder of this chapter, members of the domain Bacteria will be referred to as bacteria and this review will concentrate mainly on bacteria and fungi as they are the microbes in near-surface environments on which most work has focussed.

Defining microbial habitats

Microbes are found from the poles to the equatorial regions, and from xeric to freshwater and marine environments (Watling, 1997). This review concentrates on microbial microhabitats in the near-surface environments of soils and plant litter. The microhabitats available to microbes for colonisation are, to a large extent, dependent on the macrohabitats determined by the particular soil and vegetation types of a particular ecosystem. In the UK, truly natural habitats are relatively rare: examples of seminatural habitats are plantation forest, improved grasslands and arable land. Table 1, showing habitat losses and gains in Scotland, illustrates the increases that have occurred in habitats associated with human activity. One of the undesirable legacies of technical progress and increased human population is the creation of areas in which ecosystem processes and structures have been so damaged that the land cannot be used productively without major improvement (Urbanska et al., 1997). The causes of pollution are manifold and include organic (e.g. polyaromatic hydrocarbons, PAHs) and inorganic compounds (e.g. heavy metals, radionuclides) of industrial origin as well as biological pollution (e.g. faecal pollution, released organisms and genes). Such pollution can impact on either a local or global scale.

Certain environments afford specific habitats to terrestrial bacteria and fungi. These include habitats on leaf surfaces, in the rhizosphere, rocks and extreme environments. Pathogenic interactions between microbes and both plants and animals are discussed below. Endolithic microbial communities have been described elsewhere; for example, Amy et al. (1992) cultured a range of bacteria, predominately *Pseudomonas* species, from samples of crushed rocks (see Chapter 10, this volume).

Sampling microbial communities

Before addressing the composition and function of microbial communities, the issue of sampling from natural environments must be mentioned. There are a number of theoretical and practical issues associated with sampling microbial communities, including scale, relevance to study objective, design of sampling, statistical analyses, and sample collection and transport. Sampling strategy can therefore profoundly affect the nature of data emerging from a given experiment or survey, and in turn any information derived. Often sampling difficulties arise from the intrinsically heterogeneous nature of natural environments. For example, Frankland et al. (1990) wrote: 'All methods of quantifying soil fungi are imperfect. The heterogeneity of soil and the difficulties of separating the organisms from this complex medium, or of differentiating living from dead, cause major problems'. This is true to a large extent today. Two problems emerge: (1) how to analyse in a consistent manner variable soil types (soil is a mixture of organic matter, clay, silt and sand); and (2) how to analyse the spatial variation and microenvironments within one soil. Apart from rigour in sampling, the methodology used to analyse soil microbes is important, and cross-referencing of information from different approaches is particularly

Table 1
Habitat losses and gains in Scotland (Watling, 1997)

Losses	(%)
Pinus sylvestris woodland	51
Hedgerows	37
Lowland raised bog	23
Heather moorland	18
Broadleaved woodland	14
Unimproved grassland	9
Blanket mire	8
Grassy mire	7
Gains	(%)
Young plantation	525
Coniferous plantation	226
Recreational areas	107
Quarrying	63
Building	42
Bare ground	24
Canalised water	17
Semi-improved grassland	13
Transport corridors	5

helpful (Frankland et al., 1990). These issues are discussed in more detail in the remainder of this chapter and the following chapter.

2. Studying the microbial world

As in macroecology, it is important in microbial ecology to be able to characterise, enumerate and determine the roles of organisms within 'natural' environments (Ovreas, 2000). From these 'baselines', the impact of specific environmental perturbations can be determined. Current scientific understanding of the microbial world has been, and still is, limited in terms of what can be achieved technically. Information on microbes in 'natural' environments has been gathered in a number of different ways. These range from the microscopic counting of cells to methodologies based on molecular biology.

Absolute numbers

Frankland et al. (1990) stated that the counting of absolute numbers of organisms has traditionally been the prerogative of bacteriologists rather than mycologists. This may change however since the discovery that higher fungi often occur in the field as populations of genetically distinct mycelia.

In general, numerical surveys of fungal fruit-bodies give unreliable results unless there is repeated sampling, both within a season and over a period of several years, to offset possible bias arising from the ephemeral nature of many species and the vagaries of fruiting. A further complication is that although the occurrence of fruit-bodies does indicate the presence of a mycelium in the substratum with certainty, the lack of fruit-bodies does not necessarily reflect the absence of mycelia. Arnolds (1995) and Watling (1995) have both produced extremely readable accounts of the problems associated with estimating absolute numbers of fruit-bodies.

Because they are intimately associated with roots, enumeration of mycorrhizal fungi is particularly fraught with statistical and practical problems. Most quantitative techniques require the whole of a fraction of the root system to be recovered from the soil, and soil particles removed before an estimate of infection can be made. The VAM 'inoculum potential' of soils can be evaluated by counting infective propagules. It is usually based on the number of spores contained in a unit volume of soil, although infected root fragments may also act as sources of inoculum. The spores, because of their relatively large size, shape and wall characteristics, can be quantitatively recovered from the soil and identified.

Enumeration of the total bacterial cells in a given environment is frequently performed by direct epifluorescent microscopic examination of samples. These studies analyse soil preparations stained by fluorochromes such as acridine orange and diamidino-phenylidole (DAPI) (Kepner & Pratt, 1994). Other fluorescent dyes, such as 5-cyano-2,3-ditolyl tetrazolium chloride (CTC), have been used to determine the number of metabolically active bacteria in a sample (Rodriguez et al., 1992). The numbers of bacterial cells in soil or sediment environments can vary widely: in a study of a diverse range of soil and sediment samples, Weinbauer et al. (1998) reported values ranging from 7.8×10^6 to 1.3×10^{10} bacteria per gram of sample under investigation. Sandy soils typically have lower bacterial numbers than humus-rich soils. Despite problems associated with the technique, such as the non-homogenous distribution of soil bacteria coupled with frequently high background staining associated with soil samples (Kepner & Pratt, 1994), these data form an important first step in the characterisation of an environment in bacterial terms.

Biomass

Additional information can be gathered on the microbial component of environments by estimating biomass. On etymological grounds, the term 'biomass' should be restricted to living matter, but owing to its frequent misuse it is specified here as living or dead (Frankland et al., 1990). As discussed later, the dead microbial component is, however, an important resource, both in terms of the soil food web and genetically. The methodological details underpinning the approaches to estimating microbial biomass are discussed elsewhere (e.g. Frankland et al., 1990; Chapter 2, this volume). Briefly, biomass can be estimated by fumigating soil samples using chloroform, followed by monitoring the amount of carbon dioxide evolved by various biological or chemical procedures. Estimations of microbial biomass have been made by measuring the levels of specific biochemical components of microbial cells present in soils. The choice of component to be used as the marker is crucial: factors such as component presence/absence in different cell types, uniformity in relative proportion and rates of lability have all to be assessed. Biomass estimations therefore are constructed to meet the objectives of particular studies through the exploitation of this biochemical knowledge. For example, soil ATP levels are often used to indicate the active microbial biomass because ATP is both present in every microbial cell and disappears rapidly in dead cells. Other specific cell wall components are used to indicate the presence of bacteria (e.g. lipopolysaccharide for Gram-negative bacteria), photosynthetic pigments (for photosynthetic algae/bacteria) and fungi (chitin, ergosterol; Ruzicka et al., 2000). As with total microscopic count data, a range of values are obtained for different soil samples. Fungal biomass estimates range from 0.3 g dry weight mycelium m^{-2} in the litter layer of a tundra polygon to 91.6 g dry weight mycelium m^{-2} in the top 10 cm of a temperate woodland (Kjøller & Struwe, 1982). For fungi, biomass is equivalent to the standing crop or mycelial concentration. Fungal biomass, although small relative to the soil mass, usually exceeds that of bacteria by a ratio of approximately 3 : 1.

Cultivation

As described in the preceding sections, soil environments contain both large numbers of microbes and different microbial cell types. However, these analyses provide no genus- or species-level information on the microbes within soil environments. This information is fundamental to characterising the role of a particular microbe in soil. Historically, the first approach to this problem was the in vitro isolation and cultivation of soil microbes on solid growth media. This approach, which dates from the pioneering work by Koch in the late nineteenth century, still forms an important means of analysis today (Atlas & Bartha, 1998).

Following cultivation, pure single strains of microbes can be assigned to particular species. Given the lack of morphological detail typical of many microbes, particularly bacteria, species were defined empirically, largely on the basis of phenotypic data such as growth and biochemical characteristics. Although in many cases these species descriptions are still used to the present day, there is also active debate on the how best to define 'species' in microbial terms (see Ward, 1998; Lawrence, 1999; Lan & Reeves, 2000; and below).

A wide range of microbial genera and species have been cultured from soil samples. The following bacterial genera contain species that are cultured frequently in soil: *Acinetobacter, Agrobacterium, Alcaligenes, Arthrobacter, Bacillus, Brevibacterium, Caulobacter, Cellulomonas, Clostridium, Corynebacterium, Flavobacterium, Micrococcus, Mycobacterium, Pseudomonas, Staphylococcus, Streptococcus* and *Xanthomonas* (Atlas & Bartha, 1998). Actinomycetes, in addition, typically form a large proportion of bacteria in soil. Although these bacteria are common, the proportions of the different bacterial groups listed above differ from soil to soil. The conditions required for the growth of these bacterial genera in vitro are very varied; some genera, e.g. *Micrococcus*, require oxygen, while others, e.g. *Clostridium*, require an anaerobic environment for growth. These different genera also utilise a wide range of substrates for growth, and their presence may reflect the various microhabitats that become established within a soil and soil aggregate. No single set of growth conditions, nor single growth medium, can therefore culture every microbial strain. Much effort has been focused on defining media of different compositions for particular tasks. This has, as a consequence, led to the description of many hundreds of different growth media (Atlas, 1997).

The process of cultivation involves not only the selection of a culture medium, but also of temperature and atmospheric conditions. Every cultivation procedure therefore is in some sense selective. Moreover, for soil samples, a profound difference between the number of microbial colonies formed on agar media compared to the number of microbes visualised by direct microscopic counts was identified and described as the 'great plate count anomaly' (Staley & Konopka, 1985). The percentage of culturable bacteria has been estimated to be 0.3% in soil samples, with similar values (0.001–3%) recorded for aquatic habitats (Amann et al., 1995). There are many possible reasons for this discrepancy between plate and direct counts. For example, the response of many microbes to the frequently encountered limitation of nutrients in natural environments is to enter a viable, but non-culturable state (Edwards, 2000). In addition, obtaining pure individual cultures of microbes, by definition, removes the potential for ecological interactions (Atlas & Bartha, 1998). From these limitations emerged the drive (described below) to study soil microbes without prior cultivation, using molecular techniques based on analysis of nucleic acids and other cellular components.

Cultivation remains, however, an important first means of assessment of the microbes present within soil samples. Individual cultures of soil microbes provide information on the composition of the microbial community. Moreover, such single colonies form a resource that enables detailed assessments of the phenotype and genotype of each isolate that cannot be studied in a mixed community through molecular-based approaches. Balestra & Misaghi (1997) have shown that the number of morphologically distinct bacteria cultured from soil can be increased by using a range of different agar media as opposed to a single medium. As such, this reinforces the importance of methodology and the concept of the variety of different bacterial growth requirements.

Dilution plate counts of fungal colonies, arising as they do from hyphal fragments and spores, are meaningless in terms of either abundance or biomass of a species. Such colony counts are useful only in limited circumstances as, for example, in a comparative study of counts of fungal colony-forming units (CFUs) from a variety of biomes, ranging in the upper layer of soil from 0.005×10^4 in tundra to 100×10^4 in *Betula* woodland in Japan (Kjøller & Struwe, 1982). Heavily sporulating species of decomposer microfungi (e.g. *Penicillium*, *Trichoderma*) are typically isolated by dilution plate methods. It is probably unknown what proportions of saprotrophic or ectomycorrhizal fungi are unculturable. All VAM fungi are unable to be cultured without the plant host, although such fungi have been identified in field samples by selective enrichment of amplified DNA (Clapp et al., 1995).

Virus particles can also be recovered by 'cultivation' as visualised through plaque formation on lawns of appropriate host cell types, e.g. bacteriophage forming individual plaques following elution from the soil matrix (Hurst, 1997). Again however, most of the work in soil environments has been focused on viral pathogens, particularly those of humans (e.g. Hurst et al., 1991; Straub et al., 1995).

Molecular-based methods

Molecular-based approaches are important for two different purposes: first, to study and characterise further individual microbial cultures; and second, to analyse the microbes present in an environment without prior cultivation. The concept of the use of cellular components as markers of groups of microbes was introduced earlier. These markers can either be nucleic acids or other cellular components.

Phospholipid fatty acid (PLFA) analysis has been used to provide information on the bacterial and fungal composition simultaneously without requiring prior microbial cultivation (Frostegård & Bååth, 1996). In this approach, PLFA markers provide a broad-scale level of the microbial component within soil samples. In addition, PLFA data reflect the composition of the sample directly, and hence are not prone to potential biases in the way that exponential amplification processes, such as polymerase chain reaction (PCR), are.

Molecular analysis of nucleic acids, either for the study of single cultures or microbial communities, integrates knowledge on individual gene sequences with molecular biological procedures. Microbial phylogenetic studies have advanced greatly through the comparative sequencing of ribosomal RNA from different organisms. Ribosomal sequences were chosen because they are highly conserved yet have regions in which increased sequence variation reflects increased phylogenetic distance. As a consequence, ribosomal sequence data can be used to form a phylogenetic framework into which individual sequences of uncharacterised organisms can be placed. This phylogenetic framework enables the relationship between the three domains (Archaea, bacteria and Eukarya, including both microbial and non-microbial eukaryotes), to be described (Woese, 2000). Moreover, the sequence variation within ribosomal sequences can be exploited to distinguish different phylogenetic levels through the design of probes specific to the detection level desired, e.g. genus or species level. These can act as tools for in vitro amplification through molecular biological procedures such as PCR.

There is increasing interest in the analysis of functional genes (Costello & Lidstrom, 1999), which encode the majority of the cellular activity within microbes. Many studies have concentrated on genes which encode specific biochemical functions, such as ammonia oxidation or heavy metal resistance. However, while there is detailed knowledge of the ribosomal sequences, there is often much less information on the functional gene sequences present, which can make the design of informative probes difficult.

Genes studied for phylogenetic and 'functional' purposes can be investigated in pure cultures of microbes, with the former types of gene providing a taxonomic aid and the latter providing insight on the biochemical potential of the cell. There has also been considerable interest recently in analysing the microbial communities present in terrestrial and aquatic environments using molecular methods (Head et al., 1998). Although there are caveats (e.g. Von Wintzingerode et al., 1997), non-culture-based approaches are important given our current inability to cultivate the numerical majority of bacteria in natural environments. Using phylogenetically informative sequences, it is possible to determine the composition of natural communities of microbes without cultivation differentiating the sequences that are present and also those that are active at a given time point.

The bacteria present in a range of soils have been examined in culture-independent studies. In the first of these studies, Liesack & Stackebrandt (1992) found that α-Proteobacteria (a phylogenetic group encompassing many genera) were the predominant bacteria. Since then, such studies have routinely identified novel types of bacteria. Felske et al. (1999) reinforced this, finding no correlation between the results of culture-based as opposed to culture-independent (16S rDNA sequencing of individual clones) analysis of grassland soil bacteria. This implies that the number of bacterial taxa that cannot be cultured using standard methods is considerable.

Two notable attempts to circumvent culture techniques to examine fungal communities have been made in field samples, one in roots of *Ammophila arenaria* (Kowalchuk et al., 1997) and the other in soil from the rhizosphere of *Triticum aestivum* (Smit et al., 1999). Both studies used DNA amplified by primers for 18S rRNA gene sequences. The molecular data in the study of Kowalchuk et al. (1997) revealed fungal types that had not been detected in previous culture-based surveys, although this is understandable to some extent as the roots would contain unculturable VAM and root-pathogenic fungi. Both studies cited the incompleteness of existing genetic databases and the limited taxonomic resolution of the 18S rDNA as shortcomings. In a more recent study, Borneman & Hartin (2000) designed primers that amplify the four major phyla of fungi, Ascomycota, Basidiomycota, Chytridomycota and Zygomycota, specifically from either fungal monocultures or from DNA extracted from soil. As in many bacterial analyses, when these authors compared the fungi identified either by culture or molecular methods, different fungal types were detected by the two methods. To date, no nucleic acid-based studies of fungi in soil or litter give information about fungal abundance.

3. General importance in ecology, industry and biogeochemical cycling

Nutrient cycling

Since saprotrophic fungi are primary decomposers of organic matter, and mycorrhizas facilitate plant nutrient uptake, fungal activity is a major influence on the nutrition of plants through uptake, immobilisation (incorporation into a microorganism rendering nutrients temporarily unavailable to plants) and mineralisation of nutrients (Frankland et al., 1990). The nutrient content of fungal biomass itself is often neglected, but has been studied by Bååth & Söderström (1979) using mycelia of some common soil fungi grown on soil extract media for one month. The average values in the dry mycelium were 3.7% N and 0.7% P. Multiplying by fungal biomass estimates, these mycelial pools of N and P were relatively large, constituting up to 19.6% N and 18.2% P of the total amounts of nutrient in the soil horizon in which they occurred. The mycelium of decomposer basidiomycetes in particular has the potential to be an important accumulator of nutrients (Frankland, 1982). This has implications for the immobilisation of radionuclides and heavy metals discussed in later chapters. Bacteria are also important in the cycling of many nutrients in soil (Kennedy, 1999). Much of the research carried out on bacteria has focused on their role in specific nutrient cycles, particularly in aquatic environments, e.g. nitrogen cycling (Herbert, 1999). Highly specific symbioses between plants and bacteria, which are of great importance in nutrient cycling, can be identified. Probably the best characterised of these is the symbiosis between *Rhizobium* and leguminous plants, for which the nitrogen fixation process has been studied in fine genetic detail (Schultze & Kondorosi, 1998).

Decomposition

For many years, the received wisdom was that recycling of nutrients from dead organic matter in soils was predominantly driven by saprotrophic fungi, whereas mycorrhizas facilitated plant nutrient uptake. This is confirmed by the statement of Hering (1982) that decomposer and mycorrhizal fungi should be considered as 'two water-tight, mutually exclusive classes, both spreading their mycelium in the same material'. This view, however, has been moderated over the intervening time interval, since a range of both ericoid and ectomycorrhizal fungi has been found to possess the classes of extracellular enzymes needed to take part directly in decomposer activities (Smith & Read, 1997). The role of bacteria in decomposition processes has also been investigated. Hu et al. (1999) showed that bacteria of different nutritional strategies displayed different dynamics in response to the decomposition of crop debris. Many different bacteria have been found capable of decomposing cellulose including *Actinomycetes* and *Streptomyces* spp. (Ulrich & Wirth, 1999); certain *Streptomyces* also known to degrade chitin (De Boer et al., 1999). More generally, bacteria produce a wide range of enzymes, e.g. proteases and deaminases, that are important in the decomposition of compounds in the natural environment.

Pathogenicity

Fungi in ecosystems are not only present as decomposers or as symbionts in mycorrhizas and lichens. Some fungi live pathogenically on plants and cause a number of serious plant diseases, e.g. Dutch Elm Disease caused by *Ophiostoma ulmi*. A few fungi are parasitic on animals, including humans (e.g. *Trichophyton*, *Aspergillus fumigatus*, *Candida albicans*), although in general fungi are less significant as animal pathogens than other microbes.

Bacteria are responsible for a wide range of disease in both plant and animals. In terms of human disease, many different pathogens can be found in the terrestrial environment. Certain pathogens are considered to be typical inhabitants of soil, with other pathogens, e.g. those present in sewage or animal manures, found in certain soils as a result of human activities (Gagliardi & Karns, 2000). Sjoling & Cowan (2000), for example, detected bacteria indicative of human faecal contamination associated with base camp sites in the Antarctic. In addition to the pathogen itself, the spread of particular genes or groups of genes, which can be carried on mobile genetic elements (such as plasmids, which are described later), is matter of growing concern (Davison, 1999).

The mechanisms by which bacteria cause disease are also widely studied (see Brogden et al., 2000). Certain pathogenic mechanisms such as type III protein secretion systems are common to bacteria that are pathogens of plants and animals (He, 1998). The mechanisms by which specific pathogens cause disease will be assisted greatly by the current genomic advances that have already seen the publication of the full genome sequence for the plant pathogen *Xylella fastidiosa* (Simpson et al., 2000) and a range of animal and human pathogens including *Mycobacterium tuberculosis* (http://www.tigr.org/tdb/mdb/mdbcomplete.html).

Reservoirs for diversity

Fungi not only produce important commodities, e.g. citric acid, alcohol and edible mushrooms, but a few species produce extremely high-value biochemical products (Bills, 1995). For example, the fungal metabolites cyclosporin A and lovastatin (Mevacor®, Merck & Co., Inc.) are each pharmaceuticals with annual gross sales that exceeded one billion dollars in 1995. The anticancer drug, taxol, which had been previously supposed to occur only in the plant genus *Taxus*, has been reported in fungi of disparate genera (Strobel et al., 1996). Novel products with antifungal properties (Kang et al., 1998), or still wider antimicrobial activity (Cain et al., 2000), have been identified in soil bacteria. Industrial laboratories muster huge microbial collections in an effort to maximise chemical diversity, especially for screening for specialised metabolites and enzymes. Because of their ease of isolation, high species diversity and production of an extraordinarily wide range of metabolites, soil fungi will continue to be a mainstay of screening programmes for natural products (Bills, 1995), with other strategies emerging as important (Bull et al., 2000; Rondon et al., 2000).

Bacterial plasmids

Bacterial plasmids are extrachromosomal genetic elements that are not essential for the growth of the bacterial cell. The genes carried on these plasmids include those involved in organic compound (e.g. toluene) metabolism (Sentchilo et al., 2000), heavy metal (e.g. mercury) resistance (Lilley et al., 1996), and antibiotic resistance (Hill & Top, 1998). Often plasmids also contain other mobile genetic element structures. Many other sequences present on bacterial plasmids cannot be assigned to a particular function and plasmids can also vary enormously in size. Plasmid transfer has been documented in soil (Pukall et al., 1996) and in plant roots and leaves (Lilley et al., 1996). This is important as it demonstrates the potential for horizontal transfer of important plasmid-borne sequences and hence provides one mechanism of gene flow in the environment.

To detect movement of plasmids, typing schemes are required. Lilley et al. (1996) were able to both detect gene transfer from the native bacterial populations present on sugar beet roots and leaves into a recipient *Pseudomonas putida* strain, and also to discriminate these plasmids on the basis of restriction endonuclease digestion into five different groups. The importance of this typing is that these authors were able to detect the same plasmid groupings over successive years. There is increasing interest in developing DNA probes that will enable different plasmid types to be followed in the environment (e.g. Couturier et al., 1988; Osborn et al. 2000).

4. Community structure and diversity

A community is the biotic component of an ecosystem (Cooke & Rayner, 1984). It is implicit in this that the community comprises taxonomically diverse organisms, and has its own distinctive structure, activities and laws, including a unique internal economy which depends on relationships between the organisms that constitute it. The studies of communities should therefore be multidisciplinary, but the inevitable tendency has been to consider only those components which are relevant to any particular ecological discipline.

What is meant by structure and which methods have been used?

The identity of species, location and spatial arrangement of mycelia and single microbial cells, resource (defined as any identifiable component of detritus; Swift et al., 1979) relationships and involvement with other organisms, are all of significance in the structure of microbial assemblages in soil.

Ideally, the complete description of the structure of a microbial community would include identification of each species in situ, measurement of separate biomass values in situ for them, and a map of their locations within the soil profile, or in a leaf or twig, in relation to particular resources. From the preceding sections it is clear that by direct observation alone this may be impossible because of either the lack of morphological detail for bacteria, or the similarity in appearance of mycelia of different fungal species. It is also difficult to separate fungi from the complex medium of soil or litter, and it is demanding to differentiate living from dead mycelium by eye. It is therefore important, but difficult, to examine a microbial community in situ.

Using traditional techniques, the closest one can come to quantifying the abundance of single fungal species in soil or litter is to bring into the laboratory specific fractions of the substratum (e.g. soil particles of a particular size, or from a specific location in the profile) which are serially washed to remove 'contaminant' surface spores (after Harley & Waid, 1955) and plated into defined media (e.g. Robinson et al., 1994). Estimates of the percentage frequency of occurrence can be obtained by relating the number of observations or isolations of each species to unit amounts of the substratum. For example, presence and absence data obtained from Warcup soil plates (Warcup, 1957) are usually expressed as the percentage of plates on which the fungus has grown. An example of the type of results obtained is shown in Fig. 1 (Widden & Parkinson, 1979). This technique of isolation on defined media has the obvious problem that fungi which cannot grow (e.g. VAM fungi) will not be isolated. Much has also been written about problems of competition within the plated particle affecting fungal outgrowth (e.g. Bååth, 1988). The structure of fungal communities will change over time, and a number of these substratum successions have been described. Frankland (1992) has reviewed such successions involving decomposer (and ectomycorrhizal) fungi, and the mechanisms behind them.

Many different approaches have been used to follow the microbial community within soils. Traditional culture-based methods are being reinforced gradually by molecular methods that study either nucleic acids or some other cellular components. Of these molecular options, most studies currently use either PLFA or ribosomal sequence analysis. PLFA provides a broad-scale level of detail and has advantages in that it examines the whole microbial community (i.e. bacteria and fungi), and does so in a linear manner, although PLFA cannot be used to give species identity. PLFA has not been used to show where fungal mycelia or bacterial cells are located in soil at the fine scale. Most ribosomal studies involve the (non-linear) PCR amplification of DNA. Because of non-linearity, there can be difficulties in extrapolating back to the starting concentrations of ribosomal templates extracted from soils. When analysing microbial community structure in soil, care must be

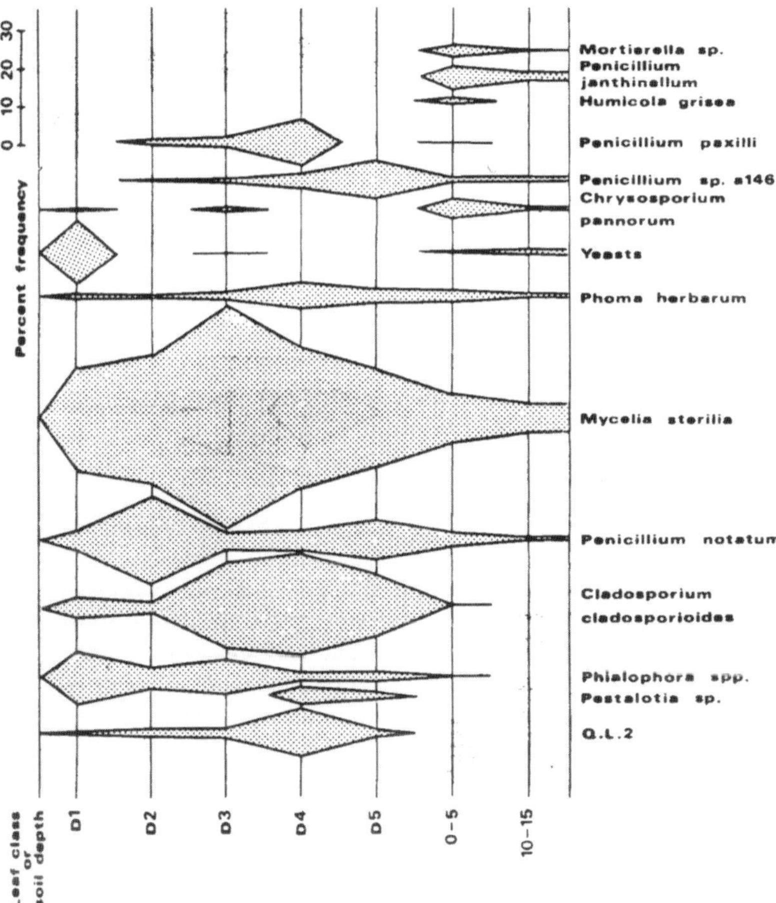

Fig. 1. Distribution of the major fungal taxa on *Dryas integrifolia* leaves, litter, and in the soil of a raised beach, Truelove Lowland, Devon Island, N.W.T., Canada (75°35′N, 84°40′W). Widths of 'kites' are proportional to the mean frequency of occurrence of the fungi. *Dryas integrifolia*: (D1) living, green leaves; (D2) brown leaves; (D3) brown leaves with a grey cast; (D4) grey leaves, entire; AND (D5) grey, fragmenting leaves. 0–15, 10–15: soil depths in cm. QL2 = unidentified fungal strain (after Widden & Parkinson, 1979).

taken at various stages not to replace biases inherent in culture-based protocols with biases specific to molecular studies. Such stages include DNA extraction (see Krsek & Wellington (1999) for a recent assessment of DNA extraction strategies) and PCR amplification.

The resulting view of bacterial community structure is greatly affected by the methodology used and the sample under investigation. Many studies have generated clone libraries of 16S rRNA genes (16S rDNA) after extracting the DNA from soil either directly (the most commonly used approach) or following an attempt to separate different microbial fractions from the soil prior to lysis. These studies typically show that proteobacteria (particularly of the α-subdivision), members of the Cytophaga-Flexibacter-Bacteroides group,

Verrucomicrobiales, the *Holophaga/Acidobacterium* cluster and both high and low (such as *Bacillus*) G+C content Gram-positive groups are common (Liesack & Stackebrandt, 1992; Borneman et al., 1996; Felske & Akkermans, 1998; Felske et al., 1998). As discussed below, these are broad groupings and as such contain a wide range of diverse bacteria. Studies that examine 16S rDNA therefore reveal which bacteria are present in an environment. Moreover, because of the lability of the molecule, such studies also identify specifically those bacteria which are active.

Data emerging from these and many other studies have shown that soil bacterial communities are complex. To elucidate the structure of the community requires some other technology and two approaches have been devised that can provide such insights. Profiling methodologies such as denaturing gradient gel electrophoresis (DGGE; Muyzer & Smalla, 1998; Muyzer, 1999), temperature gradient gel electrophoresis (TGGE; Muyzer & Smalla, 1998; Muyzer, 1999) and terminal-restriction fragment length polymorphism (T-RFLP; Liu et al., 1997; Marsh, 1999) provide a means of assessing community structures. These methodologies are often used in parallel with cloning studies (Felske et al., 1998) and as discussed below, can follow the dynamics of communities. Specific probing can also be used to examine the community structure within a soil either through extraction of nucleic acids from soils or the in situ hybridisation of the microbial community. Zarda et al. (1997) examined the bacterial community structure of soil using in situ hybridisation. Similar to those studies described above in which 16S rDNA clone libraries were analysed, these workers found that approximately half of the analysable soil bacteria belonged to the α-, β-, γ- and δ-subdivisions of Proteobacteria, high G+C Gram-positive bacteria, members of the Cytophaga-Flexibacter-Bacteroides group, and the planctomycetes.

Environmental factors are important in shaping bacterial community structures. For example, profound differences in community structure, as assessed by both measures of G+C content and 16S rDNA sequencing of clones, were observed in the transition between forest and pasture vegetation (Nusslein & Tiedje, 1999; Borneman & Triplett, 1997). Through the analysis of 16S RNA clones, Marilley & Aragno (1999) showed that bacterial communities differed with distance from the roots of *Trifolium repens* and *Lolium perenne*, with bacterial diversity decreasing with proximity to plant roots. Similar findings have been reported by Normander & Prosser (2000), who profiled distinct bacterial communities by DGGE analysis of 16S rDNA for barley plant roots and adjacent soil.

Moreover, other techniques that examine the similarities between samples of DNA extracted from different soils (cross-hybridisation), or the diversity within a soil DNA sample (reassociation kinetics) indicate that the soil 'genome' is varied (Clegg et al., 1998). That is, while the phylogenetically-informative ribosomal sequences are often similar, the majority of the other sequences (i.e. functional genes) in soil microbes can be quite different. Clegg et al. (1998) used these approaches to study the microbial communities in grassland soils that had been managed differently (and so had a vegetation cover of different plant species). These workers found that there were large differences between the DNA extracted from these soil types, with samples from 'unimproved' soil proving the most complex and 'improved' soil the least complex.

Levels of variation observed in specific communities

The bacterial lineages identified above have been identified as the most common in soils through molecular analyses. As such, these bacteria are assumed to be important in ecological terms. However it is important to note that these lineages are broad and cover a wide range of different bacterial genera. Variation within soil bacterial communities – and more widely microbial communities – must therefore be seen at a series of different levels. This is most readily seen by DNA probing studies where distinct taxonomic levels can be differentiated sequentially to identify which bacteria (or other microbes) are present. Through this nested approach, DNA probes can be designed to detect microbes at genus or subgenus levels. Hristova et al. (2000) identified such different levels within *Desulfotomaculum* spp. present within a variety of study samples including soils by using 16S rRNA hybridisation. Variation also exists at the subspecies level. Duncan et al. (1994) identified that there was extensive genotypic diversity using a variety of molecular approaches within strains of *Bacillus subtilis* and *Bacillus licheniformis* isolated from a single sample of desert soil. This provided evidence for a high degree of genetic recombination for this species within this environment.

What exactly has been found?

By horizon

Contributing a great deal towards the 'complete' picture of the structure of a fungal assemblage outlined above, fungal community structure in relation to resources for two broad species' groupings, rather than for individual species, was detailed in a painstaking study by Frankland (1982). The distribution of living and total fungal biomass in a deciduous woodland soil, estimated from hyphal length, with hyphae with cell contents assumed to be living, and classified as belonging to basidiomycetes or microfungi according to the presence or absence of clamp connections, is shown in Table 2. A relatively low quantity of non-basidiomycete mycelium (kg ha^{-1}) occurred in the lowest soil horizons. However, this reflected the sheer bulk of the subsoil. Fungal mycelium of all types was most concentrated (g g^{-1} substrate) in the thin organic horizons, and basidiomycete mycelium, excluding as far as possible that of mycorrhizal and pathogenic species on and in living roots, was almost confined to this area of the profile, dominating the fungal assemblage during the decomposition of the cell walls of plant debris.

Basidiomycetes can therefore form a significant proportion of the microbial biomass of a woodland soil, but their ecological importance in biomass terms becomes much more obvious if the large quantities of fungal mycelium in dead wood and dead roots are taken into account (Table 2). In Meathop Wood, these substrates were often packed with mycelium, and many dead tree branches and roots contained virtually a 'pure culture' of a basidiomycete, such as *Stereum hirsutum* or *Armillaria mellea*.

Little molecular work has been carried out on defining the bacterial communities present in different soil horizons. However, Dejonghe et al. (2000) have shown that the bacterial communities within the A horizon (0–30 cm deep) and the B horizon (30–60 cm deep) of a soil contaminated with 2,4-dichlorophenoxyacetic acid differed with respect to their degradative abilities.

Table 2

Comparison of the distribution of the biomass of basidiomycetes (kg ha^{-1} dry wt) with that of other microbial decomposers in the floor of a temperate deciduous woodland with mull humus (Meathop Wood, Cumbria, UK; Frankland, 1982)

	Basidiomycetes		Other fungi		Bacteria and actinomycetes	
Substrate or horizon	Living	Total	Living	Total	Living	Total
Woody debris	30.5	216.9	7.3	34.7	2.6	601.6
L	3.1	8.7	0.5	4.1		
(Oh + Ah)	8.9	31.7	3.4	12.9	37.3	8433.3
A	<1.0	<1.0	26.4	97.5		
B	<1.0	<1.0	31.4	155.6		
Dead roots	228.0	1628.5	65.1	325.7	8.0	1851.1
Total	271.5	1886.8	134.1	630.5	47.9	10866.0

A further characterisation of the structure of a fungal assemblage, in this instance by age-class of plant litter and soil depth, is shown in Fig. 1. By isolating fungi in the laboratory from leaf and litter samples of defined ages from *Dryas integrifolia*, as well as from soil from 0–5 and 10–15 cm depths, collected from a raised beach in the Canadian high arctic, Widden & Parkinson (1979) were able to show the structure of the fungal community in relation to resources at various depths. Visser and Parkinson (1975) also provided as complete a picture as possible of the fungal community by identifying species in live leaves and individual soil horizons, and measuring the total fungal biomass in each. Most biomass was located in the humus horizon (196 g wet weight m^{-2}) and least in a deep, resource-poor horizon (0.91 g m^{-2}).

Fluorescent monoclonal antibodies, specific to mycelia of particular fungi (Dewey et al., 1997), have been applied in defined environmental samples brought into the laboratory in order to pinpoint the location of specific mycelia in relation to resources. In such autoecological studies it is possible to map the structure of one, or a few species, and to estimate their biomass. Using such a technique, *Mycena galopus* mycelium was found to be concentrated in a single horizon of a *Picea sitchensis* plantation within and beyond the area of arcs of basidiomes observed around some of the trees. Biomass of *M. galopus* increased outwards from the tree bole to a maximum at the position of fruiting (Frankland et al., 1995). Unfortunately, this technique is unlikely to be applied to characterise the structure of the whole fungal community, since at least tens of fungal species are usually present, each requiring the time-consuming production of species-specific antibodies.

By different size soil particles

A fungal community could be structured according to soil particle size, rather than by soil horizon, or age class of plant litter as outlined above. Bååth (1988) washed a humus horizon from a coniferous forest soil through sieves, removing surface spores and isolating fungi found in various particle size fractions. Organic soil particles from the different fractions were plated on carboxymethylcellulose agar medium. The mean colonisation frequencies (isolates plated per particle) of the different particle sizes were 0.71 in the 50–80 μm fraction, 0.96 in 80–100 μm, 1.26 in 100–125 μm, 1.47 in 125–180 μm and 1.84 in the 180–250 μm fraction, showing there were fewer fungi associated with smaller particles. The same species were dominant in all size fractions and the species diversity values were also similar in all fractions. However, the abundance of different species differed between particles of different sizes, and the larger the difference in size of the particle, the greater the difference in the abundance of the isolated species. The fungal assemblage isolated from the smaller particles was characterised by higher abundances of slow-growing fungi (e.g. *Oidiodendron echinulatum*) compared with those from the larger particles. Those species that were preferentially isolated from larger particles were all fast-growing species (e.g. *Penicillium spinulosum*). These results suggest that, rather than this fungal assemblage being structured by particle size, slow-growers were prevented from growing out from larger organic particles because more than one, perhaps faster-growing, fungus was present. Thus, Bååth's results show that the use of smaller particles, where mostly no more than one isolate per particle was present, will overcome problems with fungal isolates interacting on the agar media. They also illustrate the difficulty in accurately characterising the structure of the fungal community, and that this assemblage from coniferous soil does not appear to be structured by organic particle size.

The community structure and activities of prokaryotes within different aggregate size fractions have also been studied. Ramakrishnan et al. (2000) found differences in methane production associated with soil aggregate size. Small aggregates (<50 and 50–100 μm) were able to generate least methane, with those of intermediate size (200–2000 μm) able to generate the most. Interestingly, there were only minor differences between the archaeal communities, analysed by T-RFLP of small subunit archaeal rRNA genes, present in each of the aggregate size fractions studied. Similarly, Gelsomino et al. (1999) identified a high degree of similarity in the bacterial 16S rDNA DGGE profiles of soil divided into a series of aggregate size fractions. Ranjard et al. (2000b), however, found that the inner portion of the soil aggregate contained the most bacteria and found additional evidence that certain bacterial populations were specific to a given location within the aggregate. Differences were also seen with respect to the response of the bacterial communities within the microenvironments of a silt loam soil to mercuric ion stress (Ranjard et al., 2000a). More mercury-resistant bacteria were identified in response to this stress in outer and coarse size fractions. These total bacterial community profiles showed a change in the dominant bacterial populations from the outer to clay size fractions.

Host specificity and 'is everything everywhere?'

Saprotrophic fungal species in soil are probably largely cosmopolitan (Fenchel et al., 1997; Finlay et al., 1997), and it is the *structure* of the decomposer fungal community, specifically the balance of dominant and rare species and the location of the mycelia, which differs between ecosystems (e.g. Swift, 1976; Frankland, 1998; Robinson et al., 1998). Evidence that 'everything is not everywhere' comes from the fact that the structure and composition of fungal assemblages, especially in mycorrhizal fungi, is affected by host specificity. For example, Newton & Haigh (1998), when examining host specificity in ectomycorrhizal fungi within the UK, found that 233 species of ectomycorrhizal fungi out of a total of 577 for which host information was collected, appeared to be specific to a single host plant species.

'Host' specificity can also occur between decomposer fungi and their substrata. Although comprehensive analysis of the fungi decaying plant debris will yield many hundreds of species, an initial partitioning of this community can be readily made on the basis of specificity for different types of plant litter, even though the nature of this specialisation is unclear (Swift, 1976). Such fungi have been termed resource-specific. Many microfungi, however, may be found on a wide variety of plant tissues, and some overlap between fungal species on different types of resources is common, resulting in the term of resource-non-specific fungi. An example of this phenomenon has been found by Robinson et al. (1994) who followed the frequency of occurrence of fungal species in stem internodes and leaves of *Triticum aestivum* buried in an arable soil at 10 cm depth for 32 weeks. *Cladorrhinum foecundissimum* was a resource-non-specific fungus found in similar frequencies on both leaves and internodes, whereas *Epicoccum nigrum* was an example of a resource-specific species, being found significantly more frequently on leaves. Thus, the decomposer fungal community is structured by 'host'-specific fungi, and species which are more widespread, or plurivorous (Ellis & Ellis, 1997). As well as host identity, the age and condition of the substrata (i.e. soil particle quality, rather than size as outlined above) can be important in determining which fungi are present in a community.

Through non-culture-based approaches, Rheims et al. (1996) have demonstrated that members of an uncultivated actinomycete group were present in peat sampled in Germany, soils sampled in Finland, Australia and Japan and ocean water of both the Atlantic and Pacific Oceans. Kuske et al. (1997) similarly reported a widespread distribution for soil bacteria studied using culture-independent techniques. Certain prokaryotes have a global distribution in marine systems, for example, the bacterial 'Sar11' cluster and group 1 marine Archaea (Garcia-Martinez & Rodriguez-Valera, 2000). Zwart et al. (1998) found three different clusters (one itself the SAR11 cluster) of highly related bacteria in lake waters in both the Netherlands and the USA, based on 16S ribosomal RNA gene sequences. In addition to these studies, Dawid (2000) has further examined the distribution of myxobacteria in soils across the world. The distribution of individual gene sequences, such as mercury resistance (Osborn et al., 1997) has also been considered at this global level.

Diversity and its interpretation

Issues of diversity, function and interactions, often referred to as the black box of soil, provide many scientific challenges (O'Donnell & Gorres, 1999; Tiedje et al., 1999). Microbial communities are enormously diverse, and this diversity has been termed both the wealth and despair of soil ecological research (Ohtonen et al., 1997). Studies have emphasised the diversity of microbes in soil: Torsvik et al. (1990) found that a single gram of soil could contain thousands of individual bacterial genome equivalents, and Gelsomino et al. (1999) have also shown that there are profound differences in the DGGE profiles of 16S rDNA generated from a wide range of different soil types. In fact, soil has been considered to contain most of the biodiversity that still has to be catalogued (Tiedje et al.,

1999), with recent DGGE profiling evidence showing, for agricultural soils at least, that bacterial diversity is greater than archaeal diversity (Nakatsu et al., 2000). When this is coupled with the estimate of Whitman et al. (1998) that there are 2.6×10^{29} prokaryotes distributed globally in soil, and the estimate of Dykhuizen (1998) that there might be over 10^{29} bacterial species, it becomes apparent that much of this problem is one of scale and of determining the means by which to interpret it.

Diversity indices

It is possible to calculate indices to estimate the species diversity of a sampled habitat, and the Shannon-Weiner and Brillouin diversity indices (Pielou, 1975) have been used for fungal assemblages (e.g. Kjøller & Struwe, 1982; Durrall & Parkinson, 1991; Robinson et al., 1994; Donnison et al., 2000). Kjøller & Struwe (1982), in their excellent review of the occurrence and activity of microfungi in soil and litter, showed that for sites worldwide, the maximum values of the diversity indices were reached in the organic layers of soil, whereas leaves and mineral soil generally showed lower diversity.

Researchers have used molecular profiling methodologies to generate diversity indices. Nubel et al. (1999) found that richness estimates and the Shannon-Weiner indices essentially provided the same answer whether based on data from molecular markers of DNA, or on other cellular components of the oxygenic phototrophic microbes present in microbial mats. In soils, Dunbar et al. (2000) stated that T-RFLP profiles of 16S rRNA genes of soil samples were not useful in generating traditional community richness or evenness values. Hedrick et al. (2000), however, found species richness estimates to be well supported, but not species evenness, by data from both PLFA and 16S rDNA DGGE profiling. Sharma et al. (1998) followed the soil microbial community dynamics, using BIOLOG plates, in response to the addition of maize litter for a year, measuring the Shannon index of diversity, the catabolic versatility and a coefficient for functional diversity. Litter amendment increased each index, with this increase most pronounced over the first 16-week period. Yang et al. (2000) used random amplified polymorphic DNA analysis of DNA extracted from four soil microbial communities to examine various indices of diversity (richness, Shannon-Weiner index). This approach indicated differences in terms of these indices between the soil and provided evidence that agricultural chemicals affected soil microbial community diversity at the DNA level.

Diversity in relation to heavy metals

There is good evidence that heavy metal-impacted soils support different microbial community structures and lower diversity than non-contaminated soils (Smit et al., 1997). Sandaa et al. (1999) have also shown by DNA reassociation analysis that bacterial diversity was much lower in contaminated as opposed to non-contaminated soil, although 16S rDNA clone sequence data showed comparatively few differences in the bacterial types present in low or highly metal-impacted soils.

Many bacterial genetic mechanisms that confer resistance to heavy metal ions have been described (Silver, 1996). Organisms with access to these mechanisms can have a selective advantage in these environments. In addition, the ability of certain microbes to grow in contaminated sites has important bioremediation possibilities (Diels et al., 1999; see Chapters 6 and 7, this volume).

Diversity in macroecology

The wider ecological community has moved beyond simple evaluations of species richness to assess the importance of functional diversity (see Section 5 below) in ecosystem processes (Zak & Visser, 1996). The suggestion is that in particular communities some species are 'redundant', i.e. could be lost with little effect on the structure and functioning of the whole community (Gitay et al., 1996). At present, it is unclear whether such redundancy occurs in microbial communities.

5. Function

What is meant by function?

As early as 1982, Kjøller and Struwe stated that a full understanding of the role of fungi in an ecosystem is not reached through independent observations on numbers, biomass, lists of species or physiological groups, but only through combined investigations where the relative occurrence of the different groups of fungi is linked to their function, that is, activity or capacity for substrate utilisation. Nearly 20 years later, the need for insight into the functional role of fungi in ecosystems is still great, partly because of the inadequate description of fungal assemblages in relation to resources. Some successful attempts, however, have been made to match organisms and their activities together (e.g. Flanagan & Scarborough, 1974; Boddy, 1986; Newsham et al., 1992b). Table 3 provides a list of functions of fungi in ecosystems.

How is function assessed, what has been found, and functional redundancy

Given the diversity of microbes within soil, it is unsurprising that there is a similarly diverse set of metabolic potentials within soil microbial communities. However, not all of this potential is expressed at any one time (White, 1995). Measuring the metabolic activity of microbes is therefore another challenge, made more complex by localised heterogeneity in physiochemical factors. Current understanding of the metabolic activities of microbes in natural environments derives in large part from in vitro assessments of microbial physiology.

Decomposition and nutrient cycling
The broad functions of fungal mycelium in soil and litter are decomposition and nutrient cycling. Decay rates (usually estimated from changes in weight, tensile strength and chemical composition) of individual litter components or pure substrates in the field have been studied extensively. However, these methods measure the activities of a decomposer community as a whole, and attempts to partition functions between specific groups are difficult. Frankland et al. (1990) stated the principal value of field samples lies in the clues

Table 3

Functions of fungi in ecosystems (after Miller, 1995)

Physiological and metabolic
Decomposition of organic matter: volatilisation of C, H and O; fragmentation
Elemental release and mineralisation of N, P, K, S and other ions
Elemental storage: immobilisation of elements
Accumulation of toxic materials
Synthesis of humic materials
Ecological
Facilitation of energy exchange between above- and belowground systems
Promotion and alteration of niche development
Regulation of successional trajectory and velocity
Mediative and integrative
Facilitation of transport of essential elements and water from soil to plant roots
Facilitation of plant-to-plant movement of essential elements and carbohydrates
Regulation of water and ion movement through plants
Regulation of photosynthetic rate of primary producers
Regulation of C allocation below ground
Increased survivability of seedlings
Protection from root pathogens
Modification of soil permeability and promotion of aggregation
Modification of soil ion exchange and water-holding capacities
Detoxification of soil (degradation, volatilisation or sequestration)
Participation in saprotrophic food chains
Instigation of parasitic and mutualistic symbioses
Production of environmental biochemicals (antibiotics, enzymes and imunosuppressants)

they can give to the functions of the predominant decomposer fungi, especially when observations are backed up by laboratory tests on isolated species. For example, Hering (1967), using pure cultures growing on γ-irradiated *Quercus* leaves, showed that lignin and hemicelluloses were most actively decomposed by *Mycena galopus* and *Collybia peronata*; cellulose by *M. galopus* and *Polyscytalum fecundissimum*. In a similar set of laboratory experiments on relatively realistic substrata, nutrient release, rather than decomposition, was followed from γ-irradiated internodes of *Triticum aestivum* inoculated with single and mixed species (Robinson et al., 1993b). Release of Na^+, K^+ and NH_4^+–N was similar from all combinations, but Ca^{2+}, Mg^{2+} and PO_4^{3-}–P release depended on species. These results suggest the fungi were functionally 'equivalent' for the monovalent cations above, but species' identity was important for the other ions analysed.

Spatial relocation of nutrients (e.g. N and P) and water by fungi are ecosystem functions likely to be of utmost importance. Such functions have been illustrated, usually in realistic substrata in the laboratory, by labelling compounds translocated by fungal mycelia that form specialised linear organs, cords and rhizomorphs (reviewed by Boddy, 1999). Two studies are especially noteworthy: one by Gray et al. (1995) chosen to link with later chapters as it demonstrated relocation of caesium in situ by *Schizophyllum commune*, and the second by Lindahl et al. (1999) who carried out an elegant microcosm demonstration of P transfer from saprotrophic to mycorrhizal mycelium, thus 'shortcutting' conventional nutrient cycling. Many more studies of the latter type are necessary, where interactions are studied between the different types of microorganism which share the same microsite.

As outlined above, 16S rRNA has been used to indicate the activity of particular bacterial species within environments. With the exception of ammonia-oxidising bacteria, few other inferences based on 16S rRNA analysis alone can be made on the types of activities with which particular bacterial species are associated. For this, the levels of expression for specific genes must be studied. In molecular studies, researchers have investigated the link between the activity of specific genes and the environmental transformation with which they are associated. For example in bacteria, the form of nucleic acid indicating activity, mRNA, for the gene encoding nitrogen reductase has also been detected in natural environments (Zani et al., 2000). Jeffrey et al. (1996) examined the expression of the *merA* gene in relation to the production of elemental mercury, and Henckel et al. (2000) detected the *pmoA* gene encoding the particulate methane monooxygenase only in soil layers that had detectable levels of methane oxidation.

Other approaches are making contributions to the understanding of the decomposer and nutrient cycling functions that bacteria carry out in natural environments. Radajewski et al. (2000) demonstrated a means by which specific metabolic processes could be studied through the acquisition of a specific ^{13}C label by microbes able to grow using the labelled substrate. Other means have been devised to study function within soil bacterial communities. Borneman (1999) proposed a related technique based on the capture of microbial DNA into which a nucleotide analogue has been incorporated by growing cells. Function can also be assessed by the use of specific bacterial sensors (Prosser et al., 1996) in which, for example, the metabolic activity of luminescent-marked strains can be detected in real time and so provide information on gene expression in specific environments.

Function can also be studied in terms of removing components of the microbial community. Griffiths et al. (2000) demonstrated this approach by reducing soil biodiversity using progressive fumigation and determining the consequent effect on a number of key soil processes. Although there was no direct relationship between diversity and function – as diversity decreased, certain functional parameters increased, e.g. the rate at which plant residues decomposed, others decreased, e.g. nitrification, and methane oxidation – this provides a way to follow the consequences of manipulating the function of soil microbial communities. Other researchers have aimed to understand the extent of functional redundancy in soil microbial communities. Yin et al. (2000) identified that bacterial functional redundancy existed along a gradient of soil reclamation from a tin mine to a forest. They found that the richness and diversity of bacterial groups (as measured by 16S–23S rRNA intergenic variation in bromodeoxyuridine-incorporated DNA following in situ growth on selected carbon substrates) increased along this gradient.

Potential decomposer activity using specific substrates

Fungal function is often tested in artificial media, which may not be good approximations to more natural substrata, but are often useful in trying to understand the potential roles of fungal isolates within a community. A wide range of tests exists to show utilisation of a C source in agar media; for example Robinson et al. (1993a) showed that, of the isolates tested, only basidiomycetes were able to clear lignin agar medium, whereas *Chaetomium globosum* and four basidiomycetes cleared cellulose agar medium. However, it is worthwhile remembering that the ability to decompose lignin and cellulose in vitro is not necessarily related to the abilities of the organism to decompose plant tissues (Swift, 1976).

Biomass increase of single species in liquid culture on various C and N sources can be measured as indications of the use of these compounds (reviewed by Paterson & Bridge, 1994). The ultimate extension of the type of substrate tests outlined above is the commercially available BIOLOG plate (Garland & Mills, 1991) used to assess functional differences among soil and water bacteria based on their patterns of utilisation of 96 substrates. The substrates include carbohydrates, carboxylic acids, amino acids, amines and amides. Lawlor et al. (2000) identified differences within the microbial communities, based on substrate utilisation profiles generated from BIOLOG plates, between untreated soils and those treated with sewage sludge. Smalla et al. (1998) examined the bacterial community, using 16S rDNA DGGE profiling, formed in each of these single carbon substrate wells. Sequence analysis indicated that only fast-growing soil bacteria were associated with growth in the individual BIOLOG wells. Dobranic & Zak (1999) have developed this method (FungiLog) to examine fungal functional diversity. Ideally, one should know which of the substrates are present in the field and their relative abundance. Studies including more ecologically-relevant substrates (e.g. root exudates) have been developed for microbes generally (e.g. Hodge et al., 1998). Fluorogenic enzyme substrates have been used to quantify specific components of fungal chitinase and cellulase activities in situ in soil (Miller et al., 1998).

Enzyme activity

Biochemical tests for a wide range of extracellular enzymes exist, but most involve addition of a substrate so that, although useful in laboratory studies, they measure potential rather than actual activity, as do the majority of the tests in the previous section. The range of enzymes produced from a single fungus can be great, for example, 18 types of extracellular enzyme from *Agaricus bisporus* (the commercial white mushroom) have been detected in compost cultures (Wood, 1998). Recently, enzyme activity has been measured using specific functional gene sequences. For example, Lamar et al. (1995) quantitatively assessed two lignin peroxidase mRNA transcripts from *Phanerochaete chrysosporium* in soil.

In summary, as seen from the scattered information above, there is still far to go before fungal function is satisfactorily described and analysed. Model studies using mixtures of decomposer/mycorrhizal fungi may help to explain the succession of functionally different fungi during decomposition. mRNA has potential as an in situ indication of enzyme activity. It is necessary to relate results of functional tests to community structure, which should be possible for 'key' species, if not for the whole community.

Workers have examined the levels of specific microbial enzymes within different soils. Tscherko & Kandeler (1999), for example, examined enzymatic activities that drive various stages within many fundamental biogeochemical cycles, e.g. N-mineralisation, in a range of different soils. For the soils studied, there appeared to be a strong influence both in terms of soil type and management.

Typically, these processes are presented in terms of individual biogeochemical cycles where the biotic and abiotic processes that link the element, and individual compounds derived from it, in a two-dimensional diagram. Given the range of compounds within environments, these diagrams can be complex. In addition to this complexity, more profound understandings of function will follow when it is possible to link different biogeochemical cycles.

6. Development and response to change

What is meant by change?

Much has been written about fungal community development (see, for example, Frankland, 1992) and it is generally accepted that the interactions involved are complex. They include the inherent individualities of each species, availability of space, availability of species of differential performance, dispersal, combative interactions and grazing by soil fauna. External changes having an effect on the fungal community could be environmental perturbations such as N or heavy metal addition, pollution by SO_2 or elevated CO_2, or extensification of agricultural grazing regimes. Most work has been carried out on the effects of perturbation on the frequency of occurrence of fungal species or, to a lesser extent, on fungal biomass. Few studies exist on the effects of perturbations on fungal community structure and function (but see Newsham et al., 1992a, b, c), partly because this type of project is so time consuming.

Community dynamics observed

The information presented to date provides in the main a ‘freeze-frame’ image of soil microbial communities. Some data show that soil bacterial communities are stable for long periods of time; Gelsomino et al. (1999) identified few changes in the bacterial community structure within a silt loam soil profiled over the course of a year by DGGE of 16S rDNA. More generally, however, microbial communities have the potential for dynamic interactions with either other microbes or their wider environment.

Dynamics in relation to seasonal, environmental and agricultural processes

Assessments of microbial community dynamics have been based largely on biomass. The microbial biomass has been found in studies to be higher in winter months in an alpine dry meadow (Lipson et al., 1999), and in leaf litter within a pine forest (Berg et al., 1998). Lipson et al. (1999) further showed that immobilised nitrogen within alpine meadow soil was greatest in the winter months. Protein released by microbes, coupled with soil protease activity, was vital in providing plants with amino acids for the growing season.

It has already been stated that vegetation cover is correlated with specific microbial communities. However, Duineveld et al. (1998) found that the bacterial community in the rhizosphere of the chrysanthemum sampled at different time points was quite constant when measured by 16S rDNA DGGE profiling, but identified variation according to the stage of plant maturation for the majority of the single carbon utilisation assays.

The impact of agriculturally-important processes have also been considered in a dynamic sense. Frey et al. (1999) determined the impact on soil microbes of tillage. While bacterial biomass did not vary consistently in the study samples, fungal biomass was greater in the non-tilled soil. This was linked to moisture availability which encouraged fungal growth. Frostegård et al. (1997) examined the dynamics of microbial communities in 'hot spots' of manure within soil over the course of 21 days. PLFA analyses indicated that the microbial biomass had doubled and that this was associated with a change in the microbial community structure because of the dissolved organic carbon from the manure.

Bruce et al. (2000) examined the bacterial community response to elevated carbon dioxide in a set of replicated terrestrial ecosystems using DGGE of 16S rDNA. No evidence was obtained for a shift in bacterial community structure resulting from elevated atmospheric carbon dioxide levels. However, although the overall bacterial community DGGE profiles were conserved, these profiles became more diverse with time.

Griffiths et al. (1999) found continual change in the microbial community – measured by community DNA hybridisation, %G+C profiling and PLFA analyses – when a synthetic root exudate was applied constantly for 2 weeks, with the fungal fraction becoming more dominant than bacteria at high substrate levels.

Resistance and resilience

Resistance is a measure of the amount of change that can be applied to a system before it is displaced from its equilibrium (Calow, 1998), whereas resilience measures if and when a system falls back, after a change or perturbation, to its previous realm of variation, or if it is reset to a new long-term dynamic (Collins & Benning, 1996). The applicability of the concepts of resistance and resilience to soil microbial communities was investigated by Griffiths et al. (2000), who imposed both transient and persistent stress to soil samples through heating and addition of heavy metals respectively. Soil microbial communities showed resilience to the transient stress, with resilience proportional to biodiversity, although the communities were not resilient to the persistent stress.

7. Future research

In terms of future research directions, new insights will be gained in a number of ways. First, there is a requirement for increasing interaction between specialists in different subject fields (Lavelle, 2000). One goal of this research is the drive towards modelling the responses of the soil microbial community accurately so that the consequences of environmental perturbations can be predicted. These factors need to be studied from the micro level (i.e. within a soil aggregate) to the global level. This is a hard task, particularly when the number and diversity of microbes, and by extension, the complexity of the interactions which are possible at both biotic and abiotic levels, are considered. As this predictive ability is currently lacking, important questions on the soil microbial response to environmental perturbations are hard to address.

Although already an important factor, it is likely that soil microbes will be viewed increasingly as an exploitable genetic resource. In genetic terms, there is already concern over the loss from the soil microbial community gene pool (Hagvar, 1998). The function of many sequences within even well-characterised organisms cannot currently be identified and, as the complex soil microbial community is examined, it is likely that the percentage of currently unidentified sequences lodged within databases will increase. Understanding and exploiting these sequences present an exciting bioinformatic and practical challenge. Microbial ecology has often been led by technical developments. Much of the current excitement stems from the increasing sequence information that is known for completed and partially finished genomes. In addition to the resource they provide for further genomics-based study, the introduction of gene arrays and gene chips provide ever increasing opportunities to examine the interaction of microbes with the environment. However, it is essential that these techniques are applied carefully to complex environmental systems, with strong hypotheses being clearly tested and experiments robustly designed.

References

Amann, R. I., Ludwig, W. & Schleifer, K. H. (1995). Phylogenetic identification and in situ detection of individual microbial cells without cultivation. *Microbiological Reviews, 59*, 143–169.

Amy, P. S., Haldeman, D. L., Ringelberg, D., Hall, D. H. & Russell, C. (1992). Comparison of identification systems for classification of bacteria isolated from water and endolithic habitats within the deep subsurface. *Applied and Environmental Microbiology, 58*, 3367–3373.

Aravalli, R. N., She, Q. & Garrett, R. A. (1998). Archaea and the new age of microorganisms. *Trends in Ecology and Evolution, 13*, 190–194.

Arnolds, E. (1995). Problems on measurements of species diversity of macrofungi. In D. Allsopp, R. R. Colwell & D. L. Hawksworth (Eds), *Microbial Diversity and Ecosystem Function* (pp. 337–353). Wallingford, UK: CAB International.

Ashelford, K. E., Day, M. J., Bailey, M. J., Lilley, A. K. & Fry, J. C. (1999). In situ population dynamics of bacterial viruses in a terrestrial environment. *Applied and Environmental Microbiology, 65*, 169–174.

Atlas, R. M. (1997). *Handbook of Microbiological Media*. Boca Raton, FL, USA: CRC Press.

Atlas, R. M. & Bartha, R. (1998). *Microbial Ecology: Fundamentals and Applications* (4th edn). Menlo Park, CA, USA: Benjamin/Cummings Science Publishing.

Bååth (1988). A critical examination of the soil washing technique with special reference to the size of of the soil particles. *Canadian Journal of Botany, 66*, 1566–1569.

Bååth, E. & Söderström, B. (1979). Fungal biomass and fungal immobilisation in Swedish coniferous forest soils. *Revue d'Écologie et de Biologie du Sol, 16*, 477–489.

Balestra, G. M. & Misaghi, I. J. (1997). Increasing the efficiency of the plate counting method for estimating bacterial diversity. *Journal of Microbiological Methods, 30*, 111–117.

Berg, M. P., Kniese, J. P. & Verhoef, H. A. (1998). Dynamics and stratification of bacteria and fungi in the organic layers of a Scots pine forest soil. *Biology and Fertility of Soils, 26*, 313–322.

Bills, G. F. (1995). Analyses of microfungal diversity from a user's perspective. *Canadian Journal of Botany, 73*, S33-S41.

Boddy, L. (1986). Water and decomposition processes in terrestrial ecosystems. In P. G. Ayres & L. Boddy (Eds), *Water, Fungi and Plants* (pp. 375–398). Cambridge, UK: Cambridge University Press.

Boddy, L. (1999). Saprotrophic cord-forming fungi: meeting the challenge of heterogeneous environments. *Mycologia, 91*, 13–32.

Borneman, J. (1999). Culture-independent identification of microorganisms that respond to specified stimuli. *Applied and Environmental Microbiology, 65*, 3398–3400.

Borneman, J. & Hartin, R. J. (2000). PCR primers that amplify fungal rRNA genes from environmental samples. *Applied and Environmental Microbiology, 66*, 4356–4360.

Borneman, J. & Triplett, E. W. (1997). Molecular microbial diversity in soils from eastern Amazonia: evidence for unusual microorganisms and microbial population shifts associated with deforestation. *Applied and Environmental Microbiology, 63*, 2647–2653.

Borneman, J., Skroch, P. W., O'Sullivan, K. M., Palus, J. A., Rumjanek, N. G., Jansen, J. L., Nienhuis, J. & Triplett, E. W. (1996). Molecular microbial diversity of an agricultural soil in Wisconsin. *Applied and Environmental Microbiology, 62*, 1935–1943.

Brogden, K. A., Roth, J. A, Stanton, T. B., Bolin, C. A., Minion, F. C. & Wannemuehler, M. J. (2000). *Virulence Mechanisms of Bacterial Pathogens*. Portland, OR, USA: ASM Press.

Bruce, K. D., Jones, T. H., Bezemer, T. M., Thompson, L. J. & Ritchie, D. A. (2000). The effect of elevated atmospheric carbon dioxide levels on soil bacterial communities. *Global Change Biology, 6*, 427–434.

Bull, A. T., Ward, A. C. & Goodfellow, M. (2000). Search and discovery strategies for biotechnology: the paradigm shift *Microbiology and Molecular Biology Reviews, 64*, 573.

Cain, C. C., Henry, A. T., Waldo, R. H., Casida, L. J. & Falkinham, J. O. (2000). Identification and characteristics of a novel *Burkholderia* strain with broad-spectrum antimicrobial activity. *Applied and Environmental Microbiology, 66*, 4139–4141.

Calow, P. (1998). *The Encyclopaedia of Ecology and Environmental Management*. Oxford, UK: Blackwell Science.

Clapp, J. P., Young, J. P. W., Merryweather, J. W. & Fitter, A. H. (1995). Diversity of fungal symbionts in arbuscular mycorrhizas from a natural community. *New Phytologist, 130*, 259–265.

Clegg, C. D., Ritz, K. & Griffiths, B. S. (1998). Broad-scale analysis of soil microbial community DNA from upland grasslands. *Antonie van Leeuwenhoek International Journal of General and Molecular Microbiology, 73*, 9–14.

Collins, S. L. & Benning, T. L. (1996). Spatial and temporal patterns in functional diversity. In K. J. Gaston (Ed.), *Biodiversity: A Biology of Numbers and Difference* (pp. 253–280). Oxford, UK: Blackwell Science.

Cooke, R. C. & Rayner, A. D. M. (1984). *Ecology of Saprotrophic Fungi*. London, UK: Longman.

Costello, A. M. & Lidstrom, M. E. (1999). Molecular characterization of functional and phylogenetic genes from natural populations of methanotrophs in lake sediments. *Applied and Environmental Microbiology, 65*, 5066–5074.

Couturier, M., Bex, F., Bergquist, P. L. & Maas, W. K. (1998). Identification and classification of bacterial plasmids. *Microbiological Reviews, 52*, 375–395.

Davison, J. (1999). Genetic exchange between bacteria in the environment. *Plasmid, 42*, 73–91.

Dawid, W (2000). Biology and global distribution of myxobacteria in soils. *FEMS Microbiology Reviews, 24*, 403–427.

De Boer, W., Gerards, S., Gunnewiek, P J. A. & Modderman, R. (1999). Response of the chitinolytic microbial community to chitin amendments of dune soils. *Biology and Fertility of Soils, 29*, 170–177.

Dejonghe, W., Goris, J., El Fantroussi, S., Hofte, M., De Vos, P., Verstraete, W. & Top, E. M. (2000). Effect of dissemination of 2,4-dichlorophenoxyacetic acid (2,4-D) degradation plasmids on 2,4-D degradation and on bacterial community structure in two different soil horizons. *Applied and Environmental Microbiology, 66*, 3297–3304.

De Long, E. F. (1998). Everything in moderation: Archaea as 'non-extremophiles'. *Current Opinion in Genetics and Development, 8*, 649–654

Dewey, F. M., Thornton, C. R. & Gilligan, C. A. (1997). Use of monoclonal antibodies to detect, quantify and visualise fungi in soils. *Advances in Botanical Research, 24*, 275–308.

Diels, L., De Smet, M., Hooyberghs, L. & Corbisier, P. (1999). Heavy metals bioremediation of soil. *Molecular Biotechnology, 12*, 149–158.

Dobranic, J. K. & Zak, J. C. (1999). A microtiter plate procedure for evaluating fungal functional diversity. *Mycologia, 91*, 756–765.

Donnison, L. M., Griffith, G. S., Hedger, J., Hobbs, P. J. & Bardgett, R. D. (2000). Management influences on soil microbial communities and their function in botanically diverse haymeadows of northern England and Wales. *Soil Biology and Biochemistry, 32*, 253–263.

Duineveld, B. M., Rosado, A. S., Van Elsas, J. D. & Van Veen, J. A. (1998). Analysis of the dynamics of bacterial communities in the rhizosphere of the chrysanthemum via denaturing gradient gel electrophoresis and substrate utilization patterns. *Applied and Environmental Microbiology, 64*, 4950–4957.

Dunbar, J., Ticknor, L. O. & Kuske, C. R. (2000). Assessment of microbial diversity in four southwestern United States soils by 16S rRNA gene terminal restriction fragment analysis. *Applied and Environmental Microbiology, 66*, 2943–2950.

Duncan, K. E., Ferguson, N., Kimura, K., Zhou, X. & Istock, C. A. (1994). Fine scale genetic and phenotypic structure in natural populations of *Bacillus subtilis* and *Bacillus licheniformis* – implications for bacterial evolution and speciation. *Evolution, 48*, 2002–2025.

Durrall, D. M. & Parkinson, D. (1991). Initial fungal community development on decomposing timothy (*Phleum pratense*) litter from a reclaimed coal-mine spoil in Alberta, Canada. *Mycological Research, 95*, 14–18.

Dykhuizen, D. E. (1998). Santa Rosalia revisited: why are there so many species of bacteria? *Antonie van Leeuwenhoek International Journal of General and Molecular Microbiology, 73*, 25–33.

Edwards, C. (2000). Problems posed by natural environments for monitoring microorganisms. *Molecular Biotechnology, 15*, 211–223.

Ellis, M. B. & Ellis, J. P. (1997). *Microfungi on Land Plants* (2nd edn). Slough, UK: The Richmond Publishing Co.

England, L. S., Holmes, S. B. & Trevors, J. T. (1998). Persistence of viruses and DNA in soil. *World Journal of Microbiology and Biotechnology, 14*, 163–169.

Felske, A. & Akkermans, A. D. L. (1998). Prominent occurrence of ribosomes from an uncultured bacterium of the Verrucomicrobiales cluster in grassland soils. *Letters in Applied Microbiology, 26*, 219–223.

Felske, A., Wolterink, A., Van Lis, R. & Akkermans, A. D. L. (1998). Phylogeny of the main bacterial 16S rRNA sequences in Drentse A grassland soils (the Netherlands). *Applied and Environmental Microbiology, 64*, 871–879.

Felske, A., Wolterink, A., Van Lis R., De Vos, W. M. & Akkermans, A. D. L. (1999). Searching for predominant soil bacteria: 16S rDNA cloning versus strain cultivation. *FEMS Microbiology Ecology, 30*, 137–145.

Fenchel, T., Esteban, G. F. & Finlay, B. J. (1997). Local versus global diversity of microorganisms: cryptic diversity of ciliated protozoa. *Oikos, 80*, 220–225.

Finlay, B. J., Maberly, S. C. & Cooper, J. I. (1997). Microbial diversity and ecosystem function. *Oikos, 80*, 209–213.

Flanagan, P. W. & Scarborough, A. M. (1974). Physiological groups of decomposer fungi on tundra plant remains. In A. J. Holding, O. W. Heal, S. F. MacLean Jr. & P. W. Flanagan (Eds), *Soil Organisms and Decomposition in Tundra* (pp. 159–181). Stockholm, Sweden: Tundra Biome Steering Committee.

Frankland, J. C. (1982). Biomass and nutrient cycling by decomposer basidiomycetes. In J. C. Frankland, J. N. Hedger & M. J. Swift (Eds), *Decomposer Basidiomycetes: Their Biology and Ecology*. Cambridge, UK: Cambridge University Press.

Frankland, J. C. (1992). Mechanisms in fungal succession. In G. C. Carroll & D. T. Wicklow (Eds), *The Fungal Community* (2nd edn), (pp. 383–401). New York: Marcel Dekker.

Frankland, J. C. (1998). Presidential Address. Fungal succession – unravelling the unpredictable. *Mycological Research, 102*, 1–15.

Frankland, J. C., Dighton, J. & Boddy, L. (1990). Methods for studying fungi in soil and forest litter. In R. Grigorova & J. R. Norris (Eds), *Methods in Microbiology* (Vol. 22), (pp. 343–404). London, UK: Academic Press.

Frankland, J. C., Poskitt, J. M. & Howard, D. M. (1995). Spatial development of populations of a decomposer fungus, *Mycena galopus. Canadian Journal of Botany, 73*, S1399–S1406.

Frey, S. D., Elliott, E. T. & Paustian, K. (1999). Bacterial and fungal abundance and biomass in conventional and no-tillage agroecosystems along two climatic gradients. *Soil Biology and Biochemistry, 31*, 573–585.

Frostegård, A. & Bååth, E. (1996). The use of phospholipid fatty acid analysis to estimate bacterial and fungal biomass in soil. *Biology and Fertility of Soils, 22*, 59–65.

Frostegård, A., Petersen, S. O., Bååth, E. & Nielsen, T. H. (1997). Dynamics of a microbial community associated with manure hot spots as revealed by phospholipid fatty acid analyses. *Applied and Environmental Microbiology, 63*, 2224–2231.

Gadd, G. M. (2000). Bioremedial potential of microbial mechanisms of metal mobilization and immobilization. *Current Opinion In Biotechnology, 11*, 271–279.

Gagliardi, J. V. & Karns, J. S. (2000). Leaching of *Escherichia coli* O157 : H7 in diverse soils under various agricultural management practices. *Applied and Environmental Microbiology, 66*, 877–883.

Garcia-Martinez, J. & Rodriguez-Valera, F. (2000). Microdiversity of uncultured amrine prokaryotes: the SAR11 cluster and the marine Archaea of Group I. *Molecular Ecology, 9*, 935–948.

Garland, J. L. & Mills, A. L. (1991). Classification and characterisation of heterotrophic microbial communities on the basis of patterns of community-level sole-carbon-source utilisation. *Applied and Environmental Microbiology, 57*, 2351–2359.

Gelsomino, A., Keijzer-Wolters, A. C., Cacco, G. & Van Elsas, J. D. (1999). Assessment of bacterial community structure in soil by polymerase chain reaction and denaturing gradient gel electrophoresis. *Journal of Microbiological Methods, 38*, 1–15.

Gitay, H., Wilson, J. B. & Lee, W. G. (1996). Species redundancy: a redundant concept? *Journal of Ecology, 84*, 121–124.

Gray, S. N., Dighton, J., Olsson, S. & Jennings, D. H. (1995). Real-time measurement of uptake and translocation of 137Cs within mycelium of *Schizophyllum commune* Fr. by autoradiography followed by quantitative image analysis. *New Phytologist, 129*, 449–465.

Griffiths, B. S., Ritz, K., Ebblewhite, N. & Dobson, G. (1999). Soil microbial community structure: effects of substrate loading rates. *Soil Biology and Biochemistry, 31*, 145–153.

Griffiths, B. S., Ritz, K., Bardgett, R. D., Cook, R., Christensen, S., Ekelund, F., Sorensen, S. J., Bååth, E., Bloem, J., De Ruiter, P. C., Dolfing, J. & Nicolardot, B. (2000). Ecosystem response of pasture soil communities to fumigation-induced microbial diversity reductions: an examination of the biodiversity-ecosystem function relationship. *Oikos, 90*, 279–294.

Hagvar, S. (1998). The relevance of the Rio-Convention on biodiversity to conserving the biodiversity of soils. *Applied Soil Ecology, 9*, 1–7.

Harley, J. L. & Waid, J. S. (1955). A method of studying active mycelia on living roots and other surfaces in the soil. *Transactions of the British Mycological Society, 38*, 104–118.

He, S. Y. (1998). Type III protein secretion systems in plant and animal pathogenic bacteria. *Annual Review of Phytopathology, 36*, 363–392.

Head, I. M., Saunders, J. R. & Pickup, R. W. (1998). Microbial evolution, diversity, and ecology: a decade of ribosomal RNA analysis of uncultivated microorganisms. *Microbial Ecology, 35*, 1–21.

Hedrick, D. B., Peacock, A., Stephen, J. R., Macnaughton, S. J., Bruggemann, J. & White, D. C. (2000). Measuring soil microbial community diversity using polar lipid fatty acid and denaturing gradient gel electrophoresis data. *Journal of Microbiological Methods, 41*, 235–248.

Henckel, T., Jackel, U., Schnell, S. & Conrad, R. (2000). Molecular analyses of novel methanotrophic communities in forest soil that oxidize atmospheric methane. *Applied and Environmental Microbiology, 66*, 1801–1808.

Herbert, R. A. (1999). Nitrogen cycling in coastal marine ecosystems. *FEMS Microbiology Reviews, 23*, 563–590.

Hering, T. F. (1967). Fungal decomposition of oak leaf litter. *Transactions of the British Mycological Society, 50*, 267–273.

Hering, T. F. (1982). Decomposing activity of basidiomycetes in forest litter. In J. C. Frankland, J. N. Hedger. & M. J. Swift (Eds), *Decomposer Basidiomycetes: Their Biology and Ecology* (pp. 213–225). Cambridge, UK: Cambridge University Press.

Hill, K. E. & Top, E. M. (1998). Gene transfer in soil systems using microcosms. *FEMS Microbiology Ecology, 25*, 319–329.

Hodge, A., Grayston, S. J., Campbell, C. D., Ord, B. G. & Killham, K. (1998). Characterisation and microbial utilisation of exudate material from the rhizosphere of *Lolium perenne* grown under CO_2 enrichment. *Soil Biology and Biochemistry, 30*, 1033–1043.

Horn, C., Paulmann, B., Kerlen, G., Junker, N. & Huber, H. (1999). In vivo observation of cell division of anaerobic hyperthermophiles by using a high-intensity dark-field microscope. *Journal of Bacteriology, 181*, 5114–5118.

Hristova, K. R., Mau, M., Zheng, D., Aminov, R. I., Mackie, R. I., Gaskins, H. R. & Raskin, L. (2000). *Desulfotomaculum* genus- and subgenus-specific 16S rRNA hybridization probes for environmental studies. *Environmental Microbiology, 2*, 143–159.

Hu, S. J., Van Bruggen, A. H. C. & Grunwald, N. J. (1999). Dynamics of bacterial populations in relation to carbon availability in a residue amended soil. *Applied Soil Ecology, 13*, 21–30.

Hurst, C. J. (1997). Sampling viruses from soil. In *Manual of Environmental Microbiology* (pp. 381–382). Washington DC, USA: ASM Press.

Hurst, C. J., Schaub, S. A., Sobsey, M. D., Farrah, S. R., Gerba, C. P., Rose, J. B., Goyal, S. M., Larkin, E. P., Sullivan, R., Tierney, J. T., O'Brien, R. T., Safferman, R. S., Morris, M. E., Wellings, F. M., Lewis, A. L., Berg, G., Britton, P. W. & Winter, J. A. (1991). Multilaboratory evaluation of methods for detecting enteric viruses in soils. *Applied and Environmental Microbiology, 57*, 395–401.

Jeffrey, W. H., Nazaret, S. & Barkay, T. (1996). Detection of the merA gene and its expression in the environment. *Microbial Ecology, 32*, 293–303.

Kang, Y. W., Carlson, R., Tharpe, W. & Schell, M. A. (1998). Characterization of genes involved in biosynthesis of a novel antibiotic from *Burkholderia cepacia* BC11 and their role in biological control of *Rhizoctonia solani*. *Applied and Environmental Microbiology, 64*, 3939–3947.

Kennedy, A. C. (1999). Bacterial diversity in agroecosystems. *Agriculture Ecosystems and Environment, 74*, 65–76.

Kepner, R. L. & Pratt, J. R. (1994). Use of fluorochromes for direct enumeration of total bacteria in environmental samples: past and present. *Microbiological Reviews, 58*, 603–615.

Killham, K. (1994). In *Soil Ecology* (pp. 34–61). Cambridge, UK: Cambridge University Press.

Kjøller, A. & Struwe, S. (1982). Microfungi in ecosystems: fungal occurrence and activity in litter and soil. *Oikos, 39*, 391–422.

Kowalchuk, G. A., Gerards, S. & Woldendorp, J. A. (1997). Detection and characterisation of fungal infections of *Ammophila arenaria* (marram grass) roots by Denaturing Gradient Gel Electrophoresis of specifically amplified 18S rDNA. *Applied and Environmental Microbiology, 63*, 3858–3865.

Krsek, M. & Wellington, E. M. H. (1999). Comparison of different methods for the isolation and purification of total community DNA from soil. *Journal of Microbiological Methods, 39*, 1–16.

Kuske, C. R., Barns, S. M. & Busch, J. D. (1997). Diverse uncultivated bacterial groups from soils of the arid southwestern United States that are present in many geographic regions. *Applied and Environmental Microbiology, 63*, 3614–3621.

Lamar, R. T., Schoenike, B., Vanden Wymelenberg, A., Stewart, P., Dietrich, D. M. & Cullen, D. (1995). Quantitation of fungal mRNAs in complex substrates by reverse transcription PCR and its application to *Phanerochaete chrysosporium*-colonised soil. *Applied and Environmental Microbiology, 61*, 2122–2126.

Lan, R. & Reeves, P. R. (2000). Intraspecies variation in bacterial genomes: the need for a species genome concept. *Trends in Microbiology, 8*, 396–401.

Lawlor, K., Knight, B. P., Barbosa-Jefferson, V. L., Lane, P. W., Lilley, A. K., Paton, G. I., McGrath, S. P., O'Flaherty, S. M. & Hirsch, P. R. (2000). Comparison of methods to investigate microbial populations in soils under different agricultural management. *FEMS Microbiology Ecology, 33*, 129–137.

Lawrence, J. G. (1999). Gene transfer, speciation, and the evolution of bacterial genomes. *Current Opinion in Microbiology, 2*, 519–523.

Lavelle, P. (2000). Ecological challenges for soil science. *Soil Science, 165*, 73–86.

Liesack, W. & Stackebrandt, E (1992). Occurrence of novel groups of the domain Bacteria as revealed by analysis of genetic material isolated from an Australian terrestrial environment. *Journal of Bacteriology, 174*, 5072–5078.

Lilley, A. K., Bailey, M. J., Day, M. J. & Fry, J. C. (1996). Diversity of mercury resistance plasmids obtained by exogenous isolation from the bacteria of sugar beet in three successive years. *FEMS Microbiology Ecology, 20*, 211–227.

Lindahl, B., Stenlid, J., Olsson, S. & Finlay, R. (1999). Translocation of ^{32}P between interacting mycelia of a wood-decomposing fungus and ectomycorrhizal fungi in microcosm systems. *New Phytologist, 144*, 183–193.

Lipson, D. A., Schmidt, S. K. & Monson, R. K. (1999). Links between microbial population dynamics and nitrogen availability in an alpine ecosystem. *Ecology, 80*, 1623–1631.

Liu, W. T., Marsh, T. L., Cheng, H. & Forney, L. J. (1997). Characterization of microbial diversity by determining terminal restriction fragment length polymorphisms of genes encoding 16S rRNA. *Applied and Environmental Microbiology, 63*, 4516–4522.

Madigan, M. T., Martinko, J. M. & Parker, J. (2000). *Brock Biology of Microorganisms* (9th edn). New Jersey, USA: Prentice Hall.

Marilley, L. & Aragno, M. (1999). Phylogenetic diversity of bacterial communities differing in degree of proximity of *Lolium perenne* and *Trifolium repens* roots. *Applied Soil Ecology, 13*, 127–136.

Marsh, T. L. (1999). Terminal restriction fragment length polymorphism (T-RFLP): an emerging method for characterizing diversity among homologous populations of amplification products. *Current Opinion in Microbiology, 2*, 323–327.

Meharg, A. A. & Cairney, J. W. G. (2000). Ectomycorrhizas – extending capabilities of rhizosphere remediation? *Soil Biology & Biochemistry, 32*, 1475–1484.

Miller, M., Palojärvi, A., Rangger, A., Reeslev, M. & Kjøller, A. (1998). The use of fluorogenic substrates to measure fungal presence and activity in soil. *Applied and Environmental Microbiology, 64*, 613–617.

Miller, S. L. (1995). Functional diversity of fungi. *Canadian Journal of Botany, 73*, S50–S57.

Muyzer, G. (1999). DGGE/TGGE a method for identifying genes from natural ecosystems. *Current Opinion in Microbiology, 2*, 317–322.

Muyzer, G. & Smalla, K. (1998). Application of denaturing gradient gel electrophoresis (DGGE) and temperature gradient gel electrophoresis (TGGE) in microbial ecology. *Antonie van Leeuwenhoek International Journal of General and Molecular Microbiology, 73*, 127–141.

Nakatsu, C. H., Torsvik, V. & Ovreas, L. (2000). Soil community analysis using DGGE of 16S rDNA polymerase chain reaction products. *Soil Science Society of America Journal, 64*, 1382–1388.

Newsham, K. K., Frankland, J. C., Boddy, L. & Ineson, P. (1992a). Effects of dry-deposited sulphur dioxide on fungal decomposition of angiosperm tree leaf litter. I, Changes in communities of fungal saprotrophs. *New Phytologist, 122*, 97–110.

Newsham, K. K., Ineson, P., Boddy, L. & Frankland, J. C. (1992b). Effects of dry-deposited sulphur dioxide on fungal decomposition of angiosperm tree leaf litter. II, Chemical content of litters. *New Phytologist, 122*, 111–126.

Newsham, K. K., Boddy, L., Frankland, J. C. & Ineson, P. (1992c). Effects of dry-deposited sulphur dioxide on fungal decomposition of angiosperm tree leaf litter. III, Decomposition rates and fungal respiration. *New Phytologist, 122*, 127–140.

Newton, A. C. & Haigh, J. M. (1998). Diversity of ectomycorrhizal fungi in Britain: a test of the species-area relationship, and the role of host specificity. *New Phytologist, 138*, 619–627.

Normander, B. & Prosser, J. I. (2000). Bacterial origin and community composition in the barley phytosphere as a function of habitat and presowing conditions. *Applied and Environmental Microbiology, 66*, 4372–4377.

Nubel, U., Garcia Pichel, F., Kuhl, M. & Muyzer, G. (1999). Quantifying microbial diversity: morphotypes, 16S rRNA genes, and carotenoids of oxygenic phototrophs in microbial mats. *Applied and Environmental Microbiology, 65*, 422–430.

Nusslein, K. & Tiedje, J. M. (1999). Soil bacterial community shift correlated with change from forest to pasture vegetation in a tropical soil. *Applied and Environmental Microbiology, 65*, 3622–3626.

O'Donnell, A. G. & Gorres, H. E. (1999). 16S rDNA methods in soil microbiology. *Current Opinion in Biotechnology, 10*, 225–229.

Ohtonen, R., Aikio, S. & Vare, H. (1997). Ecological theories in soil biology. *Soil Biology and Biochemistry, 29*, 1613–1619.

Osborn, A. M., Bruce, K. D., Strike, P. & Ritchie, D. A. (1997). Distribution, diversity and evolution of the bacterial mercury resistance (*mer*) operon. *FEMS Microbiology Reviews, 19*, 239–262.

Osborn, A. M., Pickup, R. W. & Saunders, J. R. (2000). Development and application of molecular tools in the study of IncN-related plasmids from lakewater sediments. *FEMS Microbiology Letters, 186*, 203–208.

Ovreas, L. (2000). Population and community level approaches for analysing microbial diversity in natural environments. *Ecology Letters, 3*, 236–251.

Paterson, R. R. M. & Bridge, P. D. (1994). *Biochemical Techniques for Filamentous Fungi*. International Mycological Technical Handbook. Wallingford, UK: CAB International.

Pielou, E. C. (1975). *Ecological Diversity*. New York, USA: Wiley-Interscience.

Prosser, J. I., Killham, K., Glover, L. A. & Rattray, E. A. S. (1996). Luminescence-based systems for detection of bacteria in the environment. *Critical Reviews in Biotechnology, 16*, 157–183.

Pukall, R., Tschape, H. & Smalla, K. (1996). Monitoring the spread of broad host and narrow host range plasmids in soil microcosms. *FEMS Microbiology Ecology, 20*, 53–66.

Radajewski, S., Ineson, P., Parekh, N. R. & Murrell, J. C. (2000). Stable-isotope probing as a tool in microbial ecology. *Nature, 403*, 646–649.

Ramakrishnan, B., Lueders, T., Conrad, R. & Friedrich, M. (2000). Effect of soil aggregate size on methanogenesis and archaeal community structure in anoxic rice field soil. *FEMS Microbiology Ecology, 32*, 261–270.

Ranjard, L., Nazaret, S., Gourbiere, F., Thioulouse, J., Linet, P. & Richaume, A. (2000a). A soil microscale study to reveal the heterogeneity of Hg(II) impact on indigenous bacteria by quantification of adapted phenotypes and analysis of community DNA fingerprints. *FEMS Microbiology Ecology, 31*, 107–115.

Ranjard, L., Poly, F., Combrisson, J., Richaume, A., Gourbiere, F., Thioulouse, J. & Nazaret, S. (2000b). Heterogeneous cell density and genetic structure of bacterial pools associated with various soil microenvironments as determined by enumeration and DNA fingerprinting approach (RISA). *Microbial Ecology, 39*, 263–272.

Rheims, H., Sproer, C., Rainey, F. A. & Stackebrandt, E. (1996). Molecular biological evidence for the occurrence of uncultured members of the actinomycete line of descent in different environments and geographical locations. *Microbiology, 142*, 2863–2870.

Robinson, C. H., Dighton, J. & Frankland, J. C. (1993a). Resource capture by interacting fungal colonisers of straw. *Mycological Research, 97*, 547–558.

Robinson, C. H., Dighton, J., Frankland, J. C. & Coward, P. A. (1993b). Nutrient and carbon dioxide release from straw by interacting species of fungi. *Plant and Soil, 151*, 139–142.

Robinson, C. H., Dighton, J., Frankland, J. C. & Roberts, J. D. (1994). Fungal communities on decaying wheat straw of different resource qualities. *Soil Biology and Biochemistry, 26*, 1053–1058.

Robinson, C. H., Fisher, P. J. & Sutton, B. C. (1998). Fungal biodiversity in dead leaves of fertilised plants of *Dryas octopetala* from a high Arctic site. *Mycological Research, 102*, 573–576.

Rodriguez, G. G., Phipps, D., Ishiguro, K. & Ridgway, H. F. (1992). Use of a fluorescent redox probe for visualization of actively respiring bacteria. *Applied and Environmental Microbiology, 58*, 1801–1808.

Rondon, M. R., Goodman, R. M. & Handelsman, J. (1999). The Earth's bounty: assessing and accessing soil microbial diversity. *Trends in Biotechnology, 17*, 403–409.

Rondon, M. R., August, P. R., Bettermann, A. D., Brady, S. F., Grossman, T. H., Liles, M. R., Loiacono, K. A., Lynch, B. A., MacNeil, I. A., Minor, C., Tiong, C. L., Gilman, M., Osburne, M. S., Clardy, J., Handelsman, J. & Goodman, R. M. (2000). Cloning the soil metagenome: a strategy for accessing the genetic and functional diversity of uncultured microorganisms. *Applied and Environmental Microbiology, 66*, 2541–2547.

Ruzicka, S., Edgerton, D., Norman, M. & Hill, T. (2000). The utility of ergosterol as a bioindicator of fungi in temperate soils. *Soil Biology and Biochemistry, 32*, 989–1005.

Sandaa, R. A., Torsvik, V., Enger, O., Daae, F. L., Castberg, T. & Hahn, D. (1999). Analysis of bacterial communities in heavy metal-contaminated soils at different levels of resolution. *FEMS Microbiology Ecology, 30*, 237–251.

Schultze, M. & Kondorosi, A. (1998). Regulation of symbiotic root nodule development. *Annual Review of Genetics, 32*, 33–57.

Sentchilo, V. S., Perebituk, A. N., Zehnder, A. J. B. & Van der Meer, J. R. (2000). Molecular diversity of plasmids bearing genes that encode toluene and xylene metabolism in *Pseudomonas* strains isolated from different contaminated sites in Belarus. *Applied and Environmental Microbiology, 66*, 2842–2852.

Sharma, S., Rangger, A., Von Lutzow, M. & Insam, H. (1998). Functional diversity of soil bacterial communities increases after maize litter amendment. *European Journal of Soil Biology, 34*, 53–60.

Silver, S. (1996). Bacterial resistances to toxic metal ions – A review. *Gene, 179*, 9–19.

Simpson, A. J. G., Reinach, F. C., Arruda, P., Abreu, F. A., Acencio, M., Alvarenga, R., Alves, L. M. C., Araya, J. E., Baia, G. S., Baptista, C. S., Barros, M. H., Bonaccorsi, E. D., Bordin, S., Bove, J. M., Briones, M. R. S., Bueno, M. R. P., Camargo, A. A., Camargo, L. E. A, Carraro, D. M., Carrer, H., Colauto, N. B., Colombo, C., Costa, F. F., Costa, M. C. R., Costa-Neto, C. M., Coutinho, L. L., Cristofani, M., Dias-Neto, E., Docena, C., El-Dorry, H., Facincani, A. P., Ferreira, A. J. S., Ferreira, V. C. A., Ferro, J. A., Fraga, J. S., Franca, S. C., Franco, M. C., Frohme, M., Furlan, L. R., Garnier, M., Goldman, G. H., Goldman, M. H. S., Gomes, S. L., Gruber, A., Ho P. L., Hoheisel, J. D., Junqueira, M. L., Kemper, E. L., Kitajima, J. P., Krieger, J. E., Kuramae, E. E., Laigret, F., Lambais, M. R., Leite, L. C. C., Lemos, E. G. M., Lemos, M. V. F., Lopes, S. A., Lopes, C. R., Machado, J. A., Machado, M. A., Madeira, A. M. B. N., Madeira, H. M. F., Marino, C. L., Marques, M. V., Martins, E. A. L., Martins, E. M. F., Matsukuma, A. Y., Menck, C. F. M., Miracca, E. C., Miyaki, C. Y., Monteiro-Vitorello, C. B., Moon, D. H., Nagai, M. A., Nascimento, A. L. T. O., Netto, L. E. S., Nhani, A., Nobrega, F. G., Nunes, L. R., Oliveira, M. A., De Oliveira, M. C., De Oliveira, R. C., Palmieri, D. A., Paris, A., Peixoto, B. R., Pereira, G. A. G., Pereira, H. A., Pesquero, J. B., Quaggio, R. B., Roberto, P. G., Rodrigues, V., Rosa, A. J. D., De Rosa, V. E., De Sa, R. G., Santelli, R. V., Sawasaki, H. E., Da Silva, A. C. R., Da Silva, A. M., Da Silva, F. R., Silva, W. A., Da Silveira, J. F., Silvestri, M. L. Z., Siqueira, W. J., De Souza, A. A., De Souza, A. P., Terenzi, M. F., Truffi, D., Tsai, S. M., Tsuhako, M. H., Vallada, H., Van Sluys, M. A., Verjovski-Almeida, S., Vettore, A. L., Zago, M. A., Zatz M., Meidanis, J. & Setubal, J. C. (2000). The genome sequence of the plant pathogen *Xylella fastidiosa*. *Nature, 406*, 151–157.

Sjoling, S. & Cowan, D. A. (2000). Detecting human bacterial contamination in Antarctic soils. *Polar Biology, 23*, 644–650.

Smalla, K., Wachtendorf, U., Heuer, H., Liu W. T. & Forney L. (1998). Analysis of BIOLOG GN substrate utilization patterns by microbial communities. *Applied and Environmental Microbiology, 64*, 1220–1225.

Smit, E., Leeflang, P. & Wernars, K. (1997). Detection of shifts in microbial community structure and diversity in soil caused by copper contamination using amplified ribosomal DNA restriction analysis. *FEMS Microbiology Ecology, 23*, 249–261.

Smit, E., Leeflang, P., Glandorf, B., Van Elsas, J. D. & Wernars, K. (1999). Analysis of fungal diversity in the wheat rhizosphere by sequencing of cloned PCR-amplified genes encoding 18S rRNA and temperature gradient gel electrophoresis. *Applied and Environmental Microbiology, 65*, 2614–2621.

Smith, S. E. & Read, D. J. (1997). *Mycorrhizal Symbiosis*. London, UK: Academic Press.

Staley, J. T. & Konopka, A. (1985). Measurement of in situ activities of nonphotosynthetic microorganisms in aquatic and terrestrial habitats. *Annual Review of Microbiology, 39*, 321–346.

Straub, T. M., Pepper, I. L. & Gerba, C. P. (1993). Virus survival in sewage sludge amended desert soil. *Water Science and Technology, 27*, 421–424.

Straub, T. M., Pepper, I. L. & Gerba, C. P. (1995). Comparison of PCR and cell culture for detection of enteroviruses in sludge-amended field soils and determination of their transport. *Applied and Environmental Microbiology, 61*, 2066–2068.

Strobel, G. A., Hess, W. M., Ford, E., Sidhu, R. S. & Yang X. (1996). Taxol from fungal endophytes and the issue of biodiversity. *Journal of Industrial Microbiology and Biotechnology, 17*, 417–423.

Swift, M. J. (1976). Species diversity and the structure of microbial communities in terrestrial habitats. In J. M. Anderson & A. MacFadyen (Eds), *The Role of Aquatic and Terrestrial Organisms in Decomposition Processes* (pp. 185–222). Oxford, UK: Blackwell Scientific Publications.

Swift, M. J., Heal, O. W. & Anderson, J. M. (1979). *Decomposition in Terrestrial Ecosystems*. Oxford, UK: Blackwell Scientific Publications.

Tiedje, J. M., Asuming-Brempong, S., Nusslein, K., Marsh, T. L. & Flynn, S. J. (1999). Opening the black box of soil microbial diversity. *Applied Soil Ecology, 13*, 109–122.

Torsvik, V., Goksoyr, J. & Daae, F. L. (1990). High diversity in DNA of soil bacteria. *Applied and Environmental Microbiology, 56*, 782–787.

Tscherko, D. & Kandeler, E. (1999). Classification and monitoring of soil microbial biomass, N-mineralization and enzyme activities to indicate environmental changes. *Bodenkultur, 50*, 215–226.

Ulrich, A. & Wirth, S. (1999). Phylogenetic diversity and population densities of culturable cellulolytic soil bacteria across an agricultural encatchment. *Microbial Ecology, 37*, 238–247.

Urbanska, K. M., Webb, N. R. & Edwards, P. J. (1977). *Restoration Ecology and Sustainable Development*. Cambridge, UK: Cambridge University Press.

Visser, S. & Parkinson, D. (1975). Fungal succession on aspen poplar leaf litter. *Canadian Journal of Botany, 53*, 1640–1651.

Von Wintzingerode, F., Gobel, U. B. & Stackebrandt, E. (1997). Determination of microbial diversity in environmental samples: pitfalls of PCR-based rRNA analysis. *FEMS Microbiology Reviews, 21*, 213–229.

Warcup, J. H. (1957). Studies on the occurrence and activity of fungi in a wheat-field soil. *Transactions of the British Mycological Society, 40*, 237–262.

Ward, D. M. (1998). A natural species concept for prokaryotes. *Current Opinion in Microbiology, 1*, 271–277.

Watling, R. (1995). Assessment of fungal diversity: macromycetes, the problems. *Canadian Journal of Botany, 73*, S15–S24.

Watling, R. (1997). Pulling the threads together: habitat diversity. *Biodiversity and Conservation, 6*, 753–763.

Weinbauer, M. G., Beckmann, C. & Höfle, M. G. (1998). Utility of green fluorescent nucleic acid dyes and aluminum oxide membrane filters for rapid epifluorescence enumeration of soil and sediment bacteria. *Applied and Environmental Microbiology, 64*, 5000–5003.

White, D. C. (1995). Chemical ecology: possible linkage between macro- and microbial ecology. *Oikos, 74*, 177–184.

Whitman, W. B., Coleman, D. C. & Wiebe, W. J. (1998). Prokaryotes: the unseen majority. *Proceedings of the National Academy of Sciences USA, 95*, 6578–6583.

Widden, P. & Parkinson, D. (1979). Populations of fungi in a high arctic ecosystem. *Canadian Journal of Botany, 57*, 2408–2417.

Woese, C. R. (2000). Interpreting the universal phylogenetic tree. *Proceedings of the National Academy of Sciences, 97*, 8392–8396.

Woese, C. R., Kandler, O. & Wheelis, M. L. (1990). Towards a natural system of organisms: proposal for the domains Archaea, Bacteria, and Eucarya. *Proceedings of the National Academy of Sciences*, USA, *87*, 4576–4579.

Wood, D. A. (1998). Extracellular enzymes of *Agaricus bisporus. Proceedings of Sixth International Symposium of the Mycological Society of Japan*. UK/Japan Joint Symposium, Chibo, November 1998.

Yang, Y. H., Yao, J., Hu, S. & Qi, Y. (2000). Effects of agricultural chemicals on DNA sequence diversity of soil microbial community: a study with RAPD marker. *Microbial Ecology, 39*, 72–79.

Yin, B., Crowley, D., Sparovek, G., De Melo, W. J. & Borneman, J. (2000). Bacterial functional redundancy along a soil reclamation gradient. *Applied and Environmental Microbiology, 66*, 4361–4365.

Zak, J. C. & Visser, S. (1996). An appraisal of soil fungal biodiversity: the crossroads between taxonomic and functional biodiversity. *Biodiversity and Conservation, 5*, 169–183.

Zani, S., Mellon, M. T., Collier, J. L. & Zehr, J. P. (2000). Expression of *nifH* genes in natural microbial assemblages in Lake George, New York, detected by reverse transcriptase PCR. *Applied and Environmental Microbiology, 66*, 3119–3124.

Zarda, B., Hahn, D., Chatzinotas, A., Schonhuber, W., Neef, A., Amann, R. I. & Zeyer, J. (1997). Analysis of bacterial community structure in bulk soil by in situ hybridization. *Archives of Microbiology, 168*, 185–192.

Zavarzin, G. A. (1995). The microorganisms: a concept in need of clarification or one now to be rejected? In D. Allsopp, R. R. Colwell & D. L. Hawksworth (Eds), *Microbial Diversity and Ecosystem Function* (pp. 17–26). Wallingford, UK: CAB International.

Zillig, W., Prangishvilli, D., Schleper C., Elferink, M., Holz, I., Albers, S., Janekovic, D. & Gotz, D. (1996). Viruses, plasmids and other genetic elements of thermophilic and hyperthermophilic Archaea. *FEMS Microbiology Reviews, 18*, 225–236.

Zwart, G., Hiorns, W. D., Methe, B. A., Van Agterveld, M. P., Huismans, R., Nold, S. C., Zehr, J. P. & Laanbroek, H. J. (1998). Nearly identical 16S rRNA sequences recovered from lakes in North America and Europe indicate the existence of clades of globally distributed freshwater bacteria. *Systematic and Applied Microbiology, 21*, 546–556.

Miranda J. Keith-Roach and Francis R. Livens (Editors)
© 2002 Elsevier Science Ltd. All rights reserved

Chapter 2

The characterisation of microbial communities in environmental samples

Nisha R. Parekh[a], Richard D. Bardgett[b]

[a]*Centre for Ecology and Hydrology – Merlewood, Grange-over-sands, Cumbria LA11 6JU, UK*
[b]*Institute of Environmental and Natural Sciences, Department of Biological Sciences, Lancaster University, Lancaster LA1 4YQ, UK*

1. Introduction

A major challenge in environmental microbiology is the understanding of microbial processes, such as decomposition, nutrient cycling, pollutant degradation and energy flow, at the organism level. There are two major goals here: first, to determine what controls the distribution, diversity, structure and abundance of natural microbial communities; and second, to understand how microbial dynamics and interactions at the population and community level influence process rates in the environment. Until recently, progress in these research areas was hampered by the inability of methods to characterise the vast diversity of microbial communities in the environment. This is due largely to the non-culturability of most microbial cells – for example, in soil, the portion which can be cultured in the laboratory has been estimated to be 0.3% of the total number of cells observed microscopically – but also to problems of extracting micoorganisms from complex and variable matrices. In this chapter, we report on how recently developed molecular methods, which do not rely on the cultivation of microorganisms prior to analysis, offer the potential to address some of these contemporary research issues and questions in environmental microbiology. The methods described in this chapter can be separated into three categories: (1) those that generate phenotypic data through the analysis of lipids and other microbial biomarkers; (2) those that provide genotypic data through the analysis of extracted nucleic acids; and (3) those that make mechanistic linkages between phenotypic data on microorganisms and natural processes, through measuring the assimilation of isotopically labelled substrates by biomarkers of the microbial community. We purposefully do not provide recipes for the methods described; these are available elsewhere (Akkermans et al., 1995; Van Elsas et al., 1997; Roberston et al., 1999).

2. Phenotypic analysis: phospholipid fatty acid analysis

The direct derivation of cellular components from entire microbial communities represents one important method for quantitative assessment of microbial abundance and composition. Cellular components such as the fatty acid methyl esters (FAME) and phospholipid fatty acids (PLFA) are frequently cited as a means of performing this kind of community profiling. These techniques are based on the knowledge that microorganisms contain a wide range of different types of fatty acids in their lipid structures and, furthermore, that certain fatty acids are found only in particular microbial taxa and groups. These approaches, therefore, provide a means of quantitative discrimination of different microbial taxa and groups within mixed microbial communities.

Although FAME analysis has been used to study microbial communities in situ (Cavigelli et al., 1995), most workers have opted for analysis of PLFA, which provides a fast and reliable assessment of microbial biomass and composition in environmental samples. Indeed, in the late 1990s there has been an explosion in the use of PLFA analysis to study the effects of environmental perturbations on soil microbial communities, including: the effects of changes in land management and plant community composition on microbial communities of agricultural soils (Zelles et al., 1992; Yeates et al., 1997; Bardgett et al., 1999a; Bardgett & McAlister, 1999; Wardle et al., 1999); the effects of climate change (e.g. elevated atmospheric carbon dioxide and temperature) on soil microorganisms and ecosystem processes (Zak et al., 1996; Zogg et al., 1997; Bardgett et al., 1999b); the effects of soil pollution (e.g. heavy metals and liming) on microbial communities of forest soils (Frostegård et al., 1993), and; studies of the dynamics of microbial communities of marine sediments (Findlay, 1989, 1990; Rajendran et al., 1992, 1994, 1999). The technique has also been employed in various microcosm experiments to elucidate effects of plant manipulations on rhizosphere microbial communities (Bardgett et al., 1999c; Denton et al., 1999; Griffiths et al., 2000).

PLFA methodology

The PLFA technique is based largely on two facts: first, that phospholipids exist in membranes of all living cells; and second, that different subsets of the microbial community contain different PLFAs, or at least differ in their fatty acid composition (Tunlid & White, 1992). Furthermore, PLFA analysis measures only the living part of the microflora; phospholipids are not used as storage material, and when released upon cell death they are used as substrate by living micoorganisms, and within minutes to hours are metabolised to diglyceride and PO_4^{3-} (White et al., 1979). The assessment of PLFA pattern of an environmental sample, therefore, can be viewed as an integrated measurement of all living organisms present in that sample. However, it should be emphasised that since most PLFAs exist in different concentrations in a taxonomically wide range of microorganisms (Ratledge & Wilkinson, 1988), it cannot be used to measure specific species or genera within entire microbial communities; it can be used only to determine changes in the relative abundance of very broad taxonomic groups.

The method for PLFA analysis is outlined in Fig. 1, and further experimental details can be found in Frostegård et al. (1993). Using gas chromatography mass-spectrometry (Tunlid

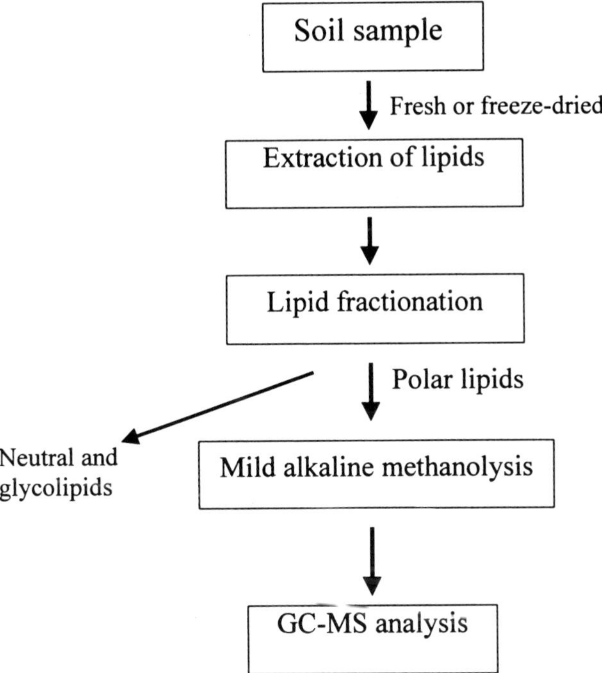

Fig. 1. Schematic diagram showing the methodology for PLFA.

& White, 1992), the relative abundance of some 30–35 individual PLFAs can be assessed, ranging from C12 to C20. Several of these can be used as broad-scale taxonomic markers (Table 1): branched fatty acids (e.g. i15:0, a15:0, i16:0, i17:0, a17:0) are commonly found in phospholipids of Gram-positive bacteria, and the cyclopropane fatty acids, cy17:0 and cy19:0, are known to be synthesised by Gram-negative bacteria (Wilkinson, 1988; O'Leary & Wilkinson, 1988). Studies of marine sediments have used 16:1ω9, 19:1 (Rajendran et al., 1992) and 16:1ω7, 18:1ω9 (Findlay et al., 1990) as indicators of aerobic bacteria, and the fatty acid 16:0 branched to distinguish sulfate-reducing and other anaerobic bacteria (Findlay et al., 1990, Rajendran et al., 1992). The tuberculostearic acid 10Me18:0 is considered to be unique to several actinomycete genera (Lechevalier, 1977), and has thus been used to study population dynamics of this group of organisms (Macnaughton & O'Donnell, 1994; Frostegård et al., 1993).

The fatty acid 18:2ω6 is a major PLFA in many saprophytic fungi, and has been used widely as a measure of fungal biomass in soil (Federle, 1986). Furthermore, the relative proportion of 18:2ω6 to the total amount of bacterial fatty acids (although opinions as to which fatty acids should be considered specific to bacteria vary, the following are commonly used to measure total bacterial PLFA in soil: i15:0, a15:0, 15:0, i16:0, 16:1ω9, 16:1ω7t, i17:0, a17:0, 17:0, cy17:0, 18:1ω7, and cy19) has been used to measure broad-scale shifts in the relative abundance of fungi and bacteria within the entire microbial community (Frostegård & Bååth, 1996; Bardgett et al., 1999a, b, c; Bardgett & McAlister, 1999). The fatty acid 16:1ω5 is found in arbuscular mycorrhizae (AM) (Olsson et al.,

Table 1

Commonly used indicator phospholipid fatty acids (PLFA)

Taxonomic group	PLFA
Gram-positive bacteria	i14:0, i15:0, a15:0, i16:0, i17:0, a17:0, 18:1ω9
Gram-negative bacteria	cy17:0, cy19:0, 18:1ω7
Fungi	18:2ω6
Arbuscular mycorrhizae	16:1ω5
Actinomycete	10Me18:0

Note: Fatty acids are designated in terms of the total number of carbon atoms:number of double bonds, followed by the position of the double bond from the methyl end of the molecule. The prefixes 'a' and 'i' indicate anteiso- and isobranching, cy refers to cyclopropane fatty acids, and 10Me indicates a methyl group on the tenth carbon atom from the carboxyl end of the molecule.

1995) and has been used to differentiate biomass responses to land management of AM fungi from those of saprophytic fungi in the field (Olsson et al., 1999).

Some limitations of PLFA

There are two main problems associated with the use of PLFA for the quantitative assessments of broad taxonomic groups: first, several fatty acids are not entirely specific to particular taxonomic groups; and second, the PLFA composition of various microorganisms varies with species and with factors such as growth phase, carbon source, pH, temperature (Federle, 1986). In relation to the first problem, a certain amount of caution is needed when making judgements about the origin of certain fatty acids. For example, although the fungal fatty acid 18:2ω6 has been shown to account for a large proportion (43%) of the total PLFA composition of a wide range of soil fungi (Federle, 1986), it is also known to occur at high concentrations in plant cells and especially in plant roots (Tunlid et al., 1985). Similarly, the PLFA 16:1ω5, which has proved suitable for estimating the biomass of AM fungi in roots (Olsson et al., 1999), is also present in other organisms, such as bacteria (Olsson et al., 1995); a non-mycorrhizal control is therefore needed to quantify the amount of PLFA in organisms other than AM fungi. Despite these problems of non-specificity, the common finding that concentrations of 18:2ω6 and 16:1ω5 are strongly correlated with other measures of fungal (e.g. ergosterol: Frostegård & Bååth, 1996; Donnison et al., 2000) and AM fungal biomass (AM hyphal length: Olsson et al., 1997), respectively, suggests that they can be used to make quantitative assessments of changes in the relative abundance of these organisms within diverse microbial communities. With these problems in mind, it has been recommended that PLFAs (and especially the fungal fatty acid) be used as a relative, rather than absolute measure of biomass in environmental samples (Frostegård & Bååth, 1996).

Variations in the PLFA composition of membranes that occur in response to environmental change may also be exploited to determine the metabolic status of an organism. The presence of *trans* fatty acids have been associated with the physiological status of the microorganism (Guckert et al., 1986) and the ratio of *trans* to *cis* isomers of monosaturated fatty acids has been used as an indicator of physiological stress. For example, Bååth et al. (1992) suggested that a lower ratio of *trans* to *cis* isomers of the fatty acid 16:1ω7 in unlimed than limed coniferous forest soils was indicative of nutrient stress in the former. Other studies, however, have shown no effect of nutritional stress on the *trans* to *cis* ratio of the microbial community (Petersen & Klug, 1994); it is therefore likely that stress-induced changes in these PLFA ratios only occur in certain species of bacteria and that these changes are not evident at the whole community scale.

3. Genotypic analysis: nucleic acid-based techniques

Over the past 15 years, molecular biological techniques based on analysis of the heterogeneity of RNA and DNA sequences have revised microbial taxonomy and greatly expanded our knowledge of the relationships between environmental organisms. These techniques have been made possible by our ability to extract and purify DNA and RNA from various environments including soil. Genes providing information on the presence and diversity of a single physiological group (via a functional gene) or indicating the phylogenetic relationship of detected microorganisms with other organisms (via universally distributed, conserved genes) have been used. Among the most widely used phylogenetic genes are those of the RNA of the small subunit of the ribosome, the 16S rRNA and 18S rRNA genes of prokaryotic and eukaryotic microorganisms respectively. This is due, in large part, to the accepted use of a phylogenetic system based on rRNA sequences, where the extent of similarity between ribosomal sequences provides an estimate of phylogenetic or evolutionary relationships (Woese, 1987). The 16S rRNA gene, and homologous 18S rRNA gene, contains conserved regions, present in all microorganisms, and variable regions which can be used to distinguish between sequences from different species or strains. The rRNA sequence database was initially compiled from well characterised strains from culture collections (Larsen et al., 1993) and now represents a dynamic and ever increasing data set against which new sequences may be compared. The data base includes sequences from cultivated strains as well as from cloned gene sequences from genetic material extracted directly from environmental samples (Pace et al., 1986). The genetic diversity of soil bacterial communities has been studied by a variety of nucleic acid hybridisation techniques, including gene probe analysis, and restriction enzyme analysis. Development of the polymerase chain reaction (PCR) has revolutionised studies of microbial ecology and led to new methods for the analysis of bacterial community structure and diversity in soil samples. Since the mid-1980s, the majority of PCR-based studies of microbial ecology have centred on the analysis of bacterial 16S rRNA. Recently, the 18S rRNA molecule of fungi has also been used as a target in ecological studies (Kowalchuck et al., 1997; Anderson & Kohn, 1998). In addition to rRNA-based analyses, the detection and analysis of specific functional genes can also be used to provide useful information on the microbial ecology of environmentally important processes.

Extraction of nucleic acids from complex environmental matrices

Soil is the most challenging environmental matrix for the extraction of nucleic acids and DNA can be extracted from soil either by cell extraction (indirect method) before lysis or direct lysis within the soil matrix. In the former procedure bacterial cells are first extracted from soil and then lysed to release DNA (Faegri et al., 1977; Ramsay, 1984; McDonald, 1986; Hopkins et al., 1991). Although these indirect methods yield a fairly pure DNA product, they may also selectively separate bacteria that are easily dislodged, leaving behind those that are more strongly attached to soil particles. To overcome these problems the direct lysis method is more widely used and comprises the following steps: (1) lysis of microbial cells; (2) separation of DNA from the other cell constituents (e.g. proteins and polysaccharides) and soil particles; and (3) precipitation of DNA. The original procedures described by Ogram et al. (1987) yielded 8–26 μg DNA g^{-1} soil or sediment sample approximately 10 kb in length. Various protocols have since been developed to simplify and increase the efficiency of these procedures (e.g. Tsai & Olson, 1991; Picard et al., 1992; Holben, 1994; Zhou et al., 1996; Frostegård et al., 1999; Miller et al., 1999). The direct approach can yield DNA from both live and dead cells (prokaryotic and eukaryotic) in addition to extracellular DNA. The efficacy of both cell lysis and recovery of DNA from the extraction milieu and the quality of the recovered nucleic acid are factors that should be considered when choosing a method for the extraction of nucleic acids from a particular environmental sample. For the detection of microbial populations which are present in low numbers within a given sample, it is preferable to extract and analyse ribosomal RNA (rRNA) as a cell may contain thousands of ribosomes but only one molecule of DNA (Olsen et al., 1986).

DNA hybridisation techniques

DNA hybridisation is the binding together, or annealing, of two single strands of DNA from the same or different organisms. The sequence similarity of the two strands determines the rate of hybridisation, similar strands will reanneal quickly whereas disparate strands will anneal more slowly. The genetic diversity of soil bacterial communities has been studied by a variety of DNA hybridisation techniques namely, DNA reassociation kinetics, DNA:DNA hybridisation and gene probe analysis. The binding or reassociation kinetics of denatured DNA can be used to provide an estimate of community complexity; the longer the reannealing period the more complex the mixture of DNA. Using this method, Torsvik et al. (1990) found a high level of genetic diversity, up to 4000 bacterial genome equivalents, in a gram of forest soil. Bej et al. (1991) similarly used reassociation kinetics to study the effect of treatment with a herbicide, 2,4,5-trichlorophenoxacetic acid, on the community diversity and showed that the overall genetic diversity was reduced by herbicide treatment.

DNA:DNA hybridisation determines the extent of annealing of one DNA sample to another prepared as a probe and therefore indicates the similarity between communities. Ritz & Griffiths (1994) found this technique to be most successful in the study of DNA isolated from the extractable bacterial fraction of the soil microbial community. Griffiths et al. (1996) further developed the technique so that the cross-hybridisation was done using random prime labelled probes (labelling of random nucleotides within the probe) prepared following digestion of DNA using restriction endonuclease enzymes. The diversity in DNA isolated from samples ranging in genetic complexity, from single bacterial cultures to the extractable bacterial fraction from soil was studied. Griffiths et al. (1996) also demonstrated differences in the diversity present in various agricultural soil samples. Clay loam, loamy sand, sandy clay loam and sandy loam soil samples were studied and the biological diversity was greatest in the latter soil type.

Gene probes analysis is the most commonly used technique for detecting and monitoring genetically defined populations within a complex microbial community (Holben et al., 1988; Jain et al., 1988). A nuclear probe is generally a single strand of either DNA or RNA that binds, or hybridises, with its complementary sequence of the target nucleic acid. The probe can be specific for either a functional or structural gene sequence or for a phylogenetic sequence, it can be radiolabelled (^{32}P) or labelled by incorporating biotinylated or digoxigenin nucleotide analogues. The target nucleic acid, extracted and purified from an environmental sample, is transferred and immobilised to a solid support (e.g. nylon or nitrocellulose membranes) and the probe is added in solution. Hybridisation or binding will occur if the target nucleic acid has genetic homology with the DNA or RNA probe. At the end of the hybridisation period, the unbound probe is washed off and the intensity of binding between the target nucleic acid and the probe is assessed by detection of the bound probe (Ogram & Bezdicek, 1994). Figure 2 shows an outline of the steps involved in nucleic acid hybridisation and probe detection. The conditions of hybridisation and subsequent washing (temperature and salt concentration) determine the degree of homology between the probe and target DNA or the 'hybridisation stringency'.

Although interpretation of gene probe analysis can be relatively simple with prokaryotes, there are serious limitations to the analysis of eukaryotes due to the more complex molecular genetics of these organisms. Gene probe techniques can be very effective if the molecular genetics of both the probe and the target sequences are known. Artefacts may arise if more than one type of gene is responsible for a given function or there is sequence homology between the probe and non-target genes. The former may result in an underestimation of the total number of organisms capable of performing the function of interest and the latter in an overestimation of the sequence.

DNA-DNA probe hybridisation analysis for functional genes provide measurements of the potential rather than the actual activity of the target gene (Ogram & Bezdicek, 1994). The target material used to obtain information on the actual activity of a specific function is mRNA. However, analysis of mRNA is problematic because it is adsorbed by humic molecules and easily degraded by extracellular and intracellular ribonucleases during soil extraction (Felske et al., 1996). A better understanding of the distribution and extent of microbial-mediated processes can be obtained by detecting metabolic genes transcripts (mRNAs) in conjunction with assays of the relative metabolites, metabolic or enzymatic rates. An example of this approach was shown by Fleming et al. (1993) who found that soil bacterial nahA mRNA (responsible for the synthesis of naphthalene dioxygenase) levels were significantly correlated with [^{14}C] naphthalene mineralisation rates.

Whole cell in situ hybridisation, with fluorescently-labelled oligonucleotide probes or fluorescent in situ hybridisation (FISH) is a powerful technique for identifying individual microbial cells within environmental samples (DeLong et al., 1989). The sample is fixed to permeabalise the cells whilst retaining their morphological integrity, cells are attached to gelatin coated microscope slides which are immersed in hybridisation solution or hybridised in suspension to the appropriate probe. The probe is bound as detailed above

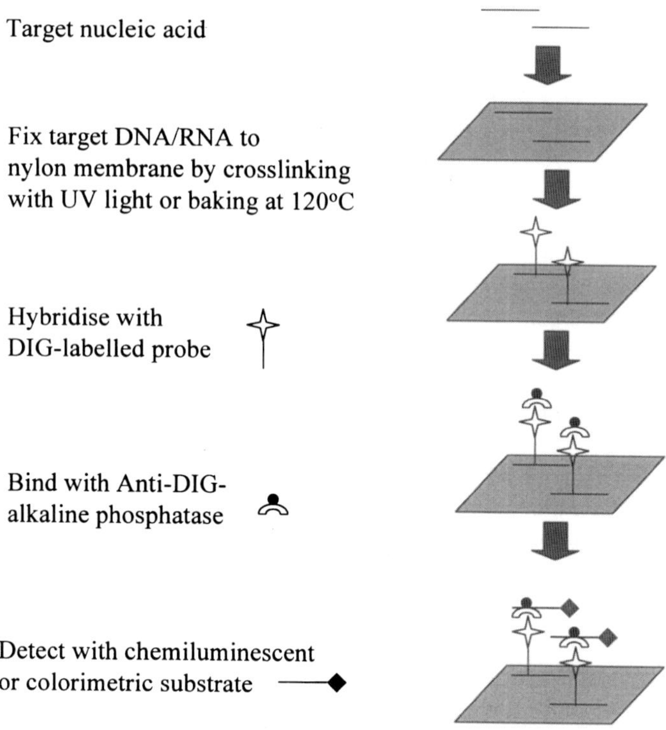

Fig. 2. Steps involved in nucleic acid hybridisation and detection using the digoxigenin system as an example.

with conditions of stringency being determined by temperature and/or concentration of formamide in the hybridisation solution. The sample is then washed to remove unbound probe. Methods for detection of the fluorescent probe include epifluorescence microscopy (DeLong et al., 1999), laser scanning confocal microscopy (Wagner et al., 1994), and flow cytometry (Amman et al., 1990; Fuchs et al., 2000). The FISH approach has been successfully used for phylogenetic identification of individual cells in plankton, sediments and soils (Snaidr et al., 1997; Christensen et al., 1999; DeLong et al., 1999). The method has also been used for estimation of physiological activity in biofilms (Poulsen et al., 1993) and is particularly useful for spatial localisation of microorganisms along environmental gradients (Schramm et al., 1996) or in symbiotic associations (Polz et al., 1994). Although the approach has great potential for studies of microbial ecology, difficulties in methodology are encountered when applying FISH to complex environmental samples (Head et al., 1998; DeLong et al., 1999).

Restriction enzyme analysis

Restriction enzyme analysis has been used for the analysis of community diversity. DNA endonucleases recognise and cleave DNA at specific base sequences (restriction sites) and can be used to recognise different bacterial strains or communities. DNA from genetically different samples will have a characteristic number of restriction sites and a distinctive profile of DNA fragments on an electrophoretic gel (Sadowsky, 1994). This type of DNA fingerprinting gives an impression of differences between samples but profiles from soil DNA can be highly complex and difficult to interpret (Holben et al., 1988; Selenska & Klingmùller, 1991; Tsai & Olson, 1991). Restriction Fragment Length Polymorphism (RFLP) is an extension of DNA fingerprinting technique where DNA or RNA probes are used to selectively bind (hybridise) to complementary restriction fragments from the sample DNA. Small differences between samples of single species or community DNA can be determined by RFLP analysis (Sadowsky, 1994).

PCR amplification of nucleic acids

PCR is catalysed by the Taq polymerase, a heat stable DNA polymerase enzyme that is still active at 98°C (Saiki et al., 1988). Knowledge of the nucleotide sequence at the distal and proximal ends of the sequence to be amplified is a prerequisite to use of the PCR technique. The double-stranded target DNA is denatured and a short oligonucleotide *primer* is annealed to the single strands. The primer sequences are then extended using DNA polymerase to complete the synthesis of strands complementary to the original target single strands. This cycle of denaturation, annealing and elongation is repeated many times and results in an exponential increase in the number of copies of the specific target region. Figures 3 and 4 show the steps involved in PCR amplification of DNA and the temperature changes during an amplification reaction. Thus, of the multitude of sequences in the target nucleic acid, only one sequence is selectively amplified. The small amounts of DNA needed as template enable PCR or amplification of selected sequences from a few bacterial cells or eukaryotic organelles or minute amounts of soil.

Initial studies involving PCR amplification of soil DNA were limited by the presence of substances such as humic acids which coextract with DNA and inhibit the reaction (Tsai & Olson, 1991). Several methods to analyse the soil bacterial communities have since been published including protocols for the extraction of DNA and RNA suitable for PCR amplification (Liesack & Stackebrandt, 1992; Bruce et al., 1992; Rochelle et al., 1992; Pepper & Pillai, 1994; Felske et al., 1996). Highly conserved nucleotide tracts in dispersed regions of the 16S rRNA gene and the availability of a large number of sequences has enabled the design of primers that amplify nearly the full 16S rRNA gene (Lane, 1991). As the template for these primers is the rRNA gene the PCR product is often referred to as rDNA. When amplified from a complex community, as is generally present in soil, the pool of PCR products generated will contain a number of different 16S rDNA sequence types. The complexity of the profile of ribosomal bands has been taken to be representative of the bacterial diversity in the sample, that is, the more complex the profile the greater the diversity of bacteria in the environment under examination. PCR primers that allow the specific amplification of fungal 18S rDNA sequences, even in the presence of non-fungal DNA, have recently been designed (Kowalchuck et al., 1997; Smit et al., 1999). Whilst these techniques have mostly been used to investigate simple host-fungus relationships (Anderson & Kohn, 1998), they have considerable potential for a broad analysis of fungal

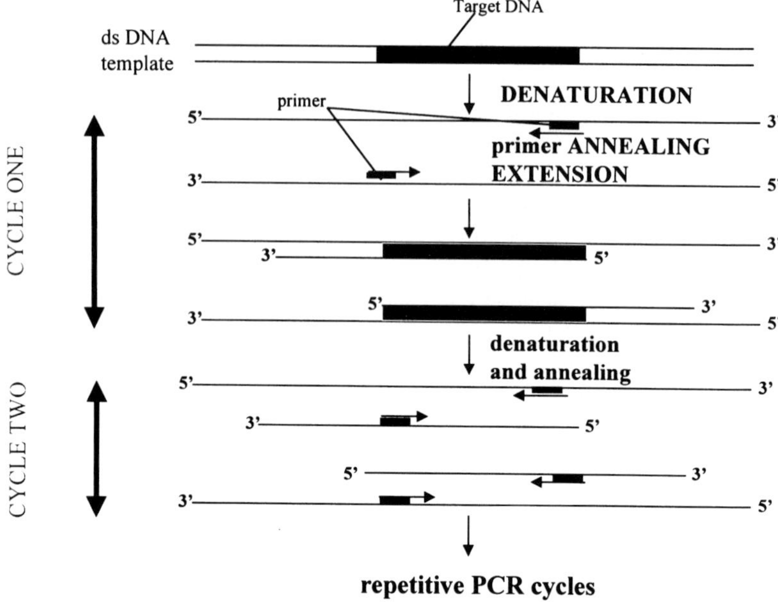

Fig. 3. Schematic showing the steps involved in PCR amplification of DNA.

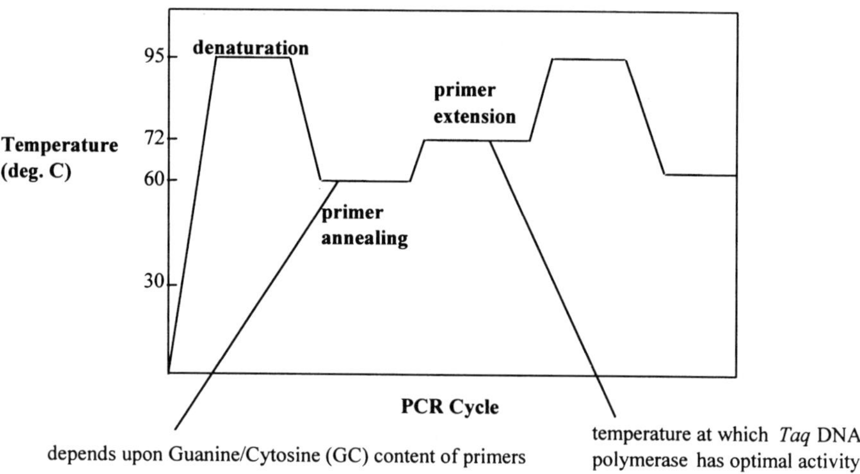

Fig. 4. Temperature changes during a PCR amplification cycle.

communities in a range of different environmental samples. Recently, increased specificity has been attained by reverse transcriptase-dependent amplification (RT-PCR) of specific rRNA and mRNA genes (Lamar et al., 1995; Xu & Tabita, 1996; Teske et al., 1996) combined with the quantitative accuracy of competitive primer methodology (Lamar et al., 1995; Bogan et al., 1996). The ability to use new techniques allows us to sensitively and accurately monitor, in situ, the gene coding for enzymes involved in ecologically important processes and determine the likely phylogenetic position of the active organisms.

Several factors must be considered when using PCR-based methods for community analysis. DNA can persist in soil after cell lysis and 'false' signals may be generated from minute quantities of extracellular DNA (Trevors & Van Elsas, 1995). There is also the possibility of bias introduced by the lysis procedure; harsh methods are needed to lyse dwarf cells and spores but these methods result in the excessive fragmentation of easily extractable DNA from large cells (Bakken & Olsen, 1989). Another potential concern, given that the PCR only requires one target molecule to enable amplification, is the introduction of 'contaminating' DNA sequences during the DNA isolation or PCR amplification steps. Generation of chimeric sequences from the coamplification of homologous genes should also be considered, this can lead to the description of non-existent species or even the misinterpretation of the level of bacterial diversity (Wang & Wang, 1996). The use of PCR amplification and analysis of 16SrDNA segments in community diversity analysis is based on the assumption that amplification of homologous but non-identical rDNA sequences occurs with equal efficiency. However, there have been reports of preferential or no amplification of sequences from certain species (Reysenbach et al., 1992). Another example of PCR bias was encountered by Farrelly et al. (1995) when the amount of 16S rDNA amplification products differed from that which was predicted for mixtures of DNA from two species. Hansen et al. (1998) presented evidence that the observed PCR bias may occur because the genomic DNA of some species contain segments outside of the amplified sequence that inhibit the initial steps of the PCR. The number of cycles used for amplification and the template concentration can also determine the representation of the ecosystem obtained by analysis of the subsequent amplification products (Wilson & Blitchington, 1996; Chandler et al., 1997).

Analysis of PCR products – clone libraries

The pool of mixed PCR products generated from community DNA can be analysed by cloning of amplified fragments followed by RFLP and/or sequence analysis of the resulting clone library. Each clone within a library contains a 16S rDNA sequence amplified from a 16S rRNA gene within one bacterial cell. From the sequence data, for a clone library generated from DNA extracted from an environmental sample, inferences can be made on the composition and phylogenetic diversity of the microbial community within the sample. Liesack & Stackebrandt (1992) used this protocol to examine non-cultivated bacteria in an Australian soil and identified a large percentage of clones related to the nitrogen-fixing bacteria in the α_2 subclass of the class Proteobacteria. Ueda et al. (1995) identified 16S rDNA sequences from soybean field soil with similarity to the high G+C content subdivision of Gram-positive bacteria, green sulfur bacteria and Proteobacteria. Borneman et al. (1996) found 16S rDNA clones from *Proteobacteria*, the *Cytophaga-Flexibacter-Bacteroides* group, and low-G+C-content Gram-positive bacteria in soil from an American clover-grass pasture. Other studies have examined archaeal diversity in sediment taken from a hot spring (Barns et al., 1994), peat bog material (Rheims et al., 1996) and in marine sediments (Grey & Herwig, 1996). The majority of cloned DNA sequences examined in the above studies did not match those from cultivated isolates in the ribosomal database or those from the same samples. This indicates that bacterial lineages other than those recognised to date may be present in the environment. Whilst DNA sequence characterisation provides highly specific information on individual members of the bacterial community within soil, the 16S rDNA clone library approach has drawbacks. The cloning stage itself may bias the structure of the mixed PCR pool and the frequency of clones may not be a reliable basis for the estimation of community composition (Amman et al., 1995). Further, as the method is highly labour intensive, relatively few sequences can be studied.

Analysis of PCR products – detection of specific groups

In addition to characterising the 'entire' diversity of a community, PCR-based analyses can also be used for detection of specific members within the community. In general, such studies use specific 16S rDNA oligonucleotide primers to amplify the 16S rRNA genes from the bacterial species/genera of interest (see Head et al., 1998). The specificity of amplification can be confirmed by oligonucleotide probe hybridisation of the 16S rDNA PCR products using a probe specific for an internal region of the amplified fragment. This approach has been used to detect ammonia-oxidising bacteria of the *Nitrosospira genus* (Hiorns et al., 1995), *Nitrobacter* populations (Degrange & Bardin, 1995), *Mycobacterium chlorophenolicum* (Briglia et al., 1996) and ammonia-oxidising β-proteobacteria (Stephen et al., 1996) in soils.

Analysis of PCR products – rapid profiling techniques

More rapid methods, also based on PCR amplification of 16S rDNA or other genes of interest, which allow display of a profile of the community diversity in a single electrophoretic profile, are now available. These methods, each with its own limitations, allow a rapid estimation of the diversity and composition of complex microbial communities and thus allow spatial and temporal comparisons to be made. Profiling methods include RFLP of amplified DNA (Porteous & Armstrong 1993), fluorescent or terminal restriction fragment length polymorphism (Flu-RFLP or T-RFLP) (Scheinert et al., 1996; Bruce, 1997; Liu et al., 1997), denaturing or temperature gradient gel electrophoresis (DGGE or TGGE) (Muyzer et al., 1993; Felske et al., 1997; Heuer et al., 1997), and single-strand conformation polymorphism (SSCP) (Lee et al., 1993).

Restriction fragment length polymorphism (RFLP) of amplified 16S rDNA fragments demonstrated that the DNA extracted from five different soil samples provided similar RFLP banding profiles, yet differences in both the presence and intensity of certain bands could be identified in different samples (Porteous & Armstrong, 1993; Porteous et al., 1994). This method when applied to the 16S rDNA gene is also referred to as amplified ribosomal DNA restriction analysis (ARDRA) (Hiorns et al., 1995; Martinez-Murcia et al., 1995). Smit et al. (1997) evaluated the use of the ARDRA method for assessing shifts in microbial communities from copper contaminated soils. More detailed information can be obtained by restriction analysis of cloned 16S rRNA genes and clones of particular interest can be sequenced (Hiorns et al., 1995; Smit et al., 1997).

Fluorescent dye labelled terminal primers (Flu-RFLP or T-RFLP) are also being used in the genetic fingerprinting of microbial communities (Scheinert et al., 1996). T-RFLP analysis is based in the restriction endonuclease digestion of fluorescent end-labelled PCR products where either one or both primers used for the PCR can be labelled (with different fluorescent dyes if necessary). The digested product is mixed with a DNA size standard, with a distinct dye label, and either gel or capillary electrophoresis can be used to separate the fragments. The internal standards facilitate gel to gel comparison and thus enhance reproducibility. The combination of florescent PCR with capillary electrophoresis allows automated laser analysis of fragments and thus reduces analysis time and leads to improved resolution and sensitivity (Osborne et al., 2000). Bruce (1997) used T-PCR to analyse the diversity of mercury resistance genes in polluted soils and Liu et al. (1997) studied bacterial populations in natural habitats. Both authors discussed the possibility that T-RFLP may permit semiquantitative analysis of relative proportions of dominant genotypes within a microbial community.

Denaturating gradient gel electrophoresis (DGGE) or temperature gradient gel electrophoresis (TGGE) can be used to analyse PCR-amplified DNA fragments to study the structural diversity of microbial communities (Muyzer & Smalla, 1998). These methods separate DNA fragments of the same length but of different sequences according to their melting properties. During electrophoresis the DNA fragment remains double-stranded until it reaches the conditions causing melting of the lower temperature melting domains. This partial melting causes the branching of the molecule with a decrease in the mobility of the DNA fragment in the gel. With DGGE, PCR-amplified fragments are separated by parallel gel electrophoresis in a linearly increasing denaturating gradient of urea/and formamide at 60°C (increasing gradient from top to bottom). TGGE is based on a linearly increasing temperature gradient in the presence of a constant high concentration of urea and formamide. The melting behaviour of the DNA fragments needs to be determined prior to DGGE or TGGE analysis so that the gradient and duration of the electrophoresis can be optimised for ideal separation of the DNA fragments. This is determined experimentally with perpendicular gradient gels which have an increasing gradient of denaturants or temperature perpendicular to the direction of electrophoresis i.e. from left to right (Muyzer & Smalla, 1998). Electrophoresis of the sample in a perpendicular gel will result in a sigmoid shaped migration curve with double-stranded DNA on the side of low denaturant (left), single-stranded DNA on the side of high denaturant (right) and molecules with different degrees of melting at intermediate concentrations of denaturant. As well as the melting behaviour of DNA fragments, the optimal gradient for multilane analysis in parallel DGGE or TGGE gels can also be determined from these perpendicular gels.

Denaturing gradient gel electrophoresis (DGGE) was first applied to the study of bacterial community diversity in natural environments by Muyzer et al. (1993) who examined marine microbial mats and bacterial biofilms from wastewater treatment plants and demonstrated that rDNA amplified from the bacterial community could be separated by this method. DGGE profiles were then blotted onto nylon membranes and hybridised with an oligonucleotide probe specific for sulfate-reducing bacteria (Amman et al., 1992) and community members were identified by sequence analysis of excised DGGE bands (Muyzer & De Waal, 1994). Subsequent studies have profiled bacterial communities within a microbial mat in the USA (Ferris et al., 1996). The diversity of sulfur-oxidising bacteria in hydrothermal vents (Muyzer et al., 1995) and sulfate-reducing bacteria in a stratified marine water column (Teske et al., 1996) have also been examined using DGGE. The latter study used DGGE to compare 16s rDNA fragments obtained by PCR and RT-PCR from environmental DNA and RNA from different depths. This gave a measure of diversity of microorganisms at each depth as well as an indication of the active populations. Similarly, Felske et al. (1997) showed that the TGGE profiles of rRNA and rDNA from the same soil sample were different. These profiling techniques have been used extensively over the past 5 years to characterise the diversity of microbial communities important in many different environments and to study community shifts as a result of perturbation (Ferris et al., 1997; Eichner et al., 1999). Microbial communities and environments studied using DGGE or TGGE include cyanobacteria in hot springs, bacterioplankton in estuaries, bacteria in biodegraded oil paintings and microbial communities in zinc contaminated soils (see Muyzer & Smalla, 1998).

Single-strand conformation polymorphism (SSCP) analysis is based on the fact that a single-base modification can change the confirmation of single-strand DNA molecules leading to a different electrophoretic mobility in a non-denaturing gel (Orita et al., 1989; Lee et al., 1996). As with DGGE and TGGE, DNA fragments of the same size but with different base composition can be separated into different bands in polyacrylamide gel electrophoresis because of different mobilities of their folded structure. Lee et al. (1996) used SSCP to study the structure and diversity of bacterial communities in an aquatic ecosystem. Zumstein et al. (2000) adapted a PCR-based SSCP analysis to an automated sequencer to study the evolution of bacterial and archael communities over a 2-year period in a laboratory anaerobic bioreactor. One strand of the V3 16S rDNA variable region was fluorescently labelled and after PCR amplification, the product is denatured and subjected to electrophoresis in a polyacrylamide gel using an automated DNA sequencer. Each peak in the resulting pattern from a microbial community is correlated with the V3 16S rDNA sequence of a single microorganism and rapid shifts in structure of the bacterial community were seen throughout the study period where as the archael community remained stable in the bioreactor environment. Although most studies using profiling techniques have concentrated on analysis of the 16S rRNA gene, analysis of functional genes can also provide useful information on the microbial ecology of specific populations (Wawer & Muyzer, 1995).

Quantification of PCR products

Two approaches have been used for the quantification of PCR amplification products from environmental DNA, most probable number-PCR (MPN-PCR) and competitive PCR. MPN-PCR involves serial dilution of extracted DNA with detection in replicate samples at each dilution using PCR amplification (Picard et al., 1992; Degrange & Bardin, 1995). Numbers of target sequences are then calculated from published tables. With competitive PCR the target DNA is coamplified with an internal standard, or competitor DNA, which is similar to but distinct from the target DNA (Jansson & Leser, 1996; Phillips et al., 2000). Quantitative RT-PCR has also been used to describe microbial activity in complex natural communities (Lamar et al., 1995). However, greater understanding of PCR and RT-PCR at low template concentrations is required for accurate implementation of competitive-PCR and RT-PCR assays for the quantification of low abundance templates in environmental samples.

4. Linking environmental processes with microorganisms

Recently developed techniques involving radio and stable isotope labelling of cells and biomarkers allow us to identify, for the first time, the specific microorganisms involved with ecologically important processes. Techniques with potential for studies of microbial ecology include FISH combined with microautoradiography and tracking of labelled substrates (Ouverney & Fuhrman, 1999; Lee et al., 1999), rRNA probe-based extraction of bacteria (Stoffels et al., 1999), detection of PLFAs labelled with radio or stable isotopes (Roslev et al., 1998; Bull et al., 2000) and stable isotope probing of community DNA (Radajewski et al., 2000).

FISH combined with microautoradiography allows for the analysis of the metabolic activities of prokaryotic cells under ecologically relevant conditions. Microorganisms are first exposed to a radiolabelled substrate and transferred onto autoradiographic emulsion-gelatin coated slides, the slides are then probed with group or species specific fluorescent rRNA-targeted oligonucleotide probes and finally stained with a vital stain (e.g. DAPI) (Lee et al., 1999). The total number of viable cells is determined with UV excitation. Cells that take up the radiolabelled substrate can be directly visualised using the UV and transmission mode of a microscope by observation of silver grains in autoradiographic film covering the sample. Within the same microscopic field, cells that are labelled with specific oligonucleotide probes can be visualised using green or red excitation filters as appropriate. Lee et al. (1999) suggest that development of the autoradiographic film after the FISH procedure and observation with a confocal laser scanning microscope improve the cellular signal intensities observed. The FISH-autoradiography combination of techniques allows for the direct analysis of in vivo substrate uptake by phylogenetically identified organisms under different environmental conditions although the results are not quantitative.

Immunomagnetic separation of cells has been used for the enrichment and isolation of cells from different environments (Christensen et al., 1992). Paramagnetic beads coated with antibodies against the target bacteria are used to label the cells. Following incubation, the labelled target cells can be removed from the background of unlabelled cells in a strong magnetic field. This technique has been shown to be efficient and sensitive but it is reliant on the availability of pure cultures for the production of specific antibodies. Recently, Stoffels et al. (1999) developed a similar method that combines the advantages of in situ hybridisation with that of magnetic separation and tested it with mixtures of pure cultures. Briefly, target cells were labelled with biotinylated transcript probes targeted against the hypervariable V3 region of the 23S rRNA. These cells were then incubated with streptavidin-coated paramagnetic beads and separated from the mixture of pure cultures in a strong magnetic field. The transcript probe acts as an anchor between the hybridised target cells and the paramagnetic beads. Enriched bacterial cells extracted with this method can be directly observed under a fluorescence or epifluorescence microscope, for identification and quantification of rRNA-probed cells, and used for subsequent molecular, genomic and metabolic analyses.

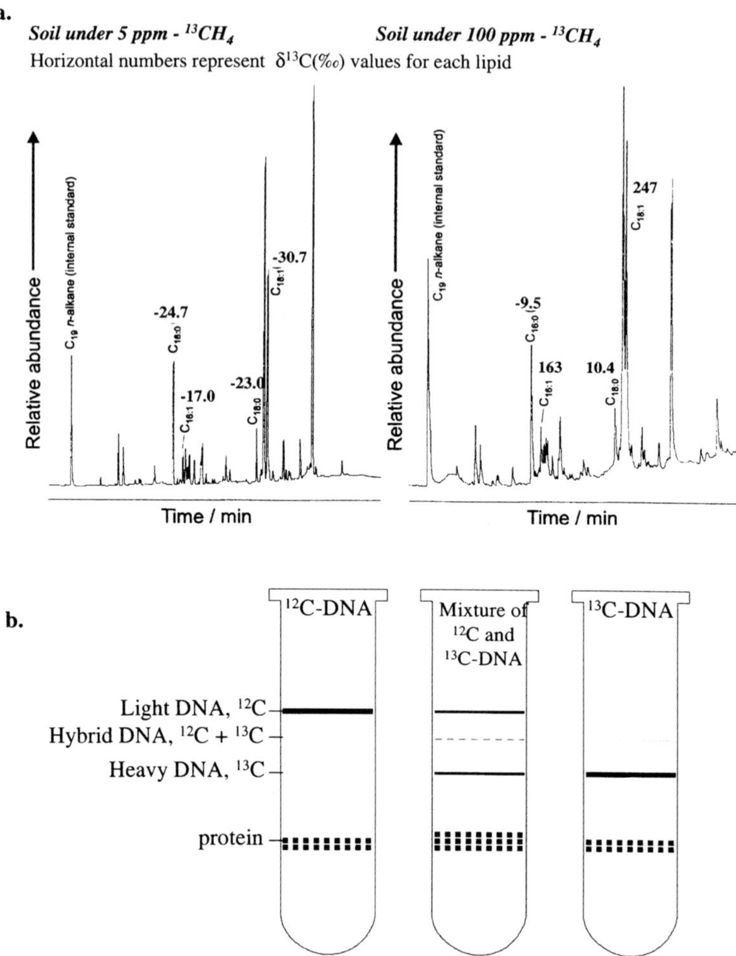

Fig. 5. Separation of isotopically labelled biomarkers by (a) gas chromatography coupled with isotope ratio mass spectrometry, and (b) density gradient centrifugation.

Novel methods based on radio and stable isotope labelling of microorganisms have recently been applied to the identification of microorganisms involved in key processes (Boschker et al., 1998; Pelz et al., 1998; Roslev et al., 1998; Bull et al., 2000; Radajewski et al., 2000). Active microbial populations utilising a substrate with a radio or stable isotopic label (e.g. ^{14}C- or ^{13}C-methane) will incorporate the labelled isotope unambiguously into cell biomarkers such as PLFAs, amino acids or DNA. Since both ^{13}C and ^{14}C are 'heavier' than ^{12}C, the labelled microbial cells and their constituent biomarkers would be expected to have a higher mass than those grown on conventional unlabelled substrates. These isotopically labelled biomarkers can be separated from unlabelled material by gas chromatography coupled with isotope ratio mass spectrometry or by density gradient centrifugation (see Fig. 5). Radiolabelled PLFAs can also be analysed by scintillation counting of $^{14}CO_2$ produced by combustion after gas chromatographic separation of individual lipids as demonstrated by Roslev et al. (1998) who analysed the incorporation of ^{14}C-methane by soil bacteria.

The presence of a ^{13}C-label in specific biomarkers can be analysed by isotope ratio mass spectrometry coupled to a gas chromatograph with a combustion interface (GC-C-IRMS). This approach was used by Pelz et al. (1998) to trace carbon sources in bacterial cells and cell wall amino acids in estuarine waters and soil and to identify carbon sources that support microbial growth. Boschker et al. (1998) used ^{13}C-labelling of PLFAs to link microbial populations to the process of methane oxidation in aquatic sediments. The method was used by Bull et al. (2000) to unambiguously identify, for the first time, organisms involved in the oxidation of atmospheric levels of methane in a forest soil. This study highlighted the sensitivity of the GC-IRMS approach for detection of isotopically labelled compounds over that of radiocarbon labelling. PLFA analysis coupled with ^{13}C-labelling was used by Hanson et al. (1999) to measure the response of a soil microbial community to toluene, a widespread organic pollutant.

Recently, ^{13}C-DNA from bacterial cells which were exposed to ^{13}C-methanol was separated from unlabelled material by density gradient centrifugation (Radajewski et al., 2000). By extending this technique to PCR amplification, cloning and sequencing of functional and taxonomically important genes (e.g. 16S rDNA), the authors propose a novel and powerful method to directly link microbially mediated processes to the phylogenetic identity of the organisms involved.

Compound specific stable isotope analysis has significant implications for the future of microbial ecological studies as it provides an extremely sensitive and non-hazardous means of investigating the role of microorganisms involved in ecologically important processes in the laboratory and directly in the field environment.

All of the new and developing techniques described above have great potential for the analysis of both cultivable and previously uncultured microorganisms within complex environmental samples. Combined application of these new techniques will allow us to gain insights into the ecological roles of microorganisms in situ and assess the impacts of environmental perturbation and pollutant stress on microorganisms involved in ecologically important processes.

References

Akkermans, A. D. L., Van Elsas, J. D. & De Bruijn, F. J. (1995). *Molecular Microbial Ecology Manual.* Dordtrecht, The Netherlands. Kluwer Academic Publishers.

Amman, R. I., Binder, B. J., Olsen, R. J., Chisholm, S. W., Devereux, R. & Stahl, D. (1990). Combination of 26S rRNA-targeted oligonucleotide probes with flow cytometry for analysing mixed microbial populations. *Applied and Environmental Microbiology, 56*,1919–1925.

Amman, R., Zarda, B., Stahl, D. A. & Schleifer, K. H. (1992). Identifcation of individual prokaryotic cells by using enzyme-labelled, rRNA targeted oligonucleotide probes. *Applied and Environmental Microbiology, 58*, 3007–3011.

Amman, R. I., Ludwig, W. & Schleifer, K. H. (1995). Phylogenetic identification and in situ detection of individual microbial cells without cultivation. *Microbiology Review, 59*, 143–169.

Anderson, J. B. & Kohn, L. M. (1998). Genotyping gene genealogies and genomics bring fungal population genetics above ground. *Trends in Ecology and Evolution, 13*, 444–449.

Atlas, R. M., Horowitz, A., Krichevsky, M. & Bej, A. K. (1991). Response of microbial-populations to environmental disturbance. *Microbial Ecology, 22*, 249–256.

Bååth, E., Frostegård, Å. & Fritze, H. (1992). Soil bacterial biomass, activity, phopsholipid fatty acid pattern, and pH tolerance in an area polluted with alkaline dust deposition. *Applied and Environmental Microbiology, 58*, 4026–4031.

Bakken, L. R. & Olson, R. A. (1987). The relationship between cell size and viability of soil bacterial. *Microbial Ecology, 13*, 103–114.

Bakken, L. R. & Olsen, R. A. (1989). DNA-content of soil bacteria of different cell-size. *Soil Biology and Biochemistry, 21*, 789–793.

Bardgett, R. D. & McAlister, E. (1999). The measurement of soil fungal:bacterial biomass ratios as an indicator of ecosystem self-regulation in temperate grasslands. *Biology and Fertility of Soils, 19*, 282–290.

Bardgett, R. D., Lovell, R. D., Hobbs, P. J. & Jarvis, S. C. (1999a). Dynamics of below-ground microbial communities in temperate grasslands: influence of management intensity. *Soil Biology and Biochemistry, 31*, 1021–1030.

Bardgett, R. D., Kandeler, E., Tscherko, D., Hobbs, P. J., Jones, T. H., Thompson, L. J. & Bezemer, T. M. (1999b). Below-ground microbial community development in a high temperature world. *Oikos, 85*, 193–203.

Bardgett, R. D., Mawdsley, J. L., Edwards, S., Hobbs, P. J., Rodwell, J. S. & Davies, W. J. (1999c). Plant species and nitrogen effects on soil biological properties of temperate upland grasslands. *Functional Ecology, 13*, 650–660.

Barns, F. M., Fundyga, R. E., Jefferies, M. W. & Pace, N. R. (1994). Remarkable archaeal diversity detected in a Yellowstone National Park hot spring environment. *Proceedings of the National Academy of Sciences USA, 91*, 1609–1613.

Bej, A. K., Perlin, M. & Atlas, R. M. (1991). Effect of introducing genetically engineered microorganisms on soil microbial diversity. *FEMS Microbiology Ecology, 86*, 169–175.

Bogan, B. W., Schoenike, B., Lamar, R. T. & Cullen, D. (1996). Manganese peroxidase mRNA and enzyme activity levels during bioremediation of polycyclic aromatic hydrocarbon-contaminated soil with *Phanerochaete chrysosporium. Applied Environmental Microbiology, 62*, 2381–2386.

Borneman, J., Skroch, P. W., O' Sullivan, K. M., Palus, J. A., Rumjanek, N. G., Jansen, J. L., Nienhuis, J. & Triplett, E. W. (1996). Molecular microbial diversity of an agricultural soil in Wisconsin. *Applied Environmental Microbiology, 62*, 1935–1943.

Boschker, H. T. S., Nold, S. C., Wellsbury, P., Bos, D., De Graaf, W., Pel, R., Parkes, R. J. & Cappenberg, T. E. (1998). Direct linking of microbial populations to specific biogeochemical processes by ^{13}C-labelling of biomarkers. *Nature, 392*, 810–805.

Briglia, M., Eggen, R. I. L., De Vos, W. M. & Van Elsas, J. D. (1996). Rapid and sensitive method for the detection of *Mycobacterium chlorophenolicum* PCP–1 in soil based on 16S rRNA gene-targeted PCR. *Applied and Environmental Microbiology, 62*, 1478–1480.

Bruce, K. D. (1997). Analysis of mer gene subclasses within bacterial communities in soils and sediments resolved by fluorescence-PCR-restriction fragment length polymorphism profiling. *Applied and Environmental Microbiology, 63*, 4914–4919.

Bruce, K. D., Hiorns, W. D., Hobman, J. L., Osborn, A. M., Strike, P. & Ritchie, D. A. (1992). Amplification of DNA from native populations of soil bacteria by using the polymerase chain-reaction. *Applied and Environmental Microbiology, 58*, 3413–3416.

Bull, I. D., Parekh, N. R., Hall, G., Ineson, P. & Evershed, R. P. (2000). Detection and classification of atmospheric methane-oxidising bacteria in soil. *Nature, 405*, 175–178.

Cavigelli, M. A., Robertson, G. P. & Klug, M. J. (1995). Fatty acid methyl ester (FAME) profiles as measures of soil microbial community structure. *Plant and Soil, 70*, 99–113.

Chandler, D. P., Fredrickson, J. K. & Brockman, F. J. (1997). Effect of PCR template concentration on the composition and distribution of total community 16S and rDNA clone libraries. *Molecular Ecology, 6*, 475–482.

Christensen, B., Torsvik, T. & Lien, T. (1992). Immunomagnetically captured thermophilic sulfate-reducing bacteria from the North Sea oil field waters. *Applied and Environmental Microbiology, 8*, 1244–1248.

Christensen, H., Hansen, M. & Sorensen, J. (1999). Counting and size classification of active soil bacteria by fluorescence in situ hybridization with an rRNA oligonucleotide probe. *Applied and Environmental Microbiology, 65*, 1753–1761.

Degrange, V. & Bardin, R. (1995). Detection and counting of *Nitrobacter* populations in soil by PCR. *Applied and Environmental Microbiology, 61*, 2093–209.

DeLong, E. F., Wickham, G. S. & Pace, N. R. (1989). Phylogenetic stains: Ribosomal RNA-based probes for the identification of single cells. *Science, 243*, 1360–1363.

DeLong, E. F., Taylor, L. T., Marsh, T. L. & Preston, C. M. (1999). Visualization and enumeration of marine planktonic archaea and bacteria by using polyribonucleotide probes and fluorescent in situ hybridization. *Applied and Environmental Microbiology, 65*, 5554–5563.

Denton, C. S., Bardgett, R. D., Cook, R. & Hobbs, P. J. (1999). Low amounts of root herbivory positively influences the rhizosphere microbial community of a temperate grassland soil. *Soil Biology and Biochemistry, 31*, 155–165.

Donnison, L. M., Griffith, G. S., Hedger, J., Hobbs, P. J. & Bardgett, R. D. (2000). Management influences on soil microbial communities and their function in botanically diverse haymeadows of northern England and Wales. *Soil Biology and Biochemistry, 32*, 253–263.

Eichner, C. A., Erb, R. W., Timmis, K. N. & Wagner-Dobler, I. (1999). Thermal gradient gel electrophoresis analysis of bioprotection from pollutant shocks in the activated sludge microbial community. *Applied and Environmental Microbiology, 65*, 102–109.

Faegri, A., Torsvik, V. L. & Goksòyr, J. (1977). Bacterial and fungal activities in soil: Separation of bacteria and fungi by a rapid fractionated centrifugation technique. *Soil Biology and Biochemistry, 9*, 105–112.

Farrelly, V., Rainey, F. A. & Stackebrandt, E. (1995). Effect of genome size and rRNA gene copy number on PCR amplification of 16S rRNA genes from a mixture of bacterial species. *Applied and Environmental Microbiology, 61*, 2798–2801.

Federle, T. W. (1986). Microbial distribution in soil – new techniques. In F. Megusar & M. Gantar (Eds), *Perspectives in Microbial Ecology* (pp. 493–498). Ljubljana: Slovene Society for Microbiology.

Felske, A., Engelen, B., Nübel, U. & Backhaus, H. (1996). Direct ribosomal isolation from soil to extract bacterial rRNA for community analysis. *Applied and Environmental Microbiology, 62*, 4162–4167.

Felske, A., Rheims, H., Wolterink, A., Stackebrandt, E. & Akkermanns, A. D. L. (1997). Ribosome analysis reveals prominent activity of an uncultured member of the class *Actinobacteria* in grassland soils. *Microbiology, 143*, 2983–2989.

Ferris, M. J., Muyzer, G. & Ward, D. M. (1996). Denaturing gradient gel electrophoresis profiles of 16S rRNA-defined populations inhabiting a hot spring microbial mat community. *Applied and Environmental Microbiology, 62*, 340–346.

Ferris, M. J., Nold, S. C., Revsbech, N. P. & Ward, D. M. (1997). Population structure and physiological changes within a hot spring microbial mat community following disturbance. *Applied and Environmental Microbiology, 63*, 1367–1374.

Findlay, R. H., King, G. M. & Waitling, L. (1989). Efficacy of phospholipid analysis in determining microbial biomass in sediments. *Applied and Environmental Microbiology, 55*, 2888–2893.

Findlay, R. H., Trexler, M. B., Guckert, J. B. & White, D. C. (1990). Laboratory study of disturbance in marine sediments: response of a microbial community. *Marine Ecology Programme Serial*, 62, 121–133.

Fleming, J. T., Sanseverino, J. & Sayler, G. S. (1993). Quantitative relationship between naphthalene catabolic gene frequency and expression in predicting PAH degradation in soils at town gas manufacturing site. *Environmental Science and Technology, 27*, 1068–1074.

Frostegård, Å. & Bååth, E. (1996). The use of phospholipid fatty acid to estimate bacterial and fungal biomass in soil. *Biology and Fertility of Soils, 22*, 59–65.

Frostegård, Å., Bååth, E. & Tunlid, A. (1993). Shifts in the structure of soil microbial communities in limed forests as revealed by phospholipid fatty acid analysis. *Soil Biology and Biochemistry, 25*, 723–730.

Frostegård, Å., Courtois, S., Ramisse, V., Clerc, S., Bernillon, D., Le Gall, F., Jeannin, P., Nesme, X. & Simonet, P. (1999). Quantification of bias related to the extraction of DNA directly from soils. *Applied and Environmental Microbiology, 12*, 5409–5420.

Fuchs, B. M., Zubkov, M. V., Sahm, K., Burkill, P. H. & Amann, R. (2000). Changes in community composition during dilution cultures of marine bacterioplankton as assessed by flow cytometric and molecular biological techniques. *Environmental Microbiology, 2*, 191–201.

Grey, J. P. & Herwig, R. P. (1996). Phylogenetic analysis of the bacterial communities in marine sediments. *Applied and Environmental Microbiology, 62*, 4049–4059.

Griffiths, B. S., Ritz, K. & Glover, L. A. (1996). Broad-scale approaches to the determination of soil microbial community structure: application of the community DNA hybridization technique. *Microbial Ecology, 31*, 269–280.

Griffiths, B. S., Ritz, K., Bardgett, R., Cook, R., Christensen, S., Ekelund, F., Sørensen, S., Bååth, E., Bloem, J., De Ruiter, P., Dolfing, J. & Nicolardot, B. (2000). Stability of soil ecosystem-level processes following the experimental manipulation of soil microbial community diversity. *Oikos, 90*, 279–294.

Guckert, J. B., Hood, M. A. & Whit, D. C. (1986). Phospholipid ester-linked fatty acid profile changes during nutrient deprivation of Vibrio cholerae: increases in the trans/cis ratio and proportions of cyclopropyl fatty acids. *Applied and Environmental Microbiology, 52*, 794–801.

Hansen, M. C., Tolker-Nielsen, T., Givskov, M. & Molin, S. (1998). Biased 16S rRNA PCR amplification caused by interference from DNA flanking the template region. *FEMS Microbiology Ecology, 26*, 141–149.

Hanson, J. R., Macalady, J. L., Harris, D. & Scow, K. M. (1999). Linking toluene degradation with specific microbial populations in soil. *Applied and Environmental Microbiology, 65*, 5403–5408.

Head, I. M., Saunders, J. R., and Pickup, R. W. (1998). Microbial evolution, diversity, and ecology: a decade of ribosomal RNA analysis of uncultivated microorganisms. *Microbial Ecology, 35*, 1–12.

Heuer, H., Krsek, M., Baker, P., Smalla, K. & Wellington, E. M. H. (1997). Analysis of actinomycete communities by specific amplification of genes encoding 16S rRNA and gel-electrophoretic separation in denaturing gradients. *Applied and Environmental Microbiology, 63*, 3233–3241.

Hiorns, W. D., Hastings, R. C., Head, I. M., McCarthy, A. J., Saunders, J. R., Pickup, R. W. & Hall, G. H. (1995). Amplification of 16S ribosomal RNA genes of autotrophic ammonia oxidizing bacteria demonstrates the ubiquity of *Nitrosospiras* in the environment. *Microbiology, 141*, 2793–2800.

Holben, W. (1994). Isolation and purification of bacterial DNA from soil. In R. W. Weaver, S. Angle, P. Bottomley, D. Bezdicek, S. Smith, A. Tabatabai & A. Wollum (Eds), *Methods of Soil Analysis*, Part 2. *Microbiological and Biochemical Properties* (pp. 729–751). Soil Science Society of America.

Holben, W. E., Jansson, J. K., Chelm, B. K. & Tidje, J. M. (1988). DNA probe method for the detection of specific microorganisms in the soil bacterial community. *Applied and Environmental Microbiology, 54*, 703–711.

Hopkins, D. W., Macnaughton, S. J. & O'Donnell, A. G. (1991). A dispersion and differential centrifugation, technique for representative sampling of microorganisms from soil. *Soil Biology and Biochemistry, 23*, 217–225.

Jain, R. K., Burlage, R. S. & Sayler, G. S. (1988). Methods for detecting recombinant DNA in the environment. *CRC Critical Reviews in Biotechnology, 8*, 33–84.

Jansson, J. K., and Leser, T. D., (1996). Quantitative PCR of environmental samples. In A. D. L. Akkermans, J. D. Van Elsas & F. J. De Bruijn (Eds), *Molecular Microbial Ecology Manual*, 2.7.4. (pp. 1–19). Dordtrecht, The Netherlands: Kluwer Academic Publishers.

Kowalchuck, G. A., Stephen, J. R., De Boer, W., Prosser, J. I., Embley, T. M. & Woldendorp, J. W. (1997). Analysis of ammonia-oxidising bacteria of the γ-subdivision of the class *Proteobacteria* in coastal sand dunes by denaturing gradient gel electrophoresis and sequencing of PCR-amplified 16S ribosomal DNA fragments. *Applied and Environmental Microbiology, 63*, 1489–1497.

Lamar, R., Schoenike, B., Wymelenberg, V., Stewart, P., Dietrich, M. & Cullen, D. (1995). Quantitation of fungal mRNAs in complex substrates by reverse transcription PCR and its application to *Phanerochaete chrysosporium*-colonized soil. *Applied and Environmental Microbiology, 61*, 2122–2126.

Lane, D. J. (1991). 16S/23S rRNA sequencing. In E. Stackebrandt & M. Goodfellow (Eds), *Nucleic Acid Techniques in Bacterial Systematics* (pp. 115–175). New York: John Wiley.

Larsen, N., Olsen, G. J., Maidak, B. L., McCaughey, M. J., Overbeek, R., Macke, T. J., Marsh, T. L. & Woese, C. R. (1993). The ribosomal database project. *Nucleic Acid Research, 21*, 3021–3023.

Lechevalier, M. P. (1977). Lipids in bacterial taxonomy – a taxonomist's view. *Critical Reviews in Microbiology, 5*, 109–210.

Lee, S., Malone, C. & Kemp, P. F. (1993). Use of multiple 16S rRNA targeted fluorescent probes to increase signal strength and measure cellular RNA from natural planktonic bacteria. *Marine Ecology Programme*, Serial 101, 193–201.

Lee, D-H., Zo, Y. G. & Kim, S. J. (1996). Non-radioactive method to study genetic profiles of natural bacterial communities by PCR-single-stranded-conformation polymorphism. *Applied and Environmental Microbiology, 62*, 3112–3120.

Lee, N., Nielsen, P. H., Andeasen, K. H., Juretschko, S., Nielsen, J. L., Schleifer, K. H. & Wagner, M. (1999). Combined microautoradiography – a new tool for structure function analysis in microbial ecology. *Applied and Environmental Microbiology, 65*, 1289–1297.

Liesack, W. & Stackebrandt, E. (1992). Occurrence of novel groups of the domain Bacteria as revealed by analysis of genetic material isolated from an Australian terrestrial environment. *Journal of Bacteriology, 174*, 5072–5078.

Liu, W. T., Marsh, T. L., Cheng, H. & Forney, L. J. (1997). Characterisation of microbial diversity by determining terminal restriction fragment length polymorphism of genes encoding 16S rRNA. *Applied and Environmental Microbiology, 63*, 4516–4522.

Macnaughton, S. J. & O'Donnell, A. G. (1994). Tuberculostearic acid as a means of estimating the recovery (using dispersion and differential centrifugation) of actinomycetes from soil. *Journal of Microbiological Methods, 20*, 69–77.

Martinez-Murcia, A. J., Acinas, S. G. & Rogriguez-Valera, F. (1995). Evaluation of prokaryotic diversity by restrictase digestion of 16S rDNA directly amplified from hypersaline environments. *FEMS Microbiology Ecology, 17*, 247–256.

McDonald, R. M. (1986). Sampling soil soil microflora: dispersion of soil by ion exchange and extraction of specific microorganisms from suspension by elutriation. *Soil Biology and Biochemistry, 18*, 399–406.

Miller, D. N., Bryant, J. E., Madsen, E. L. & Ghiorse, W. C. (1999). Evaluation and optimization of DNA extraction and purification procedures for soil and sediment samples. *Applied and Environmental Microbiology, 65*, 4715–4724.

Muyzer, G. & De Waal, E. C. (1994). Determination of the genetic diversity of microbial communities using DGGE analysis of PCR-amplified 16S rRNA. *NATO ASI*, Series G35, 207–214.

Muyzer, G., De Waal, E. C. & Uitterlinden, A. G. (1993). Profiling of complex microbial populations by denaturing gradient gel electrophoresis analysis of polymerase chain reaction-amplified genes coding for 16S rRNA. *Applied and Environmental Microbiology, 59*, 695–700.

Muyzer, G. & Smalla, K. (1998). Application of denaturing gradient gel electrophoresis (DGGE) and temperature gradient gel electrophoresis (TGGE) in microbial ecology. *Antonie van Leeuwenhoek, 73*, 127–141.

Muyzer, G., Teske, A. & Wirsen, C. O. (1995). Phylogenetic relationships of *Thiomicrospira* species and their identification in deep-sea hydrothermal vent samples by denaturing gradient gel electrophoresis of 16S rDNA fragments. *Archives of Microbiology, 164*, 165–172.

Ogram, A. V. & Bezdicek, D. F. (1994). Nucleic acid probes. In R. W. Weaver, S. Angle, P. Bottomley, D. Bezdicek, S. Smith, A. Tabatabai & A. Wollum (Eds), *Methods of Soil Analysis*. Part 2. *Microbiological and Biochemical properties* (pp. 665–687). Madison, WI: American Society of Agronomy

Ogram, A., Sayler, G. S. & Barkay, T. J. (1987). The extraction and purification of microbial DNA from sediments. *Journal of Microbiological Methods, 7*, 57–66.

O'Leary, W. M. & Wilkinson, S. G. (1988). Gram-positive bacteria. In C. Ratledge & S. G. Wilkinson (Eds), *Microbial Lipids* (Vol. 1) (pp. 117–185). New York: Academic Press.

Olsen, G. J., Lane, D. J., Giovannoni, S. J. & Pace, N. R. (1986). Microbial ecology and sediments. *Annual Reviews of Microbiology, 40*, 337–365.

Olsson, P. A., Thingstrup, I., Jakobsen, I. & Bååth, E. (1999). Estimation of the biomass of arbuscular mycorrhizal fungi in a linseed field. *Soil Biology and Biochemistry, 31*, 1879–1887.

Olsson, P. A., Bååth, E., Jakobsen, I. & Söderström, B. (1995). The use of phospholipid and neutral lipid fatty acids to estimate biomass of arbuscular mycorrhizal fungi. *Mycological Research, 99*, 623–629.

Olsson, P. A., Bååth, E. & Jakobsen, I . (1997). Phosphorus effects on the mycelium and storage structures of an arbuscular mycorrhizal fungus as studied in the soil and root by analysis of fatty acid signatures. *Applied and Environmental Microbiology, 63*, 3531–3538.

Orita, M., Iwahana, H., Kanazawa, H., Hayashi, K. & Sekiya, T. (1989). Detection of polymorphisms of human DNA by gel-electrophoresis as single-strand conformation polymorphisms. *Proceedings of the National Academy of Sciences of the USA, 86*, 2766–2770.

Osborn, M. A., Moore, E. R. B. & Timmis, K. N. (2000). An evolution of terminal-restriction fragment length polymorphism (T-RFLP) analysis for the study of microbial community structure and dynamics. *Environmental Microbiology, 2*, 39–50.

Ouverney, C. C. & Fuhrman, J. A. (1999). Combined microautoradiography-16S rRNA probe technique for determination of radioisotope uptake by specific microbial cell types in situ. *Applied and Environmental Microbiology, 65*, 1746–1752.

Pace, N. A., Stahl, D. A., Lane, D. J. & Olson, G. (1986). The analysis of natural microbial populations by ribosomal RNA sequences. *Advances in Microbial Ecology, 21*, 59–68.

Pelz, O., Cifuentes, L. A., Hammer, B. T., Kelley, C. A. & Coffin, R. B. (1998). Tracing the assimilation of organic compounds using ^{13}C analysis of unique amino acids in the bacterial peptidoglycan cell wall. *FEMS Microbiology Ecology, 25*, 229–240.

Pepper, I. L. & Pillai, S. D. (1994). Detection of specific DNA sequences in environmental samples via polymerase chain reaction. In R. W. Weaver, S. Angle, P. Bottomley, D. Bezdicek, S. Smith, A. Tabatabai & A. Wollum (Eds), *Methods of Soil Analysis*. Part 2. *Microbiological and Biochemical properties* (pp. 707–726). Madison, WI: American Society of Agronomy.

Petersen, S. O. & Klug, M. J. (1994). Effects of sieving, storage, and incubation temperature on the phospholipid fatty acid profile of a soil microbial community. *Applied and Environmental Microbiolgy, 60*, 2421–2430.

Phillips, C. J., Paul, E. A. & Prosser, J. I. (2000). Quantitative analysis of ammonia-oxidising bacteria using competitive PCR. *FEMS Microbiology Ecology, 32*, 167–175.

Picard, C., Ponsonnet, C., Nesme, X. & Simonet, P. (1992). Detection and enumeration of bacteria in soil by direct DNA extraction and polymerase chain reaction. *Applied and Environmental Microbiology, 58*, 2717–2722.

Polz, M. F., Distel, D. L., Zarda, B., Amann, R., Felbeck, H., Ott, J. A. & Cavanaugh, C. M. (1994). Phylogenetic analysis of a highly specific association between ectosymbiotic, sulfur-oxidizing bacteria and a marine nematode. *Applied and Environmental Microbiology, 60*, 4461–4467.

Porteous, L. A. & Armstrong, J. L. (1993). A simple mini-method to extract DNA directly from soil for the use with polymerase chain reaction amplification. *Current Microbiology, 27*, 115–118.

Porteous, L. A., Armstrong, J. L., Seidler, R. J. & Watrud, L. S. (1994). An effective method to extract DNA from environmental samples for polymerase chain-reaction amplification and DNA fingerprint analysis. *Current Microbiology, 29*, 301–307.

Poulsen, L. K., Ballard, G. & Stahl, D. A. (1993). Use of rRNA fluorescence in situ hybridization for measuring the activity of single cells in young and established biofilms. *Applied and Environmental Microbiology, 59*, 1354–1360.

Radajewski, S., Ineson, P., Parekh, N. R. & Murrell, J. C. (2000). Stable-isotope probing as a tool in microbial ecology. *Nature, 403*, 646–649.

Rajendran, N. & Nagatomo, Y. (1999). Seasonal changes in sedimentary microbial communities of two eutrophic bays as estimated by biomarkers. *Hydrobiologia, 393*, 117–125.

Rajendran, N., Matsuda, O., Imamura, N. & Urushigawa, Y. (1992). Variation in microbial biomass and community structure in sediments of eutrophic bays as determined by phospholipid ester-linked fatty acids. *Applied and Environmental Microbiology, 58*, 562–571.

Rajendran, N., Matsuda, O., Urushigawa, Y. & Simidu, U. (1994). Characterisation of microbial community structure in the surface sediment of Osaka Bay, Japan, by phospholipid fatty acid analysis. *Applied and Environmental Microbiology, 60*, 248–257.

Ramsay, A. J. (1984). Extraction of bacteria from soil: efficiency of shaking or ultrasonication as indicated by direct count autoradiography. *Soil Biology and Biochemistry, 16*, 475–481.

Ratledge, C. & Wilkinson, S. G. (1988). *Microbial lipids* (Vol. 1). New York: Academic Press.

Reysenbach, A-L., Giver, L. J., Wickenham, G. S. & Pace, N. R. (1992). Differential amplification of rRNA genes by polymerase chain reaction. *Applied and Environmental Microbiology, 58*, 3417–3418.

Rheims, H., Sproer, C., Rainey, F. A. & Stackebrandt, E. (1996). A molecular approach to search for diversity among bacteria in the environment. *Journal of Industrial Microbiology, 17*, 159–169.

Ritz, K. & Griffiths, B. S. (1994). Potential application of a community hybridization technique for assessing changes in the population-structure of soil microbial communities. *Soil Biology and Biochemistry, 26*, 963–971.

Rochelle, P. A., Fry, J. C., Parkes, R. J. & Weightman, A. J. (1992). DNA extraction for 16S rRNA gene analysis to determine genetic diversity in deep sediment communities. *FEMS Microbiology, 100*, 59–66.

Roberston, G. P., Coleman, D. C., Bledsoe, C. S. & Sollis, P. (1999). *Standard Soil Methods for Long-Term Ecological Research*. Oxford, UK: Oxford University Press.

Roslev, P., Iversen, N. & Henriksen, K. (1998). Direct fingerprinting of metabolically active bacteria in environmental samples by substrate specific radiolabelling and lipid analysis. *Journal of Microbiological Methods, 31*, 99–111.

Sadowsky, M. J. (1994). DNA fingerprinting and restriction fragment length polymorphism analysis. In R. W. Weaver, S. Angle, P. Bottomley, D. Bezdicek, S. Smith, A. Tabatabai & A. Wollum (Eds), *Methods of Soil Analysis*. Part 2. *Microbiological and Biochemical Properties* (pp. 119–144). Madison, WI: American Society of Agronomy.

Saiki, R. K., Gelfand, D. H., Stoffel, S. J., Scharf, S. J., Higuchi, R., Horn, G. T., Mullis, K. B. & Erlich, H. A. (1988). Primer-directed enzymatic amplification of DNA with thermostable DNA polymerase. *Science, 239*, 487–491.

Sayler, G. S., Fleming, J., Applegate, B., Werner, C. & Nikbakht, K. (1989). Microbial community analysis using environmental nucleic acid extracts. In T. Hattori, Y. Ishida, Y. Maruyama, Y. Morita & A. Ushida (Eds), *Recent Advances in Microbial Ecology, Proceedings of the 5th International Symposium on Microbial Ecology* (pp. 658–662).

Scheinert, P., Krausse, R., Ullmann, U., Soller, R. & Krupp, G. (1996). Molecular differentiation of bacteria by PCR amplification of the 16S–23S rRNA spacer. *Journal of Microbiological Methods, 26*, 103–117.

Schramm, A., Larsen, L. H., Revsbech, N. P., Ramsing, N. B., Amann, R. & Schleifer, K. H. (1996). Structure and function of a nitrifying biofilm as determined by in situ hybridization and the use of microelectrodes. *Applied and Environmental Microbiology, 62*, 4641–4647.

Selenska, S. & Klingmúller, W. (1991). DNA recovery and direct detection of Tn5 sequences in soil. *Letters in Applied Microbiology, 13*, 21–24.

Smit, E., Leeflang, P. & Wernars, K. (1997). Detection of shifts in microbial community structure and diversity in soil caused by copper contamination using amplified ribosomal DNA restriction analysis. *FEMS Microbial Ecology, 23*, 249–261.

Smit, E., Leeflang, P., Glandorf, B., Van Elsas, D. J. & Wernars, K. (1999). Analysis of fungal diversity in the wheat rhizophere by sequencing of cloned PCR-amplified genes encoding 18S rRNA and temperature gradient gel electrophoresis. *Applied and Environmental Microbiology, 6*, 2614–2621.

Snaidr, J., Amann, R., Huber, I., Ludwig, W. & Schleifer, K. H. (1997). Phylogenetic analysis and in situ identification of bacteria in activated sludge. *Applied and Environmental Microbiology, 63*, 2884–2896.

Stephen, J. R., McCaig, A. E., Smith, Z., Prosser, J. I. & Embley, T. M. (1996). Molecular diversity of soil and marine 16S rDNA sequences related to the γ-subgroup ammonia-oxidising bacteria. *Applied and Environmental Microbiology, 62*, 4147–4154.

Stoffels, M., Ludwig, W. & Schleifer, K. H. (1999). rRNA probe-based cell fishing of bacteria. *Environmental Microbiology, 1*, 259–271.

Teske, A., Sigalevich, P., Cohen, Y. & Muyzer, G. (1996). Molecular identification of bacterial from a co-culture by denaturing gradient gel electrophoresis of 16S ribosomal DNA fragments as a tool for isolation in pure culture. *Applied and Environmental Microbiology, 62*, 776–781.

Torsvik, V., Goksoyr, J. & Daae, F. L. (1990). High diversity of DNA of soil bacteria. *Applied and Environmental Microbiology, 56*, 782–787.

Trevors, J. T. & Van Elsas, J. D. (1995). *Nucleic Acids in the Environment*. Berlin, Germany: Springer-Verlag.

Tsai, Y. L. & Olson, B. H. (1991). Rapid method for direct extraction of DNA from soil and sediments. *Applied and Environmental Microbiology, 57*, 1070–1074.

Tunlid, A. & White, D. C. (1992). Biochemical analysis of biomass, community structure, nutritional status, and metabolic activity of microbial communities in soil. In G. Stotzky & J. M. Bollag (Eds), *Soil Biochemistry* (Vol. 7) (pp. 229–262). New York: Marcel Dekker.

Tunlid, A., Baird, B. H., Trexler, M. B., Olsson, S., Findlay, R. H., Oldham, G. & White, D. C. (1985). Determination of phospholipid ester-linked fatty acid and poly α-hyroxybutyrate for the stimulation of bacterial biomass and activity in the rhizosphere of the rape plant *Brassica napus* (L.). *Canadian Journal of Microbiology, 31*, 1113–1119.

Ueda, T., Suga, Y. & Matsuguchi, T. (1995). Molecular phylogenetic analysis of a soil microbial community in a soybean field. *European Journal of Soil Science, 46*, 415–421.

Van Elsas, J. D., Trevors, J. T. & Wellington, E. M. H. (1997). *Modern Soil Microbiology*. New York: Marcel Dekker.

Wagner, M., Erhart, R., Manz, W., Amann, R., Lemmer, H., Wedi, D. & Schleifer, K. H. (1994). Development of an rRNA-targeted oligonucleotide probe specific for the genus *Actinetobacter* and its application for in situ monitoring in activated sludge. *Applied and Environmental Microbiology, 60*, 792–800.

Wang, G. C. Y. & Wang, Y. (1996). The frequency of chimeric molecules as a consequence of PCR co-amplification of 16S rRNA genes from different bacterial species. *Microbiology, 142*, 1107–1114.

Wardle, D. A., Bonner, K. L., Barker, G. M., Yeates, G. W., Nicholson, K. S., Bardgett, R. D., Watson, R. N. & Ghani, A. (1999). Plant removals in perennial grassland: vegetation dynamics, decomposers, soil biodiversity, and ecosystem properties. *Ecological Monographs, 69*, 535–568.

Wawer, C. & Muyzer, G. (1995). Genetic diversity of *Desulfovibro* spp. In environmental samples analyzed by denaturing gradient gel electrophoresis of [NiFe) hydrogenase gene fragments. *Applied and Environmental Microbiology, 61*, 2203–2210.

White, D. C., Davis, W. M., Nickels, J. S., King, J. C. & Bobbie, R. J. (1979). Determination of the sedimentary microbial biomass by extractable lipid phosphate. *Oecologia, 40*, 51–62.

Wilkinson, S. G. (1988). Gram-negative bacteria. In C. Ratledge & S. G. Wilkinson (Eds), *Microbial Lipids* (Vol. 1) (pp. 299–457). New York: Academic Press.

Wilson, K. H. & Blitchington, R. B. (1996). Human colonic biota studied by ribosomal DNA sequence analysis. *Applied and Environmental Microbiology, 62*, 2273–2278.

Woese, C. R. (1987). Bacterial evolution. *Microbiology Review, 51*, 221–271.

Yeates, G. W., Bardgett, R. D., Cook, R., Hobbs, P. J., Bowling, P. J. & Potter, J. F. (1997). Faunal and microbial diversity in three Welsh grassland soils under conventional and organic management regimes. *Journal of Applied Ecology, 34*, 453–470.

Xu, H. H. & Tabita, F. R. (1996). Ribulose-1,5-bisphosphate carboxylase/oxygenenase gene expression and diversity of Lake Erie planktonic microorganisms. *Applied and Environmental Microbiology, 62*, 1913–1921.

Zak, D. R., Ringelberg, D. B., Pregitzer, K. S., Randlett, D. L., White, D. C. & Curtis, P. S. (1996). Soil microbial communities beneath *Populus grandidenata* grown under elevated atmospheric CO_2. *Ecological Applications, 6*, 257–262.

Zelles, L., Bai, Q. Y., Beck, T. & Beese, F. (1992). Signature fatty acids in phospholipids and lipopolysaccharides as indicators of microbial biomass and community structure in agricultural soils. *Soil Biology and Biochemistry, 24*, 317–323.

Zhou, J., Bruns, M. A. & Tiedje, J. M. (1996). DNA recovery from soil of diverse composition. *Applied and Environmental Microbiology, 62*, 316–322.

Zogg, G. P., Zak, D. K., Ringleberg, D. B., MacDonald, N. W., Pregitzer, K. S. & White, D. C. (1997). Compositional and functional shifts in microbial communities due to soil warming. *Soil Science Society of America, 61*, 475–481.

Zumstein, E., Moletta, R. & Godon, J. J. (2000). Examination of 2 years of community dynamics in an anaerobic bioreactor using fluorescence polymerase chain reaction (PCR) single-strand conformation polymorphism analysis. *Environmental Microbiology, 2*, 69–78.

Miranda J. Keith-Roach and Francis R. Livens (Editors)
© 2002 Elsevier Science Ltd. All rights reserved

Chapter 3

The role of microorganisms during sediment diagenesis: implications for radionuclide mobility

Kurt O. Konhauser[a], Robert J.G. Mortimer[a], Katherine Morris[b], Vicky Dunn[a]

[a]*School of Earth Sciences, University of Leeds, Leeds, LS2 9JT, UK*
[b]*School of the Environment, University of Leeds, Leeds, LS2 9JT, UK*

1. Introduction

Microbial reactions control the chemical composition of the sedimentary environment during early diagenesis. These reactions involve the breakdown and hydrolysis of solid organic matter, the destruction or formation of inorganic compounds and the chemical modification of pore-waters. Although the changes in organic and inorganic chemistry are complex, transient and involve numerous recycling reactions, they produce characteristic patterns which reflect the dominant microbial assemblages growing at a particular depth of sediment. Organic byproducts (e.g. hydrogen gas, fermentation products) may in turn activate dormant microbial consortia elsewhere in the sediment profile, while inorganic byproducts (e.g. HCO_3^-, Mn^{2+}, Fe^{2+}, NH_3, NO_2^-, HS^- and HPO_4^{2-}) may trigger important abiotic reactions between the solid and dissolved phases.

In addition to their influence on the major elemental cycles (C, S, N, P and Fe), microbial activity can also greatly influence the fate and mobility of contaminants introduced into the sediments through anthropogenic activities. Radionuclides, for example, may become mobilised through direct enzymatic microbial reduction, adsorption onto free-living microbes or chelation with biogenically-generated organic ligands (see also Chapters 6, 7 and 11, this volume). Microorganisms can also cause their direct immobilisation by adsorbing contaminants to attached populations or promoting the indirect precipitation of authigenic phases that can either adsorb or co-precipitate the metals. Microbial transformation of these host phases may further mobilise the radionuclides, and their partitioning between the sediment and pore-water will be governed largely by the prevailing redox conditions. In this chapter we review some of the principal microbially-driven chemical reactions within the sediment, and consider the relationship between these and the mobility of some environmentally-important radionuclides, namely Cs, Tc, U, Np and Pu.

2. Organic geochemical signals

Organic matter input to sediments is predominantly in the form of complex polymeric and macromolecular solids. Hydrolysis and fermentation transform these polymers into simple, soluble organic molecules such as amino acids, fatty acids and hydrogen gas (H_2), which are then used by a variety of chemoheterotrophic bacteria – microorganisms that use organic molecules as both a carbon and energy source. This both decreases the total organic carbon (TOC) content of the sediment, and mobilises O, H, N, P and S associated with the organic matter (Santschi et al., 1990a). Release of organic N and P is particularly important since these nutrients drive the sedimentary N and P cycles (Martens et al., 1978; Jørgensen, 1983a; see also Chapter 1, this volume). For example, organic N released into solution as amino acids is deaminated, producing NH_4^+, which then drives the microbial nitrification-denitrification couple (Jenkins & Kemp, 1984; Santschi et al., 1990a).

Characteristic and reproducible organic geochemical signals are generated in the solid phase and pore-waters during organic matter degradation. This is particularly evident in undisturbed sediments that receive a continued input of fresh organic material to the surface. Under such conditions there is a pronounced decrease in solid phase labile organic material with depth (e.g. Hulth et al., 1996; Mortimer and Rae, 2000). However, this steady-state profile may be obscured by a number of factors, including physical mixing, variations in sedimentation rate or short-term changes in organic matter input. For instance, Hedges et al. (1988) have shown that plankton-derived organic matter exhibited approximately 5 times the reactivity of land-derived organic material, so seasonality in planktonic blooms could have a significant bearing on the organic flux to the ocean sediment. The other solid phase signal which may be evident is a change in the C:N:P ratio. Although the C:N ratio of the organic matter input will depend on the relative contribution of terrestrial and marine material (Hulth et al., 1996), there may be recognisable trends with depth in a given sediment. For example, the average elemental composition of organic matter from plankton entering marine sediment is 106C:16N:1P (Redfield, 1958), yet preferential stripping of the N into solution causes the C:N ratio of the sediment to increase with depth (Jørgensen, 1983a). This is partially counteracted by the assimilation of some of the NH_4^+ into growing bacterial cells (which are included in measurements of sediment TOC).

Blackburn (1980) showed that the actual C:N ratio of organic matter mineralised in the upper sediment was lower than that of the average organic detritus present at that depth, whereas the opposite case was true deeper in the sediment. This serves to illustrate how microbial cells themselves should be viewed as an integral component of the sedimentary package. Although there is clear evidence for changes in the C:N ratio during early diagenesis, there is debate as to whether or not preferential stripping of P also occurs, increasing C:P or whether this is partially counteracted by the assimilation of PO_4^{3-} into cells (Jørgensen, 1983a; Ramirez & Rose, 1992).

Degradation of organic matter produces soluble products which are rapidly utilised by other bacteria. Acetate is probably the most important fermentation product (Lovley & Phillips, 1989; Parkes et al., 1989), and hence attempts have been made to determine its concentration in pore-water profiles (Balba & Nedwell, 1982). Nevertheless, current sampling methods produce elevated acetate concentrations that overestimate bioavailability (Shaw & McIntosh, 1990; Wellsbury & Parkes, 1995). Measurements of organic degradation products are therefore limited to turnover rates (Reeburgh, 1983), rather than concentration-depth profiles.

3. Major inorganic geochemical signals

Pore-water and mineralogical changes in sediments are dominated by the activity of chemoheterotrophic microorganisms in all environments at near-neutral pH and ambient surface temperature. The types of organic compounds that can be oxidised vary for different microorganisms; some microorganisms use hydrolytic enzymes to break down complex molecules into simple monomers such as sugars, amino acids and fatty acids which they can then utilise, while others are restricted to simple fermentation products. The electron acceptors used by bacteria in dissimilative metabolism depend on both what candidates are available and, in the situation when multiple terminal electron acceptors (TEAs) are present, on the energy yield of the reaction (Lovley & Chapelle, 1995). For example, using acetate as the organic substrate at pH 7, most energy is obtained via aerobic respiration (reaction 1). As soon as oxygen is depleted, anaerobic respiration takes over, with nitrate reduction (reaction 2); manganese reduction (reaction 3); iron reduction (reaction 4); and sulfate reduction (reaction 5) being used successively.

$$CH_3COO^- + 2O_2 \rightarrow H_2O + 2CO_2 + OH^- \quad \Delta G = -856 \text{ kJ/mol } CH_3COO^- \tag{1}$$

$$CH_3COO^- + 1.6NO_3^- \rightarrow 2CO_2 + 0.8N_2 + 2.6OH^- + 0.2H_2O \quad \Delta G = -806 \text{ kJ/mol } CH_3COO^- \tag{2}$$

$$CH_3COO^- + 4MnO_2 + 3H_2O \rightarrow 4Mn^{2+} + 2HCO_3^- + 7OH^- \quad \Delta G = -569 \text{ kJ/mol } CH_3COO^- \tag{3}$$

$$CH_3COO^- + 8Fe(OH)_3 \rightarrow 8Fe^{2+} + 2HCO_3^- + 15OH^- + 5H_2O \quad \Delta G = -361 \text{ kJ/mol } CH_3COO^- \tag{4}$$

$$CH_3COO^- + SO_4^{2-} \rightarrow HS^- + 2HCO_3^- \quad \Delta G = -48 \text{ kJ/mol } CH_3COO^- \tag{5}$$

Chemoautotrophs have evolved the metabolic capacity to oxidise inorganic substances for cellular energy, with carbon dioxide as their source of carbon. The reduced chemical species used include H_2S, elemental S, NH_3, NO_2^-, Mn^{2+} and Fe^{2+}. Their oxidation by some of the terminal electron acceptors illustrated above produces characteristic chemical and mineralogical signals, particularly in sediments where environmental parameters (e.g. pH, oxygen fugacity) deviate from normal conditions, as for example, in acid mine drainage waters (Ledin & Pedersen, 1996; Freise et al., 1998).

Signals caused by chemoheterotrophs

Biogeochemical zones

Organic matter degradation in sediments is controlled predominantly by terminal electron accepting processes, which in turn are based on the potential thermodynamic yield of the various metabolic processes. Biogeochemical zones of differing terminal electron accepting pathways form the framework for describing early diagenesis of aquatic sediments. Evidence for the existence of these sequential microbial processes has been known for more than half a century (Mortimer, 1941), but a formalised depth-related scheme was not developed until the 1970s (Claypool & Kaplan, 1974; Froelich et al., 1979). This scheme, which represents the microbial degradation of organic matter using successively less energy efficient electron acceptors, is shown in Fig. 1. During the last 20 years, this sequential framework has been used successfully to describe the degradation of organic matter in a variety of aquatic environments, from hypertrophic systems (Barica & Mur, 1980; Klump & Martens, 1981; Canfield et al., 1993) to the oligotrophic deep ocean (Froelich et al., 1979, Christensen et al., 1987; Hulth et al., 1997). Although the depths over which these zones occur may vary from a few mm or cm in coastal sediments with a high organic carbon flux to several metres in the deep ocean where the carbon flux is lower, they occur in any system where the supply of labile organic matter outpaces diffusion of oxygen into the sediment (Jørgensen, 1983a). The reactions which occur in these biogeochemical zones are important not only because they are responsible for the degradation and transformation of organic matter, but also because they play a crucial role in the formation of early diagenetic minerals (Coleman, 1985). In simple terms, for any given system, the depth range of each zone can be determined from a characteristic sequence of chemical changes in the sediment pore-water which occur as electron acceptors are used up and reduced products are formed.

Aerobic respiration

Aerobic respiration is the most energy efficient mechanism for the degradation of organic matter, and therefore represents the first biogeochemical zone. Microorganisms that use oxygen as their terminal electron acceptor completely oxidise a wide variety of natural and synthetic organic compounds to carbon dioxide, water and cell biomass. Aerobic metabolism involves enzymes that selectively degrade individual classes of compounds. For example, the degradation of cellulose involves the activity of a number of hydrolase enzymes, each specific to the breakage of a particular bond (Senior & Balba, 1990). The breakdown of more refractory compounds, such as lignin, is made possible by the use of highly reactive oxygen-containing radicals, such as the superoxide radical (O^{2-}), hydrogen peroxide (H_2O_2) and hydroxyl radicals (OH); each generated through the biochemical reduction of O_2 (Senior & Balba, 1990).

Aerobic respiration has been shown to be the most important terminal electron accepting process in many environments, oxidising >90% of the organic carbon in pelagic marine sediments (Bender & Heggie, 1984) and approximately 45% in marginal marine and freshwater lake environments (Jørgensen, 1983a; Jones, 1985). Oxygen consumption rates in sediments typically range from 3 to 4500 mg m^{-2} day^{-1} for freshwater and marine environments (Reeburgh, 1983), and depend on the carbon source, oxygen concentration

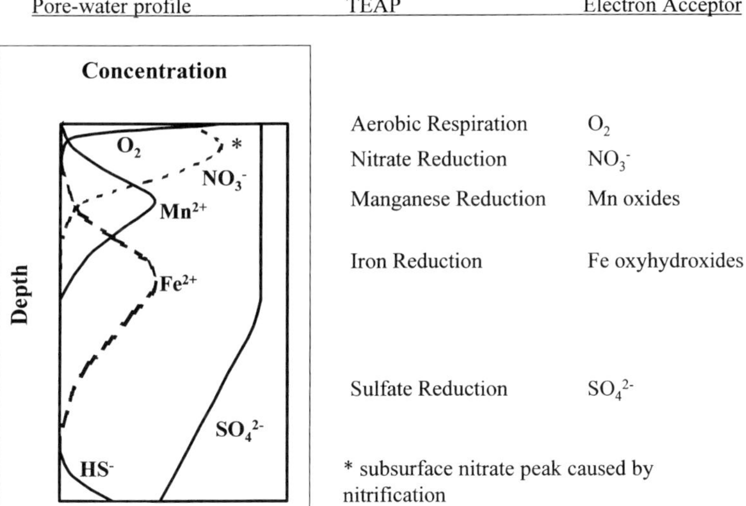

Fig. 1. Hypothetical pore-water profiles produced by successive terminal electron accepting processes (TEAPs) during decomposition of organic matter by chemoheterotrophs. Note also the subsurface nitrate peak caused by chemoautotrophic nitrification (modified from Burdige, 1993).

in the overlying water, the physical regime of the environment and the sedimentation rate (Burdige, 1993; Canfield, 1994). In particular, aerobic respiration rates show a good correlation with sedimentation rates, whereby carbon oxidation is reduced by a factor of 4.8 for every factor of 10 increase in sedimentation rate (Canfield, 1993). Under the low sedimentation rates associated with pelagic marine sediments, little organic matter is deposited and enough time exists for almost complete decomposition via aerobic respiration. At high deposition rates (e.g. continental margins), abundant organic matter is buried and the organic-rich sediment accumulates rapidly enough to pass through the oxic zone. It is estimated that only 45% of the carbon oxidised in these marine settings occurs by aerobic processes, an observation supported by models which suggest that aerobic carbon oxidation dominates when the sedimentation rate is below 0.01 cm y^{-1}, whereas anaerobic carbon oxidation dominates at higher deposition rates (Boudreau & Canfield, 1993; Canfield, 1994). Oxygen is typically used up within the upper few mm of sediment in hypertrophic environments where consumption is faster than resupply (Jørgensen & Revsbech, 1985), but may persist to half a metre or more in pelagic environments (Murray & Grundmaris, 1980). Bioirrigation by macrofauna may introduce oxygenated water to depth, thereby locally increasing the depth of the oxic zone (Hammond et al., 1985).

Nitrogen cycling

Nitrogen cycling in aquatic sediments is relatively complex, with nitrification and dissimilatory nitrate reduction mainly controlling nitrate pore-water profiles. The former is a chemoautotrophic process, and hence will be discussed later. Dissimilatory nitrate reduction most closely resembles aerobic respiration in that nitrate-reducing bacteria (which reduce nitrate to ammonia) or denitrifying bacteria (which reduce nitrate to N_2) are capable of completely degrading complex organic matter to carbon dioxide, with the concomitant depletion of pore-water nitrate. Dissimilatory nitrate reduction generally occurs at the depth where oxygen is used up, although in some instances O_2 can substitute when it is available, thereby reducing both O_2 and nitrate simultaneously (Tiedje et al., 1982). In coastal sediments, dissimilatory nitrate reduction reduces nitrate to negligible levels within the first couple of centimetres of the pore-water profile. No nitrate occurs below this depth except where burrows introduce oxygenated water which oxidises any ammonia present (Hammond et al., 1985; Mortimer et al., 1999). Nitrate may also be reduced to organic nitrogen by assimilatory nitrate reduction (Jørgensen, 1983a), although in most environments this process is of minor importance (Tiedje et al., 1982). Rates of nitrate reduction range over several orders of magnitude in marine sediments, between values as low as 1 μmol m^{-2} day^{-1} in deep-sea sediments (Bender & Heggie, 1984) to over 10 mmol m^{-2} day^{-1} in coastal sediments (Seitzinger, 1988). One of the main factors controlling the relative significance of nitrate reduction is the NO_3^- to O_2 concentration ratio (Canfield, 1993), and in estuarine sediments where water column nitrate levels are similar to dissolved O_2 levels, nitrate reduction is responsible for more carbon oxidation than aerobic respiration (Jørgensen & Sørensen, 1985). In freshwater lake sediments with high nitrate, nitrate reduction may be responsible for the oxidation of up to 20% of the organic carbon (Jones, 1985).

Neither aerobic respiration or dissimilatory nitrate reduction have a significant impact on sediment mineralogy. However, the subsequent terminal electron accepting processes described below involve important mineral dissolution and precipitation reactions (Fig. 2). The pathways for organic matter oxidation by most anaerobic microbial processes are very different from those for aerobic respiration or dissimilatory nitrate reduction. Anaerobic decomposition of organic matter is not generally accomplished by a single organism, but instead by a consortium of interdependent microorganisms (see Chapter 1, this volume). These are very limited in the types of organic compounds that they can oxidise, but together can exploit simple organic acids, long chain fatty acids and monoaromatic compounds (Lovley & Chapelle, 1995). Thus in suboxic to anoxic sediments, much of the organic matter released from hydrolysis of complex organic matter is first metabolised via fermentative microorganisms. Some of the primary fermentative byproducts include acetate, lactate, propionate, butyrate and H_2. Because fermentative microorganisms do not completely oxidise organic matter to CO_2, other types of microorganisms operate in conjunction to bring about complete oxidation of organic matter. Fermentation can in fact be inhibited by a high concentration of fermentable byproducts (Schink, 1988; Senior & Balba, 1990) because their accumulation makes the reactions thermodynamically unfavourable (Canfield, 1994).

Reduction of Fe and Mn

Following dissimilatory nitrate reduction, reduction of manganese oxides becomes the most energy-efficient bacterial respiratory process (Santschi et al., 1990a). The principal mineralogical change associated with manganese reduction is dissolution of Mn oxides to produce dissolved Mn^{2+} (Myers & Nealson, 1988), which often diffuses upwards and reprecipitates as fresh Mn oxides at the sediment surface (Burdige, 1993). Downward diffusion of Mn^{2+} most commonly results in the formation of rhodocrosite ($MnCO_3$)

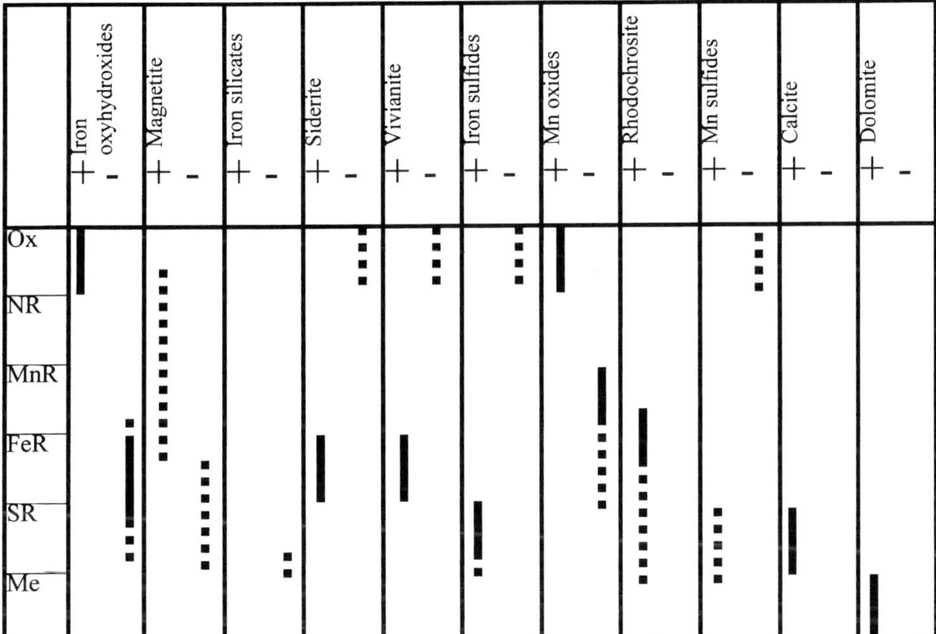

Fig. 2. Hypothetical depth profiles illustrating zones where principal authigenic minerals may precipitate (+) or be dissolved (–) within organic-rich sediments. Both microbial/inorganic precipitation and reductive dissolution reactions have been included. Solid lines indicate depth intervals over which minerals are commonly precipitated or dissolved. Dashed lines indicate other depths where these processes may occur. Ox = aerobic respiration; NR = nitrate reduction; MnR = manganese reduction; FeR = iron reduction; SR = sulfate reduction; Me = methanogenesis.

(Emerson, 1976; Burdige, 1993) or other Mn-carbonates such as kutnahorite and manganoan calcite (Aller & Rude, 1988). Although rhodocrosite has been seen to form in cultures of Mn-reducing bacteria (Lovley & Phillips, 1988) and its formation in lakes and marine sediments has been attributed to bacterial activity (Sokolova-Dubina & Deryugina, 1967; Suess, 1979), it is not known whether the bacteria play a role in mineral formation beyond supplying the ions necessary for precipitation. In environments where pore-water Mn^{2+} concentrations are relatively high, such as in the Baltic Sea basins, MnS may form (Suess, 1979). Because of the limited amount of manganese deposited in sediment, Mn reduction is of minor importance in terms of the amount of total organic carbon oxidised, generally <5% in most environments (Jørgensen, 1983a; Bender & Heggie, 1984; Jones, 1985).

Below the zones of dissimilatory nitrate reduction and manganese reduction, when nitrate has been exhausted and Eh drops accordingly, iron reduction occurs. Dissimilatory iron reduction is broadly distributed amongst several known bacterial genera (Fredrickson & Gorby, 1996). *Geobacter metallireducens* (Lovley & Phillips, 1988) and *Shewanella putrefaciens* (Lovley et al., 1989a) were among the first bacteria studied in pure culture that could gain energy from coupling the oxidation of organic matter and/or H_2 to Fe(III) reduction. Since then, many more species, including some known sulfate reducers, such as *Desulfuromonas* sp. (Roden & Lovley, 1993) and *Desulfovibrio* sp. (Coleman et al., 1993), have shown the ability to reduce ferric iron. Iron reduction has potentially the most significant impact on sediment mineralogy. Amorphous to poorly-crystalline iron hydroxides, such as ferrihydrite and lepidocrocite, are considered the most important source of Fe(III) for iron reduction, with the rates of reduction declining rapidly with depth as they become depleted (Lovley & Phillips, 1986a, b, 1987). More crystalline Fe(III) oxides (e.g. hematite and goethite) are also microbially reducible, and recent experimental observations suggest that these minerals may provide energy for cellular growth comparable to that derived from the poorly crystalline phases (Roden & Zachara, 1996). Other studies have shown that a variety of microorganisms can even reduce Fe^{3+} in both magnetite (Kostka & Nealson, 1995) and smectite clays (Kostka et al., 1999), clearly indicating that a wide array of iron minerals can be used by Fe(III)-reducing bacteria. The rates of iron reduction are dependent upon a number of factors, including the surface area and site concentration of the solid iron phase, the composition of the aqueous solution in which the microbes grow and the amount of Fe^{2+} sorbed to the oxide surface (Roden & Zachara, 1996; Fredrickson et al., 1998; Urrutia et al., 1998).

Removal of Fe from solution

The reduction of ferric iron minerals produces an increase in the concentration of ferrous iron in the pore-water. Some of this ferrous iron may diffuse upwards to be reoxidised, but most is removed from solution by precipitation of metal sulfides at depth. There is often a pore-water ferrous iron maximum at the boundary between the iron reduction and sulfate reduction (SR) zones (Jørgensen, 1983a; Burdige, 1993). Below this depth, the iron is removed from solution by reaction with sulfide produced by sulfate-reducing bacteria (SRB) below. This forms metastable iron monosulfide minerals such as greigite and mackinawite which are precursors to pyrite (FeS_2) (Postma, 1982; Berner, 1984). The precise mechanism for the transformation of monosulfides into pyrite, however, remains poorly understood (Benning et al., 2000).

In addition to sulfide formation, iron may be removed from solution by the abiogenic or biogenic precipitation of hydroxide, mixed valence oxide, carbonate, phosphate and silicate minerals, depending on the chemical conditions of the particular sediment (Santschi et al., 1990a; Lovley, 1991). The biologically-induced formation of epicellular iron hydroxide by bacteria can occur either passively or actively (Konhauser, 1997, 1998; Ferris et al., 1999). In the first instance, the oxidation and hydrolysis of cell-bound ferrous iron and/or the alteration of local pH and redox conditions around the cell due to metabolic activity can all induce transformation to insoluble hydroxide forms. Alternatively, ferrous iron diffusing upwards into more oxidising sediment spontaneously reacts with dissolved oxygen (or potentially NO_3^- or MnO_2) to precipitate rapidly (abiotically) as ferric hydroxide on available nucleation sites (Burdige, 1993; Canfield, 1993). Bacteria passively act as such sites, and over a short period of time the microbes can become completely encrusted in amorphous iron as abiological surface catalysis accelerates the rate of mineral precipitation. Indeed, ferrihydrite develops on the organic remains of dead cells, implying that iron mineralisation can occur independent of cell morphology, trophic classification or physiological state (Ferris et al., 1989). It has also been suggested that under circumneutral pH any bacterium that produces acidic, extracellular polymers will nonspecifically adsorb positively charged Fe-hydroxides (Ghiorse, 1984) since the point of zero charge (pH where the mineral has zero charge) of amorphous Fe-hydroxides lies in the range of 5.3–7.5 for natural samples (Schwertmann & Fechter, 1982). Reactive organic sites can therefore scavenge ferric iron from the surrounding waters. The active process by which iron hydroxides form stems from the ability of Fe(II)-oxidising bacteria to oxidise ferrous iron as an energy source (see below).

In suboxic sediments there is clear evidence for authigenic magnetite (Fe_3O_4) formation (Karlin et al., 1987), although its actual origin is uncertain because a variety of potential precipitation mechanisms exist (Lovley, 1991). The biogenic magnetite fraction is commonly single domain, with a high natural magnetic remanence, and is known to proceed under both 'biologically controlled' and 'biologically induced' conditions (see Blakemore & Frankel, 1989). In the first instance, magnetotactic bacteria produce intracellular chains of single, pure magnetite crystals (35–120 nm in diameter), under precise biochemical, chemical and probably genetic control (Mann et al., 1990; Bazylinski, 1996). Magnetotactic bacteria are common in aquatic habitats, their greatest abundance being at the oxic–anoxic boundary (Bazylinski, 1995). Obligately microaerophilic bacteria, such as *Aquaspirillum magnetotacticum* strain MS-1, use oxygen, nitrate (Bazylinski & Blakemore, 1983) and possibly ferric iron (Blakemore & Frankel, 1989) as terminal electron acceptors. In contrast, pure cultures and consortia of iron reducers have been shown to produce magnetite crystals quite different from those formed under controlled conditions. This magnetite is extracellular, there is no evidence of cellular material associated with it, and the crystals typically consist of a mixture of round and oval particles that range in size from 10–50 nm (Lovley et al., 1987). It remains unclear whether the iron reducers produce magnetite enzymatically or whether they simply create the necessary chemical conditions for abiotic precipitation to proceed (Lovley, 1991). However, since the magnetite produced is crystalline, and not particularly reducible, iron-reducing bacteria effectively convert part of the labile iron oxide pool into a more refractory one.

Other minerals which have been observed to form in cultures of iron-reducing bacteria are siderite ($FeCO_3$) (Lovely & Phillips, 1986a; Mortimer & Coleman, 1997; Mortimer et al., 1997) and vivianite ($Fe_3(PO_4)_2 \cdot 8H_2O$) (Fredrickson et al., 1998; Zachara et al., 1998). As with magnetite, it is unclear whether microorganisms enzymatically catalyse the precipitation reactions or simply produce chemical conditions to make abiotic precipitation favourable. Siderite formation has been linked to the activity of sulfate-reducing bacteria and methanogens, as well as iron reducers (Coleman, 1985; Lovley, 1991; Coleman et al., 1993). Since iron reduction produces both ferrous iron and bicarbonate, conditions may become favourable for the precipitation of siderite. However, in marine sediments, sulfate reduction produces sulfide which will preferentially react with any ferrous iron to precipitate iron monosulfide minerals. Therefore, siderite is more common in freshwater environments (Postma, 1982); the precise juxtaposition of iron reduction and sulfate reduction in marine sediments controls whether siderite and/or pyrite may form (Coleman, 1985; Coleman et al., 1993). Vivianite forms under anoxic conditions where sufficient phosphate and iron are present and the sulfide activity is low enough to preclude formation of iron sulfides (Hearn et al., 1983). Precipitation is extremely slow and may be inhibited by the presence of ferric hydroxides which will react with phosphate to form insoluble ferric hydroxy-phosphates (Emerson & Widmer, 1978; Postma, 1981). Vivianite has been observed in the upper iron reduction zone in anaerobic lake sediments (Emerson & Widmer, 1978), in freshwater swamps (Postma, 1982) and in river sediments (Hearn et al., 1983). Indirect evidence has also shown that it may form under marine conditions before the onset of sulfate reduction (Martens et al., 1978).

When the sediment receives significant input of detrital material and/or biogenic silica (e.g. diatom frustrules) their dissolution may lead to reprecipitation of authigenic aluminosilicates. Biogenic iron silicate minerals are ubiquitous in freshwater sediments, and precipitate when dissolved silica and aluminum react with cellularly bound iron via hydrogen bonding between the hydroxyl groups in the bound iron and the hydroxyl groups in the silica and aluminum (Konhauser & Urrutia, 1999). This is not unusual, as ferrihydrite commonly adsorbs large quantities of silica in natural mineral deposits (Carlson & Schwertmann, 1981). Over time, these hydrous compounds dehydrate, with some phases converting to more stable crystalline forms. Authigenic formation of iron silicates in the marine environment has also been recently reported from the Amazon delta, where Michalopoulos & Aller (1995) reported authigenic K–Fe–Mg clays contributing >3% of sediment weight.

With respect to the percentage of total organic carbon oxidised, the importance of iron reduction compared to other terminal electron accepting pathways varies according to the environment. It is also difficult to assess because methods used to estimate iron reduction are generally flawed and tend to underestimate rates (Lovley, 1991). However, iron reduction is known to be an extremely important process in soils (Ponnamperuma, 1972) and in tropical sediments rich in ferric oxyhydroxide, such as those along the Amazon Shelf (Aller et al., 1986, 1998). It has also been shown to be important in temperate marine and freshwater environments (Jones et al., 1984; Sørensen & Jørgensen, 1987; Lovley et al., 1990; Coleman et al., 1993; Hines et al., 1997).

Other electron acceptors

In addition to Fe(III) and Mn(IV), some bacteria may use other metals as electron acceptors. In cultures of *G. metallireducens*, the *c*-type cytochromes were oxidised by gold, silver and mercury, although whether the reduction of these metals can function to support growth was not determined (Lovley et al., 1993). The microbial reduction of selenate (SeO_4^{2-}) to selenite (SeO_3^{2-}) or elemental selenium has been shown in anoxic sediments and culture (Macy et al., 1989; Oremland et al., 1989). Other microbial taxa have been shown to reduce As(V) to As(III) (Ahmann et al., 1994; Newman et al., 1997); Cr(VI) to Cr(III) (Wang et al., 1989); V(V) to V(IV) (Yurkova & Lyalikova, 1991); Mo(VI) to Mo(V) (Sugio et al., 1988); and Cu(II) to Cu(I) (Sugio et al., 1990). The likely effects of microbially mediated reduction on the solubility of radionuclides will be discussed in more detail below.

Interestingly, many of the microbially-mediated metal reductions can be coupled to the oxidation of organic contaminants or humic compounds. Lovley et al. (1989b) and Lovley & Lonergan (1990) have shown that, in contaminated groundwater, *G. metallireducens* could couple the oxidation of synthetic aromatic hydrocarbons, such as benzoate, toluene, phenol and *p*-cresol, to Fe(III) reduction. Purified enrichment cultures metabolised other aromatics such as syringic acid, ferulic acid, nicotinic acid, *m*-cresol and a variety of mono-

and dihydroxybenzoates (Lonergan & Lovley, 1991). The reductive dehalogenation of chlorinated organic compounds, e.g. tetrachloromethane, has similarly been demonstrated for pure cultures of *S. putrefaciens* (Picardal et al., 1993). Kazumi et al. (1995) have suggested that a consortium of microorganisms may be involved in the degradation of chlorinated aromatics, with one or more microbes being responsible for dehalogenating the substrate, while other microbes utilise the aromatic ring and degradation products. Iron reducers can, however, also contribute indirectly to organic oxidation reactions by forming chemically reactive forms of Fe(II). Klausen et al. (1995) have shown that surface-bound Fe(II) species, because of their significantly lower reduction potential compared with aqueous Fe(II), contributed to the reduction of nitroaromatic compounds to amines and anilines.

Until recently, humic substances were considered resistant to bacterial degradation. However, Lovley et al. (1996) established that cell suspensions of *G. metallireducens* and *Shewanella alga* were able to use humic substances as electron acceptors for the anaerobic oxidation of organic compounds. Microbial humic reduction also mediated the reduction of Fe(III) oxides. First, the bacteria oxidised acetate or lactate with the humics as intermediate electron acceptors. Second, the reduced humic acids donated electrons to Fe(III), thereby regenerating the humic substances in an oxidised form to accept, once again, electrons from the humic-reducing bacteria. In essence, the humics served as electron shuttles between the microbes and the solid iron oxide. In a subsequent study, Coates et al. (1998) demonstrated that humic-reducing bacteria of the *Geobacteraceae* family could be recovered from a variety of sediment types, suggesting that the potential for humic substances to actively enhance microbial reduction processes was widespread.

Sulfate reduction and sulfidation

Below the zones described by Froelich et al. (1979) as 'suboxic' (nitrate-reducing, manganese-reducing and iron-reducing), is the 'anoxic' sulfate reduction zone. Sulfate reduction only occurs when all other available terminal electron acceptors have been exhausted. In marine environments, where there is a high concentration of dissolved sulfate, it is quantitatively a very important process of organic matter degradation and may occur over a considerable thickness of sediment (Jørgensen, 1983a). It has been estimated that sulfate reduction is responsible for nearly all organic carbon oxidation in euxinic sediments (Canfield, 1993), approximately 50% of the organic carbon oxidation in coastal marine sediments (Jørgensen, 1983a), but <10% in pelagic and freshwater environments (Bender & Heggie, 1984; Jones, 1985). Some freshwater environments such as swamps, which are very organic-rich and hence have substantial releases of S from organic matter, may support increased sulfate reduction (Berner, 1980). The rate of sulfate reduction is proportional to the quantity and reactivity of organic matter entering the sulfate reduction zone, which in turn is a function of the extent to which this organic matter was first degraded in the oxic zone (Westrich & Berner, 1984). Sulfate reduction rates in marine sediments vary by six orders of magnitude and are highest in rapidly deposited continental shelf sediments, with intermediate levels in more slowly accumulating hemipelagic sediments, and lowest levels in deep-sea pelagic sediments (Canfield, 1993).

Sulfate reduction results in an increase in pore-water sulfide at the expense of sulfate. Dissolved H_2S can be extremely toxic to sulfate-reducing bacteria because it combines with the iron of cytochromes (Postgate, 1984). On the other hand, reaction with extracellular iron is a common detoxification mechanism for sulfide in the environment through formation of insoluble iron monosulfides (FeS), and eventually pyrite. Initially, this process is driven by reaction of the biogenic sulfide with dissolved ferrous iron diffusing down from the iron-reducing zone above. However, as the sulfide produced overwhelms the ferrous iron available, sulfide accumulates in the pore-water up to the base of the iron-reducing zone and is removed more slowly by abiotic reaction with ferric iron minerals (Canfield, 1989). This liberates further ferrous iron, leading to the formation of additional iron sulfide, and eventually pyrite. The amount and reactivity of iron phases dictates how much pyrite will form (Canfield, 1989; Canfield et al., 1992). In very rapidly accumulating sediments (>1 mm y^{-1}), there may be insufficient interstitial dissolved sulfide at depth to allow reaction with all detrital iron minerals. In other words, there is an over abundance of iron minerals deposited. By contrast, at lower sedimentation rates, potentially all of the Fe supplied to the sediments may become pyritised because the supply of dissolved sulfide exceeds the amount of iron initially deposited, although this does not generally happen because some Fe phases react only very slowly. For example, Canfield et al. (1992) have shown that ferrihydrite has a half-time for reaction of 4 hours. Lepidocrocite, goethite, hematite, magnetite and sheet silicates are increasingly less reactive, with the latter having a reaction half-time of 100,000 years. Those minerals which are most reactive to sulfide can also be readily utilised by iron-reducing bacteria, suggesting that the two processes may act in competition. In addition to reacting with iron minerals, sulfide may reduce manganese oxides in a similar fashion (Aller & Rude, 1988; Santschi et al., 1990a). These reactions highlight the importance of sulfate reduction with respect to the sediment mineralogy; along with iron reduction, it is responsible for the predominant mineralogical transformations during early diagenesis and the oxidation of most, if not all, remaining organic matter left in the sediments.

Methanogenesis

In some instances, the terminal step in the anaerobic degradation of organic material is methanogenesis. The methanogens are strictly limited in the types of compounds that they can metabolise (Madigan et al., 1997) and most use CO_2 (as HCO_3^- at circumneutral pH) as both their terminal electron acceptor and carbon source, while H_2 serves as the electron donor (reaction 6). A second class of reactions involves the reduction of compounds containing the methyl functional group, such as methanol (CH_3OH) and methylamine (CH_3NH_2) to methane (reaction 7). Other methanogens disproportionate acetate to give methane and bicarbonate (reaction 8). Methanogens cannot use long chain fatty acids and aromatic compounds so, under anoxic conditions where methanogens grow, the homoacetogenic bacteria, are also required. These bacteria convert fatty acids and aromatics to acetate and H_2, which are then consumed by the methanogens.

$$4H_2 + CO_2 \rightarrow CH_4 + 2H_2O \quad (6)$$
$$\Delta G = -131 \text{ kJ/mol } CH_4$$

$$4CH_3OH \text{ (methanol)} \rightarrow 3CH_4 + CO_2 + 2H_2O \quad (7)$$
$$\Delta G = -106 \text{ kJ/mol } CH_4$$

$$CH_3COO^- + H_2O \rightarrow CH_4 + HCO_3^- \quad (8)$$
$$\Delta G = -31 \text{ kJ/mol } CH_4$$

Methanogenic activity is particularly abundant in fresh- and brackish-water sediments, such as peat deposits, tidal estuaries and eutrophic lakes (Smith & Klug, 1981; Williams & Crawford, 1984; Avery & Martens, 1999), or in some rapidly depositing shallow water marine sediments (Canfield, 1993), where the accumulation of organic material exceeds decay by conventional terminal electron acceptors, particularly sulfate. By contrast, marine waters contain sufficiently high levels of sulfate such that SRBs effectively out compete the methanogens for available acetate or H_2 in the sediments (Lovley et al., 1982; Lovley & Klug, 1986). This pattern largely stems from the thermodynamic and kinetic advantages sulfate-reducing bacteria have over methanogens, as indicated by the standard Gibbs free energy change of the oxidation of the substrates (Rinzema & Lettinga, 1985) and the higher affinity sulfate reducing bacteria have for the oxidisable substrates (Lovley & Klug, 1983). Because of this competitive disadvantage, the major precursors of methane production in marine environments are compounds which are inefficiently used by sulfate-reducing bacteria, i.e. the methylated compounds (King, 1984).

Carbonate minerals and sediment hydrogen concentrations

The pore-water and mineralogical signatures described in the previous sections outline the major chemical and mineralogical changes that occur with depth through successive biogeochemical zones. However, both carbonate mineral formation and microbial oxidation of hydrogen may occur in any biogeochemical zone, and these are worth describing separately.

The saturation state of sediment pore-waters with respect to calcium carbonate is dependent on organic matter oxidation during diagenesis (Van Cappellen & Gaillard, 1996). Aerobic respiration results in the complete oxidation of organic carbon to CO_2 and hence may promote the dissolution of biogenic carbonate (Archer et al., 1989). Anaerobic respiration, on the other hand, generates bicarbonate and potentially drives the precipitation of early diagenetic carbonate minerals (Boudreau & Canfield, 1993). These minerals are relatively stable once formed and are not subject to further rapid recycling by redox reactions in the same way as sulfides and oxides. Therefore, a characteristic sequence produced by different terminal electron accepting processes can be distinguished on both mineralogical and stable isotopic grounds in environments where precipitation of carbonate minerals is possible, e.g. organic-rich muds (Berner, 1981, Maynard, 1982; Coleman, 1985; Fig. 3). Aerobic respiration produces non-ferroan calcite because there is no dissolved iron present under oxic conditions. This calcite has a stable carbon isotopic composition ($\delta^{13}C_{PDB}$) of 0‰ (as do all primary marine carbonates). Nitrate reduction does not produce carbonate minerals with characteristic chemistry. Manganese and iron reduction produce rhodocrosite and siderite, respectively, with increasingly negative $\delta^{13}C_{PDB}$ (typically –2 and –10‰, respectively) due to the incorporation of an increasing component of organogenetic carbonate which has a $\delta^{13}C_{PDB}$ signature of –20 to –30‰. Sulfate reduction produces non-ferroan

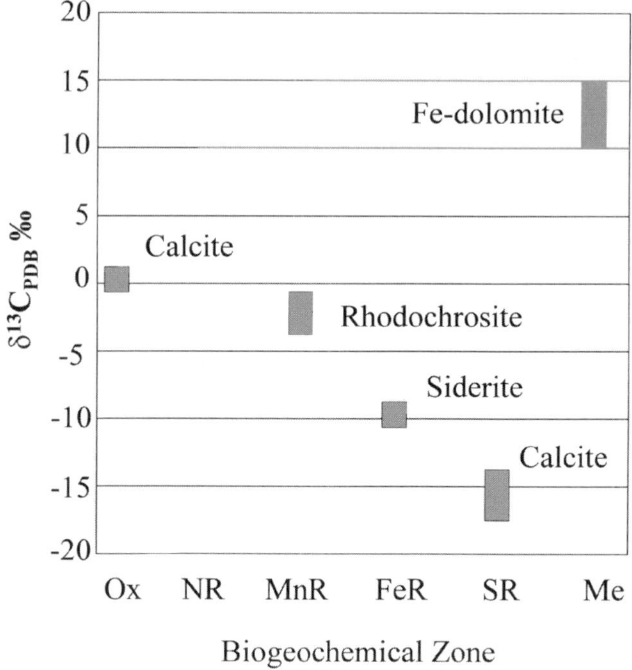

Fig. 3. Characteristic chemical and carbon isotopic composition of authigenic carbonate minerals produced in different biogeochemical zones (after Coleman, 1985).

calcite because any available iron reacts preferentially with sulfide. These calcites have an even more negative $\delta^{13}C_{PDB}$ signature (typically –15‰). Finally, below the zone of sulfate reduction, methanogenesis leads to the production of ferroan carbonates (siderite or ferroan dolomite) with a characteristic positive $\delta^{13}C_{PDB}$ signature due to coupling of methane oxidation with reduction of residual iron minerals (Irwin et al., 1977; Coleman, 1985).

As discussed above, the dominant terminal electron accepting pathways are generally segregated into distinct zones in sediments based on the potential thermodynamic yield of the various metabolic processes. However, reactions yielding less energy should also take place as long as they are energetically favourable and segregation can perhaps be more accurately explained on the basis of competition between different types of microorganisms for electron donors (Lovley & Phillips, 1987; Lovley et al., 1994; Lovely & Chapelle, 1995). For example, microorganisms may couple the oxidation of hydrogen to the reduction of nitrate, manganese, iron, sulfate or carbon dioxide and this is the predominant control on the concentration of dissolved hydrogen gas in aquatic sediments. Lovley & Goodwin (1988) also showed that microorganisms oxidising H_2 with the reduction of more electrochemically positive terminal electron acceptors can maintain lower hydrogen concentrations than microorganisms using terminal electron acceptors which yield less energy. Therefore, sediments in which nitrate and/or manganese reduction were the dominant terminal electron accepting pathway had the lowest hydrogen concentrations

(<0.05 nM), and these levels increased through iron reduction (0.2 nM), sulfate reduction (1–1.5 nM) and methanogenesis (7–10 nM). This pattern of H_2 concentrations can be ascribed to the physiological capabilities of the microorganisms growing in the sediments. Microorganisms that use Fe(III) for H_2 oxidation can metabolise it to concentrations lower than those that can be utilised by sulfate reducers, and the sulfate reducers can metabolise the same substrates to concentrations below those usable by methanogens. These findings are consistent with observations of sediments where sulfate is not reduced, and methane not produced until the reducible Fe^{3+} has been converted to Fe^{2+}. Thus, when the availability of Fe(III) does not limit the rates of microbial Fe(III) reduction, sulfate reduction and methane production are inhibited because of the iron reducers' competitive advantage in using available substrates at much lower concentrations (Lovley & Phillips, 1987). Trace metals are subject to similar forms of competition, with for example, the presence of nitrate inhibiting As(V) reduction, while sulfate reduction is inhibited by the presence of As(V) (Dowdle et al., 1996).

Problems with the biogeochemical zone scheme

There are a variety of potential problems with the simple zonal scheme. Natural aquatic environments contain complex microbial communities rather than a simple succession of 'pure cultures', and hence there is potential for overlap of the processes described above (Canfield et al., 1993, Postma & Jakobsen, 1996). In addition, microenvironments may form, allowing the coexistence of suboxic and anoxic processes at the same sediment depth (Goldhaber et al., 1977). There may also be complex and continuous recycling reactions occurring which are simply not seen because their net effects cancel each other out.

Our understanding of the individual microbial reduction reactions which form the basis of the biogeochemical zone scheme is incomplete, particularly in the case of iron and manganese reduction. For instance, we have known for more than a decade that ferric oxides persist at depth below the iron reduction zone, even in the presence of labile organic matter and microorganisms (Canfield, 1989; Lovley et al., 1990), but we do not know why iron reduction ceases. There is growing evidence for complex cycling of Mn–Fe–S rather than a simple succession of manganese reduction, iron reduction and sulfate reduction. Canfield et al. (1993) challenged the view that the sequence of Mn^{2+} and then Fe^{2+} in pore-water was indicative of a change from manganese reduction to iron reduction because Mn oxides may oxidise Fe^{2+} to Fe^{3+} oxides, releasing Mn^{2+} into solution. Similarly, reduced sulfur species, specifically S^0, organic S and/or FeS, can be used by chemoautotrophic bacteria to reduce Mn-oxides completely under anoxic conditions, thereby complicating measurement of the actual sulfate reduction rate (Aller & Rude, 1988). Canfield (1989) also showed that it is possible to have pore-waters rich in Fe^{2+}, but low in H_2S, despite active sulfate reduction. This is due to the reaction of ferric oxyhydroxides with H_2S which forms iron sulfides, leaving Fe^{2+} in solution. He invoked microenvironments to explain the contradictory findings that iron reduction was the major source of pore-water Fe^{2+} but most ferric oxides were reduced by H_2S. In one microenvironment, iron reduction released Fe^{2+} into solution, whereas in the other, ferric oxides were reduced by H_2S, with the Fe^{2+} generated quickly reacting with additional H_2S to precipitate iron sulfides. More recently, it has been shown that sulfate reduction may occur in Fe-rich environments (Jakobsen & Postma, 1994), and that sulfate-reducing bacteria are able to switch to the reduction of ferric iron (Coleman et al., 1993). It has also been observed that certain nitrate-reducing bacteria are capable of iron reduction (Sørensen, 1982, 1987; Lovley & Phillips, 1988; Ehrlich, 1990) and that fermenters can reduce manganese and ferric iron (Jones, 1985).

Recent advances in sampling both pore-water geochemistry and microbiological populations also suggest that the sequential scheme of biogeochemical zones is an oversimplification. Techniques described in Chapter 2 of this volume, such as signature lipid biomarkers, can now be used to distinguish in situ microbial biomass, community structure and nutritional status (White, 1993). Nucleic acid sequence analysis of genes also allows characterisation of community structure and function (Muyzer et al., 1993). This has led to the development of 'gene probes' to sample for specific groups of microorganisms. These technologies have been used in redox stratified environments (Coleman et al., 1993; Duan et al., 1996) to show that siderite-iron monosulfide concretions, present in anoxic salt-marsh sediments, were caused by direct enzymatic reduction of Fe(III) by sulfate-reducing bacteria (*Desulfovibrio*), rather than the iron-reducing populations.

New methods have also been devised to sample pore-water chemistry at higher resolution than previously possible. The two principal new approaches are microelectrodes (Brendel & Luther, 1995) and gel probes (Davison et al., 1991, 1994; Krom et al., 1994; Zhang et al., 1995; Mortimer et al., 1998). Microelectrode profiling of redox species in cores from continental margin sites off the coast of Nova Scotia has recently revealed evidence for complex interactions between the manganese and nitrogen cycles (Luther et al., 1997, 1998), whereby Mn^{2+} did not appear until many millimetres below the oxic layer. Those studies used field data and laboratory evidence to suggest that dinitrogen could be produced by the oxidation of ammonia and organic-N with MnO_2 in air, and that dissolved Mn^{2+} could abiotically reduce nitrate to dinitrogen. Direct evidence for close coupling between Mn and N has also recently been obtained in laboratory experiments (Hulth et al., 1999).

Signals caused by chemoautotrophs

During chemoheterotrophic activity, reduced byproducts such as NH_3, Mn^{2+}, Fe^{2+}, H_2S and CH_4 are generated below the oxic–suboxic boundary. Physical and macrofaunal processes will cause the net transport of these reduced species from the deeper layers towards the sediment surface. As a result, a fraction of the oxygen consumed in the surface layers is diverted away from aerobic respiration toward the reoxidation of reduced species by chemoautotrophic bacteria (Van Cappellen & Gaillard, 1996). The impact of chemoautotrophic bacteria on sediment mineralogy and pore-water chemistry is variable, and generally site specific. Bacterial nitrification often occurs in sediments between the zones of aerobic respiration and nitrate reduction. Nitrifying bacteria develop especially well in lakes and streams that receive inputs of ammonia-rich sewage or at sites beneath planktonic blooms where extensive protein decomposition occurs (ammonification). Methanotrophs occur wherever stable sources of methane are present. In more extreme environments, such as those associated with acid mine waters or under low oxygen concentrations, iron and sulfur-oxidising bacteria play a major role in the localised Fe and S cycles.

Just below the sediment–water interface, ammonia diffusing up from depth comes into contact with oxygen in the oxic layer and is converted to nitrate by the process of nitrification (Hansen et al., 1981). The ammonium is first oxidised to nitrite by ammonium-oxidising bacteria (e.g. *Nitrosomonas*) (reaction 9), and this is then further oxidised to nitrate by nitrite oxidisers (e.g. *Nitrobacter*) (reaction 10); no bacterium is known that will carry out the complete oxidation of ammonia to nitrate (Madigan et al., 1997). Combined, these processes cause a subsurface peak of nitrate in the pore-water (Fig. 1). In coastal sediments with a high carbon flux and relatively low nitrate concentration in the overlying water, this nitrification peak is often very close to the sediment–water interface (Mortimer et al., 1999).

$$NH_3 + 1.5O_2 \rightarrow NO_2^- + H_2O + H^+ \quad (9)$$
$$\Delta G = -96\ \text{kJ/0.5 mol } O_2$$

$$NO_2^- + 0.5O_2 \rightarrow NO_3^- \quad (10)$$
$$\Delta G = -74\ \text{kJ/0.5 mol } O_2$$

Although methane is a relatively stable compound, a variety of bacteria, the methanotrophs, utilise it as an electron donor for energy generation and as their sole source of carbon. These bacteria are all obligate aerobes and they grow abundantly at the chemocline in lakes and sediment, where methane from the underlying anoxic zone diffuses upwards to the overlying oxic zone. Methanotrophs are also able to oxidise ammonia, although they cannot grow with it as the sole electron donor (Madigan et al., 1997).

In acid mine drainage, *Thiobacillus ferrooxidans* and *Leptospirillum ferrooxidans* catalyse the reaction between iron and oxygen, leading to the formation of ferric iron minerals (reaction 11). Because very little energy is generated in the oxidation of ferrous to ferric iron (with oxygen as the electron acceptor), these bacteria must oxidise large quantities of iron in order to grow. For example, it has been estimated that a consumption of 90.1 mol of Fe^{2+} is required to assimilate only 1.0 mol of carbon (Ehrlich, 1990). Consequently, even a small number of bacteria can be responsible for precipitating vast amounts of iron.

$$2Fe^{2+} + 0.5O_2 + 2H^+ \rightarrow 2Fe^{3+} + H_2O \quad (11)$$
$$\Delta G = -54\ \text{kJ/0.5 mol } O_2$$

At neutral pH, Fe(II) oxidation by *Sphaerotilius natans*, *Leptothrix ochracea* and *Gallionella ferruginea* occurs under partially reduced conditions, with an Eh range of +200 to +320 mV and oxygen levels of 0.1–1.0 mg of O_2 per litre (Ehrlich, 1990). The existence of these microaerophilic bacteria depends on their oxidising efficiency relative to abiotic oxidation at low oxygen fugacity. Although there is no conclusive evidence that iron bacteria other than the acidophiles derive energy from Fe(II) oxidation, *Gallionella ferruginea*, for example, can grow autotrophically with Fe^{2+} as its sole energy source at a pH just below 7 (Hallbeck & Pedersen, 1991). Straub et al. (1996) also demonstrated that the biological oxidation of ferrous iron in the absence of oxygen was possible using nitrate instead as the electron acceptor. This observation, that nitrate reducers, which had never previously been grown in iron media, exhibited a capacity for ferrous iron oxidation, implies that this form of microbial oxidation of ferrous iron may be common in the suboxic zone of aquatic environments.

The role of sulfur-oxidising bacteria in controlling sediment mineralogy and pore-water chemistry is also limited to environments where the spontaneous reaction of sulfide with oxygen is inhibited (e.g. where the oxygen concentration is low). This includes being just above the oxic/anoxic boundary, or living at hot springs, where the solubility of oxygen at elevated temperatures is low. The oxidation of hydrogen sulfide normally occurs in stages, with sulfate generally the final end product. Microaerobic bacteria such as *Beggiatoa* sp., *Thiovulum* sp. and *Thiothrix* sp. are specifically dependent on H_2S and are commonly found in sediments at the transition between O_2 and H_2S (Jørgensen, 1983a; Jørgensen & Revsbech, 1983; Nelson & Jannasch, 1983). *Beggiatoa* sp. oxidise H_2S to elemental sulfur (reaction 12), which may then be further oxidised to SO_4^{2-} (reaction 13) by other bacteria, such as *Thiobacillus thiooxidans*, which grows attached to the mineral surface and uses it as an electron donor (Madigan et al., 1997). It is unclear whether *Beggiatoa* gain energy or whether these H_2S oxidisers produce intracellular sulfur for other reasons, e.g. to protect the cell from harmful oxygen compounds such as hydrogen peroxide (Kuenen & Beudeker, 1982) or as an electron acceptor, whereby it may be reduced back to H_2S if conditions become anaerobic (Jørgensen & Revsbech, 1983). Another interesting type of metabolism is shown by *Thioploca* sp., which grow as thick mats on the ocean floor. These cells can migrate upwards as much as 10 cm above the surface to accumulate NO_3^- intracellularly within vesicles and then migrate back downwards where they reduce it with the concomitant oxidation of hydrogen sulfide (Fossing et al., 1995).

$$H_2S + 0.5O_2 \rightarrow S^0 + H_2O \qquad (12)$$
$$\Delta G = -209 \text{ kJ/0.5 mol of } O_2$$

$$S^0 + H_2O + 1.5O_2 \rightarrow SO_4^{2-} + 2H^+ \qquad (13)$$
$$\Delta G = -195 \text{ kJ/0.5 mol of } O_2$$

The sulfur oxidisers also include those bacteria that inhabit the low pH environments associated with acid mine drainage environments. *Thiobacillus thiooxidans* enzymatically oxidises sulfide, sulfur, thiosulfate and other reduced sulfur compounds for cellular growth (Kuenen & Beudeker, 1982; Jørgensen, 1983b).

4. Radionuclide contaminants

The biogeochemical behaviour of radionuclides is becoming increasingly important due to the issues of their disposal as nuclear wastes, their long-term containment and ultimately their movement through the environment. Important radionuclides in the environment include high yield fission products such as radiocaesium and technetium, as well as the actinide elements, including uranium, plutonium and neptunium. The nuclear properties of significant isotopes of these elements, as well as an indication of their oxidation states and typical environmental concentrations of some contaminated areas are included in Table 1.

Table 1
Nuclear properties, oxidation states and environmental concentrations of key environmental radionuclides of Cs, Tc, U, Pu and Np

Element	Isotope	Half-life (y)	Environmental oxidation state	Major decay mode	Sample type	Concentration range
Caesium	134	2.06	I	Gamma	[a]Seawater (Irish Sea)	5.0×10^{-3}–100 Bq l^{-1}
	*137	30.17		Gamma	[b,c]Pore waters	2.8×10^{-3}–2.3×10^{-1} Bq l^{-1}
					[c,d]Sediment	0.1– 3.5×10^{4} Bq kg^{-1}
Technetium	99	2.15×10^{5}	IV, VII**	Beta	[e]Seawater (Irish Sea)	1–340 $\times 10^{-3}$ Bq l^{-1}
					[f]Surface water (fallout)	5.0×10^{-5}–6.3×10^{-4} Bq l^{-1}
					[d,g]Sediment	0.1–60.4 Bq kg^{-1}
Uranium	238	4.47×10^{9}	IV, VI**	Alpha	[h]Seawater (average)	3 μg l^{-1}
					[h]Surface waters	0.1–500 μg l^{-1}
					[i]Sediment	1.2–120 μg g^{-1}
Neptunium	237	2.14×10^{6}	IV, V**, VI	Alpha	[j]Seawater (Irish Sea)	8.0×10^{-5}–1.4×10^{-3} Bq l^{-1}
					[k]Sediment (Irish Sea)	1.2×10^{-2}–13 Bq kg^{-1}
Plutonium	238	87.7	III, IV**, V, VI	Alpha	[l]Seawater	1.0×10^{-5}–5.0×10^{-2} Bq l^{-1}
	*239, 240	2.41×10^{4}, 6.55×10^{3}		Alpha	[l]Surface waters	3.7×10^{-6}–4.8×10^{-3} Bq l^{-1}
	241	14.4		Beta	[m]Porewaters (Irish Sea)	2.0×10^{-3}–6.4×10^{-2} Bq l^{-1}
					[l]Sediment	1.6×10^{-1}–1.3×10^{4} Bq kg^{-1}

* Isotopes for which concentration data are quoted; ** Indicates the most common oxidation state in the natural environment.
[a] Kershaw et al., 1992; [b] Comans et al., 1992; [c] Sholkovitz & Mann, 1984; [d] Morris et al., 2000; [e] Leonard et al., 1997; [f] cited in Wildung et al., 1979; [g] Morita et al., 1993; [h] cited in Murphy & Shock, 1999; [i] Langmuir, 1997; [j] Pentreath & Harvey, 1981; [k] Assinder et al., 1991; [l] cited in Sholkovitz, 1983; [m] Livens et al., 1994.

Caesium

Large amounts of radiocaesium (predominantly ^{137}Cs and ^{134}Cs) have been deposited in the environment from nuclear weapons testing, nuclear accidents and in waste effluents associated with nuclear fuel reprocessing (Choppin et al., 1995). Radiocaesium has a relatively short half-life, and is a major contributor to the radiotoxicity of spent nuclear fuel in the first 300 years or so after removal from the nuclear power station. The extent of incorporation of $^{134,137}Cs$ into sediment burial is limited by their relatively short half-lives. For particle associated radiocaesium and other short lived (half-life <30 years) radionuclides, high sedimentation rates are required if the radionuclides are to be fully incorporated into biogeochemical cycles associated with sediment deposition.

The majority of radiocaesium (oxidation state +1) behaves conservatively in the environment, although it interacts strongly with clay minerals in soils and sediments (Evans et al., 1983; Cremers et al., 1988). Thus, radiocaesium may be removed from the water column during settling of suspended sediment, resulting in reduced mobility and bioavailability (Santschi et al., 1988; Santschi et al., 1990b). In certain sediments, this situation may be complicated by changes in pore-water chemistry during sediment burial and early diagenesis resulting in a remobilisation of radiocaesium (Sholkovitz, 1985). For example, Comans et al. (1989) reported in-situ solid/liquid distribution coefficients (Kd values) for radiocaesium in lake sediment contaminated by Chernobyl derived ^{137}Cs. They showed that the Kd values for radiocaesium in a sediment core decreased with depth, suggesting a decline in sediment-associated ^{137}Cs with burial. The binding mechanism for caesium in natural systems is regulated by a small number of highly selective ion exchange sites located at the edges of clay minerals (Sawhney, 1970). In natural waters, monovalent ions with a low ionic radius such as K^+ and NH_4^+ are expected to compete with Cs for these sites (Sawhney, 1972) and, in the sediment profile described by Comans et al. (1989), NH_4^+ showed a significant negative linear relationship with the Kd values for ^{137}Cs. The authors concluded that the decrease in the observed ^{137}Cs Kd values with depth was caused by the ion exchange of radiocaesium bound to sediments for pore-water NH_4^+. Therefore, production of NH_4^+ via microbial metabolic activity may well lead to post depositional remobilisation of ^{137}Cs from sediments (Zwolsman et al., 1993; Kaminski et al., 1994).

Technetium

There are no stable isotopes of Tc, but one isotope, ^{99}Tc, is a long-lived (half-life 2.1×10^5 y), high yield fission product which is produced in kilogram quantities in nuclear reactors. Technetium is important as an environmental contaminant because it is one of the most mobile radionuclides (Garland et al., 1983). It has been released to the environment primarily as a result of nuclear weapons testing and fuels reprocessing and, due to its long half-life, will be an important component of high level wastes when they are finally disposed (Wildung et al., 1979; Bird & Evenden, 1996).

Technetium is present in the environment in one of two oxidation states, Tc(IV) and Tc(VII), the latter forming the pertechnetate (TcO_4^-) anion (Wildung et al., 1979). Tc(IV) is only formed under neutral pH and conditions where Eh values lie below approximately +220 mV (Brookins, 1988; Sparkes & Long, 1988). Studies on the behaviour of ^{99}Tc in the natural environment have been limited due to the low concentrations generally found in environmental materials and due to numerous difficulties associated with its analysis (Holm, 1993). In oxic waters, the majority of technetium exists as pertechnetate and is expected to remain in solution until it decays (Beasley & Lorz, 1986). This is because the TcO_4^- anion is highly soluble and poorly sorbed by soils and sediments due to its predominantly negatively-charged surfaces (Walton et al., 1986; Elwear et al., 1992). In fact, TcO_4^- is often used as a conservative tracer for water transport studies (Leonard et al., 1997). In contrast, Tc(IV) reacts readily with mineral surfaces, i.e. iron oxyhydroxides and sulfides, or insoluble humic substances (Lieser & Bauscher, 1988; Sparkes & Long 1988), both of which may lead to decreased solubility. Only when Tc(IV) reacts with low molecular weight, soluble, organic ligands may its solubility be increased (Wildung et al., 1979; Sheppard et al., 1990).

The variability in the chemical behaviour of Tc is highlighted by the observed Kd values. Values range from <10 ml g^{-1} (Kershaw et al., 1992) to, rather more infrequently, values as large as 1000 ml g^{-1} reported for anaerobic sorption experiments (Lieser & Bauscher, 1987). The low values reflect the limited retention of Tc in oxic environments, whereas the higher values point to immobilisation under reducing conditions. For example, a soil core profile containing ^{99}Tc and other artificial radionuclides from Sellafield waste effluents, including ^{137}Cs, was taken at an intertidal salt marsh in west Cumbria (Morris et al., 2000). The marsh is relatively well characterised and due to its hydrology, remains sub-oxic throughout the year (Keith-Roach et al., 2000). The ^{99}Tc profile in the core strongly resembled both the ^{137}Cs profile and the Sellafield discharge history for ^{137}Cs. From this, Morris et al. (2000) concluded that the majority of ^{99}Tc input to the marsh was particle associated, and that the profile had a qualitative relationship with historical discharges of Tc from Sellafield. The mechanisms by which ^{99}Tc became incorporated into the particulate input remain unidentified, but microbially-mediated reductive sorption in reducing microenvironments prior to transport to the marsh may explain the pattern of retention seen in these suboxic sediments.

The association of technetium with anaerobic microorganisms can be demonstrated by considering the reported concentration ratios (Bq g^{-1} dry weight in organism/Bq ml^{-1} in solution) for a range of microbes and for aquatic organisms. For sulfate-reducing bacteria, reported concentration ratios range from 5750 ml g^{-1} (for indirect sulfide precipitation; Lloyd et al., 1998) to 12,850 ml g^{-1} (for dissimilatory TcO_4^- reduction; Lloyd et al., 1999). These compare to concentration ratio maxima for aquatic organisms of 1000–1500 ml g^{-1} (Masson et al., 1989) and Kd values for bulk sediments of 10–1000 ml g^{-1}. The concentration ratios for sulfate-reducing bacteria are at least an order of magnitude higher than values reported for aquatic organisms or sediments suggesting a very strong association (either directly or indirectly) of technetium with sulfate-reducing bacteria.

The actinide elements

The actinide elements can be treated as a group in terms of their geochemical behaviour, but the sources of these elements in the natural environment are very different. Uranium is a primordial radioactive element which is ubiquitous in the environment and has a crustal abundance of 2.3 mg kg^{-1} (Krauskopf, 1988). Naturally occurring uranium comprises of three isotopes, ^{234}U, ^{235}U and ^{238}U, all of which decay by α-emission. Uranium is exploited in the nuclear industry, and uranium milling and mining has led to locally enhanced concentrations of uranium and U-series radionuclides (Krauskopf, 1988; IAEA, 1999). In addition, significant quantities of uranium associated with the nuclear fuel cycle will eventually be disposed of in nuclear repositories. Other environmentally significant actinides, e.g. Np and Pu, are produced in nuclear reactors via neutron capture and β^--decay reactions. They have been released to the natural environment via nuclear weapons and accidents, as well as in effluents associated with nuclear fuel reprocessing (Hanson, 1980). The principal isotopes of the transuranic elements are α-emitters, with half-lives ranging from 8.77×10^1 to 2.14×10^6 years (Table 1).

Extensive reviews of the complex geochemical behaviour of the actinides are included in Morse & Choppin (1991) and Dozol et al. (1993) and an overview is given in Chapter 4 of this volume. Briefly, oxidation states of +3 to +6 inclusive may be encountered in the environment (Dozol et al., 1993). The lower valence states form simple cations, M^{3+} and M^{4+}, which are easily hydrolysed due to their high charge density, and are readily lost from solution by reaction with solid surfaces (Nelson & Lovett, 1978). The higher valence states form the dioxo- cations MO_2^+ and MO_2^{2+}, which are more soluble due to both their lower charge density and their ready complexation with common oxygen-containing ligands, such as CO_3^{2-}, to give neutral and anionic carbonate-complexes (Morse & Choppin, 1991; Clark et al., 1995). Thus, for the actinide elements, there is a clear distinction between 'reduced', relatively insoluble species, and 'oxidised', relatively soluble species (Nelson & Lovett, 1978).

The relative solubilities of U, Np and Pu in the environment are primarily governed by the dominant oxidation state(s) of the element (Table 1). In most geochemical systems, the dominant form of uranium is expected to be the relatively soluble uranyl ion, UO_2^{2+}. The relatively insoluble U(IV) is only expected in strongly reducing conditions where it forms uraninite (UO_2) (Suzuki & Banfield, 1999). The transuranic elements display different behaviour depending on their accessible range of oxidation states. Plutonium potentially has access to all four oxidation states in redox conditions found in the natural environment (Dozol et al., 1993), although in sedimentary environments, Pu^{4+} is thought to dominate, with PuO_2^+ present as the metastable soluble form in seawaters (Nelson & Lovett 1978; Morse & Choppin, 1991). Neptunium displays an intermediate range of oxidation states with Np^{4+}, NpO_2^+ and NpO_2^{2+} all possible species at environmental redox potentials. In oxic to mildly reducing environments NpO_2^+ is thought to dominate (Lieser & Muhlenweg, 1988; Morris & Livens, 1996); indeed, Np(V) is not effectively removed from waste effluents via typical physicochemical or biotechnological processes (Lloyd et al., 2000). Thus, neptunium is the most mobile of the transuranic elements in oxic environments, with plutonium displaying lower solubility as it is more prone to reductive sorption (Hursthouse & Livens, 1993). This general environmental behaviour is reflected in published Kd values for Np and Pu of 10^4 and 10^5 ml g^{-1}, respectively (Morse & Choppin, 1991; Kershaw et al., 1992).

5. Biogeochemical processes affecting radionuclide mobility

The fate of radionuclides in the sedimentary environment is a complex function of both abiotic and biotic factors. Abiotic factors such as pH, ionic strength and the presence of competing and complexing ions are important in controlling radionuclide speciation. These factors, in turn, control their mobility in the environment (Santschi & Honeyman, 1989; Pedersen, 1993). Microbial activity may affect actinide mobility via a number of processes including (i) direct mechanisms, such as enzymatic reduction or sorption on to microbial biomass, and (ii) indirect mechanisms, such as mineral dissolution/precipitation reactions or interactions with organic and inorganic metabolites. The biogeochemical behaviour of a radionuclide will also depend on the form in which it is released and upon its initial associations. For example, different radionuclides may be released as refractory oxides or in the elemental form, or upon release, they may become preferentially associated with clay mineral fractions, organic matter and iron/manganese hydroxides (Choppin & Stout, 1989; Santschi et al., 1990b).

Direct enzymatic reduction

Microorganisms can, in principle, obtain energy to support growth from the dissimilatory reduction of redox active radionuclides including Tc, U, Np, and presumably Pu (Lovley, 1995; Lloyd & Macaskie, 1996; Lloyd et al., 2000; see also Chapters 7, 8, 11 and 12, this volume). In one of the first studies that examined the potential for soil microorganisms to affect technetium solubility, Henrot (1989), found that in aerobic cultures there was little bioaccumulation of Tc (as TcO_4^-) on bacterial surfaces. By contrast, in a mixed anaerobic inoculum, a substantial fraction of the Tc became associated with the biomass in the growth medium, presumably as Tc(IV). If pertechnetate was added to a mixed anaerobic culture medium after bacteria had been removed by filtration, there was no removal of TcO_4^-. This suggested that direct enzymatic reduction of Tc(VII), and subsequent adsorption on to microbial surfaces was responsible for the changes in Tc solubility. Lloyd & Macaskie (1996) subsequently demonstrated that the dissimilatory metal-reducing bacteria *G. metallireducens* and *S. putrefaciens* could enzymatically reduce pertechnetate in the presence of acetate and lactate. In those pure culture experiments, the reduced Tc products were species-specific: *S. putrefaciens* produced soluble reduced Tc species in the supernatant, whereas *G. metallireducens* precipitated appreciable quantities of technetium as a low valence, insoluble oxide, hydroxide or oxohydroxo compound (e.g. Tc_2O_5 or TcO_2). Lloyd et al. (1997) further established that anaerobic cultures of *Escherichia coli* were able to couple the oxidation of formate or hydrogen directly to the reduction of pertechnetate, while Lloyd et al. (1999) reported that a range of electron donors could be utilised by *D. desulfuricans*, with TcO_4^- as the sole electron acceptor in non-sulfidogenic cultures.

Enzymatic reduction of U(VI) has been suggested as a pathway for U(VI) immobilisation in sediments (Barnes & Cochran, 1993) and aquifers (Lovley et al., 1993). Direct dissimilatory reduction of U(VI) coupled to the oxidation of organic substrates, and the subsequent extracellular precipitation of the insoluble U(IV) mineral uraninite (UO_2), has been demonstrated for a number of pure culture species, including *Clostridium*, *Desulfovibrio*, *G. metallireducens* and *S. putrefaciens*, (Lovley et al., 1991, 1993, Gorby & Lovley, 1992; Lovley, 1993, Francis, 1994). The enzymatic reduction of U(VI), as the soluble uranyl ion, UO_2^{2+}, by *G. metallireducens* and *S. putrefaciens*, occurs via the following reactions (equations modified from Lovley et al., 1991; Lovely & Phillips, 1992):

$$CH_3COO^- + 2UO_2^{2+} + H_2O \rightarrow 2UO_2 + HCO_3^- + 4H^+ \quad (14)$$
$$\Delta G = -158 \text{ kJ/mol } CH_3COO^-$$

$$H_2 + UO_2^{2+} \rightarrow 2H^+ + UO_2 \quad (15)$$
$$\Delta G = -82 \text{ kJ/mol } H_2$$

Aqueous uranyl can also be reduced by mixed cultures of Fe- and sulfate-reducing bacteria (Ganesh et al., 1997; Abdelouas et al., 2000), although attempts to grow SRB's with UO_2^{2+} as the sole electron acceptor have been largely unsuccessful (Suzuki & Banfield, 1999). The free energy yield per mole of acetate, coupled to U(VI) reduction, is –158 kJ mol^{-1}, a value which lies between the yields for the reduction of Mn(IV) and Fe(III).

Dissimilatory reduction of Np(V) in the presence of *S. putrefaciens* has recently been observed in a series of batch experiments (Lloyd et al., 2000). The majority of the reduced Np(IV) remained soluble in the culture medium (>85%) and was thought to be present in a microcolloidal form. These results were not surprising considering that *S. putrefaciens* has been shown to reduce Fe(III), U(VI) and Tc(VII) enzymatically, and the redox couple for Np(V)/Np(IV) (+0.74 V) lies between the redox couples for U(VI)/U(IV) (+0.32 V) and Fe(III)/Fe(II) (+0.77 V). Following reduction, indirect bioprecipitation of Np(IV)-phosphate was then induced by Citrobacter sp. in the presence of glycerol-2-phosphate as the source of PO_4^{3-}. Lloyd et al. (2000) also performed a bioreduction experiment in conditions related to the behaviour of Np(V) in natural waters (i.e. in carbonate buffer and at a reduced concentration of neptunium). In these experiments a substantial fraction of the Np(V) appeared to be directly reduced by *S. putrefaciens* to an insoluble, unidentified Np species.

The biological reduction of plutonium has also been described. Iron-reducing *Bacillus* strains were able to solubilise up to 90% of Pu(IV) hydrous oxide in the presence of nitrilotriacetic acid (NTA); only 4.5% of the plutonium was solubilised in uninoculated NTA media (Rusin et al., 1994). Analysis of the solution phase, however, did not reveal the presence of a Pu(III)-NTA complex. Instead, the authors suggested that the soluble Pu(III)-NTA formed during microbial reduction spontaneously reoxidised to Pu(IV)-NTA, and thus remained in solution. Although the mechanism of Pu(IV) solubilisation in the presence of NTA and iron reduction remains unidentified, the results are of importance as they suggest that the microbially mediated reduction of highly insoluble Pu(IV) oxyhydroxide (K_{sp} $10^{-56.8}$ to $10^{-57.8}$) to the more soluble Pu(III) hydroxide (K_{sp} = $10^{-22.6}$) may increase the solubility of plutonium in the environment.

Biosorption

At the normal growth pH (between 5 and 8), structural polymers that reside in the cell wall and surrounding fabric of bacteria are naturally anionic (Beveridge, 1989). By virtue of their small size, bacteria also have the largest surface area to volume ratio of cellular life forms. This property offers bacteria a remarkable potential to sequester and accumulate an assortment of metals onto their surfaces. Furthermore, bacteria are found in every environment where liquid water is freely available. They are not only ubiquitous, but are also found in vast numbers, more than 10^8 cells g^{-1} in marine muds and garden soils, 10^5 cells ml^{-1} in river waters and 10^8 cells cm^{-2} in biofilms (Geesey et al., 1978; Ehrlich, 1990). The combination of cell reactivity and abundance implies that the biomass associated with microorganisms can have a direct effect on contaminant solubility.

Microorganisms have been shown to biosorb radionuclides (Francis, 1990; Avery, 1995), a property that lends itself to bioremediation strategies for the removal of such contaminants (Voleski & Holan, 1995; Lovley & Coates, 1997; Eccles, 1998). The similarity in ionic radius of Cs^+ to that of other monovalent cations, particularly K^+, leads to higher levels of uptake than would be expected from passive sorption processes alone (Avery, 1995). Potassium is an essential macronutrient, and microorganisms accumulate K^+ intracellularly via metabolism-dependent transport systems (Avery et al., 1992). Therefore, caesium may be incorporated into the uptake cycle for potassium, although the extent of incorporation is variable for different microorganisms and it can be affected by the presence of other monovalent cations. Biomass accumulations of ^{137}Cs have been reported for bacteria, cyanobacteria, algae, yeast and fungi (Avery, 1995), with particularly high Cs accumulations of 56 g kg^{-1} (dry weight) associated with bacteria (*Rhodococcus* sp.) isolated from ^{137}Cs-enriched soils (Tomioka et al., 1992). Similarly, unexpectedly strong retention of radiocaesium derived from Chernobyl was attributed to microbial activity in organic-rich layers of upland soils (Johnson et al., 1991).

For the actinide elements, accumulation in cells is thought to occur predominantly via metabolism-independent biosorption (Pentreath, 1981; Scoppa, 1984; Suzuki & Banfield, 1999). Biosorption is defined as the passive sorption and complexation of metal ions by microbial biomass or material derived from this. Indeed, the metal biosorption capacity of non-living biomass can be greater than that of living cells: cellular degradation increases the availability of functional groups capable of binding metals (Ferris et al., 1988) and the protons generated by membrane respiration in living cells no longer compete with dissolved metals for binding sites (Urrutia et al., 1992). In some cases, microbial biomass may even have a greater capacity for metal adsorption than inorganic fractions such as ferric oxides or clays (Berthelin et al., 1995). Studies on the biosorption of uranium have reported uptake by different microorganisms ranging from 23–9000 g kg^{-1} biomass (Eccles, 1998). For neptunium, biosorption levels for aerobic microorganisms are low (11–15 g kg^{-1} biomass) suggesting that, as expected from the geochemical behaviour of NpO_2^+, the adsorption of Np(V) is less effective than that of other radionuclides (Strandberg & Arnold, 1988). Wahlgren et al. (1980) tentatively related the observed pattern of plutonium cycling in Lake Michigan to seasonal differences in accumulation in phytoplankton. Recently, Keith-Roach (2000) used lipid biomarkers to examine the variation in microbial population in a marsh where Am, Pu and Np concentrations were measured in pore-waters over an annual cycle. The study suggested that seasonal variations in transuranic element concentrations in pore-waters were related to changes in aerobe biomass in the marsh. For example, the concentrations of plutonium and americium were at a minimum in periods where microbial biomass was at a maximum, presumably due to passive sorption. As the relative biomass concentration declined in the marsh due to cell degradation, the plutonium and americium concentrations in pore-waters subsequently increased. Although this pattern was complicated in winter months by other factors, this study suggests that direct sorption of actinide elements by microbial biomass may have a significant short term effect on actinide solubility in the environment.

Organic metabolites

Fermentation products formed during diagenesis may act as ligands with actinide ions (Munier-Lamy & Berthelin, 1987; Berry et al., 1991; Gadd, 1997). Although such reactions are expected to increase actinide solubility, in sedimentary environments many of the organic byproducts of microbial metabolism are rapidly utilised by other bacteria and their effect on actinide solubility may be considered transitory. The consumption of hydrogen, or other fermentation products, during diagenesis also regulates the redox potential of the local environment, and subsequently has the potential to shift the oxidation state distribution of the multivalent actinides (Silva & Nitsche, 1995; Choppin & Bond, 1996). This ultimately leads to differences in their environmental behaviour. For example, the greater susceptibility of Pu(V) to redox processes implies that it may be reduced to Pu(IV) more readily than Np(V) or U(VI) (Morse & Choppin, 1991). Generally, as sediment burial occurs, and a more reducing environment develops, the actinides are expected to become reduced and therefore less soluble.

In addition to the organic degradation products produced in diagenesis, bacteria and fungi may, under stress of iron deficiency, produce specific iron chelators such as siderophores. These highly specific, low molecular weight chelating agents allow Fe(III) to be sequestered and delivered into cells via active transport systems. Although virtually specific for Fe(III), siderophores can complex certain other metals including the tetravalent actinides (Birch & Bachofen, 1990; Macaskie & Dean, 1990; Brainard et al., 1992; Neu et al., 2000). Some siderophores are also capable of complexing U(VI) (Macaskie & Dean, 1990). The binding constants for the tetravalent actinides are similar to those for Fe(III) (e.g. desferrioxamine B-Fe(III) = $10^{30.6}$; desferrioxamine B-Pu(IV) = $10^{30.8}$) (Neu et al., 2000). The concentrations of these chelating agents in the environment are estimated to be very low, typically 0.01–3.0 μmol kg^{-1} in soils (Brainard et al., 1992; Neu et al., 2000). This is approximately 3–4 orders of magnitude lower than concentrations of other organic complexing agents such as organic acids or humic substances. However, Brainard et al. (1992) showed that a specific siderophore (enterobactin), at low concentrations (0.1 mM), was equal to or better than a series of strong chelators such as oxalic acid, citric acid or EDTA at concentrations three orders of magnitude higher at solubilising hydrous PuO_2. In addition to the environmental significance of these sequestering agents, biotechnological actinide remediation treatments using siderophores have also been suggested (Birch & Bachofen, 1990).

Inorganic metabolites

During diagenesis, inorganic metabolites such as bicarbonate, phosphate, ferrous iron and hydrogen sulfide are released into sediment pore-waters. These metabolites may in turn affect the speciation and mobility of the actinide elements. For example, in a closed aqueous system, increased levels of carbon dioxide produced through organic degradation may increase the solubility of actinides by forming neutral or anionic actinide-carbonate complexes such as $M(VI)O_2(CO_3)_3^{4-}$ and $M(V)O_2(CO_3)^-$ in solutions with pH > 6 (Clark et al., 1995; Banaszak 1998; Suzuki & Banfield, 1999). Some cells also produce phosphatase enzymes that release phosphate (HPO_4^{2-}) from organic compounds

(Suzuki & Banfield, 1999). This has biotechnological applications because the phosphate generated potentially binds and precipitates large quantities of metals including actinides. For instance, a copper tolerant bacterium, *Citrobacter* sp., which produced three times more phosphatase than the parent strain, accumulated up to 9 g U kg^{-1} dry cell weight as insoluble hydrogen autinite (HUO_2PO_4) (Macaskie, 1990; Macaskie et al., 1990). The biomineralisation of insoluble actinide phosphates via *Citrobacter* sp. has also been demonstrated for Pu(IV) and Np(IV) (Macaskie et al., 1994; Lloyd et al., 2000; see Chapters 11 and 12, this volume).

Ferrous iron also plays a role in the reduction of U(VI). In a series of batch experiments, uranium reduction rates ranged from greater than 3 days in the presence of 1 mM soluble Fe^{2+} to less than a few hours in the presence of both soluble Fe^{2+} and colloidal hematite (Liger et al., 1999). In the case of the latter experiments, the pseudo first order rate constants for U(VI) reduction were of the same order of magnitude as the highest corresponding rate constants for direct enzymatic reduction of U(VI). This suggests that abiotic surface catalysed reduction of U(VI) may be a major pathway for reductive immobilisation during diagenesis. The inorganic formation of a reduced TcS_2-like phase via co-precipitation of TcO_4^- with FeS (mackinawite), has also recently been observed, (Wharton et al., 2000). Interestingly, on reoxidation of the TcS_2-FeS phase, the technetium remained in an insoluble TcO_2-like phase, i.e. the Tc(IV) failed to reform into Tc(VII) on reoxidation and remained insoluble. This implies that when technetium is reduced to insoluble TcS_2 in a sulfidic environment it may remain insoluble as Tc(IV)-oxide even if the sediment is subsequently exposed to aerobic conditions (Wharton et al., 2000).

A number of workers have demonstrated that dissolved sulfide reacts only slowly with U(VI) (Kochenov et al., 1977; Mohagheghi et al., 1985). In particular, Lovley et al. (1993) showed that at circumneutral pH, the rate of abiotic uranyl reduction in the presence of HS^- was very much slower than the measured rates of direct enzymatic uranium reduction. Uranium(VI) reduction may also be catalysed by mineral, and even bacterial surfaces (Kochenov et al., 1977; Mohagheghi et al., 1985). Sulfide minerals such as pyrite and galena can reduce U(VI) under anoxic conditions (Wersin et al., 1994), and interestingly, are frequently found in close association with supergene uranium deposits (Brookins, 1988). The production of H_2S in sulfidic environments also effects technetium speciation. Henrot (1989) reported that in single culture experiments the sulfate-reducing bacteria *Desulfovibrio vulgaris* and *Desulfovibrio gigas* converted up to 70% of pertechnetate to adsorbed and/or insoluble forms. Similarly, Lloyd et al. (1998) observed that the production of H_2S by cultures of *Desulfovibrio desulfuricans* led to the extracellular precipitation of insoluble technetium sulfides. The Tc:S ratio of 1.5 in the precipitates suggested that either TcS_2 or Tc_2S_7 were formed. Moreover, the lack of intracellular precipitation of Tc noted in the sulfidogenic cultures indicated that it was unlikely that Tc was directly reduced within the cell.

Mineral precipitation/dissolution reactions

The solubility of radionuclides is affected by the precipitation and/or dissolution of oxidised mineral phases. In the first instance, the solubility of actinide elements is often limited by adsorption onto solid surfaces such as iron oxyhydroxides (Keeny-Kennicutt

& Morse, 1984, 1985; Hsi & Langmuir, 1985; Hursthouse et al., 1991). The accumulation of metals by bacteriogenic iron oxides has clearly been demonstrated (Clarke et al., 1997), and recently, Ferris et al. (1999, 2000) suggested that the relative proportion of bacterial organic matter in biogenic Fe oxides affected the solid phase uranium enrichment. A number of studies have also presented evidence that scavenging of plutonium and americium by ferric hydroxides occurs at sediment/water interfaces in marine environments (Sholkovitz & Mann, 1984; Malcolm et al., 1990), although the possible role of microorganisms in the precipitation of these ferric iron phases was not discussed.

Secondly, ferric hydroxide reduction may affect actinide solubility by releasing adsorbed radionuclides from ferric oxide phases (Sholkovitz, 1983). Barnes & Cochran (1993) reported that uranium was desorbed from sediments during the reduction of Fe/Mn hydroxides. The subsequent exposure to a reducing environment may result in the actinide ions becoming reduced, and subsequently adsorbing on to new mineral surfaces. Morris et al. (2001) reported that plutonium underwent seasonal cycling in salt marsh pore-waters, and that maximum plutonium solubility appeared to coincide with minimum iron and manganese solubility in marsh pore-waters. Interestingly, Eh measurements at the site were relatively constant over an annual cycle, suggesting that the observed cycling of Pu, Fe and Mn was complex. In other environments, no obvious link between actinide solubility and Fe/Mn oxide cycles has been observed (Sholkovitz & Mann, 1984; Malcolm et al., 1990).

6. Summary

Microbial metabolism is the predominant control on sediment mineralogy and pore-water chemistry in aquatic environments. Although changes in organic and inorganic chemistry are complex, transient and involve numerous recycling reactions, both produce characteristic patterns. Chemoheterotrophic bacteria use a sequence of terminal electron accepting processes which produce successive changes in the pore-water chemistry and associated mineralogy. These processes are integrally linked to bulk changes in sediment and soluble organic matter. The contribution of chemoautotrophic bacteria in most aquatic sediments is limited to nitrification, but they become more important at low pH and oxygen levels, or in organic-rich sediments that generate methane.

Environmental radioactivity and an understanding of the fate of radionuclides in the natural environment are of great importance when considering the legacy of nuclear waste and the future potential exploitation of nuclear power. As we have discussed, the fate of radionuclides in the natural environment cannot be considered without examining the role that microbial activity plays in either enhancing their mobility or causing their immobilisation. Because of the low concentrations of radionuclides in the natural environment, and the radiotoxicity of these elements, few studies on biogeochemical cycles of the artificial radionuclides have been performed. However, it is clear that a further understanding of the rates of movement, the pathways and the reservoirs of contaminant radionuclides in the environment continues to be one area of biogeochemical research which has contemporary relevance.

References

Abdelouas, A., Lutze, W., Gong, W., Nuttall, E. H., Streitelmeier, B. A. & Travis, B. J. (2000). Biological reduction of uranium in groundwater and subsurface soil. *The Science of the Total Environment, 250*, 21–35.

Ahmann, D., Roberts, A. L., Krumholz, L. R. & Morel, F. M. M. (1994). Microbe grows by reducing arsenic. *Nature, 371*, 750–750.

Aller, R. C. & Rude, P. D. (1988). Complete oxidation of solid phase sulphides by manganese and bacteria in anoxic marine sediments. *Geochimica et Cosmochimica Acta, 52*, 751–765.

Aller, R. C., Macklin, J. E. & Cox, R. T. J. (1986). Diagenesis of Fe and S in Amazon inner shelf muds: apparent dominance of Fe reduction and implications for genesis of ironstones. *Continental Shelf Research, 6*, 263–289.

Aller, R. C., Hall, P. O. J., Rude, P. D. & Aller, J. Y. (1998). Biogeochemical heterogeneity and suboxic diagenesis in hemipelagic sediments of the Panama Basin. *Deep-Sea Research, 45*, 133–165.

Archer, D. E., Emerson, S. & Reimers, C. (1989). Dissolution of calcite in deep-sea sediments: pH and O_2 microelectrode results. *Geochimica et Cosmochimica Acta, 53*, 2831–2846.

Avery, G. B. Jr. & Martens, C. S. (1999). Controls on the stable carbon isotopic composition of biogenic methane produced in a tidal freshwater estuarine sediment. *Geochimica et Cosmochimica Acta, 63*, 1075–1082.

Avery, S. V. (1995). Microbial interactions with caesium – implications for biotechnology. *Journal of Chemical Technology and Biotechnology, 62*, 3–16.

Avery, S. V., Codd, G. A. & Gadd, G. M. (1992). Replacement of cellular potassium by caesium in *Chlorella emersonii* – differential sensitivity of photoautotrophic and chemoheterotrophic growth. *Journal of General Microbiology, 138*, 69–76.

Banaszak, J. E., VanBriesen, J. M., Rittmann, B. E. & Reed, D. T. (1998). Mathematical modelling of the effects of aerobic and anaerobic chelate biodegradation on actinide speciation. *Radiochimica Acta, 82*, 445–451.

Balba, M. T. & Nedwell, D. B. (1982). Microbial metabolism of acetate, propionate and butyrate in anoxic sediment from the Colne Point Saltmarsh, Essex, UK. *Journal of General Microbiology, 128*, 1415–1422.

Barica, J. & Mur, L. R. (1980). Developments in hydrobiology 2. Hypertrophic ecosystems. *S.I.L. Workshop on Hypertrophic Ecosystems, Vaxjo*, September 10–14 (1979). The Hague (348pp). Boston: W. Junk BV.

Barnes, C. E. & Cochran, K. J. (1993). Uranium geochemistry in estuarine sediments: controls on removal and release processes. *Geochimica et Cosmochimica Acta, 57*, 555–569.

Bazylinski, D. A. (1995). Structure and function of the bacterial magnetosome. *ASM News, 61*, 337–343.

Bazylinski, D. A. (1996). Controlled biomineralization of magnetic minerals by magnetotactic bacteria. *Chemical Geology, 132*, 191–198.

Bazylinski, D. A. & Blakemore, R. P. (1983). Denitrification and assimilatory nitrate reduction in *Aquaspirillum magnetotacticum*. *Applied and Environmental Microbiology, 46*, 1118–1124.

Beasley, T. M. & Lorz, H. V (1986). A review of the biological and geochemical behaviour of Tc in the marine environment. In G. Desmet & C. Mytennaere (Eds), *Technetium in the Environment* (pp. 197–216). London: Elsevier.

Bender, M. L. & Heggie, D. T. (1984). Fate of organic carbon reaching the deep sea floor: a status report. *Geochimica et Cosmochimica Acta, 48*, 977–986.

Benning, L. G., Wilkin, R. T. & Barnes, H. L. (2000). Reaction pathways in the Fe-S system below 100°C. *Chemical Geology, 167*, 25–52.

Berner, R. A. (1980). *Early Diagenesis. A Theoretical Approach* (241pp). Princeton NJ. Princeton University Press.

Berner, R. A. (1981). A new geochemical classification of sedimentary environments. *Journal of Sedimentary Petrology, 51*, 359–365.

Berner, R. A. (1984). Sedimentary pyrite formation. an update. *Geochimica et Cosmochimica Acta, 48*, 605–615.

Berry, J. A., Bond, K. A., Ferguson, D. R. & Pilkington, N. J. (1991). Experimental studies of the effect of organic materials on the sorption of uranium and plutonium. *Radiochimica Acta, 52/53*, 201–209.

Berthelin, J., Munier-Lamy, C. & Leyval, C. (1995). Effect of microorganisms on mobility of heavy metals in soils. In P. M. Huanag (Ed.), *Environmental Impacts of Soil Component Interactions* (pp. 3–17). Boca Raton, FL: CRC Press.

Beveridge, T. J. (1989). Role of cellular design in bacterial metal accumulation and biomineralization. *Annual Review of Microbiology, 43*, 147–171.

Birch, L. & Bachofen, R. (1990). Complexing agents from microorganisms. *Experientia, 46*, 827–834.

Bird, G. A. & Evenden, W. G. (1996). Transfer of ^{60}Co, ^{65}Zn, ^{95}Tc, ^{134}Cs and ^{238}U from water to organic sediment. *Water Air and Soil Pollution, 86*, 251–261.

Blackburn, T. H. (1980). Seasonal variation in the rate of organic-N mineralisation in anoxic sediments. *Colloque International C.N.R.S. (Marseilles)*, No. 293 (pp. 173–183).

Blakemore, R. P. & Frankel, R. B. (1989). Biomineralization by magnetogenic bacteria. In R. K. Poole & G. M. Gadd (Eds), *Metal-Microbe Interactions* (pp. 85–98). Oxford, UK: IRL Press.

Boudreau, B. P. & Canfield, D. E. (1993). A comparison of closed- and open-system models of porewater pH and calcite saturation state. *Geochimica et Cosmochimica Acta, 57*, 317–334.

Brainard, J. R., Streitelmeier, B. A., Smith, P. H., Langston-Unkofer, P. J., Barr, M. E. & Ryan, R. R. (1992). Actinide binding and solubilization by microbial siderophores. *Radiochimica Acta, 58/59*, 357–363.

Brendel, P. J. & Luther, G. W., III. (1995). Development of a gold amalgam voltammetric microelectrode for the determination of dissolved iron, manganese, O_2, and S(-II) in porewaters of marine and freshwater sediments. *Environmental Science and Technology, 29*, 751–761.

Brookins, D. G. (1988). *Eh/pH Diagrams for Geochemistry* (176pp). Berlin: Springer-Verlag.

Burdige, D. J. (1993). The biogeochemistry of manganese and iron reduction in marine sediments. *Earth Science Reviews, 35*, 249–284.

Canfield, D. E. (1989). Reactive iron in marine sediments. *Geochimica et Cosmochimica Acta, 53*, 619–632.

Canfield, D. E. (1993). Organic matter oxidation in marine sediments. In R. Wollast, L. Chou & F. Mackenzie (Eds), *Interactions of C, N, P and S Biogeochemical Cycles* (pp. 333–363). NATO-ARW (North Atlantic Treaty Organization-Advanced Research Workshop), Berlin.

Canfield, D. E. (1994). Factors influencing organic carbon preservation in marine sediments. *Chemical Geology, 114*, 315–329.

Canfield, D. E., Raiswell, R. & Bottrell, S. (1992). The reactivity of sedimentary iron toward sulfide. *American Journal of Science, 292*, 659–683.

Canfield, D. E., Thamdrup, B. & Hansen, J. W. (1993). The anaerobic degradation of organic matter in Danish coastal sediments. Iron reduction, manganese reduction, and sulphate reduction. *Geochimica et Cosmochimica Acta, 57*, 3867–3883.

Carlson, L. & Schwertmann, U. (1981). Natural ferrihydrites in surface deposits from Finland and their association with silica. *Geochimica et Cosmochimica Acta, 45*, 421–429.

Choppin, G. R. & Stout, B. E. (1989). Actinide behaviour in natural waters. *The Science of the Total Environment, 83*, 203–216.

Choppin, G. R. & Bond, A. H. (1996). Actinide oxidation state speciation. *Journal of Analytical Chemistry, 51*, 1129–1138.

Choppin, G. R., Liljenzin, J. O. & Rydburg, J. (1995). *Radiochemistry and Nuclear Chemistry – Theory and Applications* (707pp). Oxford, UK: Butterworth Heinman.

Christensen, J. P., Smethie, W. M. & Devol, A. H. (1987). Benthic nutrient regeneration and denitrification on the Washington Continental Shelf. *Deep Sea Research, 34*, 1027–1048.

Clark, D. L., Hobart, D. E. & Neu, M. P. (1995). Actinide carbonate complexes and their importance in actinide environmental chemistry. *Chemical Reviews, 95*, 25–48.

Clarke, W. A., Konhauser, K. O., Thomas, J. C. & Bottrell, S. H. (1997). Ferric hydroxide and ferric hydroxysulphate precipitation by bacteria in an acid mine drainage lagoon. *FEMS Microbiology Reviews, 20*, 351–361.

Claypool, G. E. & Kaplan, I. R. (1974). The origin and distribution of methane in marine sediments. In I. R. Kaplan (Ed.), *Natural Gases in Marine Sediments* (pp. 99–139). New York: Plenum.

Coates, J. D., Ellis, D. J., Blunt-Harris, E. L., Gaw, C. V., Roden, E. E. & Lovely, D. R. (1998). Recovery of humic-reducing bacteria from a diversity of environments. *Applied and Environmental Microbiology, 64*, 1504–1509.

Coleman, M. L. (1985). Geochemistry of diagenetic non-silicate minerals. Kinetic considerations. *Philosophical Transactions of the Royal Society of London, A, 315*, 39–56.

Coleman, M. L., Hedrick, D. B., Lovley, D. R., White, D. C. & Pye, K. (1993). Reduction of Fe(III) in sediments by sulphate-reducing bacteria. *Nature, 361*, 436–438.

Comans, R. N. J., Middelburg, J. J., Zonderhuis, J., Woittiez, J. R. W., Delange, G. J., Das, H. A. & Vanderweijden, C. H. (1989). Mobilisation of radiocaesium in pore waters of lake sediment. *Nature, 339*, 367–369.

Cremers, A., Elsen, A., De Preter, P. & Maes, A. (1988). Quantitative analysis of radiocaesium in soils. *Nature, 335*, 247–249.

Davison, W., Grime, G. W., Morgan, J. A. W. & Clarke, K. (1991). Distribution of dissolved iron in sediment pore waters at submillimetre resolution. *Nature, 352*, 323–324.

Davison, W., Zhang, H. & Grime, G. W. (1994). Performance characteristics of gel probes used for measuring pore waters. *Environmental Science and Technology, 28*, 1623–1632.

Dowdle, P. R., Laverman, A. M. & Oremland, R. S. (1996). Bacterial dissimilatory reduction of arsenic(V) to arsenic(III) in anoxic sediments. *Applied and Environmental Microbiology, 62*, 1664–1669.

Dozol, M., Hagemann, R., Hoffman, D. C., Adloff, J. P., Vongunten, H. R., Foos, J., Kasprzak, K. S., Liu, Y. F., Zvara, I., Ache, H. J., Das, H. A., Hagemann, R. J. C., Herrmann, G., Karol, P., Maenhaut, W., Nakahara, H., Sakanoue, M., Tetlow, J. A., Baro, G. B., Fardy, J. J., Benes, P., Roessler, K., Roth, E., Burger, K., Steinnes, E., Kostanski, M. J., Peisach, M., Liljenzin, J. O., Aras, N. K., Myasoedov, B. F. & Holden, N. E. (1993). Radionuclide migration in groundwaters: review of the behaviour of actinides. *Pure and Applied Chemistry, 65*, 1081–1102.

Duan, W. M., Hedrick, D. B., Pye, K., Coleman, M. L. & White, D. C. (1996). A preliminary study of the geochemical and microbiological characteristics of modern sedimentary concretions. *Limnology and Oceanography, 41*, 1404–1414.

Eccles, H. (1998). Nuclear waste management: a bioremediation approach. In G. R. Choppin & M. K. Khankhasayev (Eds), *NATO ASI Chemical Separation Technologies and Related Methods of Nuclear Waste Management: Applications, Problems, and Research Needs*. Dubna, Russia (pp. 187–208). Dordrecht, The Netherlands: Kluwer Academic Publishers.

Ehrlich, H. L. (1990). *Geomicrobiology* (2nd edn) (646pp). New York: Marcel Dekker.

Elwear, S., German, K. E. & Peretrukin, V. F. (1992). Sorption of technetium onto inorganic sorbents and natural minerals. *Journal of Radioanalytical and Nuclear Chemistry Articles, 157*, 3–14.

Emerson, S. (1976). Early diagenesis in anaerobic lake sediments: chemical equilibria in interstitial waters. *Geochimica et Cosmochimica Acta, 40*, 925–934.

Emerson, S. & Widmer, G. (1978). Early diagenesis in anaerobic lake sediments – II. Thermodynamic and kinetic factors controlling the formation of iron phosphate. *Geochimica et Cosmochimica Acta, 42*, 1307–1316.

Evans, D. W., Alberts, J. J. & Clarke, R. A. (1983). Reversible ion exchange fixation of Cs-137 leading to mobilisation from sediments. *Geochimica et Cosmochimica Acta, 47*, 1041–1049.

Ferris, F. G., Fyfe, W. S. & Beveridge, T. J. (1988). Metallic ion binding by *Bacillus subtilis*: Implications for the fossilization of microorganisms. *Geology, 16*, 149–152.

Ferris, F. G., Tazaki, K. & Fyfe, W. S. (1989). Iron oxides in acid mine drainage environments and their association with bacteria. *Chemical Geology, 74*, 321–330.

Ferris, F. G., Konhauser, K. O., Lyven, B. & Pedersen, K. (1999). Accumulation of metals by bacteriogenic iron oxides in a subterranean environment. *Geomicrobiology Journal, 16*, 181–192.

Ferris, F. G., Hallberg, R. O., Lyven, B. & Pedersen, K. (2000). Retention of strontium, caesium, lead and uranium by bacterial iron oxides from a subterranean environment. *Applied Geochemistry, 15*, 1035–1042.

Fossing, H., Gallardo, V. A., Jørgensen, B. B., Huttel, M., Nielsen, L. P., Schulz, H., Canfield, D. E., Forster, S., Glud, R. N., Gundersen, J. K., Kuver, J., Ramsing, N. B., Teske, A., Thamdrup, B. & Ulloa,

O. (1995). Concentration and transport of nitrate by the mat-forming sulphur bacterium *Thioploca. Nature, 374*, 713–715.
Francis, A. J. (1990). Microbial dissolution and stabilisation of toxic metals and radionuclides in mixed wastes.*Experientia, 46*, 840–851.
Francis, A. J. (1994). Microbial transformations of radioactive wastes and environmental restoration through bioremediation. *Journal of Alloys and Compounds, 213/214*, 226–231.
Fredrickson, J. K. & Gorby, Y. A. (1996). Environmental processes mediated by iron-reducing bacteria. *Current Opinions in Biotechnology, 7*, 287–294.
Fredrickson, J. K., Zachara, J. M., Kennedy, D. W., Dong, H. L., Onstott, T. C., Hinman, N. W. & Li, S. M. (1998). Biogenic iron mineralisation accompanying the dissimilatory reduction of hydrous ferric oxide by groundwater bacterium. *Geochimica et Cosmochimica Acta, 62*, 3239–3257.
Friese, K., Wendt-Potthoff, K., Zachmann, D. W., Fauville, A., Mayer, B. & Veizer, J. (1998). Biogeochemistry of iron and sulphur in sediments of an acidic mining lake in Lusatia, Germany. *Water, Air and Soil Pollution, 108*, 231–247.
Froelich, P. N., Klinkhammer, G. P., Bender, M. L., Luedtke, N. A., Heath, G. R., Cullen, D., Dauphin, P., Hammond, D., Hartman, B. & Maynard, V. (1979). Early oxidation of organic matter in pelagic sediments of the eastern equatorial Atlantic. Suboxic diagenesis. *Geochimica et Cosmochimica Acta, 43*, 1075–1090.
Gadd, G. M. (1997). Roles of micro-organisms in the environmental fate of radionuclides. In J. V. Lake, G. R. Bock & G. Cardew (Eds), *Health Impacts of Large Releases of Radionuclides. Ciba Foundation Symposium, 203* (pp. 94–108). Chichester: John Wiley.
Ganesh, R., Robinson, K. G., Reed, G. & Sayler, G. S. (1997). Reduction of hexavalent uranium from organic complexes by sulphate and iron-reducing bacteria. *Applied and Environmental Microbiology, 3*, 4385–4391.
Garland, T. R., Cataldo, D. A., McFadden, K. M., Schreckheise, R. G. & Wildung, R. E. (1983). Comparative behaviour of ^{99}Tc, ^{129}I, ^{127}I and ^{137}Cs in the environment adjacent to a fuels reprocessing facility. *Health Physics, 44*, 658–662.
Geesey, G. G., Mutch, R. & Costerton, J. W. (1978). Sessile bacteria: an important component of microbial population in small mountain streams. *Limnology and Oceanography, 23*, 1214–1223.
Ghiorse, W. C. (1984). Biology of iron- and manganese-depositing bacteria. *Annual Reviews in Microbiology, 38*, 515–550.
Goldhaber, M. B., Aller, R. C., Cochran, J. K., Rosenfeld, J. K., Martens, C. S. & Berner, R. A. (1977). Sulphate reduction, diffusion, and bioturbation in Long Island Sound sediments. Report of the FOAM Group. *American Journal of Science, 277*, 193–237.
Gorby, Y. A. & Lovley, D. R. (1992). Enzymatic uranium precipitation. *Environmental Science and Technology, 26*, 205–207.
Hallbeck, L. & Pedersen, K. (1991). Autotrophic and mixotrophic growth of *Gallionella ferruginea. Journal of General Microbiology, 137*, 2657–2661.
Hammond, D. E., Fuller, C., Harmon, D., Hartman, B., Korosec, M., Miller, L. G., Rea, R., Warren, S., Berelson, W. & Hager, S. W. (1985). Benthic fluxes in San Francisco Bay. *Hydrobiologia, 129*, 69–90.
Hansen, J. I., Henriksen, K. & Blackburn, T. H. (1981). Seasonal distribution of nitrifying bacteria and rates of nitrification in coastal marine sediments. *Microbial Ecology, 7*, 297–304.
Hanson, W. C. (1980). *Transuranic Elements in the Environment* (728pp). DOE Publication No. DOE/TIC-22800.
Hearn, P. P., Parkhurst, D. L. & Callender, E. (1983). Authigenic vivianite in Potomac river sediments: control by ferric oxyhydroxides. *Journal of Sedimentary Petrology, 53*, 165–177.
Hedges, J. I., Clark, W. A. & Cowie, G. L. (1988). Fluxes and reactivities of organic matter in a coastal marine bay. *Limnology and Oceanography, 33*, 1137–1152.
Henrot, J. (1989). Bioaccumulation and chemical modification of Tc by soil bacteria. *Health Physics, 57*, 239–245.
Hines, M. E., Figareli, J. & Planinc, R. (1997). Sedimentary anaerobic microbial biogeochemistry in the Gulf of Trieste, northern Adriatic Sea: Influences of bottom water oxygen depletion. *Biogeochemistry, 39*, 5–86.

Holm, E. (1993). Radioanalytical studies of Tc in the environment, progress and solutions. *Radiochimica Acta, 63*, 57–62.

Hsi, C. D. & Langmuir, D. (1985). Adsorption of uranyl onto ferric oxyhydroxides: application of the surface complexation site binding model. *Geochimica et Cosmochimica Acta, 49*, 1931–1941.

Hulth, S., Hall, P. O. J., Blackburn, T. H. & Landen, A. (1996). Arctic sediments (Svalbard): Pore water and solid phase distributions of C, N, P and Si. *Polar Biology, 16*, 447–462.

Hulth, S., Tengberg, A., Landen, A. & Hall, P. O. J. (1997). Mineralisation and burial of organic carbon in sediments of the southern Weddell Sea (Antarctica). *Deep Sea Research, 44*, 955–981.

Hulth, S., Aller, R. C. & Gilbert, F. (1999). Coupled anoxic nitrification/manganese reduction in marine sediments. *Geochimica et Cosmochimica Acta, 63*, 49–66.

Hursthouse, A. S. & Livens, F. R. (1993). Evidence for the remobilization of transuranic elements in the terrestrial environment. *Environmental Geochemistry and Health, 15*, 163–171.

Hursthouse, A. S., Baxter, M. S., Livens, F. R. & Duncan, H. J. (1991). Transfer of Sellafield derived Np–237 to and within the terrestrial environment. *Journal of Environmental Radioactivity, 14*, 147–174.

IAEA (1999). *Environmental Activities in Uranium Milling and Mining* (173pp). IAEA/OECD Joint Publication, ISBN 92-64-17064-2. Paris: OECD Publications.

Irwin, H., Curtis, C. D. & Coleman, M. L. (1977). Isotopic evidence for the source of diagenetic carbonates formed during burial of organic-rich sediments. *Nature, 269*, 209–213.

Jakobsen, M. E. & Postma, D. (1994). In situ rates of sulphate reduction in an aquifer (Romo, Denmark) and implications for the reactivity of organic matter. *Geology, 23*, 1103–1106.

Jenkins, M. C. & Kemp, M. (1984). The coupling of nitrification and denitrification in two estuarine sediments. *Limnology and Oceanography, 29*, 609–619.

Johnson, E. E., Odonnnel, A. G. & Ineson, P. (1991). An autoradiographic technique for selecting Cs–137 sorbing microorganisms from soil. *Journal of Microbiological Methods, 13*, 293–298.

Jones, J. G. (1985). Microbes and microbial processes in sediments. *Philosophical Transactions of the Royal Society of London, A, 315*, 3–17.

Jones, J. G., Gardener, G. S. & Simon, B. M. (1984). Reduction of ferric iron by heterotrophic bacteria in lake sediments. *Journal of General Microbiology, 130*, 45–51.

Jørgensen, B. B. (1983a). Processes at the sediment-water interface. In B. C. Bolin (Ed.), *The Major Biogeochemical Cycles and Their Interactions*, SCOPE (pp. 477–509). New York: John Wiley.

Jørgensen, B. B. (1983b). Ecology of the bacteria of the sulphur cycle with special reference to anoxic-oxic interface environments. *Philosophical Transactions of the Royal Society of London B, 298*, 543–561.

Jørgensen, B. B. & Revsbech, N. P. (1983). Colorless sulphur bacteria, *Beggiatoa* spp. & *Thiovulum* spp. in O_2 and H_2S microgradients. *Applied and Environmental Microbiology, 45*, 61–70.

Jørgensen, B. B. & Revsbech, N. P. (1985). Diffusive boundary layers and the oxygen uptake of sediments and detritus. *Limnology and Oceanography, 30*, 111–122.

Jørgensen, B. B. & Sørensen, J. (1985). Seasonal cycles of O_2, NO_3^- and SO_4^{2-} reduction in estuarine sediments: the significance of an NO_3^- reduction maximum in the spring. *Marine Ecology Progress Series, 24*, 65–74.

Kaminski, S., Richter, T., Walser, M. & Linder, G. (1994). Redissolution of caesium radionuclides from sediments of freshwater lakes due to biological degradation of organic matter. *Radiochimica Acta, 66/67*, 433–436.

Karlin, R., Lyle, M. & Heath, G. S. (1987). Authigenic magnetite formation in suboxic marine sediments. *Nature, 326*, 490–493.

Kazumi, J., Haggblom, M. M. & Young, L. Y. (1995). Degradation of monochlorinated and nonchlorinated aromatic compounds under iron-reducing conditions. *Applied and Environmental Microbiology, 61*, 4069–4073.

Keeny-Kennicut, W. L. & Morse, J. W. (1984). The interaction of $Np(V)O_2^+$ with common mineral surfaces in dilute aqueous solution and seawater. *Marine Chemistry, 15*, 133–150.

Keeny-Kennicut, W. L. & Morse, J. W. (1985). The redox chemistry of $Pu(V)O_2^+$ interaction with common mineral surfaces in dilute solutions and seawater. *Geochimica et Cosmochimica Acta, 49*, 2577–2588.

Keith-Roach, M. J., Day, J. P., Fifield, L. K., Bryan, N. D. & Livens, F. R. (2000). Seasonal variations in interstitial water transuranium element concentrations. *Environmental Science and Technology, 34*, 4237–4277.

Kershaw, P. J., Pentreath, R. J., Woodhead, D. S. & Hunt, G. J. (1992). A review of radioactivity in the Irish Sea (65pp). *Report prepared for the Marine Pollution Monitoring Group. Aquatic Environment Monitoring Report No. 32*. Lowestoft, UK: MAFF.

King, G. M. (1984). Utilization of hydrogen, acetate and 'noncompetitive' substrates by methanogenic bacteria in marine sediments. *Geomicrobiology Journal, 3*, 275–306.

Klausen, J., Trober, S. P., Haderlein, S. B. & Schwarzenbach, R. P. (1995). Reduction of substituted nitrobenzenes by Fe(II) in aqueous mineral suspensions. *Environmental Science and Technology, 29*, 2396–2404.

Klump, J. V. & Martens, C. S. (1981). Biogeochemical cycling in an organic rich coastal marine basin-II. Nutrient sediment-water exchange processes. *Geochimica et Cosmochimica Acta, 45*, 101–121.

Kochenov, A. V., Korolev, K. G., Dubinchuk, V. T. & Medvedev, Y. L. (1977). Experimental data on the precipitation of uranium from aqueous solution. *Geochemistry International, 14*, 82–87.

Konhauser, K. O. (1997). Bacterial iron biomineralisation in nature. *FEMS Microbiological Reviews, 20*, 315–326.

Konhauser, K. O. (1998). Diversity of bacterial iron mineralization. *Earth Science Reviews, 43*, 91–121.

Konhauser, K. O. & Urrutia, M. M. (1999). Bacterial clay authigenesis: implications for river chemistry. *Chemical Geology, 161*, 399–414.

Kostka, J. E. & Nealson, K. H. (1995). Dissolution and reduction of magnetite by bacteria. *Environmental Science and Technology, 29*, 2535–2540.

Kostka, J. E., Wu, J., Nealson, K. H. & Stucki, J. W. (1999). The impact of structural Fe(III) reduction by bacteria on the surface chemistry of smectite clay minerals. *Geochimica et Cosmochimica Acta, 63*, 3705–3713.

Krauskopf, K. B. (1988). *Radioactive Waste Disposal and Geology* (145pp). New York: Chapman and Hall.

Krom, M. D., Davison, P., Zhang, H. & Davison, W. (1994). High resolution pore water sampling with a gel sampler. *Limnology and Oceanography, 39*, 1967–1973.

Kuenen, J. G. & Beudeker, R. F. (1982). Microbiology of *Thiobacilli* and other sulphur-oxidising autotrophs, mixotrophs and heterotrophs. *Philosophical Transactions of the Royal Society of London B, 298*, 473–497.

Ledin, M. & Pedersen, K. (1996). The environmental impact of mine wastes – roles of microorganisms and their significance in treatment of mine wastes. *Earth Science Reviews, 41*, 67–108.

Leonard, K. S., McCubbin, D., Brown, J., Bonfield, R. & Brooks, T. (1997). Distribution of Tc-99 in UK coastal waters. *Marine Pollution Bulletin, 34*, 628–636.

Liger, E., Charlet, L. & Van Cappellen, P. (1999). Surface catalysis of uranium(VI) reduction by iron(II). *Geochimica et Cosmochimica Acta, 63*, 2939–2955.

Lieser, K. H. & Bauscher, C. (1987). Technetium in the hydrosphere and geosphere. Chemistry of technetium in natural waters and the influence of redox potential on the sorption of technetium. *Radiochimica Acta, 42*, 205–213.

Lieser, K. H. & Bauscher, C. (1988). Technetium in the hydrosphere and geosphere 2. Influence of pH, complexing agents and of some minerals on the sorption of technetium. *Radiochimica Acta, 44/45*, 125–128.

Lieser, K. H. & Muhlenweg, U. (1988). Neptunium in the hydrosphere and in the geosphere. I. Chemistry of neptunium in the hydrosphere and sorption of neptunium from groundwaters on sediments under aerobic and anaerobic conditions. *Radiochimica Acta, 43*, 27–35.

Lloyd, J. R. & Macaskie, L. E. (1996). A novel phosphorImager-based technique for monitoring the microbial reduction of technetium. *Applied and Environmental Microbiology, 58*, 850–856.

Lloyd, J. R., Harding, C. L. & Macaskie, L. A. (1997). Tc(VII) reduction and precipitation by immobilized cells of *Escherichia coli. Biotechnology and Bioengineering, 55*, 505–510.

Lloyd, J. R., Nolting, H. F., Sole, V. A., Bosecker, K. & Macaskie, L. A. (1998). Technetium reduction and precipitation by sulphate-reducing bacteria. *Geomicrobiology Journal, 15*, 45–58.

Lloyd, J. R., Ridley, J., Khizniak, T., Lyalikova, N. N. & Macaskie, L. E. (1999). Reduction of technetium by *Desulfovibrio desulfuricans*: biocatalyst characterisation and use in a flowthrough bioreactor. *Applied and Environmental Microbiology, 65*, 2691–2696.

Lloyd, J. R., Yong, P. & Macaskie, L. E. (2000). Biological reduction and removal of Np(V) by two microorganisms. *Environmental Science and Technology, 34*, 1297–1301.

Lonergan, D. J. & Lovley, D. R. (1991). Microbial oxidation of natural and anthropogenic aromatic compounds coupled to Fe(III) reduction. In R. A. Baker (Ed.), *Organic Substances and Sediments in Water* (pp. 327–338). Chelsea, MI: Lewis Publishers.

Lovley, D. R. (1991). Dissimilatory Fe(III) and Mn(IV) reduction. *Microbiology Reviews, 55*, 259–287.

Lovley, D. R. (1993). Dissimilatory metal reduction. *Annual Reviews in Microbiology, 47*, 263–290.

Lovley, D. R. (1995). Microbial reduction of iron, manganese and other metals. *Advances in Agronomy, 54*, 175–231.

Lovley, D. R. & Chapelle, F. H. (1995). Deep subsurface microbial processes. *Reviews of Geophysics, 33*, 365–381.

Lovley, D. R. & Coates, J. D. (1997). Bioremediation of metal contamination. *Current Opinion in Biotechnology, 8*, 285–289.

Lovley, D. R. & Goodwin, S. (1988). Hydrogen concentrations as an indicator of the predominant terminal electron-accepting reactions in aquatic sediments. *Geochimica et Cosmochimica Acta, 52*, 2993–3003.

Lovely, D. R. & Klug, M. J. (1983). Sulphate reducers can outcompete methanogens at freshwater sulphate concentrations. *Applied and Environmental Microbiology, 45*, 187–192.

Lovely, D. R. & Klug, M. J. (1986). Model for the distribution of sulphate reduction and methanogenesis in freshwater sediment. *Geochimica et Cosmochimica Acta, 50*, 11–18.

Lovley, D. R. & Lonergan, D. J. (1990). Anaerobic oxidation of toluene, phenol, and *p*-cresol by the dissimilatory iron-reducing organism, GS-15. *Applied and Environmental Microbiology, 56*, 1858–1864.

Lovley, D. R. & Phillips, E. J. P. (1986a). Organic matter mineralization with reduction of ferric iron in anaerobic sediments. *Applied and Environmental Microbiology, 51*, 683–689.

Lovley, D. R. & Phillips, E. J. P. (1986b). Availability of ferric iron for microbial reduction in bottom sediments of the freshwater tidal Potomac River. *Applied and Environmental Microbiology, 52*, 751–757.

Lovley, D. R. & Phillips, E. J. P. (1987). Competitive mechanisms for inhibition of sulphate reduction and methane production in the zone of ferric iron reduction in sediments. *Applied and Environmental Microbiology, 53*, 2636–2641.

Lovley, D. R. & Phillips, E. J. P. (1988). Novel mode of microbial energy metabolism: organic carbon oxidation coupled to dissimilatory reduction of iron or manganese. *Applied and Environmental Microbiology, 54*, 1472–1480.

Lovley, D. R. & Phillips, E. J. P. (1989). Requirements for a microbial consortium to completely oxidise glucose in Fe (III)-reducing sediments. *Applied and Environmental Microbiology, 55*, 3234–3236.

Lovley, D. R. & Phillips, E. J. P. (1992). Reduction of uranium by *Desulfovibrio desulfuricans. Applied and Environmental Microbiology, 58*, 850–856.

Lovely, D. R., Dwyer, D. F. & Klug, M. J. (1982). Kinetic analysis of competition between sulphate reducers and methanogens for hydrogen in sediments. *Applied and Environmental Microbiology, 43*, 1373–1379.

Lovley, D. R., Stolz, J. F., Nord, G. L. Jr. & Phillips, E. J. P. (1987). Anaerobic production of magnetite by a dissimilatory iron-reducing microorganism. *Nature, 330*, 252–254.

Lovley, D. R., Phillips, E. J. P. & Lonergan, D. J. (1989a). Hydrogen and formate oxidation coupled to dissimilatory reduction of iron or manganese by *Alteromonas putrefaciens. Applied and Environmental Microbiology, 55*, 700–706.

Lovley, D. R., Baedecker, M. J., Lonergan, D. J., Cozzarelli, I. M., Phillips, E. J. P. & Siegel, D. I. (1989b). Oxidation of aromatic contaminants coupled to microbial iron reduction. *Nature, 339*, 297–299

Lovley, D. R., Chapelle, F. H. & Phillips, E. J. P. (1990). Fe(III)-reducing bacteria in deeply buried sediments of the Atlantic Coastal Plain. *Geology, 18*, 954–957.

Lovley, D. R., Phillips, E. J. P., Gorby, Y. A. & Landa, E. R. (1991). Microbial reduction of uranium. *Nature, 350*, 413–416.

Lovley, D. R., Giovannoni, S. J., White, D. C., Champine, J. E., Phillips, E. J. P., Gorby, Y. A. & Goodwin, S. (1993). *Geobacter metallireducens* gen. nov. sp. nov., a microorganism capable of coupling the complete oxidation of organic compounds to the reduction of iron and other metals. *Archives of Microbiology, 159*, 336–344.

Lovely, D. R., Chapelle, F. H. & Woodward, J. C. (1994). Use of dissolved H_2 concentrations to determine distribution of microbially catalyzed redox reactions in anoxic groundwater. *Environmental Science and Technology, 28*, 1205–1210.

Lovley, D. R., Coates, J. D., Blunt-Harris, E. L., Phillips, E. J. P. & Woodward, J. C. (1996). Humic substances as electron acceptors for microbial respiration. *Nature, 382*, 445–448.

Luther, G. W., Sundby, B., Lewis, B. L., Brendel, P. J. & Silverberg, N. (1997). Interactions of manganese with the nitrogen cycle. Alternative pathways to dinitrogen. *Geochimica et Cosmochimica Acta, 61*, 4043–4052.

Luther, G. W., Brendel, P. J., Lewis, B. L., Sundby, B., Lefraníois, L., Silverberg, N. & Nuzzio, D. B. (1998). Simultaneous measurement of O_2, Mn, Fe, I^-, and S (-II) in marine pore waters with a solid-state voltammetric microelectrode. *Limnology and Oceanography, 43*, 325–333.

Macaskie, L. E. (1990). An immobilised cell bioprocess for the removal of heavy metals from aqueous flows. *Journal of Chemical Technology and Biotechnology, 49*, 357–379.

Macaskie, L. E. & Dean, A. C. R. (1990). Metal sequestering biochemicals. In A. Mizrahi (Ed.), *Biological Waste Treatment* (pp. 199–248). New York: Liss.

Macaskie, L. E., Empson, R. M., Cheetham, A. K., Grey, C. P. & Skarnulis, A. J. (1990). Uranium bioaccumulation by a *Citrobacter* sp. as a result of enzymatically mediated growth of polycrystalline HUO_2PO_4. *Science, 257*, 782–784.

Macaskie, L. E., Jeong, B. C. & Tolley, M. R. (1994). Enzymatically accelerated biomineralisation of heavy metals – application to removal of americium and plutonium from aqueous flows. *FEMS Microbiology Reviews, 14*, 351–367.

Macy, J. M., Michel, T. A. & Kirsch, D. G. (1989). Selenate reduction by a *Pseudomonas* species: a new mode of anaerobic respiration. *FEMS Microbiology Letters, 61*, 195–198.

Madigan, M. T., Martinko, J. M. & Parker, J. (1997). *Brock. Biology of Microorganisms* (8th edn) (986pp). Upper Saddle River, NJ: Prentice Hall.

Malcolm, S. J., Kershaw, P. J., Lovett, M. B. & Harvey, B. R. (1990). The interstitial water chemistry of $^{239,240}Pu$ and ^{241}Am in the sediments of the north-east Irish Sea. *Geochimica et Cosmochimica Acta, 54*, 29–35.

Mann, S., Sparks, N. H. C., Frankel, R. B., Bazylinski, D. A. & Jannasch, H. W. (1990). Biomineralization of ferrimagnetic greigite (Fe_3S_4) and iron pyrite (FeS_2) in a magnetotactic bacterium. *Nature, 343*, 258–261.

Martens, C. S., Berner, R. A. & Rosenfeld, J. K. (1978). Interstitial water chemistry of anoxic Long Island Sound sediments. 2. Nutrient regeneration and phosphate removal. *Limnology and Oceanography, 23*, 605–617.

Masson, M., Patti, F., Colle, C., Roucoux, P., Grauby, A. & Saas, A. (1989). Synopsis of French experimental in-situ work on the terrestrial and marine behavior of Tc. *Health Physics, 57*, 269–279.

Maynard, J. B. (1982). Extension of Berner's 'New geochemical classification of sedimentary environments' to ancient sediments. *Journal of Sedimentary Petrology, 52*, 1325–1331.

Michalopoulos, P. & Aller, R. C. (1995). Rapid clay mineral formation in Amazon Delta sediments: reverse weathering and oceanic elemental cycles. *Science, 270*, 614–617.

Mohagheghi, A., Updegraff, D. M. & Goldhaber, M. B. (1985). The role of sulphate-reducing bacteria in the deposition of sedimentary ores. *Geomicrobiology Journal, 4*, 153–173.

Morris, K. & Livens, F. R. (1996). Interactions of actinides with sediments in intertidal areas in west Cumbria, UK. *Radiochimica Acta, 74*, 195–198.

Morris, K., Butterworth, J. C. & Livens, F. R. (2000). Evidence for the remobilization of Sellafield waste radionuclides in an intertidal salt marsh, west Cumbria, UK. *Estuarine, Coastal and Shelf Science, 51*, 613–625.

Morris, K., Bryan, N. D. & Livens, F. R. (2001). Plutonium solubility in sediment pore waters. *Journal of Environmental Radioactivity, 56*, 259–267.

Morse, J. W. & Choppin, G. R. (1991). The chemistry of transuranic elements in natural waters. *Reviews in Aquatic Sciences, 4*, 1–22.

Mortimer, C. H. (1941). The exchange of dissolved substances between mud and water in lakes. *Journal of Ecology, 29*, 280–329.

Mortimer, R. J. G. & Coleman, M. L. (1997). Microbial influence on the oxygen isotopic composition of diagenetic siderite. *Geochimica et Cosmochimica Acta, 61*, 1705–1711.

Mortimer, R. J. G. & Rae, J. E. (2000). Metal speciation (Cu, Zn, Pb, Cd) and organic matter in oxic to suboxic saltmarsh sediments, Severn Estuary, southwest Britain. *Marine Pollution Bulletin, 40*, 377–386.

Mortimer, R. J. G., Coleman, M. L. & Rae, J. E. (1997). Effect of bacteria on the trace element composition of early diagenetic siderite: implications for palaeoenvironmental interpretations. *Sedimentology, 44*, 759–765.

Mortimer, R. J. G., Krom, M. D., Hall, P. O. J., Hulth, S. & Stahl, H. (1998). Use of gel probes for the determination of high resolution solute distributions in marine and estuarine pore waters. *Marine Chemistry, 63*, 119–129.

Mortimer, R. J. G., Davey, J. T., Krom, M. D., Watson, P. G., Frickers, P. E. & Clifton, R. C. (1999). The effect of macrofauna on pore-water profiles and nutrient fluxes in the intertidal zone of the Humber Estuary. *Estuarine, Coastal and Shelf Science, 48*, 683–699.

Munier-Lamy, C. & Berthelin, J. (1987). Formation of polyelectrolyte complexes with the major elements Fe and Al and the trace elements U and Cu during heterotrophic microbial leaching of rocks. *Geomicrobiology Journal, 5*, 119–147.

Murray, J. W. & Grundmaris, V. (1980). Oxygen consumption in pelagic marine sediments. *Science, 209*, 1527–1530.

Muyzer, G., Dewaal, E. C. & Uitterlindern, A. G. (1993). Profiling of complex microbial populations by denaturing gradient gel electrophoresis analysis of polymerase chain reaction amplified genes coding for 16S ribosomal RNA. *Applied and Environmental Microbiology, 59*, 695–700.

Myers, C. R. & Nealson, K. H. (1988). Bacterial manganese reduction and growth with manganese oxide as the sole electron acceptor. *Science, 240*, 1319–1321.

Nelson, D. C. & Jannasch, H. W. (1983). Chemoautotrophic growth of a marine *Beggiatoa* in sulfide-gradient cultures. *Archives of Microbiology, 136*, 262–269.

Nelson, D. L. & Lovett, M. B. (1978). Oxidation state of plutonium in the Irish Sea. *Nature, 276*, 599–601.

Neu, M. P., Matonic, J. H., Ruggiero, C. E. & Scott, B. L. (2000). Structural characterisation of a plutonium (IV) siderophore complex: single crystal structure of Pu-desferriooxamine *Angewandte Chemie International Edition, 39*, 1442–1444.

Newman, D. K., Beveridge, T. J. & Morel, F. M. M. (1997). Precipitation of arsenic trisulfide by *Desulfotomaculum auripigmentum. Applied and Environmental Microbiology, 63*, 2022–2028.

Oremland, R. S., Holibaugh, J. T., Maest, A. S., Presser, T. S., Miller, L. G. & Culbertson, C. W. (1989). Selenate reduction to elemental selenium by anaerobic bacteria in sediments and culture: biogeochemical significance of a novel, independent-independent respiration. *Applied and Environmental Microbiology, 55*, 2333–2343.

Parkes, R. J., Gibson, G. R., Mueller-Harvey, I., Buckingham, W. J. & Herbert, R. A. (1989). Determination of the substrates for reducing-reducing bacteria within marine and estuarine sediments with different rates of sulphate reduction. *Journal of General Microbiology, 135*, 175–187.

Pedersen, K. (1993). The deep subterranean biosphere. *Earth Science Reviews, 34*, 243–260.

Pentreath, R. J. (1981). The biological availability to marine organisms of transuranium and other long lived nuclides (pp. 241–272). IAEA-SM-248/102.

Picardal, F. W., Arnold, R. G., Couch, H., Little, A. M. & Smith, M. E. (1993). Involvement of cytochromes in the anaerobic biotransformation of tetrachloromethane by *Shewanella putrefaciens* 200. *Applied and Environmental Microbiology, 59*, 3763–3770.

Ponnamperuma, F. N. (1972). The chemistry of submerged soils. *Advances in Agronomy, 24*, 29–96.

Postgate, J. R. (1984). *The Sulphate-Reducing Bacteria* (2nd edn) (208pp). Cambridge, UK: Cambridge University Press.

Postma, D. (1981). Formation of siderite and vivianite and the pore-water composition of a recent bog sediment in Denmark. *Chemical Geology, 31*, 225–244.

Postma, D. (1982). Pyrite and siderite formation in brackish and freshwater swamp sediments. *American Journal of Science, 282*, 1151–1183.

Postma, D. & Jakobsen, R. (1996). Redox zonation: equilibrium constraints on the Fe(III)/SO_4^{2-} reduction interface. *Geochimica et Cosmochimica Acta, 60*, 3169–3175.

Ramirez, A. J. & Rose, A. W. (1992). Analytical geochemistry of organic phosphorus and its correlation with organic carbon in marine and fluvial sediments and soils. *American Journal of Science, 292*, 421–454.

Redfield, A. C. (1958). The biological control of chemical factors in the environment. *American Journal of Science, 46*, 206–226.

Reeburgh, W. S. (1983). Rates of biogeochemical processes in anoxic sediments. *Annual Review of Earth and Planetary Science, 11*, 269–298.

Rinzema, A. & Lettinga, G. (1985). Anaerobic treatment of containing-containing waste water. In D. L. Wise (Ed.), *Biotreatment Systems* (pp. 65–109). Boca Raton, FL: CRC Press.

Roden, E. E. & Lovley, D. R. (1993). Dissimilatory Fe(III) reduction by the marine microorganism *Desulfuromonas acetooxidans*. *Applied and Environmental Microbiology, 59*, 734–742.

Roden, E. E. & Zachara, J. M. (1996). Microbial reduction of crystalline iron(III) oxides: influence of oxide surface area and potential for cell growth. *Environmental Science and Technology, 30*, 1618–1628.

Rusin, P. A., Quintana, L., Brainard, J. R., Strietelmeier, B. A., Tait, C. D., Ekberg, S. A., Palmer, P. D., Newton, T. W. & Clark, D. L. (1994). Solubilization of plutonium hydrous oxide by iron-reducing bacteria. *Environmental Science and Technology, 28*, 1686–1690.

Santschi, P. H. & Honeyman, B. D. (1989). Radionuclides in aquatic environments. *Radiation Physics and Chemistry, 34*, 213–240.

Santschi, P. H., Bollhalder, S., Farrenkothen, K., Lueck, A., Zingg, S. & Sturm, M. (1988). Chernobyl radionuclides in the environment, tracers for the tight coupling of atmospheric, terrestrial and aquatic geochemical processes. *Environmental Science and Technology, 22*, 510–516.

Santschi, P., Hohener, P., Benoit, G. & Buchholtz-Ten Brink, M. (1990a). Chemical processes at the sediment-water interface. *Marine Chemistry, 30*, 269–315.

Santschi, P. H., Bollhalder, S., Zingg, S., Lueck, A. & Farrenkothen, K. (1990b). The self cleaning capacity of surface waters after radioactive fallout. Evidence from European waters after Chernobyl. *Environmental Science and Technology, 24*, 519–524.

Sawhney, B. L. (1970). Potassium and caesium ions selectivity in relation to clay mineral structure. *Clays and Clay Minerals, 18*, 47–52.

Sawhney, B. L. (1972). Selective sorption and fixation of cations by clay minerals: a review. *Clays and Clay Mineralogy, 20*, 93–100.

Schink, B. (1988). Principles and limits of anaerobic degradation: environmental and technological aspects. In A. J. B. Zehnder (Ed.), *Biology of Anaerobic Microorganisms* (pp. 771–846). New York: John Wiley.

Schwertmann, U. & Fechter, H. (1982). The point of zero charge of natural and synthetic ferrihydrites and its relation to adsorbed silicate. *Clay Minerals, 17*, 471–476.

Scoppa, P. (1984). Environmental behaviour of trans-uranium actinides: availability to marine biota. *Inorganica Chimica Acta, 95*, 23-27.

Seitzinger, S. P. (1988). Denitrification in freshwater and coastal marine ecosystems: ecological and geochemical significance. *Limnology and Oceanography, 33*, 702–724.

Senior, E. & Balba, T. M. (1990). Refuse decomposition. In E. Senior (Ed.), *Microbiology of Landfill Sites* (pp. 18–57). Boca Raton, FL: CRC Press.

Shaw, D. G. & McIntosh, D. J. (1990). Acetate in recent anoxic sediments. Direct and indirect measurements of concentration and turnover rates. *Estuarine and Coastal Shelf Science, 31*, 775–788.

Sheppard, S. C., Sheppard, M. I. & Evenden, W. G. (1990). A novel method used to examine variation in Tc sorption among 34 soils, aerated and anoxic. *Journal of Radioactivity, 11*, 215–233.

Sholkovitz, E. R. (1983). The geochemistry of plutonium in fresh and marine water environments. *Earth Science Reviews, 19*, 95–161.

Sholkovitz, E. R. (1985). In W. Stumm (Ed.), *Chemical Processes in Lakes* (pp. 119–142). New York: John Wiley.

Sholkovitz, E. R. & Mann, D. R. (1984). The porewater chemistry of $^{239,240}Pu$ and ^{137}Cs in the sediments of Buzzards Bay, Massachusetts. *Geochimica et Cosmochimica Acta, 48*, 1107–1114.

Silva, R. J. & Nitsche, H. (1995). Actinide environmental chemistry. *Radiochimica Acta, 70/71*, 377–396.

Smith, R. L. & Klug, M. J. (1981). Electron donors utilized by reducing-reducing bacteria in eutrophic lake sediments. *Applied and Environmental Microbiology, 42*, 116–121.

Sokolva-Dubina, G. A. & Deryugina, Z. P. (1967). On the role of microorganisms in the formation of rhodocrosite in Punus-Yarvi Lake. *Microbiology, 36*, 445–451.

Sørensen, J. (1982). Reduction of ferric iron in anaerobic, marine sediment and interaction with reduction of nitrate and sulphate. *Applied and Environmental Microbiology, 43*, 319–324.

Sørensen, J. (1987). Nitrate reduction in marine sediment. Pathways and interactions with iron and sulphur cycling. *Geomicrobiology Journal, 5*, 401–421.

Sørensen, J. & Jørgensen, B. B. (1987). Early diagenesis in sediments from Danish coastal waters: microbial activity and Mn-Fe-S geochemistry. *Geochimica et Cosmochimica Acta, 51*, 1583–1590.

Sparkes, S. T. & Long, S. E. (1988). *The Chemical Speciation of Technetium in the Environment – A Literature Survey*. AERE R12743, UK Atomic Energy Authority, Harwell.

Strandberg, G. W. & Arnold, W. D. (1988). Microbial accumulation of neptunium. *Journal of Industrial Microbiology, 3*, 329 - 331.

Straub, K. L., Benz, M., Schink, B. & Widdel, F. (1996). Anaerobic, nitrate-dependent microbial oxidation of ferrous iron. *Applied and Environmental Microbiology, 62*, 1458–1460.

Suess, E. (1979). Mineral phases formed in anoxic sediments by microbial decomposition of organic matter. *Geochimica et Cosmochimica Acta, 43*, 339–352.

Sugio, T., Tsujita, Y., Katagiri, T., Inagaki, K. & Tano, T. (1988). Reduction of Mo–6+ with elemental sulphur by *Thiobacillus ferroxidans. Journal of Bacteriology, 170*, 5956–5959.

Sugio, T., Tsujita, Y., Inagaki, K. & Tano, T. (1990). Reduction of cupric ions with elemental sulphur by *Thiobacillus ferrooxidans. Applied and Environmental Microbiology, 56*, 693–696.

Suzuki, Y. & Banfield, J. F. (1999). Geomicrobiology of uranium. In P. C. Burns & R. Finch (Eds), *Reviews in Mineralogy* (Vol. 38) *Uranium – Mineralogy, Geochemistry and the Environment* (pp. 393–431). Washington DC: Mineralogical Society of America.

Tiedje, J. M., Sexstone, A. J., Myrold, D. D. & Robinson, J. A. (1982). Dentrification: ecological niches, competition and survival. *Antonie van Leeuwenhoek. Journal of Microbiology, 48*, 569–583.

Tomioka, N., Uchiyama, H. & Yagi, O. (1992). Isolation and characterisation of caesium accumulating bacteria. *Applied and Environmental Microbiology, 58*, 1019–23.

Urrutia, M. M., Kemper, M., Doyle, R. & Beveridge, T. J. (1992). The membrane-induced proton motive force influences the metal binding ability of *Bacillus subtilis* cell walls. *Applied and Environmental Microbiology, 58*, 3837–3844.

Urrutia, M. M., Roden, E. E., Fredrickson, J. K. & Zachara, J. M. (1998). Microbial and surface chemistry controls on reduction of synthetic Fe(III) oxide minerals by the dissimilatory iron-reducing bacterium *Shewanella alga. Geomicrobiology Journal, 15*, 269–291.

Van Cappellen, P. & Gaillard, J-F. (1996). Biogeochemical dynamics in aquatic sediments. In P. C. Lichtner, C. I. Steefel & E. H. Oelkers (Eds), *Reactive Transport in Porous Media. Reviews in Mineralogy* (Vol. 34) (pp. 335–376). Washington, DC: Mineralogical Society of America.

Voleski, B. & Holan, Z. R. (1995). Biosorption of heavy metals. *Biotechnology Progress, 11*, 235–250.

Walton, F. B., Paquette, J. & Ross, J. P. M. (1986) Tc(IV) and Tc(VII) interactions with iron oxyhydroxides. *Nuclear and Chemical Waste Management, 6*, 121–126.

Wang, P., Mori, T., Komori, K., Sasatsu, M., Toda, K. & Ohtakein, H. (1989). Isolation and characterization of an *Enterobacter cloacae* strain that reduces hexavalent chromium under anaerobic conditions. *Applied and Environmental Microbiology, 55*, 1665–1669.

Wellsbury, P. & Parkes, R. J. (1995). Acetate bioavailability and turnover in an estuarine sediment. *FEMS Microbial Ecology, 17*, 85–94.

Wersin, P., Hocehella, M. F., Persson, P., Redden, G., Leckie, J. O. & Harris, D. W. (1994) Interaction between aqueous uranium(VI) and sulfide minerals – spectroscopic evidence for sorption and reduction. *Geochimica et Cosmochimica Acta, 58*, 2889–2843.

Westrich, J. T. & Berner, R. A. (1984). The role of sedimentary organic matter in bacterial sulphate reduction: the G model tested. *Limnology and Oceanography, 29*, 236–249.

Wahlgren, M. A., Robbins, J. A. & Edgington, D. N. (1980). Plutonium in the great lakes. In W. C. Hanson (Ed.), *Transuranic Elements in the Environment* (pp. 659–683). DOE/TIC–22800.

Wharton, M., Atkins B., Charnock J. M., Livens F. R., Pattrick R. A. D. & Collison, D. (2000). An X-ray absorption spectroscopy study of the coprecipitation of Tc and Re with mackinawite (FeS). *Applied Geochemistry, 15*, 347–354.

White, D. C. (1993). In situ measurement of microbial biomass, community structure and nutritional status. *Philosophical Transactions of the Royal Society of London. A, 344*, 59–67.

Wildung, R. E., McFadden, K. M. & Garland, T. R. (1979). Technetium sources and behaviour in the environment. *Journal of Environmental Quality, 8*, 156–161.

Williams, R. T. & Crawford, R. L. (1984). Methane production in Minnesota peatlands. *Applied and Environmental Microbiology, 47*, 1266–1271.

Yurkova, N. A. & Lyalikova, N. N. (1991). New vanadate-reducing faculative chemolithotrophic bacteria. *Microbiology, 59*, 672–677.

Zachara, J. M., Fredrickson, J. K., Li, S. M., Kennedy, D. W., Smith, S. C. & Gassman, P. L. (1998). Bacterial reduction of crystalline Fe^{3+} oxides in single phase suspensions and subsurface materials. *American Mineralogist, 83*, 1426–1443.

Zhang, H., Davison, W., Miller, S. & Tych, W. (1995). In situ high resolution measurements of fluxes of Ni, Cu, Fe and Mn and concentrations of Zn and Cd in porewaters by DGT. *Geochimica et Cosmochimica Acta*, 59, 4181–4192.

Zwolsman, J. J. G., Berger, G. W. & Van Eck, G. T. M. (1993). Sediment accumulation rates, historical input, post depositional mobility and retention of major and trace elements in salt marsh sediments of the Scheldt Estuary, SW Netherlands. *Marine Chemistry, 44*, 73–94.

Miranda J. Keith-Roach and Francis R. Livens (Editors)
© 2002 Elsevier Science Ltd. All rights reserved

Chapter 4

Biogeochemical cycles and remobilisation of the actinide elements

Katherine Morris[a], Rob Raiswell[b]

[a]*School of the Environment, University of Leeds, Leeds, LS2 9JT, UK*
[b]*School of Earth Sciences, University of Leeds, Leeds, LS2 9JT, UK*

1. Introduction

The actinide elements uranium, neptunium, plutonium and americium are among the most important radionuclides in nuclear wastes and in contaminated soils and groundwaters. The primordial radionuclide uranium is ubiquitous in the natural environment but may also be anthropogenically enhanced, whereas the transuranic elements neptunium, plutonium and americium are essentially present in the environment solely as a result of human activities. The aim of this chapter is to review the inputs of these key actinide elements to the natural environment, to examine global cycling of uranium in the context of the global iron cycle, and to review the possible roles of microbial processes in the cycling and remobilisation behaviour of the actinides.

The mobility of the actinide elements in natural systems is controlled by their partitioning between aqueous and particulate phases, which is in turn affected by their aqueous speciation. Actinide elements which occur predominantly in the aqueous phase may be rapidly and widely dispersed via riverine discharge and ocean currents, and may remain bioavailable indefinitely. By contrast, actinide elements associated with particulates may undergo more limited transportation where sedimentation can occur near the point of discharge, with estuarine, tidal flat and salt marsh sediments acting as sinks. However, storage in these sediments is not irreversible; chemical changes below the sediment water interface may remobilise actinides to the water column. Here, particulate association may reoccur, thus facilitating return to the sediments. This recycling between water and sediment introduces considerable uncertainty into the prediction of actinide behaviour where sedimentary environments are dynamic and susceptible to short-term or seasonal variations. An important aspect of actinide behaviour to be discussed in this chapter will therefore be the controls on and extent of partitioning between aqueous and particulate phases, and the nature of the sedimentary environments into which the particulate-associated actinides are deposited.

The mobility of aqueous actinide species in the natural environment is dependent on a range of competing mechanisms. Redox conditions, pH and temperature, colloid formation and availability of complexing ligands all play a role in determining the chemical speciation and cycling behaviour of the actinides (Silva & Nitsche, 1995). Microorganisms and their metabolic activities are central in enhancing or retarding these migration effects, and processes such as direct enzymatic reduction, biosorption, chelation with organic and inorganic metabolites, precipitation/dissolution reactions, and changes in subsurface chemistry have been discussed with reference to actinide behaviour in Chapter 3. Experimental studies of microbial processes affecting actinide solubility are often laboratory 'microcosm' systems based on bioremediation technologies (Ledin, 2000). This approach cannot easily be extrapolated to the natural environment as these experiments usually contain elevated metal concentrations (compared to those found in natural systems) and are often based on single cultures rather than a consortium of microbes. Hence, this chapter will focus on studies which examine the cycling and remobilisation behaviour of actinide elements within the natural environment.

2. Actinide chemistry

The environmental chemistry of the actinides has been reviewed extensively (see Chapter 3, this volume; Morse & Choppin, 1991; Dozol et al., 1993; Silva & Nitsche, 1995) and there is, in addition, a recent review by Banaszak et al. (1999) which focuses on microbial reactions which affect actinide mobility. The most important property that defines the behaviour of the actinides is the oxidation state, as precipitation, complexation, sorption and colloid formation differ considerably from one state to another (Silva & Nitsche, 1995). The early actinides display a wide range of oxidation states (+3 to +6 inclusive under environmental conditions) due to the small energy differences between the 5f, 6d and 7s valence orbitals (Katz et al., 1986; Lander & Fuger, 1989). The dominant oxidation state for any individual actinide element reflects changes in stability of these orbitals across the series (see Katz et al., 1986 for more detail), but involves the loss of all the valence electrons for the early actinides. Thus, there is an initial increase in preferred oxidation state in the early actinides from 3 (Ac) to 6 (U), which is followed by a decrease through 5 (Np) and 4 (Pu) to 3 (Am and thereafter; see Table 1).

Table 1

The oxidation states of selected actinide elements (Katz et al., 1986)

89 Ac	90 Th	91 Pa	92 U	93 Np	94 Pu	95 Am	96 Cm
3	(3)	(3)	3	3	3	**3**	**3**
	4	4	4	4	**4**	4	4
		5	5	**5**	5	5	5?
			6	6	6	6	6?
				7	(7)	7?	

Note: Bold type = most stable; ( ) = unstable; ? = claimed but unsubstantiated.

In aqueous solution at pH < 3, four structural cations exist; M^{3+}, M^{4+}, and the partially hydrolysed MO_2^+ and MO_2^{2+} forms (Katz et al., 1986). The charge on the ions induces acidity in solution and thus the ions undergo extensive hydrolytic reactions (Katz et al., 1986). The extent of hydrolysis decreases in the order $M^{4+} > M^{3+} \approx MO_2^{2+} > MO_2^+$ and, except in strong acids, the hydrolysis chemistry dominates (Katz et al., 1986; Clark et al., 1995). In natural waters, hydroxide, carbonate, sulfate, phosphate, chloride, fluoride, nitrate and silicate will be important ligands for actinides and the trend for complexation within an individual oxidation state is $OH^- > CO_3^{2-} > F^-, HPO_4^{2-}, SO_4^{2-} > Cl^-, NO_3^-$ (Silva & Nitsche, 1995). The stability order for the complexes of a particular ligand then follows the order for hydrolysis ($M^{4+} > M^{3+} \approx MO_2^{2+} > MO_2^+$; Arhland, 1986; Morse & Choppin, 1991; Silva & Nitsche, 1995). In general, the solubility of the actinides decreases with increasing ionic charge. For example, Silva & Nitsche (1995) estimated the relative solubilities of the different actinide oxidation states in 0.01 M NaCl equilibrated with atmospheric CO_2 at pH 7 as: $MO_2^+ > MO_2^{2+} > M^{3+} \gg M^{4+}$ with the solubility of M^{4+} being three orders of magnitude lower than that of M^{3+} and five orders of magnitude lower than that of MO_2^+ (Silva & Nitsche, 1995).

Redox reactions

The effect of redox potential on actinide oxidation states has been discussed by numerous authors (Morse & Choppin, 1991; Dozol et al., 1993; Silva & Nitsche, 1995; Banaszak et al., 1999). Although, in practice, attainment of redox equilibrium in the environment is often inhibited by kinetics, it is useful to consider the predicted equilibrium redox behaviour of the early actinides. In order to calculate the Eh stability field, the stabilisation of ions in solution by complex formation must be taken into account (Silva & Nitsche, 1995). Banaszak et al. (1999) combined the redox potentials associated with the major biological redox couples with the thermodynamically stable Eh regions for different oxidation states of the early actinides in a 0.01 M NaCl solution exposed to atmospheric CO_2 at pH 7, as calculated by Silva & Nitsche (1995). Thus, Fig. 1 indicates that U(VI), Pu(V), Np(V) and Am(III) are the dominant forms in low ionic strength, oxygenated waters. However, the +4 oxidation state becomes progressively dominant for plutonium (Eh $\approx$ +600 mV), neptunium (Eh $\approx$ +400 mV) and uranium (Eh $\approx$ +100 mV) as reducing conditions develop. Am(III) is thermodynamically stable over a wide Eh range at neutral pH.

Although thermodynamics can predict the general trends in oxidation states of the actinides, it is important to appreciate that equilibrium may not be attained. Furthermore, models such as HYDRAQL and PHREEQE (e.g. Bryan et al., 1994; Silva & Nitsche, 1995; Suzuki & Banfield, 1999) require extensive thermodynamic data, many of which are poorly known for the actinides (Allard, 1984; Bryan et al., 1994), although a series of standard databases are now accepted (e.g. Grenthe et al., 1992; Silva et al., 1995; Langmuir, 1997; Murphy & Shock, 1999; Suzuki & Banfield, 1999). The presence of humic substances and colloid formation in natural waters also leads to difficulties in modelling (see Chapter 5, this volume; Silva & Nitsche, 1995) and overall, such variables can lead to considerable uncertainties in the results of redox calculations (Bryan et al., 1994).

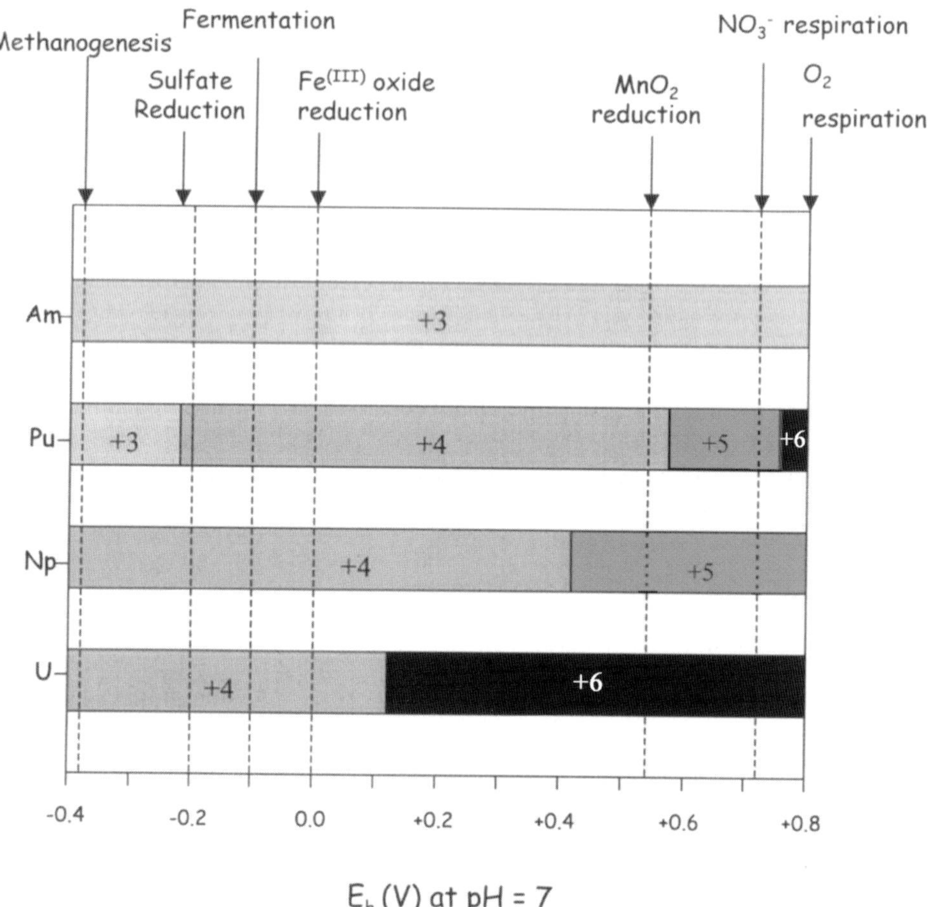

Fig. 1. Expected dominant oxidation states of the actinides as a function of standard reduction potential in pH 7 water at equilibrium with atmospheric CO_2. Arrows indicate the expected redox potentials associated with common microbial electron acceptor couples (Silva & Nitsche, 1995; Banaszak et al., 1999).

Environmental chemistry

The solubility of actinides in most natural waters is usually limited by CO_3^{2-} or OH^- compounds, although silica is important in uranium geochemistry as it can form moderately insoluble compounds such as uranophane (Choppin & Stout, 1989; Nguyen et al., 1992; Silva & Nitsche, 1995; Finch & Murakami, 1999). The actinides have limited solubility with many inorganic and organic ligands found in natural waters (e.g. $Pu(OH)_4$, log $K_{sp} \approx -54$; $Am(OH)_3$, log $K_{sp} \approx -26.6$), but free metal concentrations in the environment are generally so low that the solubility limit is not exceeded (Choppin & Stout, 1989). In this case, the radionuclides can be removed from the aqueous phase by chemisorption processes which can be irreversible (due to covalent bonding, hydrophobic bonding and hydrogen bridges) or reversible (due to longer-range Van de Waals forces and ion exchange reactions) (Silva & Nitsche, 1995). Sorption of dissolved actinide species on to mineral/soil matrices again varies for different oxidation states with the order following that for complexation and precipitation (4+ > 3+ ≈ 6+ > 5+).

Environmental measurements of solid solution distribution coefficients (K_d values) highlight general differences in the solubilities of the different transuranic elements. In the 'oxic' Irish Sea environment, average K_d values for a range of sediments are reported as 10^6, 10^5 and 10^4 l kg^{-1} for americium, plutonium and neptunium respectively. This reflects the range of predicted oxidation states for the different actinides within the Irish Sea: Am^{3+}; Pu^{4+} with limited access to PuO_2^+; and NpO_2^+ (Pentreath et al., 1986; Kershaw et al., 1992; Hursthouse & Livens, 1993). However, when radionuclides undergo sediment burial, microbially induced changes in the redox environment may lead to speciation changes and associated sorption changes for the actinides. For example, it has been reported that the adsorption ratio of neptunium increased by nearly four orders of magnitude in an anaerobic system (Eh = +80 mV) compared to an aerobic system (Eh = > +300 mV), suggesting a change from Np(V) to a less soluble Np(IV) species in anaerobic conditions (Lieser & Muhlenweg, 1988).

3. Sources of actinides in the natural environment

Uranium

Uranium is the most abundant of the naturally occurring actinides (Ac, Th, Pa and U) and ^{238}U is typically present at ppm levels in uncontaminated soils and sediments and at ppb levels in natural waters (Langmuir, 1997; Murphy & Shock, 1999). Seventy percent of economically important uranium ore deposits in the world are in one of three types: quartz-pebble conglomerates, unconformity type deposits and sandstone deposits (Nash et al., 1981), all of which generally contain uranium in the reduced U(IV) form. Microorganisms may have played a role in formation of the quartz pebble and sandstone hosted deposits (Miholic, 1952; Zaijic, 1969; Nash, 1981; Suzuki & Banfield, 1999). In the case of sandstone hosted deposits, it is thought that soluble UO_2^{2+} (as uranyl carbonate complexes) derived from leaching of granite or tuff was transported in ore forming fluids into the host sandstone soon after sedimentation, when permeability was relatively high. The soluble uranyl carbonate complexes were then reduced and precipitated as insoluble U(IV) minerals (such as uraninite and coffinite) in zones which are now associated with organic matter and pyrite. Precipitation has typically been attributed to 'indirect' inorganic U(VI) reduction by organic matter or hydrogen sulfide derived from sulfate-reducing bacteria (Zaijic, 1969; Nash et al., 1981; Nakashima et al., 1984). However, more recently it has been suggested that microorganisms may have had a direct role in the formation of these ore deposits (Pietzch et al., 1998 cited in Suzuki & Banfield, 1999) as bacteria can enzymatically reduce U(VI) (Lovley et al., 1991; see Chapter 7, this volume).

Uranium ore bodies have been worked to produce fuel for civilian power reactors and for weapons programmes. The mining process includes ore extraction via a number of different chemical and microbiological leaching techniques (Zaijic, 1969; Tuovinen & Kelly, 1974;

Eisenbud & Gesell, 1997; Beneš, 1998), with milling and chemical processing also used in order to produce 'yellow cake' (nominally U_3O_8). During these operations, large volumes of solid and liquid residues are produced. The solid residues (mill 'tailings') are sand-like in appearance and contain radioactive elements (including uranium and uranium-series radionuclides), as well as heavy metals, in low concentrations. Solid wastes from milling contain most of the radioactivity that was originally present in the uranium ore, with 70–85% of the total radioactivity, 50–100% ^{232}Th, and 93–100% ^{226}Ra retained (Beneš, 1998; Abdelouas et al., 1999). It is worth noting that much of the radioactivity associated with uranium tailings is in fact associated with uranium series radionuclides which were not extracted in processing of ores. ^{226}Ra is particularly problematic in these wastes as it emits intense alpha radiation, and forms the radioactive gas ^{222}Rn as a decay product (Krauskopf, 1988).

The disposal of tailings can, if improperly undertaken, pose a potential hazard to human health, for example by rainwater leaching of heavy metals or by wind erosion and dispersal of tailings materials (Beneš, 1998; Abdelouas et al., 1999). Most uranium is extracted during processing, but residual concentrations are typically 100–1400 mg kg^{-1} ^{238}U (Putnik, 1996; Delaney et al., 1998; Junghans & Helling, 1998); and pore-waters in contact with tailings and surface waters associated with processing can contain up to 85 ppm uranium (but are more typically between 0.3–10 ppm; Willett & Bond, 1995; Hagen & Jakubick, 1997; Delaney et al., 1998; Fernandes et al., 1998). These uranium concentrations are higher than average values for uncontaminated soils (1 ppm) and surface waters (0.1–500 ppb), see for example Langmuir (1997), Abdelouas et al. (1999) and Murphy & Shock (1999). The extent of operations associated with uranium mining and milling is highlighted by the fact that many countries have operational remediation programmes for old uranium tailings sites (Beneš, 1998; Abdelouas et al., 1999; IAEA, 1999). In the USA, it is estimated that more than 230 million tonnes of uranium tailings waste are stored at mill sites located in 10 states and, in Canada, tailings probably total of the order of 300 million tonnes (Abdelouas et al., 1999). In addition to uranium contamination associated with mining and milling operations, production of nuclear fuel for use in both civilian and military reactors and use of depleted uranium in military operations has led to the release of radioactive materials, including uranium, to the natural environment (US DoE, 1997; Francis & Dodge, 1998; Abdelouas et al., 1999; Mackenzie, 2001).

The transuranic elements

The elements beyond uranium, including the nuclides ^{237}Np and ^{239}Pu, exist naturally at vanishingly small levels in the Earth's crust (1×10^{-14} to 1×10^{-15} g kg^{-1}) due to production by neutron capture reactions with ^{238}U (Levine & Seaborg, 1951, cited in Ewing, 1999). Thus, contemporary concentrations of these elements in the environment are almost entirely due to their production for weapons manufacture or electricity generation. Transuranic elements enter the environment from three sources. Nuclear weapons testing has been responsible for by far the largest global input (Perkins & Thomas, 1980), followed by fuel reprocessing contamination, with accidental releases causing the smallest input (Pentreath, 1988).

Weapons fallout

The first nuclear device was detonated in 1945, and initiated a series of nuclear weapons tests by a number of nations (Perkins & Thomas, 1980). After the first test ban treaty was signed in 1963, atmospheric and surface tests were largely superseded by underground testing and it is estimated that a total of approximately 400 surface tests have been carried out (Facer, 1980). The majority of transuranic and other radionuclides injected to the atmosphere were generally introduced via detonation of thermonuclear devices by the USA and former Soviet Republic during 1961–62 (Perkins & Thomas, 1980). The nature of individual test conditions has an important effect on the extent to which transuranic material is dispersed in the environment (Facer, 1980). Detonations at the Earth's surface incorporate a large amount of debris into the nuclear blast and lead to localised fallout of the transuranic fraction (Pentreath, 1988). More widespread contamination has been caused by above ground tests which were of high enough yield to inject radionuclides into the stratosphere, producing a global dispersion of transuranic activity (Pentreath, 1988). The extent to which transuranic elements were dispersed from surface weapons tests is illustrated by plutonium data. It is estimated that over 14.8×10^{15} Bq (approximately 3800 kg) of $^{239,240}Pu$ were released to the natural environment by nuclear weapons testing, most of which (12.2×10^{15} Bq) was dispersed in global fallout (Pentreath, 1988). In the UK, $^{239,240}Pu$ from weapons fallout has been estimated to be 48 ± 7 Bq m^{-2} (Hardy et al., 1973) corresponding to a concentration of approximately 10^{-9} g kg^{-1}, 5–6 orders of magnitude higher than estimates of ambient concentrations associated with neutron capture in ^{238}U (see above).

Nuclear reprocessing

The global release of transuranic elements as a result of nuclear fuel reprocessing operations has been much smaller than that resulting from weapons tests (Pentreath, 1988; Choppin & Stout, 1989). However, releases from fuel reprocessing have produced localised but substantially enhanced levels of transuranic (and other) radionuclides (Sholkovitz, 1983; Pentreath, 1988; Myasoedov & Drozhko, 1998). There are a number of sites which contain globally significant quantities of radionuclides as a result of releases from reprocessing plants. These include the areas surrounding Sellafield in the UK, Cap de la Hague in France, the Hanford and Savannah River sites in the USA, and Mayak in Russia. Transuranic elements have been released from these facilities by a number of different mechanisms: (1) authorised discharges of low-level liquid wastes into coastal waters have occurred from Sellafield and Cap de la Hague (Pentreath et al., 1988; Gray et al., 1995); (2) leakage of high level nuclear waste tanks, authorised discharges of waste waters to the soil and leaching of near surface contaminants (US DoE, 1997; Banaszak et al., 1999) have allowed radionuclides to migrate into groundwater at the Hanford and Savannah River site; and (3) direct releases of radioactive wastes into the river Techa have occurred at the Mayak facility. Here releases have also arisen from the explosion of a radioactive waste tank in 1957 and the transfer of windborne particulate from the dry banks of lake which was acting as a radioactive reservoir (Akleyev & Lyubchansky, 1994; Myasoedov & Drozhko, 1998).

The scale of transuranic element releases from reprocessing facilities can be illustrated by considering the Sellafield reprocessing facility, which has made authorised discharges of low-level liquid wastes to the Irish Sea since 1951. The Sellafield site has a well documented discharge history, and has released in the order of 600×10^{12} Bq (approximately 200 kg) of ^{239,240}Pu to the local environment (Gray et al., 1995). This is significant on a global scale; indeed Sholkovitz (1983) described the site as 'presently the world's most important point source of plutonium'. Thus, numerous studies have reported elevated concentrations of transuranic radionuclides for sediments in the Irish Sea, with ^{239,240}Pu, ^{237}Np and ^{241}Am activities in the ranges 220–15,800; 0.63–13 and 250–29,300 Bq kg^{-1} respectively (Morris et al., 2000). Due to the releases of waste effluents from Sellafield to the Irish Sea, the area has been a focus for environmental studies on the behaviour of artificial radionuclides (e.g. Woodhead, 1999 and references therein).

Nuclear accidents

Prior to 1986, the major accidental contributor to global inventories of the transuranic elements was the stratospheric burn up of a plutonium powered satellite (SNAP-9A) which injected 0.63×10^{15} Bq ^{238}Pu into the upper stratosphere (Pentreath, 1988). This resulted in a three-fold increase in the ambient fallout concentration of ^{238}Pu (Hardy et al., 1973). In the spring of 1986, the Chernobyl accident in the Ukraine is estimated to have released 62×10^{12} Bq of ^{239,240}Pu (Pentreath, 1988). The percentage of the reactor core inventory of plutonium released was relatively low (approximately 3.5% compared to 20–40% for caesium radionuclides, and 100% for the noble gases) due to the refractory nature of the actinides (Eisenbud & Gesell, 1997), and plutonium contamination was mainly restricted to a relatively small fallout area.

The nuclear waste legacy

The proposed disposal of radioactive waste in mined repositories, and the storage and disposal of plutonium and uranium from weapons dismantlement and reprocessing operations, are further possible sources for actinide release to the environment (Silva & Nitsche, 1995). In the USA, approximately 3.5×10^4 tonnes of spent nuclear fuel is being stored to await disposal at the Yucca Mountain high level waste disposal facility, pending approval for operation of the site (Ewing, 1999). Globally, it has been estimated that approximately 2.3×10^5 tonnes of nuclear fuel would have been utilised in nuclear reactors by 2000 (Krauskopf, 1988). Currently, 17% of the world's electric power is generated by nuclear reactors in 31 nations and approximately 2000–3000 tonnes of spent nuclear fuel are generated in the US per year (Ewing, 1999). In addition to spent fuel, reprocessing for both military and civilian reasons has generated large volumes of high level nuclear wastes in a number of different forms. In the UK reprocessing of spent nuclear fuel from civilian and defence operations has generated approximately 1600 m^3 high level waste (HLW) and 6.1×10^4 m^3 intermediate level waste (ILW) which are stored awaiting development of a national strategy for ILW and HLW radioactive waste management (HoL, 1999). In the US, weapons production is estimated to have generated approximately 4×10^5 m^3 of high level nuclear waste, the majority of which is stored in

Table 2
Important actinide elements in two different high level wastes (Krauskopf, 1988)

Nuclide	Half-life (y)	Mass in a spent fuel assembly from a pressurised water reactor (g) Time after discharge		Mass in a borosilicate glass canister made from reprocessing waste (g) Time after discharge	
		10 years	1000 years	10 Years	1000 Years
^{234}U	2.45×10^5	88	150	2.9	16
^{238}U	4.47×10^9	440 000	440 000	9900	9900
^{237}Np	2.14×10^6	210	660	930	1000
^{238}Pu	8.87×10^1	60	260	13	0.014
^{239}Pu	2.41×10^4	2300	2300	53	68
^{240}Pu	6.56×10^3	1100	1000	36	56
^{242}Pu	3.76×10^5	210	210	4.8	5.0
^{241}Am	4.33×10^2	230	120	120	26
^{243}Am	7.38×10^3	40	36	180	160

a heterogeneous form in waste tanks at the Hanford site in Washington State (Krauskopf, 1988; Heaston et al., 1999). In addition, it is estimated that nuclear weapons production has led to radionuclide contamination of approximately 50×10^6 m^3 of soil waste, and greater than 800×10^6 m^3 of groundwaters, at numerous sites throughout the US (US DoE, 1997). The global picture for reprocessing wastes is complex and poorly-defined for a variety of reasons: different countries use different reprocessing technologies; there have been changes in reprocessing technology with time; and some countries choose to reprocess civilian fuel. The secrecy associated with weapons production also contributes to difficulties in quantifying the volumes and activities of wastes.

The half-lives and estimates of the amounts of the actinides present in spent nuclear fuel, and in high level waste that has been vitrified ready for storage and disposal, are given in Table 2 (Krauskopf, 1988). The significance of the actinide contribution to total radioactivity in spent fuel will increase with time as shorter-lived fission product isotopes decay. Indeed, a few hundred years after enriched uranium fuel is discharged from boiling or pressurised water reactors, the actinide elements uranium, neptunium, plutonium and americium will account for more than 98% of the radioactivity associated with the spent fuel (Silva & Nitsche, 1995).

Summary

From the above discussion, it can be seen that atmospheric nuclear weapons testing has resulted in widespread dispersal of transuranic elements throughout the surface environment, in freshwater, groundwater, river, estuary and seawater. In addition, the manufacture of nuclear weapons and nuclear fuels has led to localised contamination with elevated levels of transuranic elements. Uranium is present in the natural environment as a primordial radionuclide, and has been technologically enhanced at specific sites throughout the world due to its role in the nuclear fuel cycle and in the production of plutonium for nuclear weapons. Escape of the actinide elements has effected their entry into natural biogeochemical cycles. The bioavailability of these elements will be determined in the first instance by their phase associations on release (whether aqueous or particulate) and their subsequent phase changes during transportation, which control the extent of their removal into sediments. Once they are in sediments, early diagenetic reactions affect whether they are retained or remobilised back into the water column, and thus burial in sediments is not necessarily an ultimate sink. In the following sections, we shall discuss the cycling of the actinides, first on a macroscale, by examining their global cycling and fluxes, and then on a microscale, by examining their localised biogeochemical cycling.

4. Global cycling processes

Background

Global cycling processes govern the distribution of elements throughout rivers, oceans and sediments, and also redistribute long-lived radionuclides in the surface environment. It is therefore important to examine global cycling processes before focussing on more localised biogeochemical cycles although, because of the generally low environmental concentrations of the transuranic elements, uranium is the only relevant element which can be studied on this scale. Global cycling includes the physical, chemical and biological processes that affect the release, transport and deposition of uranium, from the weathering of minerals to the uptake into sinks. Research into transport and removal processes has suggested a close association between uranium and iron oxides, which have been found to be efficient sorbents for U(VI), both in laboratory and natural systems (Ho & Dourn, 1985; Hsi & Langmuir, 1985; Van der Weijden et al., 1985; Sagert et al., 1989; Sheppard & Thibault, 1992; Lienert et al., 1994; Prikryl et al., 1994; Martínez-Aguirre & Perianez, 1999; Moyes et al., 2000). However, many parts of the global cycle of uranium are poorly quantified, whereas the global iron cycle is relatively well known. Hence, a valuable insight into the global cycling of uranium may be gained by drawing analogies with the global iron cycle. Therefore, in this section we adopt this approach to the global cycling of uranium as the most widely distributed actinide element in the natural environment.

The global iron cycle

A simple model of the global iron cycle is shown in Fig. 2. The fluxes and reservoirs in this cycle are expressed in terms of total iron, and three key features are readily apparent. First, the riverine total iron particulate flux is several orders of magnitude larger than the riverine dissolved iron flux. The riverine particulate data is based on new analyses of 25 rivers (Poulton & Raiswell, 2000) supplemented by data from Canfield (1997) for a further 9 rivers. Together these 34 rivers represent approximately 23% of the global riverine particulate flux, based on the 20,000 Tg y^{-1} estimate of river sediment transport by Milliman & Syvitski (1992) for pre-dam discharges. These rivers give good coverage of

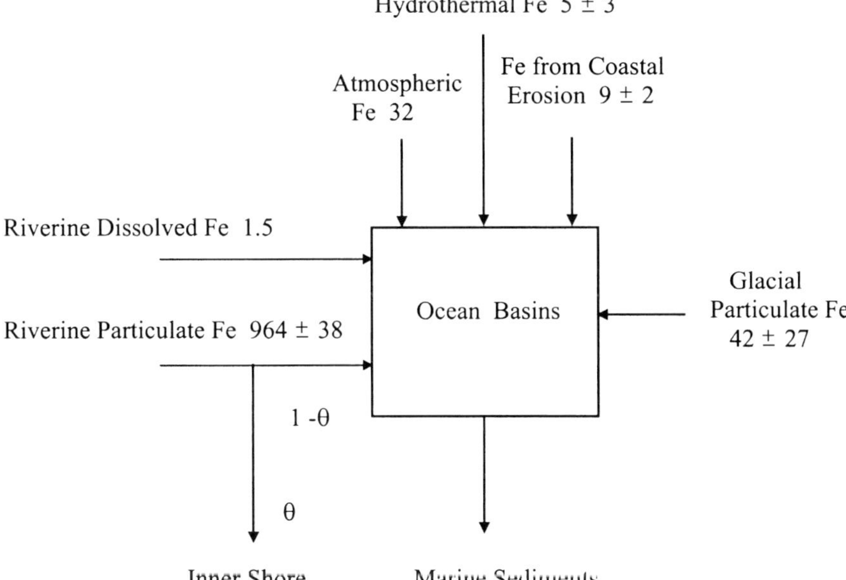

Fig. 2. The global total iron cycle, fluxes in units of 10^{12} g y^{-1}. The global rate of deposition of marine sediments is too poorly known to derive a useful removal rate for total iron into sediments. Total Fe in riverine particulates = 4.8 ± 0.2%, total Fe of marine sediments = 4.0 ± 1.0%.

most of the important sediment-discharging regions (46% of the sediment flux from North America, 73% from South America, 45% from Europe, 36% from Asia, 19% from Africa and 5% from Australia) although there are, unfortunately, no data from Central America or the Oceanic Islands.

The second key feature of Fig. 2 is that the riverine particulate flux predominates over the inputs from other sources, which are principally derived from glacial and coastal erosion, atmospheric dust and hydrothermal activity. None of these minor fluxes are well quantified but all are orders of magnitude lower than the riverine particulate flux. The third feature of Fig. 2 is the difference between the total iron contents of the riverine particulates and marine sediments (both corrected for carbonate or organic contents of greater than 1%). Thus, the differences in total iron contents cannot be reconciled by the addition of iron-poor biogenic material to the marine sediments. Poulton and Raiswell (unpublished data) propose that a particulate total iron balance is maintained because a fraction θ of the riverine particulate flux (which is enriched in oxide iron, see later) is deposited in a variety of environments such as floodplains, estuaries and salt marshes, collectively termed the innershore. This has significant implications for any trace metals associated with the riverine particulate flux as these may be accumulated in innershore environments by deposition of particulate iron.

Quantifying θ is difficult for a variety of reasons. Riverine particulate fluxes are subject to substantial temporal and spatial variability (Milliman & Syvitiski, 1992). Measuring sediment fluxes during flood conditions is often difficult or even impossible, although as much as half the annual sediment load may pass a measuring station in only a few days (Milliman & Meade, 1983). Furthermore, major flood events (100 or 1000 year floods) may remove substantial proportions of the sediment stored in the innershore reservoirs. Thus, a single snapshot measurement of discharge fluxes will approximate only crudely to the long-term sediment flux into the ocean basins, which maintains a steady state on a geological timescale. Our approach to this problem is to view the innershore as a transient reservoir, and to accept that considerable variation in θ may occur when reconciling instantaneous discharge fluxes with sediment accumulation rates (which integrate over geological timescales). Note that literature estimates for the global rate of marine sediment deposition (m_S) produce an average of 10,000 $\pm$ 3000 Tg y^{-1} and are both too few and too variable to constrain θ by comparison with the riverine flux of 20,000 Tg y^{-1} (the errors on which are also unknown).

Greater insight into the global iron cycle is obtained by partitioning the total iron flux into various operationally defined fractions. A dithionite extraction (Raiswell et al., 1994) can be used to measure the iron present as iron oxides so that the simple iron cycle can be redrawn in terms of the dithionite-extractable iron, which is here referred to as Fe_{HR} or highly reactive iron. This iron cycle (Fig. 3) shows the same key features as the total iron cycle (FeT), but note that the difference of 0.82% Fe between the total iron contents of the riverine particulates (4.82%) and marine sediments (4.00%) is roughly matched by the difference of 1.04% between the highly reactive iron contents of the riverine (2.08%) and marine (1.04%) sediments. Clearly the losses of total iron to the innershore are largely attributable to the removal of iron oxides into this reservoir. The removal of iron in the estuarine environment has been well documented on the scale of individual estuaries (Boyle et al., 1974; Holliday & Liss, 1976; Sholkovitz, 1976; Boyle et al., 1977; Sholkovitz et al., 1978), but has not previously been quantified on a global scale. Once again it is difficult to obtain a precise measure of θ from this iron model, although the ratio Fe_{HR}/FeT can be used in a mass balance approach. Using ratios is preferable to concentrations because the riverine Fe_{HR} covaries with FeT and thus there are lower errors in the ratios than in concentration data (see Fig. 3). We can therefore write

$$m_R \times R_R = \theta m_R \times R_I + (1 - \theta) \cdot m_R \times R_S$$

where m_R is the annual mass of sediment discharged globally by rivers, and R_R, R_I and R_S are respectively the values of the ratio Fe_{HR}/FeT in riverine particulates, innershore sediments and marine sediments. Using values from Fig. 3 gives

$$(0.432 \pm 0.025) = \theta\,[R_I - (0.264 \pm 0.08)] + (0.264 \pm 0.08)$$

$$\theta\,[R_I - (0.264 \pm 0.08)] = (0.168 \pm 0.083)$$

However, analyses of Fe_{HR}/FeT for innershore environments produce values of R_I of around 0.6 which suggests a minimum of 0.2 for θ. Values of R_I will be low (and thus θ will be overestimated) because significant fractions of marine sediment are carried back into estuaries and stored in innershore reservoirs (Regnier & Wollast, 1993), even if only temporarily (as discussed for the riverine particulate flux earlier). The data are sparse

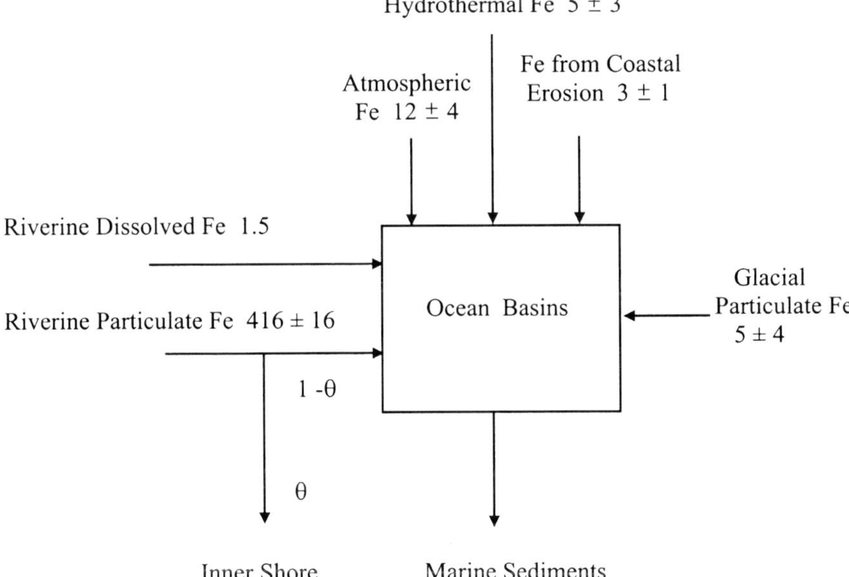

Fig. 3. The global highly reactive iron cycle, fluxes in units of 10^{12} g y^{-1}. The global rate of deposition of marine sediments is too poorly known to derive useful removal rates for any iron species into marine sediments. The highly reactive iron content of riverine particulates in 2.08 ± 0.08%, and the ratio of highly reactive iron to total iron is 0.432 ± 0.025. The highly reactive iron content of marine sediments is 1.04 ± 0.45%, and the ratio of highly reactive iron to total iron is 0.264 ± 0.083.

and geographically limited in distribution, but clearly suggest that significant fractions of riverine sediment are trapped in innershore depositional environments.

There have been few previous attempts to estimate the proportions of the riverine particulate load which are trapped prior to entry into the marine environment, although it has frequently been suggested that substantial proportions may be stored in areas such as floodplains (Milliman & Meade, 1983; Milliman & Syvitiski, 1992; Hay 1998). Probably the most clearly defined estimate of θ has been made by Allison et al. (1998), who have measured floodplain sediment accumulation rates in the Brahmaputra. Extrapolating over the whole basin, 39–71% of the river flux appears to be trapped prior to entry into the marine environment. This conclusion is consistent with the suggestion of Milliman & Syvitski (1992) that a significant proportion of the riverine particulate load of high-yield rivers in tectonically-active settings is trapped landward of the ocean. Furthermore, Halim (1991) reported that 85% of the Nile suspended sediment reached the Mediterranean prior to construction of the Aswan dam. Considerable spatial variations in θ probably occur in different river systems, and considerable temporal variations may occur within a single river system in the periods prior to, and after, major flood events. However, the possibility that significant uranium removal occurs into salt marshes suggests that estimation of θ, and the partition of sediment between the different innershore reservoirs, would be valuable for those rivers which host salt marsh environments.

Fluxes, sources and sinks for uranium

The transport and deposition of uranium derived from continental weathering and supplied via rivers to the ocean basins are poorly quantified. Two recent studies have produced estimates of the dissolved riverine flux of uranium which are in reasonably close agreement (3.7 to 4.8 $\times$ 10^7 mol y^{-1} by Palmer & Edmond, 1993; 2.1 to 3.3 $\times$ 10^7 mol y^{-1} by Windom et al., 2000). However, there have been no recent estimates of the global riverine particulate uranium flux to the oceans, although Martin & Meybeck (1979) give a median value of 3 ppm uranium for riverine particulates based on analyses of six river samples. By contrast, there have been many studies which suggest that the ocean is maintained in a steady state by the extraction of dissolved uranium from seawater followed by fixation into reducing marine sediments (Cochran et al., 1986; Anderson, 1987; Anderson et al., 1989a, b; Barnes & Cochran, 1990, 1991, 1993) or into hydrothermal ocean crust systems (Edmond et al., 1979; Klinkhammer & Palmer, 1991). More recently, salt marshes have also been implicated as another significant global sink for the removal of the dissolved uranium supplied by rivers in a similar manner as suggested earlier for Fe_{HR} (Church et al., 1996; Windom et al., 2000).

The removal of the dissolved riverine uranium flux into sediments is necessary in order to maintain a steady state ocean. One significant removal mechanism (Palmer & Edmond, 1993) is by fixation in reducing sediments, in estuaries, on the continental shelf and in deep sea hemipelagic environments. U(VI) is believed to diffuse into sediments where the presence of organic matter produces reducing conditions which transform the soluble oxidised U(VI) species into insoluble U(IV). Microbial processes may directly reduce U(VI) to U(IV) (Lovley et al., 1991; Lovley & Phillips, 1992; Barnes & Cochran, 1993) or may be involved only indirectly (by supplying reducing agents such as H_2S; Anderson et al., 1989a; Klinkhammer & Palmer, 1991). Another possibility is that uranium is removed from seawater during high temperature hydrothermal circulation at mid-ocean ridges. Edmond et al. (1979) examined hot springs (at the Galapagos Spreading Centre) which resulted from mixing of high temperature reduced hydrothermal fluids with cold oxygenated seawater. These springs were depleted in uranium, probably due to reduction of the predominant dissolved U(VI) species (a uranyl carbonate complex) followed by precipitation of insoluble U(IV).

Sarin & Church (1994) initially showed that large subtidal estuaries are seasonal summer sinks for uranium at low salinities. Subsequently, Church et al. (1996) found that intertidal salt marshes were much larger sinks for uranium at all salinities and two potential mechanisms of uranium removal were identified. Thus, uranium could be scavenged from creek waters by humic acids and iron oxides as these were flocculated during tidal mixing, or uranium might be removed by iron oxides when the marsh sediments were flooded (when the low pH produced by earlier subaerial sulfide oxidation assisted in the destabilisation of the dominant uranyl carbonate complex).

However, the global riverine flux of uranium is dominated by the particulate supply, with the particulate uranium bound to hydrous metal oxides (Payne & Waite, 1991; Plater et al., 1992; Lienert et al., 1994; Waite et al., 1994; Swarzenski et al., 1995) and/or organic matter (Swarzenski & McKee, 1998; Zielinski & Meier, 1988). Significant proportions of the riverine uranium flux may also be transported as colloids, although these have proved

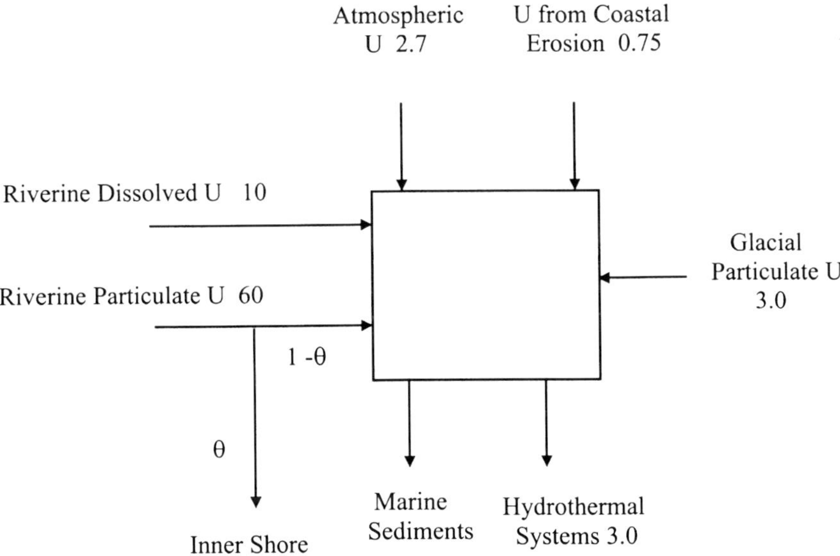

Fig. 4. The global uranium cycle, with fluxes in units of 10^9 g y^{-1}. Marine sediments contain typically 2–3 ppm uranium, but the rate of deposition of marine sediments is too poorly known to derive a useful uranium removal rate into marine sediments.

difficult to characterise (e.g. Swarzenski et al., 1995).

Global uranium cycling

A global uranium cycle (Fig. 4) can be constructed using a similar approach to that for the global iron cycle. The dissolved riverine uranium flux of ~4×10^7 mol y^{-1} (10×10^9 g y^{-1}) lies roughly midway between the estimates above (of Palmer & Edmond 1993 and Windom et al., 2000). The riverine particulate uranium flux (60×10^9 g y^{-1}) has been estimated from the data in Table 3, which is based on a total sediment discharge flux of 2571 Tg y^{-1}, about 13% of the global riverine particulate flux. These data produce a discharge-weighted particulate uranium concentration of 3 ppm.

The remaining fluxes are estimated as follows. Duce et al. (1991) have quantified the deposition of atmospheric mineral aerosol into the ocean basins as 910 Tg y^{-1}. There are no uranium data given for mineral aerosol, but it is commonly assumed that the elemental composition of mineral aerosol reflects crustal abundance (3.0 ppm for uranium; Chester, 1990). Using this value produces a particulate uranium flux of 2.7×10^9 g y^{-1} from the atmosphere (< 5% of the riverine particulate flux). The solubilities of trace elements in mineral aerosols are often large but can vary widely (e.g. Fe 1–50%, trace transition elements typically 20–80%; Duce et al., 1991) but even a high seawater solubility, such as 40%, for uranium in aerosols would produce a dissolved flux of 1×10^{-9} g y^{-1}, which is an order of magnitude less than that from rivers irrespective of the solubility. Atmospheric uranium fluxes are clearly negligible compared to riverine fluxes.

Table 3
Riverine particulate uranium contents

River	Uranium content (ppm)	Sediment discharge ($\times 10^{12}$ g)	Source
Amazon	2.5	1200	Martin & Meybeck (1979)
	2.5		Moore (1967)
	3.55		Allegre et al. (1996)
	3.0		Swarzenski et al. (1995)
Congo	3.0	43	Martin & Meybeck (1979)
	2.5		Allegre et al. (1996)
	2.9		Martin et al. (1978)
Brahmaputra	2.8	540	Martin & Meybeck (1979)
Garonne	3.6	2.2	Martin & Meybeck (1979)
Mekong	5.8	160	Martin & Meybeck (1979)
Orinoco	4.5	150	Martin & Meybeck (1979)
Gironde	3.0	2.0	Martin et al. (1978)
Charante	2.5	0.1	Martin et al. (1978)
Narbada	1.4	125	Borole et al. (1982)
Tapti	1.4		Borole et al. (1982)
Mississippi	2.0	400	Moore (1967)
Seine	2.0	1.1	Roy et al. (1999)
Savannah	2.0	0.03	Meade & Windom (1982)
Ogeechee	2.6	0.06	Meade & Windom (1982)

Estimates of the glacial particulate flux to the ocean basins are poorly constrained. Hay (1998) has reviewed previous estimates to suggest a flux of 800–5000 Tg y^{-1}, but prefers a value towards the lower end of this range. A value of 1000 Tg y^{-1}, at 3.0 ppm uranium, produces a particulate glacial flux of 3×10^9 g y^{-1} of uranium. Even if this flux were completely soluble (which is unlikely), it is still significantly smaller than the riverine dissolved flux. Coastal erosion is believed to supply approximately 250 Tg y^{-1} of sediment to the ocean basins (Garrels & Mackenzie, 1974) and a composition of 3 ppm uranium suggests a coastal erosion flux of 0.75×10^9 g y^{-1} of uranium. Again, the maximum contribution from this flux (if all the uranium is soluble) is small compared to the riverine dissolved uranium flux.

Finally, the role of hydrothermal activity is both complex and poorly quantified. Barnes & Cochran (1990) suggest that the alteration of oceanic crust may remove 3×10^9 g y^{-1} of dissolved uranium from seawater. Consistent with this, Elderfield & Schlutz (1996) show hydrothermal fluxes at ridge flanks as removing 2×10^9 g y^{-1} uranium from the oceans (see also Klinkhammer & Palmer, 1991), and the iron oxides formed in hydrothermal plumes also effectively scavenge uranium from seawater at a rate estimated as approximately 10×10^9 g y^{-1}. These removal rates are comparable to the riverine dissolved fluxes of uranium, and clearly the hydrothermal-ocean crust system plays an important role in maintaining a steady state ocean.

No further useful progress can be made from Fig. 4 in the absence of data for the unquantified fluxes, especially the removal of sediments to the innershore, and to the oceans. However, the composition of deep sea sediments (2.6 ± 1.1 ppm uranium from Mo et al., 1973; 2.0 ppm uranium from Chester, 1990) is probably not significantly different from the uranium levels found in riverine particulates. On a global basis, this implies that there is little uptake of dissolved uranium by deep sea sediments, which is consistent with their generally oxic conditions. Although uranium is fractionated into some deep sea deposits, notably iron and manganese nodules and phosphorites, the rates of accumulation of these sediments are too slow for uranium removal to be quantitatively significant (see Barnes & Cochran, 1990). Thus, in addition to the detrital flux of uranium received from riverine particulates, the innershore and continental margin sediments must be responsible for removing dissolved uranium from seawater, as a consequence of the reducing conditions established during microbial diagenesis.

We will continue by making estimates of the potential global role of suspended sediments in transporting adsorbed uranium (which is potentially the most bioavailable phase), by applying various K_d approaches to our riverine uranium and iron data. K_d values are defined as

$$K_d \text{ (units of l kg}^{-1}\text{)} = \frac{\text{(mass U adsorbed)/(mass sorbent)}}{\text{dissolved U concentration}} \tag{1}$$

Adsorption is a function of salinity, pH and fluid composition, and K_d values defined in terms of concentrations (as above) are sensitive to all these variables. Sholkovitz (1983) has drawn attention to a number of difficulties inherent to the use of K_d values; notably that they integrate the adsorption effects of particles with many different characteristics, capable of interacting with uranium in many different ways. Particles in natural systems may participate in a variety of adsorption, desorption, precipitation, ion-exchange and bioaccumulation reactions (all of which may also vary temporally and spatially). K_d values are also affected by changes in the oxidation state of the adsorbing ion, and by the presence of complexing ions. For these reasons, K_d values may not be regarded as equivalent to equilibrium constants and are not capable of accurately predicting the fractionation of uranium between solid and aqueous phases in a natural system.

Despite these problems, K_d values can provide a useful empirical picture of adsorption behaviour. Typical K_d values for uranium range from 10^4 to 10^7 l kg^{-1} (Langmuir, 1978), depending on adsorbate mineralogy and solution composition. This range of K_d can be used in the following expression (Sholkovitz, 1983), derived by recasting equation (1);

$$\%\text{ U in suspended fraction} = \frac{100}{1 + 1/(K_d \times S)} \tag{2}$$

where S = the suspended sediment concentration in kg l^{-1}. Our global sediment discharge of 20,000 Tg y^{-1} is carried by a water discharge flux of 3.2×10^{16} l y^{-1} (Garrels & Mackenzie, 1974), which indicates a global mean sediment concentration of 0.63 g l^{-1}.

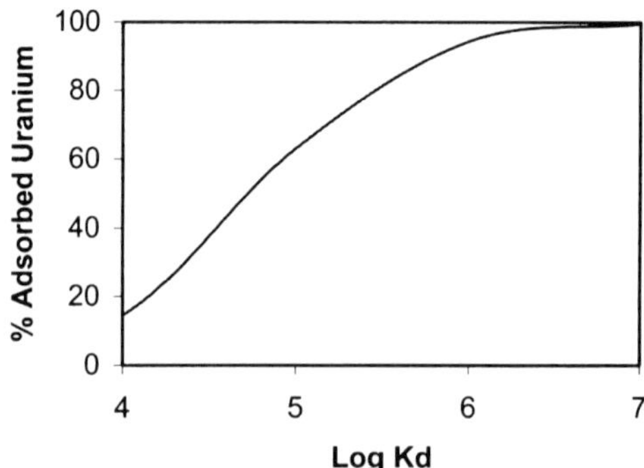

Fig. 5. Variations in the proportion of adsorbed U with K_d.

Figure 5 shows the variations in the proportion of uranium associated with this suspended sediment load for different K_d values. It is very difficult to chose a K_d value appropriate for globally-averaged riverine suspended sediment, but adsorbed uranium is significant even for the lowest K_d values. Consistent with this, uranium in rivers is predominantly present in the particulate, rather than the dissolved, load (Fig. 4).

It is possible to improve on this rather simplistic picture of uranium behaviour by making a more detailed examination of the role of iron oxides on the adsorption of uranium, and Table 4 shows literature-derived K_d values for the adsorption of uranium on to iron oxides. The values in Table 4 have been selected to approximate to the conditions found in the riverine system, in terms of concentrations of uranium, pH and presence of complexing ions. The range of values is quite large (3×10^4 to 3×10^6 l kg^{-1}) with the lower values probably the most realistic, as these reflect the adsorption behaviour in the common presence of carbonate complexes. These K_d values can be used with the data in Table 5 for mean sediment concentrations, and mean FeOOH contents (mostly less that 2×10^{-5} kg l^{-1}), of riverine particulates. The variations in the proportion of adsorbed uranium with increasing FeOOH contents are shown in Fig. 6, for a range of K_d values. Clearly even typical iron oxide contents are sufficient to adsorb a significant proportion of uranium preferentially provided $K_d > 10^5$ l kg^{-1}.

Using equation (1), the estimates of the mean dissolved uranium concentration in rivers (2×10^7 g l^{-1}; Palmer & Edmond, 1993), mean suspended sediment concentrations (0.63

Table 4

K_d values for the fractionation of uranium between solid and solution

K_d (l/kg)	Conditions	Reference
2×10^6	0.01 M NaCl, 25°C	Ames et al. (1983)
3×10^4	0.01 M $NaHCO_3$, 25°C	Ames et al. (1983)
1.1×10^6 to 2.7×10^6	Not specified	Langmuir (1978)

Table 5

Riverine fluxes of dissolved uranium and concentrations and fluxes of suspended sediment and sediment highly reactive iron content

River	Flow (10^{12} l yr^{-1})	Dissolved U (ppb)	Sediment flux (10^{12} g yr^{-1})	River sediment Fe_{HR} (%)	River sediment Fe_{HR} flux (10^{10} g yr^{-1})	Conc. of river sediment (g l^{-1})	Conc. of river sediment FeOOH ($\times 10^{-2}$ g l^{-1})
Amazon	6930	33.3	1200	2.6	3120	0.17	0.69
Changjiang	900	452	480	1.86	893	0.53	1.6
Brahmaputra	603	1690	540	0.62	335	0.90	0.88
Mississippi	530	309	120	1.27	152	0.21	0.42
Mekong	470	64.3	160	3.48	557	0.34	1.9
Yukon	195	543	60	2.57	154	0.31	1.3
Rhine	91	438	0.72	1.42	1.02	0.008	0.02
Danube	36	1083	67	2.03	136	0.78	2.5
Rhone	55	1407	31	1.61	49.9	0.56	1.4
Nile	54	14.3	0	2.40	0	0	0
Huanghe	49	4855	1100	0.96	1056	22	33.6
Seine	17	631	1.1	1.02	1.12	0.06	0.10
Brazos	10	1059	16	0.72	11.5	1.6	1.9
Rio Grande	2	219	0.8	0.68	0.54	0.4	0.27

Sediment discharge data from Miliman & Syvitski (1992) for present-day post-dam fluxes, dissolved U flux and river flow data from Palmer & Edmond (1993), Fe_{HR} data from Poulton & Raiswell (2000).

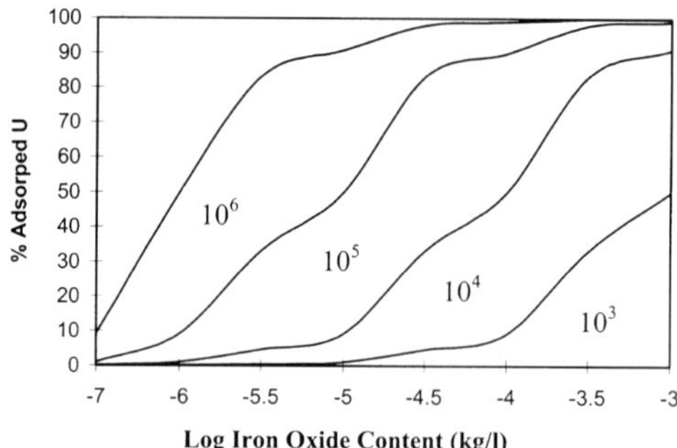

Fig. 6. Variation in percentage adsorbed uranium with iron oxide content, and for varying K_d values. Mean riverine iron oxide contents rarely exceed 2×10^{-5} kg l^{-1}.

g l^{-1}; see above) and a mean riverine suspended sediment concentration of iron from Table 5 (1.6% Fe or 2.6% FeOOH), we can calculate that adsorbed uranium concentrations lie between 0.03 ppm ($K_d = 10^4$) and 3.2 ppm ($K_d = 10^6$). As mentioned above, values towards the lower end of this range are more probable, and in this case adsorbed uranium would constitute a small but significant fraction of the 3 ppm total uranium (see Table 3).

There are very few data that can be used to examine the relationships between adsorbate concentrations, adsorbed uranium contents and dissolved uranium concentrations in rivers. However, different workers have reported on the dissolved uranium concentrations (Palmer & Edmond, 1993) and particulate iron concentrations (Poulton & Raiswell, 2000) in the same rivers, and Table 5 shows the relevant physical and chemical characteristics. Palmer & Edmond (1993) did not measure sediment discharge for these rivers at the time of sampling (although the flow state was defined), and so we have approximated for these data sets by using mean post-dam sediment discharge data (Milliman & Syvitski, 1992). Ignoring the Huanghe data, the only significant correlation between the variables in Table 5 is between dissolved uranium concentrations and suspended sediment concentrations. Figure 7 shows the straight line relationship ($r = 0.66$, significant at the $< 0.1\%$ level). The Huanghe data point roughly continues this trend but lies so far outside the ranges of the remaining data as to bias the statistics. The Huanghe is highly unusual in carrying exceedingly large concentrations of suspended sediment derived from easily-eroded loess (Poulton & Raiswell, 2000). This straight line strongly suggests that there is some type of quasi-equilibrium relationship between the dissolved uranium and the sediment load. The absence of any relationship between the iron oxide contents and the dissolved uranium concentrations suggests that concentrations adsorbed on iron oxides alone may be too small (because K_d values are low in carbonate-bearing waters) to exert any detectable control on dissolved concentrations. A detailed study of adsorbed uranium concentrations in riverine particulates (especially organics) would be able to clarify the role of solid phase adsorption.

Finally, we can attempt to place the estimates of salt marsh removal by Windom et al. (2000) within the context of a global sediment budget. Windom et al. (2000) use an estimate of the areal extent of salt marshes as 3.8×10^5 km^2 from Woodwell et al. (1973). It is difficult to estimate the mean rate of sediment accumulation in salt marshes, but Jickells &

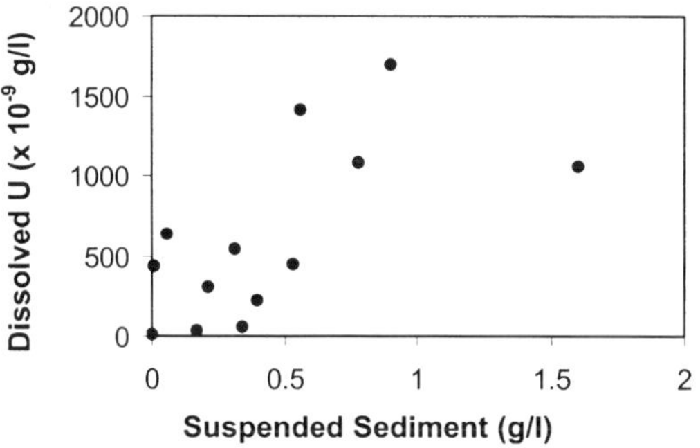

Fig. 7. Variations in dissolved uranium concentrations from Palmer & Edmond (1993) with river suspended sediment contents (sediment discharge data from Millimand & Syvitski, 1992).

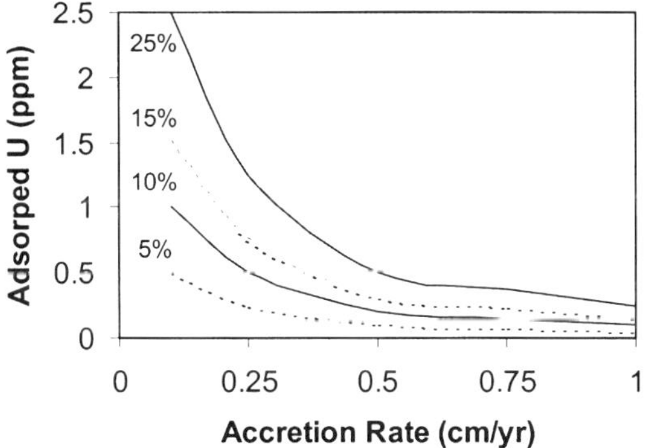

Fig. 8. Variations in the concentrations of uranium in salt marshes with accretion rate for different percent removal rates of the riverine dissolved uranium flux. Accretion rates assume 50% porosity, density 2700 kg m^{-3}, and a salt marsh area of 3.8 × 10^5 km^3. Mean accretion rates of 0.1–0.2 cm y^{-1} correspond to removal of 5-10% of the riverine particulate load.

Rae (1997) suggest that maximum marsh vertical accretion rates appear to be 1–2 cm y^{-1}, which are similar to maximum predicted rates of sea-level rise.

The accretion rate data are used in conjunction with the areal estimate in Fig. 8, but the highest vertical accretion rates produce sediment masses which seem improbably large in relation to the best estimates of the riverine particulate load (2 × 10^{16} g y^{-1}; see earlier) and mean accretion rates of 0.1–0.2 cm y^{-1} of compacted sediment (roughly equivalent to 5–10% of the riverine sediment flux) represent a more reasonable upper limit. Figure 8 shows that an accretion rate of 0.1 cm y^{-1} would produce salt marsh sediments with an adsorbed uranium content of 0.5 ppm if 5% of the dissolved riverine flux of uranium were removed, and 2.5 ppm uranium if 25% of the dissolved riverine flux were removed. Note that these adsorbed uranium concentrations would be added to the background level of 3 ppm present in riverine particulates and derived from continental weathering (Table 3). Church et al. (1981) reported uranium contents of 2.7 ± 0.9 ppm for the Delaware salt marshes, but their maximum uranium concentrations of 5 ppm might indicate the presence of adsorbed uranium at levels consistent with significant removal of the riverine flux, when judged on a global basis. There is clearly considerable scope to refine these estimates via improved data for the sediment contents and uranium concentrations of the innershore reservoirs. This should be an important focus for further study.

Summary

Overall, this attempt to link the global iron and uranium cycles has highlighted several areas where further research is required. First, the global behaviour of both elements would be significantly improved by quantitative studies of riverine sediment partitioning through estuaries and into the ocean basins. Second, such studies should also utilise chemical extractions to identify the phase associations of each metal in the riverine particulate flux, and in the different innershore reservoirs. The present discussion also suggests that such studies could be usefully extended to other radionuclides with high K_d values (e.g. Am 2×10^6 l kg^{-1}, Pu 10^5 l kg^{-1}; Morse & Choppin, 1991), which favour adsorption on sediments and storage in innershore reservoirs. Thirdly, the K_d approach can only provide a very generalised picture of element partitioning which is subject to considerable between-site variations. This is a fundamental weakness for elements such as uranium where the K_d values may vary significantly with speciation and ionic strength. Surface complexation models offer a promising way forward but are not yet at a level of understanding which encourages quantitative extrapolation to natural systems.

Storage in innershore reservoirs may be transient on the geological time scale, but not on the much shorter ones of surface biogeochemical processes. Thus, the principal threat arising from storage in the innershore lies in the potential impact on wildlife and human activities. These reservoirs are important nursery grounds for fish and invertebrates, and are widely used for food supply and recreation. Radionuclides with phase associations which favour fractionation into the innershore are thus stored in a bioavailable form in a biologically active environment. The following section examines the biogeochemical cycling of the actinides in innershore environments, along with riverine and marine systems, to complete this picture of actinide cycling.

5. The influence of microbial processes on actinide behaviour in the natural environment

Seasonal cycling of the actinide elements

A number of studies have reported seasonal cycling of actinide elements and have suggested that microbial activity may be a factor in controlling actinide solubility. Mechanisms include: redox reactions between the more soluble 'oxidised' MO_2^+ and MO_2^{2+} species and the particle reactive 'reduced' M^{3+} and M^{4+} species; indirect redox reactions whereby seasonal iron/manganese oxyhydroxide dissolution may cause the dissolution of associated actinide elements; and scavenging/release of actinides associated with growth/decomposition of microbial 'biomass'. A number of studies have shown that solution phase concentrations of the actinide elements are typically up to 6 orders of magnitude lower than solid phase activities. Thus, changes in solution concentration of the actinides provide a sensitive approach for examining mechanisms which change the solubility of these radionuclides (Sholkovitz, 1983; Klinkhammer & Palmer, 1991; Livens et al., 1994; Keith-Roach et al., 2000). However, it is important to note that the measurement of 'ultra trace' levels of transuranic elements (and of uranium) in natural waters is difficult. Uranium is present at 0.1–500 $\times$ 10^{-6} g l^{-1} in natural waters, and the transuranic elements are present at much lower concentrations (2 $\times$ 10^{-15} to 3 $\times$ 10^{-11} g l^{-1}). Typical minimum sample volumes for plutonium pore-water analyses are of the order of tens to hundreds of litres, and only in relatively contaminated sites can sample volumes be reduced to perhaps 200–300 ml (Livens et al., 1994). These sampling difficulties have deterred environmental studies, and there are only a relatively small number of such studies which focus on seasonal cycling of the actinide elements.

Uranium

In rivers, there is evidence that uranium may be seasonally influenced by microbial cycling (Lienert et al., 1994). Uranium infiltration from the River Glatt, Switzerland into a hydraulically connected saturated aquifer was examined over a five year period. The uranium in the river was thought to be derived from Lake Greifen and/or its catchment, a 'very eutrophic' lake and uranium concentrations were determined at monthly intervals in river waters and in four groundwater sampling wells. The uranium concentrations in the river varied seasonally with a maximum in summer and a decline in concentrations over the winter months. A number of mechanisms were proposed for the seasonal variation: (1) microbially-mediated oxidative decomposition of aquatic biota (containing uranium) in high productivity summer lake waters; (2) photoreductive dissolution of uranium from ferrihydrites; and (3) the dissolution of iron and manganese (hydr)oxides in low Eh sediments upstream leading to release of sorbed uranium (Lienert et al., 1994). Abiotic photoreductive dissolution of iron/manganese (hydr)oxides occurred in simulations and may therefore explain the seasonal cycling. However, decomposition of microbial biomass and/or reductive dissolution and desorption of uranium from anoxic sediments could not be ruled out and further work is required to elucidate the uranium cycling mechanism in the river.

In a study examining seasonal cycling of uranium in estuarine waters and sediment pore-waters, Shaw et al. (1994) identified two microbially driven processes that appeared to control uranium solubility in a deep, seasonally anoxic basin in Chesapeake Bay over an annual cycle: (1) productivity dependent scavenging of uranium in surface waters; and (2) redox dependent cycling of uranium in deep waters and in sediments. Seasonal anoxia in bottom waters occurred at the site in response to high organic carbon inputs as primary productivity increased during spring, and to the near-contemporaneous stratification of the water column due to freshwater runoff. The result of these processes was that deep water anoxia developed from June to September and was accompanied by relatively high dissolved manganese and iron, and very low dissolved oxygen concentrations, in the water column. Over the annual cycle, sediment pore-water profiles were analysed for dissolved uranium, iron and manganese. In addition, surface (1–3 m) and bottom waters (27–29 m) were sampled for dissolved uranium, iron, manganese and oxygen. Surface water depletion in uranium began in early April and ended by mid-November, suggesting a productivity re lated removal mechanism. Peak uranium removal in surface waters also corresponded with peak primary productivity in the bay, consistent with the removal of uranium to plankton or fresh biodetritus as suggested by Anderson et al. (1989a). Interestingly, during periods when the bottom waters at the site were anoxic, uranium was associated with particulate matter throughout the water column, indicating that reduction of U(VI) to U(IV) was occurring and causing depletion of uranium in the deep water. Under these conditions, U(IV) appeared to be transported to the sediment and ultimately buried in an authigenic phase. However, during the autumnal turnover of the bay and the resulting oxygenation of the deep waters, uranium was rapidly released from particles and sediments and a concomitant deep water uranium enrichment occurred. Uranium release in this circumstance was attributed to both the oxidation of insoluble U(IV) to soluble U(VI) and the release of biomass-bound uranium to waters due to organic matter decomposition.

The sediment pore-water uranium measurements at the site were complex. It appeared that the main flux of uranium to sediments within the Chesapeake Bay was via particulate settling during periods of anoxia, rather than by diffusive transport of dissolved uranium into sediments and reduction in-situ, as reported for deep ocean sediments by Anderson et al. (1989a) and Klinkhammer & Palmer (1991). Indeed, sediment pore-water measurements indicated that dissolved uranium was being released to overlying water at a low rate. Thus, it appears that uranium biogeochemistry at the site was controlled by a complex balance between uranium being scavenged to sediments by particulate matter, and uranium being released from sediments during burial and/or as oxygenation of bottom waters occurred during late summer. The mechanism(s) for release from sediments remained unidentified, however degradation of uranium-enriched biodetritus or reductive dissolution of iron/manganese oxyhydroxides during burial may be responsible. Overall, uranium cycling in the Chesapeake Bay appears to be affected by microorganisms via seasonal sorption on to plankton and/or fresh biodetritus, and by complex seasonal redox cycling caused by physical and biogeochemical changes in sediment and water over the annual cycle.

Transuranic elements

In one of the earliest studies to examine the biogeochemical cycling of plutonium in natural waters, Wahlgren et al. (1980) suggested that a complex seasonal cycling of fallout $^{239,240}Pu$ was occurring in the surface waters of Lake Michigan, USA, a large oxygenated lake. Up to 75% of total (suspended and dissolved) plutonium was lost from the epilimnion of the lake in summer months, and was recycled to these waters during the autumn/winter mixing period in the following year. A positive correlation was reported between the total plutonium concentration in the lake waters and SiO_2 and $CaCO_3$ concentrations, indicating that plutonium may accumulate on phytoplankton (primarily diatoms) and biogenic calcite particles. This led to a depletion in total plutonium in the lake when the particulate settled out from the epilimnion over summer months. However, in a critical review, Sholkovitz (1983) suggested that the positive correlation appeared to be 'unduly biased by one data point', casting some doubt onto the proposed scavenging mechanism. The increase in plutonium concentrations in the lake waters over winter was attributed to redissolution of the biogenic calcite and silica particles, leading to a 'delayed' release of plutonium when the thermocline began to break down. The authors suggested that the plutonium released from dissolution of biogenic silica/calcite could be scavenged by ferric and manganese oxides and then released from these fractions by an unidentified mechanism over winter.

Sholkovitz et al. (1982) reported enhanced plutonium solubility in filtered (<1 μm) oxygen-depleted bottom waters of a seasonally anoxic lake compared to oxic-surface waters. Five times as much plutonium was observed in the iron and manganese rich bottom waters of the lake compared to surface waters, and the ^{239,240}Pu concentration/depth profile in the lake followed the fallout-derived ^{55}Fe profile. This suggested that plutonium was released from bottom sediments to the water column under anoxic conditions. The exact mechanism for remobilisation of plutonium from sediments was not identified. However, the similarity between ^{239,240}Pu and ^{55}Fe in the depth profile, and the high concentrations of dissolved iron and manganese in the lake bottom waters indicated that plutonium which was adsorbed on to iron/manganese oxides may have be released to the water column as microbially driven reductive solubilisation of Fe(II) and Mn(II) occurred during development of anoxia. Although the evidence linking plutonium solubility to reductive dissolution of iron/manganese oxides was strong, the study could not rule out that formation of Pu(IV)-organic complexes in bottom waters was responsible for increased plutonium solubility with depth. Similar, enhanced levels of plutonium were reported for pore-waters from a reed bed site which was reducing (Eh = –65 to +65 mV) and contained high iron, manganese and dissolved organic carbon concentrations (Morris et al., 2001). Here, dissolved (<0.45 μm) plutonium appeared to be associated with dissolved organic carbon rather than iron or manganese dissolution. For both studies, the soluble plutonium was between 0.1–0.4% of the particle associated plutonium indicating that, even in conditions most favourable to solubilisation, the vast majority of plutonium remains associated with particulate matter (Sholkovitz et al., 1982; Morris et al., 2001). Interestingly, Sholkovitz et al. (1982) did not find any significant enrichment of uranium in the lake bottom waters discussed above, implying that reductive dissolution of uranium associated with iron/manganese oxides was not significant at that site. Unfortunately, the lake was sampled once in late summer to examine the effect of seasonal anoxia on plutonium solubility so any effect of seasonal remixing of the lake on plutonium and uranium solubility remained unidentified.

More recently, a series of studies has examined the behaviour of the transuranic elements in pore-waters at an intertidal salt marsh in west Cumbria, UK (Livens et al., 1994; Keith-Roach et al., 2000; Morris et al., 2001). The sediments within this area are radiolabelled with neptunium, plutonium and americium derived from the Sellafield reprocessing plant (Horrill, 1983; Livens et al., 1994; Morris et al., 2001). The relatively high solid phase concentrations of transuranic elements at the marsh have been shown to sustain measurable quantities of neptunium, plutonium and americium in relatively small volumes of sediment pore-waters (typically 300 ml), making detailed solution phase measurement of these elements possible. In the earliest study examining transuranic pore-water, chemistry at the site, Livens et al. (1994) utilised 'porous cup' samplers, which allow repeat extraction of pore-waters (nominally <0.45 μm filtration) over a number of months, to examine plutonium solubility at the marsh. Differences in plutonium solubility occurred in different areas of the marsh and there was considerable variation in plutonium solubility over time in pore-waters from the same zones of the marsh, suggesting that the system was not at equilibrium. However, for the relatively infrequently sampled pore-waters, there was no evidence for a consistent seasonal pattern in plutonium solubility within individual sampling zones over an 18-month period.

In a later study, Morris et al. (2001) utilised porous cup samplers emplaced at one area of the marsh to collect pore-waters at monthly intervals over an annual cycle. Dissolved $^{239,240}Pu$, iron and manganese were analysed and it was assumed that any iron and manganese in the pore-water would be present in the reduced form. The pore-waters showed a systematic change in plutonium concentrations over time. Generally, there were relatively constant dissolved plutonium values in the early part of the year, followed by an increase in June, then a decrease during July and August to a minimum in September, before concentrations returned to a second peak in December and settled back to relatively constant values in January (Fig. 9). Both iron and manganese concentrations in the water samples appeared to be inversely related to the plutonium cycling (Fig. 9). The variation in dissolved iron/manganese in pore-waters was consistent with microbial-driven reductive dissolution of iron/manganese oxides causing an increase in soluble, reduced iron and manganese in sediment pore-waters. By contrast, plutonium solubility was at a minimum during periods of high dissolved iron and manganese. This indicated that plutonium solubility might be controlled by a redox mechanism with reduced insoluble Pu^{4+} becoming more predominant during conditions which produced high dissolved iron/manganese. In periods of relatively low iron and manganese solubility, soluble plutonium displayed a three-fold increase in concentration compared to its minimum of 1.1 mBq l^{-1} (Fig. 9). An increase in the oxidised relatively soluble species PuO_2^+ during these periods may account for the observed increase in solubility. The in-situ 'redox cycling' mechanism may qualitatively explain the observed plutonium, iron and manganese fluctuations in these pore-waters. However, a number of factors were inconsistent with a direct redox cycling mechanism. Dissolved iron and manganese concentrations were very low compared to the potentially reducible iron and manganese concentrations in the marsh sediments. In addition, Eh measurements were relatively constant at the site (145–285 mV) compared to those expected for a system fluctuating between 'oxic' and manganese/iron-reducing. This suggested that direct analysis of microbial community structure, as well as analysis of radionuclide and stable element geochemistry at the site would allow further insights into the factors controlling Pu solubility at the site.

In the most recent work at the Cumbrian marsh site, Keith-Roach et al. (2000) utilised the same porous cup sampling approach to examine plutonium, americium and neptunium solubility over an annual cycle. In addition to transuranic analyses, dissolved manganese, iron, pH, Eh and temperature measurements were made on pore-waters, and changes in the microbial community were assessed by analysis of phospholipid fatty acid (PLFA) on sediment samples collected with each pore-water sample. Plutonium solubility in the marsh pore-waters displayed a remarkable similarity to the data of Morris et al. (2001), and the authors concluded that this was clear evidence that plutonium underwent seasonal cycling within the marsh (Fig. 9). However, soluble iron and manganese did not display the same pattern as earlier (i.e. there was no minimum iron or manganese concentration associated with maximum plutonium solubility), but the Eh data were similar to those observed by other workers at the site and were relatively constant throughout the annual cycle. Perhaps more importantly, the concentration of americium in the pore-waters mimicked

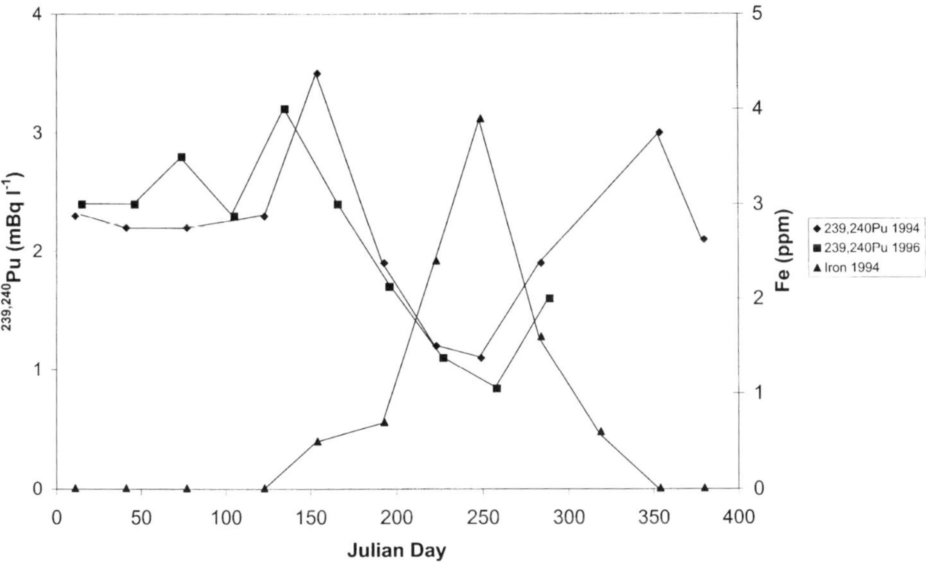

Fig. 9. Temporal Variations in ^{239,240}Pu and dissolved Fe in salt marsh pore-waters. Pu data are plotted for both 1994 and 1996 (Morris et al., 2000b; Keith-Roach et al., 2000).

plutonium solubility during the late summer. This suggested that: (1) a single process was responsible for the increase in plutonium/americium solubility in late summer; and (2) that process was probably not a direct 'redox cycling' mechanism because pore-water Eh did not significantly fluctuate and because Am^{3+}, the predicted americium species, does not undergo redox cycling. However the PLFA data suggested an alternate mechanism to explain the plutonium/americium solubility minimum in late summer. The microbial community structure was relatively constant over the annual cycle, and was dominated by aerobic organisms. The major change in the microbial community over the year was a six-fold increase in total PLFA between the minimum in mid-winter, and the maximum in late summer, i.e., the minimum in plutonium/americium solubility corresponds to the maximum sediment 'biomass' as indicated by PLFA concentrations . Keith-Roach et al. (2000) suggested that passive sorption to biomass may be responsible for the reduced americium and plutonium concentrations observed in the pore-waters. After the peak in biomass during late summer, the americium/plutonium concentrations began to increase in pore waters, suggesting that cell decomposition may release the bound radionuclides back into solution. The passive sorption of actinide elements to bacterial cells may occur in the environment (Wildung & Garland, 1980) and has been suggested as a possible bioremediation technology (Francis, 1990; Eccles, 1998; Suzuki & Banfield, 1999). More recently, Mahara & Kudo (1998), proposed that plutonium is significantly retained by living cells in sedimentary environments in anoxic conditions. Interestingly, neptunium behaviour at the salt marsh was apparently unrelated to plutonium/americium cycling in pore-waters possibly because the thermodynamically predicted stable oxidation state for neptunium at the site (NpO_2^+; Keith-Roach et al., 2000) is less particle reactive than

Am(III) and Pu(IV) species. This may also explain the enhanced mobility of neptunium in a sediment core from the same salt marsh compared to the sediment profiles of both plutonium and americium (Morris et al., 2000). Although there is convincing evidence that seasonal cycling of plutonium and other transuranic elements does occur within the marsh, it remains unclear whether these processes lead to significant remobilisation of radionuclides from the site over the annual cycle and further geochemical and microbiological work is required to ascertain the mechanisms of transuranic element cycling at the site.

Active biological uptake of uranium

Studies at a permanently anoxic fjord (Framvaren, southern Norway) have identified a clear link between uranium solubility and maximum microbial activity in the fjord waters and suggest that reduction of U(VI) in these anoxic waters is not the major control on uranium solubility (McKee & Todd, 1993; Swarzenski et al., 1999). The fjord is a small, deep (180 m) basin which has a persistently stratified water column with a permanent oxic-anoxic (i.e. O_2/H_2S) interface in the photic zone. A depth profile of the uranium concentrations in the fjord revealed a sharp decline in dissolved O_2 in the waters, and at 18 m H_2S began to appear in the water, reaching a maximum of 6 mM at depth (McKee & Todd, 1993). A sharp peak in total suspended matter (>0.4 μm), N and C occurred at 24 m and this was attributed to dense populations of phototrophic, sulfur-oxidising bacteria (*Chromatium* sp. and *Chlorobium* sp.) at the O_2/H_2S redox boundary. The total water column concentrations of uranium were lower than predicted from the oceanic ^{238}U salinity relationship, indicating a net loss of uranium from the fjord waters which was particularly pronounced in the 24–36 m region. The vertical profile of dissolved (<0.22 μm) ^{238}U was linear with respect to salinity apart from a pronounced minimum observed at 24 m. Dissolved uranium did not appear to be scavenged by the iron oxyhydroxides which reached a maximum concentration at or above the O_2/H_2S interface at 18 m. The authors suggested that reduction of U(VI) was not responsible for the depletion of uranium seen at 24 m as: (1) uranium concentrations were not depleted below this depth, implying that scavenging of U(IV) by settling matter was not occurring; and (2) U(VI) has been found in similar anoxic, sulfidic waters (Todd et al., 1988; Anderson et al., 1989b). In the absence of a redox or iron oxyhydroxide related scavenging mechanism, the dissolved uranium minimum was related to the observed bacterial maximum at the site. The particulate uranium profile was reported as being a 'mirror image' of the dissolved uranium profile over the depth range of uranium depletion, consistent with the scavenging of U(VI) directly to particulate matter associated with the peak microbial density. Below the bacterial maximum, but still in strongly 'anoxic' waters, the authors suggested that U(VI) was reconverted to a dissolved form although the mechanism for this was not identified.

In a later study at the site, Swarzenski et al. (1999) confirmed that microbial activity had a significant effect on uranium solubility in the fjord, and further refined understanding of the processes controlling uranium solubility. In this study, the uranium solubility results were somewhat different to previous studies (Todd et al., 1988; McKee & Todd, 1993) and the authors suggested a complex cycling mechanism controlled uranium solubility in the fjord. Oxidation state analyses at depths to 30 m (i.e. below the O_2/H_2S interface) confirmed that the majority (>90%) of uranium within the water column was present as U(VI). However, the peak in uranium solubility in the fjord was associated with redox-, particle- and biologically- reactive elements such as manganese, iron, barium and strontium and was present at the O_2/H_2S transition zone in the fjord. Unfortunately, inadequate resolution in the profile meant that individual reactions affecting uranium solubility could not be identified. However, the authors suggested that a complex recycling mechanism involving U(VI)/U(IV)–Mn(IV)/Mn(II) redox cycling during development of anoxia, uranium-microbial associations, and uranium-dissolved organic carbon interactions, may explain uranium solubility in the fjord waters. Microbial activity is thus either directly or indirectly implicated in all of these possible cycling mechanisms and the Framvaren fjord may offer an ideal opportunity to investigate these complex cycling processes in more detail.

Processes affecting uranium solubility in early diagenesis

In early studies, valence change of U(VI) to particle reactive U(IV) was implicated as the likely mechanism for removing dissolved uranium to sediments (Koczy et al., 1957; cited in Swarzenski et al., 1999) and enriched uranium concentrations in sapropelic muds have been reported in a variety of different, reducing shelf-water environments (e.g. Baturin, 1968; Degens et al., 1977; Huh et al., 1987; Kolodny & Kaplan, 1973; Mo et al., 1973; Weber & Sackett, 1973).

Uranium reduction in these sediments was thought to be via diffusion and in-situ, abiotic reduction by hydrogen sulfide produced within anoxic marine sediments (Langmuir, 1978; Anderson et al., 1989a; Klinkhammer & Palmer, 1991). However, a number of studies report that in anoxic, sulfidic marine waters (where reduction of U(VI) is thermodynamically favourable) virtually all uranium remains as U(VI) in the water column below the O_2/H_2S interface (Todd et al., 1988; Anderson, 1989b; McKee & Todd, 1993). This is consistent with laboratory studies which report slow U(VI) reduction in waters with sulfide at concentrations typical of anoxic environments (Kochenov et al., 1977; Mohagheghi et al., 1985). Sulfide may therefore be a poor reductant for U(VI) at circumneutral pH and low temperature. However, enhanced rates of enzymatic U(VI) reduction have been reported in pure culture for several microorganisms including both dissimilatory metal-reducing bacteria and sulfate-reducing bacteria (Lovley et al., 1991; Caccavo et al., 1992; Gorby & Lovley, 1992; Lovley & Phillips, 1992; Lovley et al., 1993; Tebo & Obraztsova, 1998).

More recently, Abdelouas et al. (2000) reported on microcosm experiments where soil was inoculated with groundwaters from a range of uranium-contaminated sites, finding that U(VI), Fe(III) and Mn(IV) were concomitantly reduced by sulfate-reducing bacteria. This implies that microbial activity may be directly implicated in enhancing rates of U(VI) reduction in anoxic sediments. In addition, both mineral and microbial surfaces have been implicated in enhancing U(VI) reduction in natural conditions (Kochenov et al., 1977; Mohagheghi et al., 1985; Liger et al., 1999). Indeed, Liger et al. (1999) reported that the rate of abiotic U(VI) reduction in the presence of hematite and Fe(II) was of the same order as the highest rate reported for enzymatic reduction of U(VI) in pure culture experiments.

Barnes & Cochran (1993) reported that microbially driven uranium removal was occurring in estuarine sediments from Long Island sound and the Amazon shelf. Sediment incubation experiments indicated that the U(VI) removal rate was first order with respect to uranium concentration and that the rate constant was linearly correlated to sulfate reduction rates, implying that U(VI) reduction was directly associated with sulfate reduction. In this case U(VI)-reducing sulfate reducers (specifically *Desulfovibrio* sp.) may be directly implicated in U(VI) reduction (Barnes & Cochran, 1993; Lovley, 1995; Abdelouas et al., 2000). Interestingly, during the early stages of incubation (prior to sulfate reduction), some experiments displayed a net release of uranium from sediments to pore-waters which was correlated with increases in dissolved iron and manganese (Barnes & Cochran, 1993). This suggested that uranium was being released from sediments during dissolution of metal oxide carrier phases driven by iron and manganese reduction. In contrast, a number of studies indicate that deposition of dissolved U(VI) in marine sediments occurs within the iron reduction zone and prior to sulfate reduction (Cochran et al., 1986; Lovley et al., 1991). In this case, dissimilatory Fe(III)-reducing microorganisms (e.g. *Geobacter* sp.) may predominate over sulfate-reducing bacteria in the sediments and may be responsible for most of the U(VI) reduction in the sediment (Lovley, 1995). Lovley et al. (2000) reported that reduction of Fe(III) and U(VI) in aquifer systems is dominated by enzymatic reduction of the metals by microorganisms. Genomic characterisation of 16S rDNA in subsurface sediments from a uranium-contaminated aquifer showed that *Geobacter* related species were 'invariably' stimulated by addition of various organic growth media used to stimulate microbial reduction of U(VI) at the site. Interestingly, Wielinga et al. (2000) reported that the rate of bacterially promoted U(VI) reduction may be inhibited by the presence of amorphous iron hydrous oxide minerals such as ferrihydrite but may be unaffected by the presence of crystalline iron (hydr)oxide minerals such as goethite and hematite. In microcosm experiments, *Shewanella alga*, a dissimilatory metal-reducing bacterium, was incubated in the presence of U(VI), goethite and ferrihydrite. In the experiments with *S. alga* in the presence of U(VI) and U(VI)/goethite, enzymatic reduction of U(VI) was complete (>95%) within 10 hours. By contrast, in the presence of *S. alga*, U(VI) and ferrihydrite, only 50% of the U(VI) was reduced within 10 hours. This suggested that amorphous iron oxides (ferrihydrite) may successfully compete with U(VI) as a terminal electron acceptor under certain conditions thereby limiting the rate of U(VI) reduction. By contrast, the rate of bacterial U(VI) reduction was unaffected in the presence of crystalline iron (hydr)oxides (goethite and hematite) suggesting that the reduced energy yield for microbial reduction of goethite/hematite compared to ferrihydrite (see Chapter 3, this volume) does not allow them to compete with U(VI) as terminal electron acceptors under certain conditions (Wielinga et al., 2000).

Recent studies have demonstrated that U(VI) reduction in sediments does not always occur even under strongly anoxic conditions (Duff et al., 1997, 1999, 2000). X-ray absorption near-edge structure (XANES) techniques revealed that uranium in a sediment profile from a contaminated, highly reducing evaporation pond was largely present as U(VI) with only 25% of the uranium present as U(IV) even in the presence of acid volatile sulfides (Duff et al., 1997). However, the percentage of U(IV) did increase with depth in the sediment profile suggesting that a gradual reduction from U(VI) to U(IV) was occurring. More recently, Duff et al. (2000) found that the U(IV)/U(VI) ratios in a different pond again increased with depth in sediments, although in these sediments U(IV) became dominant at much shallower depths. In a related study, Duff et al. (1999) again applied the XANES technique to examine the effect on uranium redox chemistry of adding substrates to stimulate microbial reduction (pond algae, acetate, sucrose, alfafa shoots). In alfafa treated sediments, 95% of the uranium was removed from solution and sulfide precipitation occurred. XANES measurements again indicated that a reduction of U(VI) to U(IV) was occurring, although on exposure to air U(IV) was readily reoxidised, implying the reduction process was reversible. Incubation experiments that examined the effect of Eh and pCO_2 on uranium solubility in pond sediment/water suspensions were performed. Contaminated sediments were equilibrated at low and high pCO_2 pressures (0.22 and 5.26 kPa) and some incubations were left to become reducing. At high Eh/high pCO_2, dissolved uranium was at higher concentrations than in high Eh/low pCO_2 systems, presumably due to complexation with CO_3^{2-}. Even at low Eh, dissolved uranium concentrations only decreased under 'intense' sulfate-reducing conditions.

It is clear that microbial activity in early diagenesis may affect uranium solubility both directly and indirectly. However, the picture is complex, with different studies reporting uranium reduction in different biogeochemical zones and other studies reporting that abiotic reduction may be as important as enzymatic reduction. Further work which utilises solid phase speciation techniques and genetic sequencing as well as rigorous geochemical analysis is needed to provide further insight into the subtleties of uranium cycling in early diagenesis.

Processes affecting transuranic element solubility during early diagenesis

Nelson & Lovett (1981) reported concentrations of ^{239,240}Pu in two oxidation state groups, 'reduced' Pu(III, IV) and 'oxidised' Pu(V, VI) in pore-waters and sediment profiles from sediment cores taken within the Irish Sea. 'Oxidised' Pu in the pore-waters decreased with depth and was at lower concentrations than in the overlying seawater. This implied that plutonium was being reduced within the sediments. In a review of plutonium geochemistry, Sholkovitz (1983) pointed out that the samples had not been handled under anoxic conditions and that there may be some artefacts associated with the 'oxidised' and 'reduced' analyses. Within these constraints, Sholkovitz (1983) suggested that a number of the pore-water 'reduced' plutonium profiles displayed subsurface maxima coincident with dissolved iron maxima. In addition, solid/solution distribution coefficients were not constant with depth in the cores. This indicated that post-depositional diagenetic reactions such as dissolution of plutonium from metal oxide carrier phases during iron reduction could be increasing the pore-water concentrations of the reduced form of plutonium (Sholkovitz, 1983).

In a series of sediment incubation experiments, Sholkovitz et al. (1983) examined the behaviour of radionuclides including ^{239,240}Pu and ^{55}Fe in early diagenesis. Sediments contaminated with fallout concentrations of ^{239,240}Pu and ^{55}Fe were homogenised and allowed to develop diagenetic profiles in a large tank. Periodically cores were taken, and pore-waters were extruded under anoxic conditions. In the sediment tanks, sulfate reduction developed and large pore-water concentration gradients were reported after 4 weeks in alkalinity, ammonia and dissolved organic carbon. Subsurface maxima were also reported for stable iron and manganese indicating that bacterially-mediated redox reactions were occurring in the tanks. Pore-water profiles for ^{55}Fe displayed a clear subsurface maximum that mirrored the stable iron profile, indicating that fallout derived ^{55}Fe was in a form that underwent diagenetic reactions. By contrast, the ^{239,240}Pu pore-water profile was constant with depth, showing that early diagenetic reactions were not affecting plutonium solubility in this system.

The absence of a biogeochemical signal for ^{239,240}Pu during early diagenesis in sediment pore-waters is further supported by studies which analysed in-situ ^{239,240}Pu and stable element profiles in sediments from a number of different locales (Sholkovitz & Mann, 1984; Buesseler & Sholkovitz, 1987; Malcolm et al., 1990). Sholkovitz & Mann (1984) examined ^{239,240}Pu pore-water profiles in sediments from Buzzards Bay. The plutonium profiles displayed a clear subsurface maximum which could not be related to dissolved iron/manganese or dissolved organic carbon. The plutonium profile was explained by a simple equilibrium-mixing model whereby the subsurface maximum resulted from a balance between equilibrium with solid phase plutonium, and downward diffusion and biological mixing of overlying seawater depleted in plutonium. Below the active mixing zone, the decrease in pore-water plutonium activity was explained by a corresponding decrease in the solid phase plutonium activity in the core and this was reflected in a relatively constant K_d (0.3–1.2×10^5 l kg^{-1}) with depth in the core (Sholkovitz & Mann, 1984). Buesseler & Sholkovitz (1987) reported that plutonium pore-water profiles did not display post-depositional plutonium mobility in environments ranging from highly reducing muddy shelf sediments to carbonate-rich oxic and suboxic sediments in the deep sea. The main factor that controlled plutonium solubility in the sediments was the distribution of plutonium in the solid phase. Pore-water plutonium distributions could be represented by distribution coefficients ranging from 0.2–23×10^4 l kg^{-1}. Diffusive flux calculations at the sediment/water interface showed that an 'insignificant' amount of ^{239,240}Pu had been remobilised in the sediments sampled.

Malcolm et al. (1990) examined the early diagenetic behaviour of both plutonium and americium in sediment cores taken from the Irish Sea. Sediment profiles for ^{239,240}Pu and ^{241}Am displayed subsurface peaks which were unrelated to organic carbon, iron or manganese in the cores, but seemed to be related to historical radionuclide discharges from the Sellafield reprocessing facility, as has been reported in numerous other studies (Kershaw et al., 1990; Livens et al., 1994; Mackenzie et al., 1994; Morris et al., 2000).

Malcolm et al. (1990) analysed sediment pore-waters for both reduced (i.e. oxidation states III + IV) and oxidised (i.e. oxidation states V + VI) plutonium, and reduced (i.e. oxidation state III) and 'oxidised' americium. In all pore-waters, more than 99% of americium was present in the reduced form presumably as Am(III). Reduced plutonium was dominant (>96%) in all but the topmost sample of each core where 27–69% of plutonium was oxidised. The conversion of oxidised Pu(V, VI) present in seawater to reduced Pu(III, IV) was coincident with a change in Eh from 350–500 mV in the surface sediments to 50–250 mV at 5–10 cm depth in the sediment core. The redox cline in the sediments was close to the site of the redox-driven mobilisation of iron/manganese, although the core resolution was too low to establish a definitive link between plutonium reduction and the iron/manganese cycle. The sediment pore-water profiles for reduced plutonium and americium were unrelated to any early diagenetic indicators but were qualitatively related to the sediment radionuclide profiles, again suggesting that an adsorption/desorption process governed plutonium and americium solubility in the sediments. K_d values were between $0.17–1.2 \times 10^6$ l kg^{-1} for reduced plutonium and $3.5–8.9 \times 10^5$ l kg^{-1} for reduced americium. The highest K_d values for plutonium and americium (i.e. the largest extent of particulate association) were at 2–4 cm, the depth where iron and manganese reoxidation and reprecipitation in the core was occurring. This indicates that oxide scavenging of plutonium and americium from solution could be a sink for these radionuclides in the uppermost few centimetres of marine sediments. Sholkovitz & Mann (1984) suggested a similar sink for plutonium in sediments from Buzzards Bay.

Overall, this body of evidence suggests that the reduced forms of plutonium and americium are controlled by adsorption/desorption processes which can be described by K_d values typically in the order of 10^5 to 10^6 l kg^{-1} within marine sediments (Sholkovitz & Mann, 1984; Buesseler & Sholkovitz, 1987; Malcolm et al., 1990). However, indirect microbial processes are implicated in aspects of transuranic solubility. There is some suggestion that oxidised plutonium undergoes reduction as iron/manganese reduction develops in sediment cores from the Irish Sea (Malcolm et al., 1990), and that pore-water plutonium may be scavenged by iron and manganese (hydr)oxide precipitation in surface sediments (Sholkovitz & Mann, 1984; Malcolm et al., 1990).

6. Conclusions

The early actinide elements are ubiquitous in the natural environment, and several of their isotopes will be significant components of nuclear wastes. Our discussion highlights substantial gaps in our knowledge of the global cycling of uranium, the most studied actinide element. Indeed, we are only now becoming able to construct a detailed global cycle for iron which takes account of sediment partitioning and phase associations, contrasting with previous attempts to study the global cycling of uranium, which have focussed on the maintenance of a steady state ocean by the removal of dissolved riverine uranium into sediments. However, construction of an accurate budget requires knowledge of the phase association of uranium, so that authigenic uranium can be distinguished from detrital uranium. Such data are not commonly available, but some understanding can be obtained from total uranium data (more readily available) and consideration of the fate of riverine inputs. This cruder model still needs to be refined by addition of riverine data on the partition of river sediments between the innershore and marine sediment reservoirs, and on the uranium content of these reservoirs.

We have reviewed the limited number of studies relating actinide solubility and microbial activity in the natural environment. There is certainly evidence that microbial processes, which drive both seasonal cycling and early diagenetic reactions, do affect actinide solubility. However, it is also clear that our understanding of the role of microbiology in these cycling and remobilisation reactions is rather rudimentary as the role of microbiology in understanding subsurface geochemical cycles has only developed in recent decades. Application of novel techniques including genomic characterisation (see Chapter 2) and direct spectroscopic observation of radionuclide (and stable element) speciation as well as rigorous geochemical analysis and developments in mathematical modelling will allow further insight into the cycling and remobilisation behaviour of actinide elements and other radionuclides in the natural environment. Understanding these cycling processes and predicting the eventual fate of these elements in the natural environment are critical to development of remediation strategies for the cleanup of existing radionuclide contaminated sites and to improving understanding of radionuclide behaviour in subsurface environments, particularly in situations relevant to nuclear waste disposal.

References

Abdelouas, A., Lutze, W. & Nuttall, E. H. (1999). Uranium contamination in the subsurface: characterisation and remediation. In P. C. Burns & R. Finch (Eds), *Uranium: Mineralogy, Geochemistry and the Environment* (pp. 433–473). Washington DC: Mineralogical Society of America.

Abdelouas, A., Lutze, W., Gong, W., Nuttall, E. H., Streitelmeier, B. A. & Travis, B. J. (2000). Biological reduction of uranium in groundwater and subsurface soil. *The Science of the Total Environment, 250*, 21–35.

Ahrland, S. (1986). Solution chemistry and kinetics of ionic reactions. In J. J. Katz, G. T. Seaborg & L. R. Morss (Eds), *The Chemistry of the Actinide Elements* (2nd edn). London, UK: Chapman and Hall.

Akleyev, A. V. & Lyubchansky, E. R. (1994). Environmental and medical effects of nuclear- weapon production in the southern Urals. *The Science of the Total Environment, 142*, 1–8.

Allard, B., Olofsson, U. & Torstenfelt, B. (1984). Environmental actinide chemistry. *Inorganica Chimica Acta, 94*, 205–221.

Allegre, C. J., Dupre, D., Negrel,, P. & Gaillardet, J. (1996). Sr–Nd–Pb isotope systematics in Amazon and Congo River systems: constraints about erosion processes. *Chemical Geology, 131*, 93–112.

Allison, M. A., Kuehl, S. A., Martin, T. C. & Hassan, A. (1998). Importance of flood-plain sedimentation for river sediment budgets and terrigenous input to the oceans: insights from the Brahmaputra-Jamuna river. *Geology, 26*, 175–178.

Ames, L. L., McGarrah, J. E., Walker, B. A. & Salter, P. F. (1983). Uranium and radium sorption on amorphous ferric oxyhydroxide. *Chemical Geology, 40*, 135–148.

Anderson, R. F. (1987). Redox behaviour of uranium in an anoxic marine basin. *Uranium, 3*, 145–164.

Anderson, R. F., LeHuray, A. P., Fleisher, M. Q. & Murray, J. W. (1989a). Uranium deposition in Saanich Inlet sediments, Vancouver Island. *Geochimica et Cosmochimica Acta, 53*, 2215–2224.

Anderson, R. F., Fleisher, M. Q., LeHuray & A. P. (1989b). Concentration, oxidation state and particulate flux of uranium in the Black Sea. *Geochimica et Cosmochimica Acta, 53*, 2215–2224.

Banaszak, J. E., Rittmann, B. E. & Reed, D. T. (1999). Subsurface interactions of actinide species and microorganisms: implications for the bioremediation of actinide-organic mixtures. *Journal of Radioanalytical and Nuclear Chemistry, 241*, 385–435.

Barnes, C. E. & Cochran, J. K. (1990). Uranium removal in oceanic sediments and the oceanic U balance. *Earth and Planetary Science Letters, 97*, 94–101.

Barnes, C. E. & Cochran, J. K. (1991). Geochemistry of uranium in Black Sea sediments. *Deep-Sea Research, 38*, 51237–51254.

Barnes, C. E. & Cochran, J. K. (1993). Uranium geochemistry in estuarine sediments: controls on removal and release processes. *Geochimica et Cosmochimica Acta, 57*, 555–569.

Baturin, G. N. (1968). Geochemistry of uranium in the Baltic. *Geochemistry International, 5*, 344–348.

Beneš, P. (1998). The environmental impacts of uranium mining and milling and the methods of their reduction. In G. R. Choppin & M. K. Khankhasayev (Eds), *NATO ASI Chemical Separation Technologies and Related Methods of Nuclear Waste Management: Applications, Problems, and Research Needs* (pp. 225–246). London, UK: Kluwer Academic Publishers.

Borole, D. V., Krishnaswami, S. & Somayajulu, B. L. K. (1982). Uranium isotopes in rivers, estuaries and adjacent coastal sediments of western India: their weathering, transport and oceanic budget. *Geochimica et Cosmochimica Acta, 46*, 125–137.

Boyle, E. A., Collier, R., Dengler, A. T., Edmond, J. A., Ng, A. C. & Stallard, R. F. (1974). On the chemical mass balance in estuaries. *Geochimica et Cosmochimica Acta, 38*, 1719–1728.

Boyle, E. A., Edmond, J. A. & Sholkovitz, E. R. (1977). The mechanism of iron removal in estuaries. *Geochimica et Cosmochimica Acta, 41*, 1313–1324.

Bryan, N. D., Livens, F. R. & Horrill, A. D. (1994). Biogeochemical behaviour of plutonium and americium and geochemical modelling of the soil solution. *Journal of Radioanalytical and Nuclear Chemistry Articles, 182*, 359–366.

Buesseler, K. O. & Sholkovitz, E. R. (1987). The geochemistry of fallout plutonium in the north Atlantic I. A pore water study in shelf, slope and deep sea sediments. *Geochimica et Cosmochimica Acta, 51*, 2605–2622.

Canfield, D. E. (1997). The geochemistry of river particulates from the continental USA: Major elements. *Geochimica et Cosmochimica Acta, 61*, 3349–3367.

Caccavo, F. J., Blakemore, R. P. & Lovley, D. R. (1992). A hydrogen-oxidizing, Fe(III)-reducing microorganism from the Great Bay estuary, New-Hampshire. *Applied and Environmental Microbiology, 58*, 3211–3216.

Chester, R. (1990). *Marine Geochemistry* (698pp). London, UK: Unwin Hyman.

Choppin, G. R. & Stout, B. E. (1989). Actinide behaviour in natural waters. *Science of the Total Environment, 83*, 203–216.

Church, T. M., Lord, C. J. III & Somayajulu, B. L. K. (1981). Uranium, thorium and lead nuclides in a Delaware salt marsh sediment. *Estuarine, Coastal and Shelf Science, 13*, 267–275.

Church, T. M., Sarin, M. M., Fleischer, M. Q. & Ferdelman, T. G. (1996). Salt marshes: an important sink for dissolved uranium. *Geochimica et Cosmochimica Acta, 60*, 3879–3887.

Clark, D. L., Hobart, D. E. & Neu, M. P. (1995). Actinide carbonate complexes and their importance in actinide environmental chemistry. *Chemical Reviews, 95*, 25–48.

Cochran, J. K., Carey, A. E., Sholkovitz, E. R. & Surprenant, L. D. (1986). The geochemistry of uranium and thorium in coastal marine sediments and sediment pore waters. *Geochimica et Cosmochimica Acta, 50*, 663–680.

Degens, E. T., Khoo, R. & Michalelis, W. (1977). Uranium anomaly in Black Sea sediments. *Nature, 269*, 566–569.

Delaney, T. A., Hockley, D. E., Chapman, J. T. & Holl, N. C. (1998). Geochemical characterisation of tailings at the McArthur River Mine, Saskatchewan. In *Proceedings of Tailings and Mine Waste '98* (pp. 571–578). Rotterdam, The Netherlands: Balkema.

Dozol, M., Hagemann, R., Hoffman, D. C., Adloff, J. P., Vongunten, H. R., Foos, J., Kasprzak, K. S., Liu, Y. F., Zvara, I., Ache, H. J., Das, H. A., Hagemann, R. J. C., Herrmann, G., Karol, P., Maenhaut, W., Nakahara, H., Sakanoue, M., Tetlow, J. A., Baro, G. B., Fardy, J. J., Benes, P., Roessler, K., Roth, E., Burger, K., Steinnes, E., Kostanski, M. J., Peisach, M., Liljenzin, J. O., Aras, N. K., Myasoedov, B. F. & Holden, N. E. (1993). Radionuclide migration in ground waters: review of the behaviour of actinides. *Pure and Applied Chemistry, 65*, 1081–1102.

Duce R. A., Liss, P. S., Merrill, J. T., Atlas, E. L., Buat-Menard, P., Hicks, B. B., Miller, J. M., Prospero, J. M., Arimoto, R., Church, T. M., Ellis, W., Galloway, J. N., Hansen, L., Jickells, T. D., Knap, A. H., Reinhardt, K. H., Schneider, B., Soudine, A., Toko, J. J., Tsunogai, S., Wollast, R. & Zhou, M. (1991). The atmospheric input of trace species to the world ocean. *Global Biogeochemical Cycles, 5*, 193–259.

Duff, M. C., Amrhein, C. Bertsch, P. M. & Hunter, D. B. (1997). The chemistry of uranium in evaporation pond sediment in the San Joaquin Valley, California, USA, using X-ray fluorescence and XANES techniques. *Geochimica et Cosmochimica Acta, 61*, 73–81.

Duff, M. C., Hunter, D. B., Bertsch, P. M. & Amrhein, C. (1999). Factors influencing uranium reduction and solubility in evaporation pond sediments. *Biogeochemistry, 45*, 95–144.

Duff, M. C., Morris, D. E., Hunter, D. B. & Bertsch, P. M. (2000). Spectroscopic characterisation of uranium in evaporation basin sediments. *Geochimica et Cosmochimica Acta, 64*, 1535–1550.

Duniker, J. C., Wollast, R. & Billen, G. (1979). Behaviour of manganese in the Rhine and Scheldt estuaries. *Estuarine and Coastal Shelf Science, 9*, 727–738.

Eccles, H. (1998). Nuclear waste management: a bioremediation approach. In G. R. Choppin & M. K. Khankhasayev (Eds), *NATO ASI Chemical Separation Technologies and Related Methods of Nuclear*

Waste Management: Applications, Problems and Research Needs (pp. 225–246). London, UK: Kluwer Academic Publishers.

Edmond, J. M., Measures, C., Mangum, B., Grant, B., Scalter, F. R., Collier, R., Hudson, A., Gordon, L. I. & Corliss, J. B. (1979). On the formation of metal-rich deposits at ridge crests. *Earth and Planetary Science Letters, 46*, 19–30.

Eisenbud, M. & Gesell, T. F. (1997). *Environmental Radioactivity from Natural, Industrial and Military Sources* (pp. 201–218). London, UK: Academic Press.

Elderfield, H. & Schlutz, A. (1996). Mid-ocean ridge hydrothermal fluxes and the chemical composition of the ocean. *Annual Reviews in Earth and Planetary Science, 24*, 191–224.

Ewing, R. C. (1999). Radioactivity and the 20th century. In P. C. Burns & R. Finch (Eds), *Uranium: Mineralogy, Geochemistry and the Environment* (pp. 1–21). Washington DC: Mineralogical Society of America.

Facer, G. (1980). Quantities of transuranic elements in the environment from operations relating to nuclear weapons. In W. C. Hanson (Ed.), *Transuranic Elements in the Natural Environment* (pp. 300–335). US Department of Energy, Springfield, VA. US DOE/TIC/22800.

Fernandes, H. M., Franklin, M. R. & Veiga, L. H. (1998). Acid rock drainage and radiological environmental impacts. A case study of the uranium mining and milling facilities at Pocos de Caldas. *Waste Management, 18*, 169–181

Finch, R. & Murakami, T. (1999). Systematics and paragenesis of uranium minerals. In P. C. Burns & R. Finch (Eds), *Uranium: Mineralogy, Geochemistry and the Environment* (pp 91–179). Washington DC: Mineralogical Society of America.

Francis, A. J. (1990). Microbial dissolution and stabilisation of toxic metals and radionuclides in mixed wastes. *Experientia, 46*, 840–851.

Francis, A. J. & Dodge, C. J. (1998). Remediation of soils and wastes contaminated with uranium and toxic metals. *Environmental Science and Technology, 32*, 3993–3998.

Garrels, R. M. & Mackenzie, F. T. (1974). *Evolution of Sedimentary Rocks* (397pp). New York: Norton.

Gorby, Y. A. & Lovley, D. R. (1992). Enzymatic uranium precipitation. *Environmental Science and Technology, 26*, 205–207.

Gray, J., Jones, S. R. & Smith, A. D. (1995). Discharges to the environment from the Sellafield site 1951–1992. *Journal of Radiological Protection, 15*, 99–131.

Grenthe, I., Fuger, J., Konings, R. J. M., Lemire, R. J., Muller, A. B., Ngyuyen-Trung, C. & Wanner, H. (1992). *Chemical Thermodynamics of Uranium* (pp. 715). North-Holland: Nuclear Energy Agency.

Hagen, A. & Jakubick, A. T. (1997). The WISMUT experience in the management of large scale projects. In R. Baker, S. Slate & G. Bender (Eds), *Proceedings of the Sixth International Conference on Radioactive Waste Management and Environmental Remediation* (pp. 153–160). New York: The American Society of Mechanical Engineers.

Halim, Y. (1991). The impact of human alteration of the hydrological cycle on ocean margins. In R. F. C. Mantoura, J.-M. Martin & R. Wollast (Eds), *Ocean Margin Processes in Global Change* (pp. 301–327). New York: Wiley-Interscience.

Hardy, E. P., Krey, P. W. & Volchok, H. L. (1973). Global inventory and fallout distribution of plutonium. *Nature, 241*, 444–445.

Hay, W. H. (1998). Detrital sediment fluxes from continents to oceans. *Chemical Geology, 145*, 287–32.

Heaston, E., Poppiti, J., Sutter, H., Knutson, D. & Hunemuller, M. (1999). Everything you ever wanted to know about the Hanford waste tanks. *Radwaste Magazine, 6*, 27–34.

Ho, C. H. & Doern, D. C. (1985). The sorption of uranyl species onto a hematite sol. *Canadian Journal of Chemistry, 63*, 1100–1104.

HoL (1999). House of Lords Select Committee on Science and Technology, *Third Report on Management of Nuclear Waste* (90pp). London, UK: The Stationery Office.

Holliday, L. M. & Liss, P. S. (1976). The behaviour of dissolved iron, manganese and zinc in the Beaulieu Estuary, S. England. *Estuarine Chemistry, 4*, 53–91.

Horrill, A. D. (1983). Concentrations and spatial distribution of radioactivity in an ungrazed salt marsh. In P. J. Coughtrey (Ed.), *Ecological Aspects of Radionuclide Release* (pp. 199–215). British Ecological Society Special Publication No. 3. Oxford, UK: Blackwell.

Hsi, D. C. & Langmuir, D. (1985). Adsorption of uranyl onto ferric oxyhydroxides: application of the surface complexation site-binding model. *Geochimica et Cosmochimica Acta, 49*, 1931–1941.

Huh, C. A., Zahnle, D. L., Small, L. F. & Noshkin, V. E. (1987). Budgets and behaviours of uranium and thorium isotopes in Santa Monica Basin Sediments. *Geochimica et Cosmochimica Acta, 51*, 1743–1754.

Hursthouse, A. S. & Livens, F. R. (1993). Evidence for the remobilization of transuranic elements in the terrestrial environment. *Environmental Geochemistry and Health, 15*, 163–171.

IAEA (1999). *Environmental Activities in Uranium Milling and Mining* (173pp). Paris, France: IAEA/OECD Joint Publication, OECD Publications. ISBN 92-64-17064-2.

Jickells, T. D. & Rae, J. E. (1997). Biogeochemistry of intertidal sediments. In T. D. Jickells & J. E. Rae (Eds), *Biogeochemistry of Intertidal Sediments* (pp. 2–15). Cambridge, UK: Cambridge University Press.

Junghans, M. & Helling, C. (1998). Historical mining, uranium tailings and waste disposal at one site: can it be managed? A hydrological analysis. In *Proceedings of Tailings and Mine Waste '98* (pp. 117–126). Rotterdam: Balkema.

Katz, J. J., Seaborg, G. T. & Morss, L. R. (1986). *The Chemistry of the Actinide Elements* (2nd edn) (Vol. 2) (pp. 1197–1212). London, UK: Chapman and Hall

Keith-Roach, M. J., Day, J. P., Fifield, L. K., Bryan, N. D. & Livens, F. R. (2000). Seasonal variations in interstitial water transuranium element concentrations. *Environmental Science and Technology, 34*, 4273–4277.

Kershaw, P. J., Woodhead, D. S., Malcolm, S. J. & Hunt, G. J. (1990). A sediment history of Sellafield discharges. *Journal of Environmental Radioactivity, 12*, 201–241.

Kershaw, P. J., Pentreath, R. J., Woodhead, D. S. & Hunt, G. J. (1992). *A Review of Radioactivity in the Irish Sea*. Report prepared for the Marine Pollution Monitoring Group. Aquatic Environment Monitoring Report No. 32, MAFF, Lowestoft.

Klinkhammer, G. P. & Palmer, M. R. (1991). Uranium in the oceans: where it goes and why. *Geochimica et Cosmochimica Acta, 55*, 1799–1806.

Kochenov, A. V., Korolev, K. G., Dubinchuk, V. T. & Medvedev, Y. L., 1977. Experimental data on the precipitation of uranium from aqueous solution. *Geochemistry International, 14*, 82-87.

Koczy, F. F., Tomic, E. & Hecht, F. (1957). Zur Geochemie des Urans im Osteebecken. *Geochimica et Cosmochimica Acta, 11*, 86–102.

Kolodny, Y. & Kaplan, I. R. (1973). Deposition of uranium in the sediment and interstitial water of an anoxic fjord. In *Proceedings of A Symposium on Hydrochemistry and Biogeochemistry* (Vol. 1) (pp. 418–422). T. H. Clarke, Washington DC.

Krauskopf, K. O. (1988). *Radioactive Waste Disposal and Geology* (145pp). *Topics in the Earth Sciences* (Vol. 1). London, UK: Chapman and Hall.

Lander, G. H. & Fuger, J. (1989). Actinides: the unusual world of the 5f electrons. *Endeavour, 13*, 8–14.

Langmuir, D. (1978). Uranium solution-mineral equilibria at low temperatures with applications to sedimentary ore deposits. *Geochimica et Cosmochimica Acta, 42*, 547–569.

Langmuir, D. (1997). *Aqueous Environmental Geochemistry* (600pp). Upper Saddle River, NJ:. Prentice Hall.

Ledin, M. (2000). Accumulation of metals by microorganisms – processes and importance for soil systems. *Earth Science Reviews, 51*, 1–31

Levine, C. A. & Seaborg, G. T. (1951). The occurrence of plutonium in nature. *Journal of the American Chemical Society, 73*, 3278–3283.

Lienert, C. Short, S. A. & Von Gunten, H. (1994). Uranium infiltration from a river to a shallow groundwater. *Geochimica et Cosmochimica Acta, 58*, 5455–5463.

Lieser, K. H. & Muhlenweg, U. (1988). Neptunium in the hydrosphere and geosphere, (1) Chemistry of neptunium in the hydrosphere and sorption of neptunium from ground waters on sediments under aerobic and anaerobic conditions. *Radiochimica Acta, 43*, 27–35.

Liger, E., Charlet, L. & Van Cappellen, P. (1999). Surface catalysis of U(VI) reduction by Fe(II). *Geochimica et Cosmochimica Acta, 63*, 2939–2955.

Livens, F. R., Horrill, A. D. & Singleton, D. L. (1994). Plutonium in estuarine sediments and the associated interstitial waters. *Estuarine Coastal and Shelf Science, 38*, 479–489.

Lovley, D. R. (1995). Microbial reduction of iron manganese and other metals. *Advances in Agronomy, 54*, 175–231.

Lovley, D. R. & Phillips, E. J. P. (1992). Reduction of uranium by *Desulfovibrio desulfuricans*. *Applied and Environmental Microbiology, 58*, 850–856.

Lovley, D. R., Phillips, E. J. P., Gorby, Y. A. & Lander, E. R. (1991). Microbial reduction of uranium. *Nature, 350*, 413–416.

Lovley, D. R., Widman, P. K., Woodward, J. C. & Phillips, E. J. P. (1993). Reduction of uranium by cytochrome-c(3) of *Desulfovibrio vulgaris*. *Applied and Environmental Microbiology, 59*, 3572–3576.

Lovley, D. R., Methe, B., Maddalena, C., Nevin, K., Childers, S., Lloyd, J. & Leang, C. (2000). Genomic approach to the study of microbial reduction of iron and uranium in subsurface environments. In *Goldschmidt, 2000. Journal of Conference Abstracts, 5*, 650. Cambridge Publications.

Mackenzie, A. B., Scott, R. D., Allan, R. L., Ben Shaban, Y. A., Cook, G. T. & Pulford, I. D. (1994). Sediment radionuclide profiles: implications for mechanisms of Sellafield waste dispersal in the Irish Sea. *Journal of Environmental Radioactivity, 23*, 39–69.

Mackenzie, D. (2001). Off target. *New Scientist*, 13 January 2001, 5.

Mahara, Y. & Kudo, A. (1998). Probability of production of mobile plutonium in environments of soil and sediment. *Radiochimica Acta, 82*, 399–404.

Malcolm, S. J., Kershaw, P. J., Lovett, M. B. & Harvey, B. R. (1990). The interstitial water chemistry of $^{239,240}Pu$ and ^{241}Am in the sediments of the north-east Irish Sea. *Geochimica et Cosmochimica Acta, 54*, 29–35.

Martin, J.-M. & Meybeck, M. (1979). Elemental mass balance of material carried by major world rivers. *Marine Chemistry, 7*, 173–206.

Martin, J.-M., Meybeck, M. & Pusset, M. (1978). Uranium behaviour in the Zaire estuary. *Netherlands Journal of Sea Research, 12*, 338–344.

Martin, J. M., Nijampurkar, V. & Salvadori, F. (1978). Uranium and thorium isotope behaviour in estuarine systems. In *Biogeochemistry of Estuarine Systems* (pp. 111–127), UNESCO.

Martinez-Aguirre, A. & Perianez, R. (1999). Distribution of natural radionuclides in sequentially extracted fractions of sediments from a marsh area in Southwest Spain: U-isotopes. *Journal of Environmental Radioactivity, 45*, 67-80.

McKee, B. A. & Todd, J. F. (1993). Uranium behaviour in a permanently anoxic fjord: microbial control? *Limnology and Oceanography, 38*, 408–414.

Miholic, S. (1952). Radioactivity of waters issuing from sedimentary rocks. *Economic Geology, 47*, 543–547.

Milliman, J. D. & Meade, R. H. (1983). World-wide delivery of sediment to the oceans. *Journal of Geology, 91*, 1–21.

Milliman, J. D. & Syvitski, J. P. M. (1992). Geomorphic/tectonic control of sediment discharge to the ocean: the importance of small mountainous rivers. *Journal of Geology, 100*, 525–544.

Mo, T., Suttle, A. D. & Sackett, W. M. (1973). Uranium concentrations in marine sediments. *Geochimica et Cosmochimica Acta, 37*, 35–51.

Mohagheghi, A., Updegraff, D. M. & Goldhaber, M. B. (1985). The role of sulfate-reducing bacteria in the deposition of sedimentary ores. *Geomicrobiology Journal, 4*, 153–173.

Moore, W. S. (1967). Amazon and Mississippi river concentrations of uranium, thorium and radium isotopes. *Earth and Planetary Science Letters, 2*, 231–234.

Morris, K., Butterworth, J. C. & Livens, F. R. (2000). Evidence for the Remobilization of Sellafield Waste Radionuclides in an Intertidal Salt Marsh, West Cumbria, UK. *Estuarine Coastal and Shelf Science, 51*, 613–625.

Morris, K., Bryan, N. D. & Livens, F. R. (2001). Plutonium solubility in sediment pore waters. *Journal of Environmental Radioactivity, 56*, 259–267.

Morse, J. W. & Choppin, G. R. (1991). The chemistry of transuranic elements in natural waters. *Reviews in Aquatic Science, 4*, 1–22.

Moyes, L. N., Parkman, R. H., Charnock, J. M., Vaughan, D. J., Livens, F. R., Hughes, C. R. & Braithwaite, A. (2000). Uranium uptake from aqueous solution by interaction with goethite, lepidocrocite, muscovite and mackinawite: an X-ray absorption spectroscopy study. *Environmental Science and Technology, 34*, 1062–1068.

Murphy, W. M. & Shock, E. L. (1999). Environmental aqueous geochemistry of the actinides. In P. C. Burns & R. Finch (Eds), *Uranium: Mineralogy, Geochemistry and the Environment* (pp. 433–473). Washington DC: Mineralogical Society of America.

Myasoedov, B. F. & Drozhko, E. G. (1998). Up to date radioecological situation around the 'Mayak' nuclear facility. In G. R. Choppin & M. K. Khankhasayev (Eds), *NATO ASI Chemical Separation Technologies and Related Methods of Nuclear Waste Management: Applications, Problems, and Research Needs* (pp. 209–224). London, UK: Kluwer Academic Publishers.

Nash, J. T., Granger, H. C. & Adams, S. S. (1981). Geology and concepts of genesis of important types of uranium deposits. *Economic Geology, 75*, 63–116.

Nakashima, S., Disnar, J. R., Perruchot, A. & Trichet, J. (1984). Experimental study of the mechanisms of fixation and reduction of uranium by sedimentary organic matter under diagenetic or hydrothermal conditions. *Geochimica et Cosmochimica Acta, 48*, 2321–2329.

Nelson, D. M. & Lovett, M. B. (1981). Measurement of the oxidation state and concentration of plutonium in interstitial waters of the Irish Sea. In *Impacts of Radionuclide Releases into the Marine Environment* (pp. 105–118). IAEA Symposium. IAEA-SM-248/145.

Nguyen, S. N., Silva, R. J., Weed, H. C. & Andrews, J. E. (1992). Standard Gibbs free energies of formation at the temperature 303.15K for four uranyl silicates: soddyite, uranophane, sodium boltwoodite and sodium weeksite. *Journal of Chemical Thermodynamics, 24*, 359–376.

Palmer, M. R. & Edmond, J. M. (1993). Uranium in river water. *Geochimica et Cosmochimica Acta, 57*, 4949–4955.

Payne, T. E. & Waite, T. D. (1991). Surface complexation modelling of uranium sorption data obtained by isotope exchange techniques. *Radiochimica Acta, 52/53*, 487–493.

Pentreath, R. J., Harvey, B. R. & Lovett, M. B. (1986). Chemical speciation of long lived transuranium elements in the marine environment. In *Speciation of Fission and Activation Products in the Environment* (312pp). London, UK and New York: Elsevier Applied Science.

Pentreath, R. J. (1988). Sources of artificial radionuclides in the marine environment. In J. C. Guary, P. Guegueniat & R. J. Pentreath (Eds), *Radionuclides, a Tool for Oceanography* (pp. 12–34). London, UK: Elsevier Science.

Perkins, R. W. & Thomas, C. W. (1980). Worldwide Fallout. In W. C. Hanson (Ed.), *Transuranic Elements in the Environment* (pp. 53–82). US Department of Energy, DOE/TIC–22800, Virginia.

Pietzsch, K., Hard, B. C. & Babel, W. (1998). A simple model for determination of the ability of sulfate-reducing bacteria to reduce U(VI). In *Abstracts of Euroconference on Bacteria-Metal/Radionuclide Interaction: Basic Research and Bioremediation* (pp. 91–92).

Plater, A. J. M., Ivanovich, M. & Dugdale, R. E. (1992). Uranium series disequilibria in river sediments and waters: the significance of anomalous activity ratios. *Applied Geochemistry, 7*, 101–110.

Prikryl, J. D., Pabalan, R. T., Turner, D. R. & Leslie, B. W. (1994). Uranium sorption on α-alumina: effects of pH and surface area/solution volume ratio. *Radiochimica Acta, 66/67*, 291–296.

Poulton, S. P. & Raiswell, R. (2000). Solid phase associations, oceanic fluxes and the anthropogenic perturbation of transition metals in world river particulates. *Marine Chemistry, 72*, 17–31.

Putnik, H. (1996). Identification and radiological characterisation of contaminated sites in Estonia. In IAEA (1996). *Planning for Environmental Restoration of Radioactively Contaminated Sites in Central and Eastern Europe*, Vol. 1: *Identification and Characterisation of Contaminated Sites* (pp. 137–142). IAEA-TECDOC-865, International Atomic Energy Agency, Vienna.

Raiswell, R., Canfield, D. E. & Berner, R. A. (1994). A comparison of iron extraction methods for the determination of degree of pyritization and the recognition of iron-limited pyrite formation. *Chemical Geology, 111*, 101–111.

Regnier, P. & Wollast, R. (1993). Distribution of trace metals in suspended matter of the Scheldt estuary. *Marine Chemistry, 43*, 3–19.

Sagert, N. H., Ho, C. H. & Miller, N. H. (1989). The adsorption of uranium (VI) onto a magnetite soil. *Journal of Colloid and Interface Science, 130*, 283–287.

Sarin. M. M. & Church, T. M. (1994). Behaviour of uranium during mixing in the Delaware and Chesapeake estuaries. *Estuarine, Coastal and Shelf Science, 39*, 619–631.

Shaw, T. J., Sholkovitz, E. R. & Klinkhammer, G. (1994). Redox dynamics in the Chesapeake Bay: the effect on sediment/water uranium exchange. *Geochimica et Cosmochimica Acta, 58*, 2985–2995.

Sheppard, M. A. & Thibault, D. H. (1992). Desorption and extraction of selected heavy metals from soils. *Journal of the Soil Science Society of America, 56*, 415–423.

Sholkovitz, E. R. (1976). Flocculation of dissolved organic and inorganic matter during mixing of river water and seawater. *Geochimica et Cosmochimica Acta, 40*, 831-845.

Sholkovitz, E. R. (1983). The geochemistry of plutonium in fresh and marine water environments. *Earth Science Reviews, 19*, 95–161.

Sholkovitz, E. R. & Mann, D. R. (1984). The pore-water chemistry of $^{239,240}Pu$ and ^{137}Cs in sediments of Buzzards Bay, Massachusetts. *Geochimica et Cosmochimica Acta, 48*, 1107–1114.

Sholkovitz, E. R., Boyle, E. A. & Price, N. B. (1978). The removal of dissolved humic acids and iron during estuarine mixing. *Earth and Planetary Science Letters, 40*, 130–136.

Sholkovitz, E. R., Carey, A. E. & Cochran, J. K. (1982). Aquatic chemistry of plutonium in seasonally anoxic lake waters. *Nature, 300*, 159–161.

Sholkovitz, E. R., Cochran, J. K. & Carey, A. E. (1983). Laboratory studies of the diagenesis and mobility of $^{239,240}Pu$ and ^{137}Cs in near shore sediments. *Geochimica et Cosmochimica Acta, 47*, 1369–1379.

Silva, R. J. & Nitsche, H. (1995). Actinide environmental chemistry. *Radiochimica Acta, 70/71*, 377–396.

Silva, R. J., Bidoglio, G., Rand, M. H., Robouch, P. B., Wanner, H. & Puigdomenech, I. (1995). *Chemical Thermodynamics of Americium* (374pp). Nuclear Energy Agency. Amsterdam: Elsevier Science.

Swarzenski, P. W. & McKee, B. A. (1998). Seasonal uranium distributions in the coastal waters of the Amazon and Mississippi rivers. *Estuaries, 21*, 379–390.

Swarzenski, P. W., McKee, B. A. & Booth, J. G. (1995). Uranium geochemistry on the Amazon shelf: chemical phase partitioning and cycling across a salinity gradient. *Geochimica et Cosmochimica Acta, 59*, 7–18.

Swarzenski, P. W., McKee, B. A., Skei, J. M. & Todd, J. F. (1999). Uranium biogeochemistry across the redox transition zone of a permanently stratified fjord: Framvaren, Norway. *Marine Chemistry, 67*, 181–198.

Suzuki, Y. & Banfield, J. F. (1999). Geomicrobiology of uranium. In P. C. Burns & R. Finch (Eds), *Uranium: Mineralogy, Geochemistry and the Environment* (374pp). Washington DC: Mineralogical Society of America.

Tebo, B. M. & Obraztsova, A. Y. (1998). Sulfate-reducing bacterium grows with Cr(VI), U(VI), Mn(IV), and Fe(III) as electron acceptors. *FEMS Microbiology Letters, 162*, 193–198.

Todd, J. F., Elsinger, R. J. & Morre, W. S. (1988). The distributions of uranium, radium and thorium isotopes in two anoxic fjords: Framvaren Fjord (Norway) and Saanich Inlet (British Columbia). *Marine Chemistry, 23*, 393–415.

Tuovinen, O. H. & Kelly, D. P. (1974). Studies on the growth of *Thiobacillus ferroxidans* II. Toxicity of uranium to growing cultures and tolerance conferred by mutation, other metal cations and EDTA. *Archives of Microbiology, 95*, 153–164.

US DoE. (1997). *Linking Legacies. Connecting the Cold War Nuclear Weapons Production Processes to Their Environmental Consequences* (230pp). US Department of Energy, Washington DC.

Van der Weijden, C. H., Van Leeuwen, M. & Peters, A. F. (1985). The adsorption of U (VI) onto precipitating amorphous ferric hydroxide. *Uranium, 2*, 53–58.

Waite, T. D., Davis, J. A., Payne, T. E., Waychunas, G. A. & Xu, N. (1994). Uranium (VI) adsorption to ferrihydrite: applications of a surface complexation model. *Geochimica et Cosmochimica Acta, 58*, 5465–5478.

Weber, F. F. & Sackett, W. M. (1973). Uranium geochemistry of Orca Basin. *Geochimica et Cosmochimica Acta, 45*, 1321–1329.

Whalgren, M. A., Robbins, J. A. & Edgington, D. N. (1980). Plutonium in the Great Lakes. In W. C. Hanson (Ed.), *Transuranic Elements in the Environment* (pp. 659–683). US Department of Energy, Springfield, VA. US DOE/TIC/22800.

Wielinga, B. A., Bostick, B., Hansel, C. M., Rosenzweig, R. F. & Fendorf, S. (2000). Inhibition of bacterially promoted uranium reduction: Ferric (hydr)oxides as competitive electron acceptors. *Environmental Science and Technology, 34*, 2190–2195.

Wildung, R. E. & Garland, T. R. (1980). The relationship of microbial processes to the fate and behaviour of transuranic elements in soils, plants and animals. In W. C. Hanson (Ed.). *Transuranic Elements in the Natural Environment* (pp. 300–335). United States Department of Energy, Springfield, VA. US DOE/TIC/22800.

Willett, I. R. & Bond, W. J. (1995). Sorption of manganese, uranium and radium by highly weathered soils. *Journal of Environmental Quality, 24*, 834–845.

Windom, H., Smith, R., Niencheski, F. & Alexander, C. (2000). Uranium in rivers and estuaries of globally diverse, smaller watersheds. *Marine Chemistry, 68*, 307–321.

Woodhead, D. S. (1999). Actinides in the Irish Sea. *Journal of Environmental Radioactivity, 44*, 127–403.

Woodwell, G. M., Rich, P. H. & Hall, C. A. S. (1973). Estuaries. In G. M. Woodwell & E. V. Pecan (Eds), *Brookhaven Symposium Biological Series 24, Carbon and the Biosphere: Proceedings of the 24th Brookhaven Symposium in Biology*. Upton, NY, 16–18 May 1972. US Atomic Energy Commission, Washington, DC.

Zaijic, J. (1969). Uranium biogeochemistry. In *Microbial Biogeochemistry* (pp. 179–195). New York: Academic Press.

Zielinski, R. A. & Meier, A. L. (1988). The association of uranium with organic matter in Holocene peat: an experimental leaching study. *Applied Geochemistry, 3*, 631–643.

Miranda J. Keith-Roach and Francis R. Livens (Editors)
© 2002 Elsevier Science Ltd. All rights reserved

Chapter 5

The effects of humic substances on radioactivity in the environment

Rose E. Keepax, Dominic M. Jones, Sarah E. Pepper, Nicholas D. Bryan

Centre for Radiochemistry Research, Department of Chemistry, University of Manchester, Oxford Road, Manchester, M13 9PL, UK

1. Introduction

Humic substances are an important class of organic molecules, which play a vital role in both the carbon cycle and the biogeochemistry of virtually all metallic elements (Livens, 1991). Since most radionuclides are metallic, including Tc, Np, Pu, Am, any treatment of environmental radioactivity cannot be complete without considering humic substances and their effects. It is thought that they are the by-products of microbial activity (Stevenson, 1982). Hence, any effects due directly to humic substances may also be attributed indirectly to microbes, and it is right to consider humic effects alongside other microbial processes.

Most research relevant to environmental radioactivity has examined generic metal-humate behaviour. However, these studies are applicable to radionuclides. Hence, in the sections below, the term 'metal' can be taken to imply virtually all radionuclides. Metal-humic interactions have been studied extensively over the last few decades. Most studies have concentrated upon the initial, reversible or exchangeable interaction. However, it has become clear that, beyond the exchangeable interaction, there is also a non-exchangeable component which develops with time. This aspect of humic behaviour is still poorly understood. However, several possible explanations are advanced.

Humics research is replete with models and modelling studies, and the diversity can often be confusing. In the following sections, various types and families of models are discussed, from the models of humic structure, which tend to be conceptual, to the various metal binding models, which tend to be more mathematical. Much of the metal-humate modelling has been driven by the need to predict the behaviour of radionuclides in the environment. In particular, much of the effort has been directed at models designed to test the long-term safety of radioactive waste repositories. The discovery of slow desorption kinetics has yet to be fully incorporated into these studies. However, some preliminary studies have shown that kinetics will have a significant impact.

Some of the work discussed here was undertaken between January 1997 and December 1999 as part of the recent EU project, *HUMICS, The Effects of Humic Substances on the Migration of Radionuclides: Complexation and Transport of the Actinides.*

2. Humic substances

Humic substances are universal and will be found wherever organic matter is being decomposed (Hayes et al., 1989). In soils, plant and animal residues can remain for different periods of time, depending on their susceptibility to decomposition by microorganisms, and may accumulate over time. Materials that remain resistant to decomposition and stay in a modified form as amorphous, brown materials below the surface layers of soil, are humic substances. In total, humic substances make up approximately 60-70% of soil organic matter (Jones & Bryan, 1998). They possess certain characteristics, which make them different to the materials from which they have been formed. For example (Wood, 1995), they dissolve in alkali and certain components may dissolve in acid. Humic substances also have a high capacity for proton exchange and the capacity to adsorb heavy metals as well as pesticides and other organic chemicals. Humic substances are abundant in aquatic systems, both in the sediments and also dissolved in the water of streams, rivers, lakes and oceans. They are formed from the condensation reactions of quinones and phenolic compounds, which are themselves formed from biological transformations of plant residues (Livens, 1991), giving organic polyelectrolyte macromolecules derived from a combination of the constituent molecules of the debris (Jones & Bryan, 1998, Stevenson, 1982). Although they are the principal organic components of soils and waters throughout the world, their composition may vary greatly, depending on their geographical location, origin and history (Kononova, 1966; Stevenson, 1982).

The shape, size and solubility of these substances have been shown to be strongly dependent on pH. Their solubility increases with pH (Hayes et al., 1989), due to the progressive ionisation of carboxylic acid groups and phenolic groups causing the macromolecules to repel and separate. In addition, the molecular arrangements become smaller and better orientated (Swift, 1989a, b). At low pH values, and in the presence of high metal concentrations, the macromolecular particles tend to aggregate, forming bundles of elongated fibres, as a direct result of interactions such as hydrogen bonding, Van der Waals forces, interactions between the electrons of adjacent ring structures and homolytic reactions of free radicals (Swift, 1989a, b; Wershaw, 1986, 1993).

The most important property of humic substances is their heterogeneity, both in terms of molecular structure and weight (Livens, 1991; Jones & Bryan, 1998). In order to understand their reactions with metals it is important to consider their structures, and particularly their functional group concentrations. Unfortunately, this information is hard to obtain, because humic substances are complex mixtures, which do not exhibit precise physical and chemical characteristics (McBride, 1989). No two humic molecules have been found to be exactly the same, and as a result the properties that are measured are only averaged properties (Livens, 1991).

The classification and characterisation of humic substances

There are three classes of humic substance, which are defined by their solubilities (Oden, 1914):

(i) humin – insoluble in aqueous systems at all pH values;
(ii) humic acid – soluble at pH > 2; and
(iii) fulvic acid – soluble in water at any pH.

This classification is preferred to one based on chemical and structural criteria, because all humic substances are heterogeneous mixtures exhibiting similar macroscopic properties, but of different microscopic chemical structure (Buffle et al., 1990; Livens, 1991). However, despite this being the best, agreed method of operationally defining humic substances, it is not entirely satisfactory, since it can also include organic compounds that are not humic substances (Swift, 1999). Generally, such classification tends to oversimplify what are complex colloidal mixtures. Further subdivision of humic acid is possible. Hymatomelanic acid is any part of humic acid that can be extracted into ethanol. From the remaining humic acid, that which is precipitated at high ionic strength is called 'Grey humic acid'. The remaining humic substance is called 'Brown humic acid' (Stevenson, 1982). However, these procedures are rarely used. While some behaviour, for example solubility, is specific to the individual fractions, other properties are common to all. The terms 'humics' and 'humic substances' should be taken to refer to all fractions (both of these are used here).

There are further distinguishing features of each of the fractions. As a general rule, humic acids are of greater molecular weight than fulvic acids and have a greater aromatic composition compared to fulvic acids (40-60% and 20-50% respectively) (Choppin, 1988). Generally, fulvic acids are poorly polymerised and condensed and contain a large number of aliphatic side chains. In terms of their chemical composition, carbon and oxygen are the major elements found in humic and fulvic acids and, on average, there is little difference between the two acids in terms of hydrogen, nitrogen and sulphur content. However, the humic fraction typically contains 10% more carbon and 10% less oxygen than fulvic acid. Functional group analysis shows that the total acidity of fulvic acids is higher than that of humic acids, especially in terms of the carboxylic acid content. Fulvic acids also contain a greater proportion of alcoholic OH groups but both acid types contain roughly the same amount of phenolic, carbonyl and methoxy groups per unit weight (Schnitzer, 1978).

The extraction of humic substances

In order to study their properties, humic substances must first be extracted quantitatively from the matrix without disrupting their physical or chemical properties. However, this is difficult to achieve in practice. The most common extractant is sodium hydroxide (Livens, 1991), which deprotonates the humic material and gives high yields (Stevenson, 1982). Such an aggressive extractant may chemically alter the humic material (Gregor and Powell, 1987), since it will encourage degradation, condensation and auto-oxidation reactions (Livens, 1991). Alternatives to NaOH include $Na_4P_2O_7$ but this is generally less effective than NaOH (Kononova, 1966). Fulvic acids, which are the only soluble components at low pH, have been selectively extracted using acids (Schnitzer et al., 1958). Organic solvents (e.g. oxalic acid, formic acid, phenol, benzene, chloroform, acetylacetone, hexamethylenetetramine, dodecylsulfate and urea), have also been used (Schnitzer & Kahn, 1972).

3. Formation of humic substances

The formation of humic substances occurs via humification, which is important in the maintenance of balance in the carbon cycle, and is thought to be mediated by microbes (Swift et al., 1987). Microorganisms play an important role in many geochemical processes by mediating crucial biochemical reactions, for example, the production of petroleum, kerogen, coal, and lignite. Organic carbon is returned to the atmosphere as CO_2 when microorganisms degrade carbon sources such as biomass or petroleum. Hydrocarbons (such as those found in crude oil), synthetic hydrocarbons and pollutant hydrocarbons (for example, those accidentally spilled on soil or water) are also degraded (Manahan,1994). Under ideal conditions, when an organism dies, its remains undergo complete mineralisation by consortia of microorganisms, causing the release of CO_2. However, when mineralisation is incomplete, complex, heterogeneous and polymeric compounds (i.e. humic substances), which are relatively resistant to microbial decay, form (Hesketh, 1995). These humic substances do turn over at a finite rate (Swift et al., 1987). Indeed, if they were not subject to some decomposition, however slow, an enormous reservoir of soil organic carbon would develop. The whole series of reactions, from the decomposition of dead organisms to the enzyme reactions that form complex polymers of proteins, lipids, lignins, tannins and polyphenols which can then interact with each other, is called humification (Hatcher et al., 1983). Such reactions occur in terrestrial soils, surface and subsurface sediments and wherever organic residues are present in aquatic environments (Hesketh, 1995).

There are many precursor molecules that could be involved in the synthesis of humic substances (Hessen & Tranvik, 1998), for example chlorophyll, polysaccharides and sterol. These precursors contain a large number of carbon-carbon double bonds and in the case of chlorophyll, nitrogen containing groups. Water soluble, precursor humic and fulvic molecules, which are mobile in soils and sediment interstitial water, are derived from the action of extracellular enzymes on these molecules. Further modification can occur after transport of the precursors, and this may involve microbiological processes or fractionation by sorption and interaction with inorganic surfaces (McKnight & Aiken, 1998).

The formation of humic substances in soils and aquatic environments could be through one of many pathways, for example, directly from the lignified tissues of plant material, or through polymerisation of simple products generated in the degradation of plant material (McKnight & Aiken, 1998). The actual formation of humic substances, from plant and animal remains, is subject to debate. However, there are four possible mechanisms (Stevenson, 1982; Hesketh, 1995):

(i) the plant residues undergo microbial transformation to form reducing sugars and amino acids. These undergo non-enzymatic polymerisation to form humic substances.

(ii) The non-lignin (for example, cellulose) sources of plant residue form polyphenols as a result of microbial metabolism. The polyphenols form quinones via enzymatic oxidation which, in the presence or absence of amino compounds, polymerise to form humic substances.

(iii) The plant residues yield phenolic aldehydes and acids which are released from lignin during microbial decomposition. Via an enzymic conversion, the phenolic aldehydes and acids form quinones which then polymerise to form humic substances.

(iv) Incomplete utilisation of lignin by the microorganisms yields modified lignins, which undergo demethylation, oxidation and condensation with nitrogen containing compounds, such as proteins, to form humic substances.

A unifying theory, which incorporates the three most widely accepted mechanisms (i, ii and iv), has been put forward by Stevenson (1982). In addition to the plant/animal derivation of humic substances, it is thought that algal metabolism is also a mechanism of formation (McBride, 1989). Free radical cross-linking of unsaturated lipids, which are released into seawater during algal growth, forms humic substances (Harvey and Boran, 1985). This distinct mechanism could account for the existence of marine humic and fulvic acids.

The lignin mechanism

Lignin makes up a plant's structural material and is found in large amounts in bark as well as in lower amounts in leaf litter, grasses and stems (Kononova, 1966; Alexander, 1977). Unlike many biomolecules, it has an irregular structure, because of a lack of control in its formation at a cellular level. Lignin is principally made up of units such as coniferyl alcohol, p-hydroxycinnamyl alcohol and synapyl alcohol. It is a complex aromatic polymer in which the repeating units are linked in an irregular way by strong ether and carbon bridges (Kononova, 1966; Alexander, 1977). These are not easily broken, making lignin resistant to decomposition and giving it its irregular arrangement of cyclic and branched structures (Rheinheimer, 1974). It can be degraded by only relatively few fungal and bacterial species (see Chapter 1, this volume), and only under aerobic conditions, to release polyphenol and phenolic compounds. These can then participate in the formation of humic substances. The degradation of lignin is a non-specific process, carried out by exocellular enzymes. The products from these processes, such as quinones, contain carboxyl groups resulting from oxidation, which may, in turn increase the solubility of larger compounds formed by condensation reactions (Kirk, 1984).

The sugar-amine mechanism

Polyphenols (quinones) are important reactive monomers that are synthesised by microorganisms (Stevenson, 1982). From them, larger molecules, which are resistant to degradation, form by polymerisation. An example is the oxidative polymerisation of phenol derivatives with amino sugar units. This is an attractive theory, since it uses reactants such as sugars and amino acids, which are of abundance in natural systems through microbial activity (Steinberg & Munster, 1985). A more detailed mechanism was proposed by Stevenson (1982). The initiation involves the addition of an amine to the aldehyde group of a reducing sugar to form a Schiff base and then the n-substituted glycosylamine. An Amadori rearrangement, involving the glycosylamine, then yields a N-substituted, 1-amino-deoxy-2-ketose which undergoes fragmentation and water loss. The resulting compounds, such as acetol, glyceraldehyde, dihydroxyacetone, reductones and hydroxymethyl furfural are all highly reactive and readily polymerise in the presence of amino compounds to form humics.

The polyphenol mechanism

The polyphenol mechanism is considered to be one of the most probable routes to humic substances (Hesketh, 1995). It is based on quinones, derived from and/or synthesised by microorganisms. The first stage is the formation of polyphenols by transformation of lignin and degradation of cellulose by fungi. These processes, carried out by extracellular enzymes of fungi, occur within, or perhaps more likely, on the exposed edges of intact molecules (Martin & Haider, 1971). C6–C3 units are oxidised to yield a series of low molecular weight aromatic acids and aldehydes. These include syringaldehyde, syringic acid, p-hydroxybenzaldehyde, p-hydroxybenzoic acid, protocatechuic acid and gallic acid. In addition, polyphenols are produced enzymatically from dead plant or microbial cells, whilst autocatalytic enzymes are still functioning, but before cell walls are ruptured by microbes (Hesketh, 1995). However the precursor polyphenols are formed, quinones are then formed by enzymatic reactions and/or spontaneous oxidation. Demethylation is also thought to occur at this stage due to the large number of methoxy groups and higher phenolic hydroxyl groups present in humic and fulvic acids. Further condensation of the quinones, especially in the presence of amino acids, occurs and finally nitrogenous polymers form. The final stage occurs on the enzymatic oxidation of the quinones in the presence of amino acids, proteins and peptides (Stevenson, 1982).

4. Colloidal properties

Colloidal size and structure

Many of the special properties exhibited by humic substances are due to their size. Fulvic acids have weight average molecular weights in the region of 500 to several thousands, whilst humic acids are in the range 5000 to several tens of thousands, usually up to 50,000 (Jones & Bryan, 1998). Hence, although humic acids have sizes at the lower end of the colloidal range, fulvic acids are, by some definitions, too small to qualify as 'proper' colloids. Nevertheless, although there are differences in behaviour between humic and fulvic acids, there are enough similarities to suggest that they share many properties.

In many ways, the physical effects due to their colloidal nature, are more important in determining the impact of humic substances on environmental radioactivity than their chemical properties. Unfortunately, these colloidal properties are more difficult to understand due to the significant polydispersity (broad molecular weight distributions) of humic and fulvic acids. In fact, not only do humic substances have a range of weights, but there is also some associated chemical heterogeneity (Buffle et al., 1990).

Although it is accepted that humic substances display a broad spectrum of molecular weights, the physical and chemical nature of the species existing in solution is uncertain. The properties of humic substances themselves create experimental difficulties in their characterisation in terms of their mass, mass distribution and deviations from ideality. They have a wide mass distribution, and concentration-dependent association of humic substances makes distinguishing between intrinsically high mass species and aggregates of high mass difficult (Jones & Bryan, 1998).

In the past, humics were thought of, and modelled, as rigid, impenetrable spheres with the metal binding sites on the surface (Tipping & Hurley, 1992). However, in recent years it has become clear that this model is an unsatisfactory description of the system. In fact, there is a growing body of evidence that humics do have fairly rigid structures, but that the interior volume of the colloid is accessible to solvent molecules and counterions, such as Na^+ (Benedetti et al., 1996; Tombacz et al., 1997; Buffle et al., 1998; Chin et al., 1998). Certainly, humic substances do display broadly hydrophilic properties when charged, although they can become hydrophobic when that charge is neutralised, for example by metal binding (Jouany & Chassin, 1987; Milne et al., 1995a, b; Kaiser, 1998). More than this, it has become clear that humics undergo at least some degree of expansion and contraction as solution conditions, such as ionic strength and pH, change (Benedetti et al., 1996; Schimpf & Petteys, 1997; Chin et al., 1998). This behaviour has been attributed to the protonation/deprotonation of humic carboxylate groups and the shielding of the humic charge (Benedetti et al., 1996; Tombacz et al., 1997; Schimpf & Petteys, 1997). In addition to these intra-colloidal changes, some have suggested that an increase in ionic strength could also lead to loose aggregates of humic 'monomers' due to the reduction in electrostatic repulsion (Buffle et al., 1998).

For a macromolecule, the radius of gyration, R_G, is defined as the root mean square distance of the electrons in the particle from the centre of charge (Wershaw, 1989). For a humic in solution, R_G varies from 1.5 nm for the lowest molecular weight material to greater than 2.5 nm for the higher weight fractions (Cameron et al., 1972a). Size exclusion chromatography has suggested that humics have structures midway in character between polysaccharides, which have expanded conformations, and proteins, which are more condensed (Cameron et al., 1972b).

Two models of humic structure have been proposed. The *random coil* approach treats the humic as an essentially contiguous single molecular strand, and assumes that the loose coiling of that strand accounts for the penetrable gel-like properties. Conversely, the *self association* approach assumes that the humic species which exist in solution are in fact (loose) associations of smaller units, and that this accounts for the penetrability.

The Random Coil model

The Random Coil model treats the humic as a molecular strand, which coils randomly with respect to time and space. This strand carries, along its length, charged and hydrated functional groups. It assumes a roughly spherical conformation with a Gaussian mass distribution about its centre (that is, the mass density is greatest at the centre and decreases to zero at the edge). The solvent penetrates throughout the structure, and that at the periphery exchanges freely with the bulk solvent. As density increases, towards the centre of the colloid, the solvent might become trapped (Swift, 1999). The colloid may be tightly or loosely coiled depending upon several factors: the nature of the solvent; the extent of the solvent penetration; the charge of the colloid; the concentration of counterions; the surrounding pH (Hayes & Swift, 1978, 1990).

It has been suggested that the carboxylate groups of humic substances, which will be deprotonated at ambient pH, account for the relatively high charge of humic colloids, and the associated inter- and intramolecular forces (Swift, 1999). Typically, a soil humic substance will contain 4–8 carboxylate groups per 1000 Daltons (Swift, 1996). The variable conformation of humics is explained in terms of these charges. At neutral to alkaline pH, the charged sites will be fully dissociated, giving rise to electrostatic repulsion within the colloid. In an attempt to reduce its electrostatic free energy, it will expand (and rearrange). The species in solution will be highly charged, expanded and highly solvated with water. Intramolecular expansion and solvation forces, together with intermolecular repulsion, will keep these species in solution. They are able to rotate, flex and respond to external influences by structural rearrangement. Increasing ionic strength will reduce the repulsion and will lead to contraction. Adding specifically bound metal ions will reduce the magnitude of the charge and hence will also lead to contraction, as will protonation at lower pH. The colloid will shrink until the point where all the solvent is expelled and the flexible structure has shrunk to its most collapsed state.

In the solid state, short-range, non-electrostatic forces are dominant and only very tightly bound water is retained. Solid state humic substances may retain a certain amount of flexibility and solvent accessibility due to the presence of some entrapped water. It could be this that gives rise to rapid ion exchange reactions and other interactions at the humic surface (Swift, 1999).

The intermolecular interactions are controlled by very similar factors. Two highly charged colloids in solution will repel each other (Swift, 1989a, b). High ionic strengths or charge neutralisation suppress the electrostatic repulsive forces, which are then superseded by short-range attractive forces. Such attractive interactions include Van der Waals forces, dipolar interactions and hydrogen bonds. At the point where the molecules are so poorly solvated and/or charged, that they can no longer remain in solution, they aggregate, and eventually precipitate (Swift, 1989a).

Information about molecular size and shape may be determined with ultracentrifugation (Cameron et al., 1972a; Swift, 1989 a, b). These results have provided evidence for the Random Coil model of humic substances, since they were in good agreement with the expected theoretical relationship (Swift, 1989a, b). Further evidence for a random coiled structure, which also contradicts a hard, rigid sphere model, includes the random ternary structure, high charge density, rapid ion exchange and high water uptake of humic macromolecules (Swift, 1999).

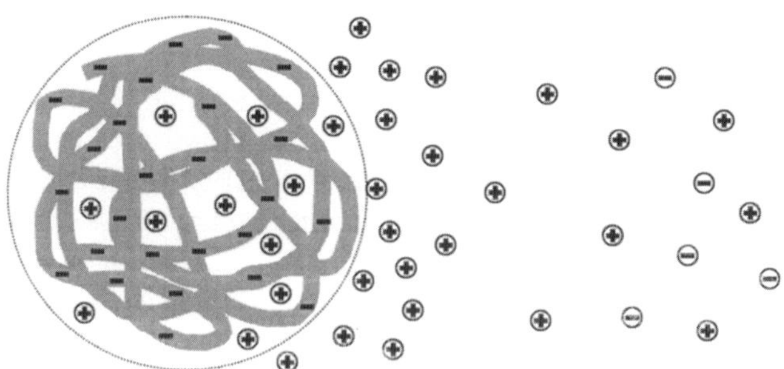

Fig. 1. Random Coil/Penetrable Gel model of humic structure showing cross-linked molecular strands carrying negative charges and penetrating counterions, plus distribution of ions in the double layer.

In its strictest sense, the Random Coil approach does treat the humic as a single molecular chain, and this does seem to be a reasonable approximation to the real system. However, given the complex and chaotic mechanisms of humic formation, it seems unlikely that humic colloids really are a single contiguous chain. In fact, Schulton & Schnitzer (1993) have found experimental evidence for significant cross-linking. Hence, it might be more realistic to think of the humic as having a sponge-like structure, which is cross-linked, but is still able to expand and contract to allow penetration by the solvent and small ions. The net effect is a penetrable gel-like structure (Fig. 1) (Benedetti et al., 1996).

The self-association model

This hypothesis holds that the humic species observed in solution are actually aggregates of smaller moieties (Wershaw, 1986, 1993; Piccolo et al., 1996; Piccolo, 1997; Conte & Piccolo, 1999). If this theory is correct, then the aggregation mechanism is not simple, since one would expect that the degree of aggregation/disaggregation would depend heavily upon the charge of the 'monomer units'. Hence, one might expect that systems with the same charge would behave very similarly. However, there is evidence that these systems are more complex than that. For example, the same concentrations of hydrochloric acid are reported to have had no effects in some cases but substantial effects in others (Piccolo et al., 1996; Conte & Piccolo, 1999). Another study (De Nobili & Chen, 1999) has subjected humics to rigorous treatments, such as acid precipitation, washing, redissolution and dialysis. However, none of this had any effect upon the weight distribution of the samples. If the humic species that exist in solution were fairly weakly bound agglomerates, then one might have expected these procedures to have an effect (Swift, 1989a, b).

The mechanism by which these smaller molecules might be held together in solution is also uncertain. Hydrogen bonds, π-bonds, charge transfer complexes and bridging metal ions have been postulated (Wershaw, 1986, 1993) in addition to hydrophobic interactions (Piccolo et al., 1996; Piccolo, 1997). Multivalent cations are capable of bridging between charged carboxylate groups in the same or different molecules, and humic substances are known to precipitate on addition of large quantities of such cations (Piccolo et al., 1996). Thus, for the humic substance to remain in solution, only a small percentage of sites can participate in bridging. Hydrophobic interactions are very weak interactions and so are unlikely to give rise to strong agglomerates. In addition to this, the strong hydration of the charged sites and other functional groups will disrupt such weak interactions. Therefore, it seems that for any molecular aggregation to survive, it would have to be held together by something stronger than hydrophobic interactions. In fact, in order to survive, the aggregates would have to be linked by interactions comparable to a covalent bond (Swift, 1999).

The double layer

Humic substances in solution will have a large negative charge due to the ionisation of their functional groups, particularly carboxylates (Jones & Bryan, 1998). This charge generates an associated potential, which attracts positive ions in the solution, and repels anions (Bartschat et al., 1992). The result is that the humic acid is surrounded by an atmosphere of cations, called a *double layer* (Fig. 1). The double layer has a significant impact upon the interaction of humics with radionuclides, which largely exist in solution as cationic species (Jones & Bryan, 1998).

The potential and double layer developed by a colloid will depend upon the colloidal charge, ionic strength and the physical state of the colloid itself. Colloids are often thought of as impenetrable, with their charges distributed over the surface (Tanford, 1961). A number of studies have treated humics in this way, either as impenetrable spheres (Nederlof et al., 1993) or cylinders (DeWit et al., 1993; Milne et al., 1995a, b). However, as discussed earlier, there is evidence that humics are not impenetrable, and that solvent and small ions are able to gain access to the interior. In fact, some authors have ceased using the impenetrable approach in favour of a penetrable one, since the later provides a better description of the system (Kinniburgh et al., 1996; van Riemsdijk et al., 1996). Hence, the theory described below is specific to a spherical penetrable humic.

The potential, ψ, around a charged colloid particle will be governed by Poisson's equation (Tanford, 1961):

$$\nabla^2\psi = \frac{-\rho}{\varepsilon} \tag{1}$$

where∇^2 is the Laplacian operator, ρ is the charge density, and ε is the permittivity. ρ will vary with distance from the centre of the colloid, and will depend upon both the humic and counterion charges (Bartschat et al., 1992). The concentration of any counterion, i, will be governed by a Boltzmann equation (Tanford, 1961),

$$[X_i^{z_i+}]_r = [X_i^{z_i+}]_{\mathrm{BULK}} e^{-\left(\frac{\psi_r z_i e}{kT}\right)} \tag{2}$$

where ψ_r is the potential at a distance, r, from the centre of the colloid, and $[X_i^{z_i+}]_{\mathrm{BULK}}$ and $[X_i^{z_i+}]_r$ are the counterion concentrations in the bulk solution and at r respectively, and z_i is the charge number of the cation.

Now, there are two distinct regions that need to be considered: region I is the colloid itself, and here there will be charge from both the counterions and the humic. In region II, only the counterions make a contribution. Combining equations (1) and (2) gives two '*Poisson-Boltzmann*' equations (Bryan et al., 2000): one for region II

$$\nabla^2\psi_{\mathrm{II}} = -\frac{N_A e}{\varepsilon}\left(\sum_{i=1}^{N_{\mathrm{ions}}} z_i [X_i^{z_i}]_{\mathrm{BULK}} \exp^{-\left(\frac{\psi_r z_i e}{kT}\right)}\right) \tag{3}$$

and the other for region I,

$$\nabla^2\psi_{\mathrm{I}} = -\frac{1}{\varepsilon}\left[\rho_{\mathrm{HUMIC}} + \beta N_A e\left(\sum_{i=1}^{N_{\mathrm{ions}}} z_i [X_i^{z_i}]_{\mathrm{BULK}} \exp^{-\left(\frac{\psi_r z_i e}{kT}\right)}\right)\right] \tag{4}$$

where ρ_{HUMIC} is the charge density of the humic and β is a factor to take account of the fact that not all of the volume is available to the solvent and counterions. The boundary conditions are (Bartschat et al., 1992, Bryan et al., 2000),

$$\left(\frac{\partial\psi}{\partial_r}\right)\Bigg|_{r=0} = 0 \tag{5}$$

and

$$\psi_r \to 0 \qquad \text{as} \qquad r \to \infty \tag{6}$$

The magnitude and shape of the potential is dependent upon the humic charge: the potential at any point increases with increasing humic charge, and decreases with increasing ionic strength. In addition, the extent of the double layer is affected, shrinking or compressing as ionic strength increases. In addition, the changing conditions will affect the degree of expansion, and hence the size of the colloid. These effects are illustrated in Fig. 2.

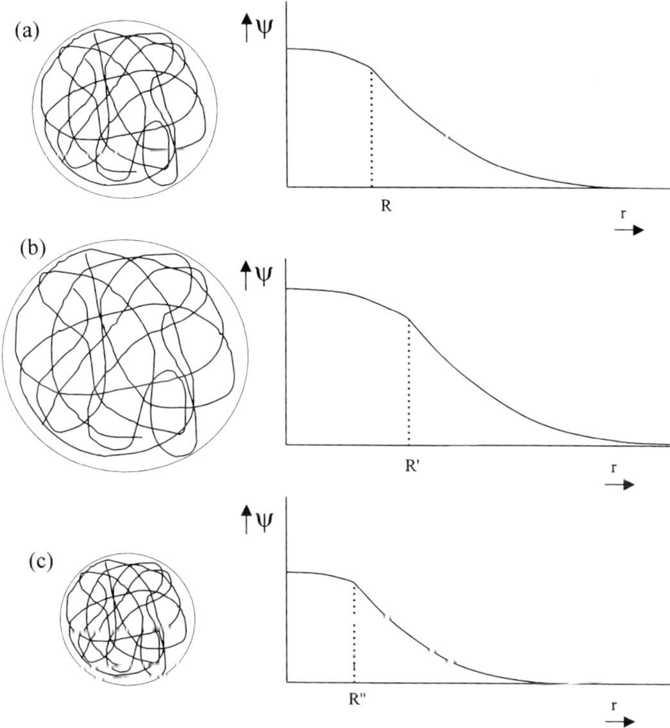

Fig. 2. Representation of colloidal structure and plots of potential generated by the colloid at distances from its centre; dotted lines represent potential at surface of colloid; (b) effect of increasing humic charge; humic expands and potential increases; (c) effect of increasing ionic strength; humic contracts, potential reduces and double layer contracts.

5. Metal binding

In recent decades, a significant effort has been put into the study of metal-humate interactions (Livens, 1991; Jones & Bryan, 1998). Humic substances interact very strongly with virtually all metal cations (Tipping & Hurley, 1992). As a result, they have an impact upon the environmental behaviour of both naturally occurring and anthropogenic metals (Jones & Bryan, 1998). It is important to study metal-humate interactions, because metals bound to humics show very different behaviours compared to the free metal ions or simple inorganic complexes (Saar & Weber, 1982). In particular, humics are suspected of enhancing the environmental mobility of metals (Khan et al., 1985; McCarthy et al., 1998; Bowell et al., 1993). Clearly, when these metals are either toxic heavy metals or radionuclides, then this is cause for concern. For example, it is thought that humic substances could interact with radionuclides in radioactive waste repositories, and thereby mobilise species which would otherwise be contained due to their low solubilities in inorganic, non-colloidal systems (Choppin, 1988; Ramsay, 1988). One field study has shown that fulvic acids mediate the environmental mobility of gold (Bowell et al., 1993). Another study has already shown that humic substances have led to the migration of radionuclides away from a radioactive waste repository (McCarthy et al., 1998). Many other studies have also demonstrated the importance of humics in the environmental chemistry of radionuclides. For example, Livens and Singleton (1991) found that the Pu specific activity of a soil from Cumbria, UK was 2860 Bq kg^{-1}, but that of the humic fraction was 27400 Bq kg^{-1}. For Am, the whole sample activity was 2650 Bq kg^{-1}, and for the humic fraction it was 18600 Bq kg^{-1}.

In the case of heavy metals, the toxicity is primarily associated with the free metal, and not with the total solution phase concentration (Buffle et al., 1990; Livens, 1991). Therefore, one might expect that humics and fulvics, which complex heavy metals and reduce the free concentration, would reduce toxicity (Petersen, 1982; Gjessing et al., 1989). However, the effect of humic substances is not always so clear-cut: in some cases, because humics can significantly increase solubility, they inhibit precipitation and retain metals in the solution phase, with the net overall effect perhaps being an increase in toxicity (Buchwalter et al., 1996). More than this, humics will affect the bioavailability of metals (Livens, 1991). It has been shown that humic substances, particularly the smaller fractions, do increase the uptake of Cd by plants (Sedlacek et al., 1989).

Clearly, humics play a crucial role in the behaviour of radionuclides in the environment (Livens, 1991). However, because of their heterogeneous and colloidal nature, the understanding of metal-humate interactions is still in its infancy, and there are still many aspects that we understand poorly (Jones & Bryan, 1998).

Binding properties

Humics will bind virtually all metal ions, and interact particularly strongly with the transition metals, the lanthanides and the actinides (Jones & Bryan, 1998). The strength of the interaction depends upon the identity of the metal cation. Ranking the following in increasing order of affinity gives (Tipping & Hurley, 1992; Read & Falck, 1996):

$$Ba(II) < Mn(II) < Mg(II) < Ca(II) < Sr(II) < Ni(II)\ Cd(II)\ Co(II) < Zn(II) < Fe(II)$$
$$< Pb(II) < Cu(II) < UO_2^{2+} \approx Al(III) \approx Eu(III) < Fe(III) < Th(IV) < U(IV)$$

Note that by 'binding strength', we mean the stability of the complex formed instantaneously between the humic and the metal. However, we shall see later that there is more to metal-humate binding than this '*exchangeable*' interaction. In common with many simple, noncolloidal ligands, the interaction is pH and ionic strength dependent; strength increases with increasing pH and decreasing ionic strength (Kim & Sekine, 1991; Carter et al., 1992; Benedetti et al., 1995; Czerwinski et al., 1996; Jones & Bryan, 1998). In the case of Cu binding, the strength of the interaction depends strongly upon the nitrogen content of the sample, because of the special affinity of Cu(II) for ligands with nitrogen donor atoms. In the case of some metals, there is evidence for humate mediated redox processes. For example, Goodman et al. (1991) found that Fe(III) was gradually reduced to Fe(II).

The key to understanding metal-humate complexation is the dual nature of the interaction (Tipping & Hurley, 1992; Jones & Bryan, 1998). The first is the chemical component, which depends upon the intrinsic chemical affinity of the metal binding site for the metal, and is familiar from studies with simple ligands. The second, which is unique to colloidal and surface interactions, is the physical component, and is due to the large negative potentials generated by the humic (Tipping & Hurley, 1992).

Humics display distinct types of behaviour towards two groups of metal cations (Devitt & Wiesner, 1998). The first group (group 1), which includes the alkali metals (Na^+, K^+ etc.), are attracted electrostatically towards the colloid, and may take part in the double layer or penetrate its structure, but which do not neutralise the humic charge, because they cannot bind specifically to the functional groups. Therefore, these metals experience the physical component only. The remainder of the metals (group 2) are still attracted by the charge, but are able to form bonds with the humate functional groups and hence, neutralise the charge (Livens, 1991; Jones & Bryan, 1998; Bryan et al., 2000). Hence, both the chemical and physical components are significant.

A very large number of studies have used a variety of techniques to determine the affinity of most metals for humic substances, under a range of solution conditions, and these have been reviewed elsewhere (Jones & Bryan, 1998). These data have been invaluable in advancing the understanding of metal-humate interactions. They have been complemented by a smaller number of studies, which have determined metal-humate reaction enthalpies (Nash & Choppin, 1980; Rao & Choppin, 1995; Samadfam et al., 1996; Bryan et al., 1998a; Warwick et al., 1998; Bryan et al., 2000). These studies have shown that the reaction enthalpies can be exo- or endothermic, but are all relatively small. In addition, it has been found that the enthalpy for a given metal depends upon the humic sample, and also ionic strength (Bryan et al., 1998a). Thus, reaction enthalpies are highly unpredictable (Bryan et al., 2000), in marked contrast to the gross binding strengths of humic substances, which are uniform over a wide range of humic samples from different origins (Read & Falck, 1996). The main conclusion of these thermodynamic studies is that, since the exchangeable binding strength is high, and the enthalpies are small, and in some cases endothermic, then the reactions must be driven by a large increase in entropy (Rao & Choppin, 1995; Bryan et al., 2000). Also, given that the binding strength is highly metal dependent, one would expect that some property of the metal has a significant influence (Bryan et al., 2000).

Coordination environment

Due to their nature, humic substances will have a distribution of binding sites (Buffle et al., 1990). This inherent heterogeneity means that it has been difficult to identify the precise coordination environment around complexed metals (Jones & Bryan, 1998). Methods such as EXAFS (extended X-ray absorption fine structure) and XANES (X-ray absorption near edge structure), which are commonly used to probe coordination environments, do not perform well with metal-humate complexes due to their heterogeneity. However, recent advances have allowed these techniques to provide some limited information (Xia et al., 1997). Ni, Co and Zn have all been found to have octahedral coordination environments, while Cu has a tetragonally distorted octahedral environment. Although the technique is currently unable to identify the donor atoms, beyond that they are probably N or O, evidence has been found for C in the second coordination shell, providing evidence for the formation of inner-sphere complexes. Ironically, the much simpler infrared spectroscopy has provided more definitive information on the coordination environment. For example, Cr has been found to be coordinated by carboxylate groups (Fukushima et al., 1995). However, the largest body of evidence for coordination environments has come from EPR (electron paramagnetic resonance) spectroscopy. Senesi et al. (1985a, b) found that the EPR spectra of Cu-fulvate complexes were consistent with the metal binding site being four coordinate in the equatorial plane with the axial positions being filled by water molecules. There were several chemically distinct binding site types. One had oxygen atoms as all four equatorial donors while others had: 3 oxygens and 1 nitrogen; 2 nitrogens and 2 oxygens; 1 oxygen and 3 nitrogens. Another study found evidence for similar sites, and concluded that the nitrogen donors were probably aliphatic, with iminodiacetate suggested as a good analogue (Lakatos et al., 1977). Boyd et al. (1981) have also suggested that the oxygen donors are carboxylates. These infrared and EPR techniques have provided our only insights into the nature of the coordination environment around the metal, and it is hard to understand why so few studies have been undertaken.

Kinetics

The equilibrium binding of metals by humics has been studied extensively (Jones & Bryan, 1998), and there are a great deal of exchangeable binding data. Metals bound in this exchangeable mode are bound very strongly, but nevertheless may still be removed instantaneously by a ligand or surface with a sufficiently high affinity for the metal (Warwick et al., 2000). However, it has become clear that the initial, exchangeable binding is followed by some degree of rearrangement, which makes the metal resistant to removal (Choppin, 1988). This '*non-exchangeably*' bound metal may be desorbed: however, the process is slow, and the rate is independent of the strength or concentration of the competing sink (Warwick et al., 2000).

A number of studies have examined this kinetic effect (Choppin, 1988; Choppin & Clark, 1991; Chakrabati et al., 1994; Cacheris & Choppin, 1998; Schussler et al., 1998; King et al., 1999; Schussler et al., 1999; Warwick et al., 2000). There are in fact a number of these non-exchangeable components, and a number of first-order rate constants are required to describe the data. Choppin et al. (Choppin, 1988; Choppin & Clark, 1991;

Table 1
Examples of first order rate constants for various metals (Bryan et al.,1999b)

Metal	Desorption rate (s^{-1})
Eu(III) (pH 4.5, 20°C)	1.2×10^{-6}
Eu(III) (pH 4.5, 40°C)	1.4×10^{-6}
Eu(III) (pH 6.5, 40°C)	5.0×10^{-7}
Co(II)	1.3×10^{-6}
Am(III)	1.1×10^{-6}

Cacheris & Choppin, 1998) have used a 'kinetic spectrum' approach, which defines a number of these components. Other studies (Schussler et al., 1998; King et al., 1999; Schussler et al., 1999; Warwick et al., 2000) have acknowledged the existence of the faster fractions, but have concentrated upon the longest lived fraction, since this is expected to be the most significant for transport in the environment (Bryan et al., 1999a). In fact, there is probably a continuum of desorption rates ranging from virtually instantaneous to this slowest fraction. The most surprising result is that this slow fraction can be identified in all humic samples. Even more surprising is that the rate is the essentially same for most metals, regardless of metal ion chemistry (Bryan et al., 1999b). Table 1 contains examples of rate constants. This behaviour is in contrast to that for exchangeable binding, where the binding strength depends very much upon the identity of the metal ion (Tipping & Hurley, 1992; Read & Falck, 1996). One question that has not yet been resolved is whether the various components or fractions are connected in series or parallel. That is, to reach the slowest fraction does the metal proceed via the faster ones, or is the progression direct? Figure 3 shows representations of the series and parallel mechanisms. Of course, these two systems are opposite extremes, and the real situation is likely to be some hybrid of the two.

Origin of the kinetic effect

There is now no doubt that humic substances bind metals non-exchangeably. However, the mechanism by which this happens is uncertain. Beyond the kinetic effect itself, there are a few clues. Von Wandruska et al. (1997) found that, although solution phase humic colloids are broadly hydrophilic, the addition of metal ions resulted in the formation of hydrophobic zones, with metals of high charge density being the most efficient at producing the effect. In addition, structural changes have been observed over some days following initial binding, which is similar to that observed for the development of non-exchangeable binding (Engebretson & Von Wandruszka, 1998).

Thinking in terms of the Self Association and Random Coil/Penetrable Gel models, one can propose mechanisms by which both could induce chemical kinetics. In the case of Self Association, metals could bind at a site on one of the smaller units, which make up the larger aggregates. Initially, that metal would remain exposed to the solution, and available for removal by surfaces or competing ligands. However, rearrangement of the humic aggregate

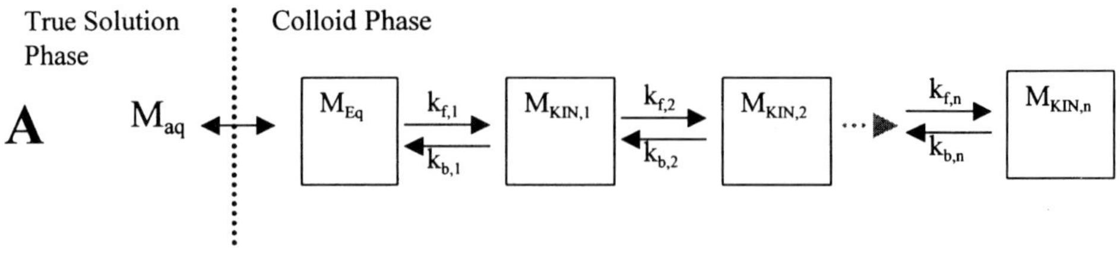

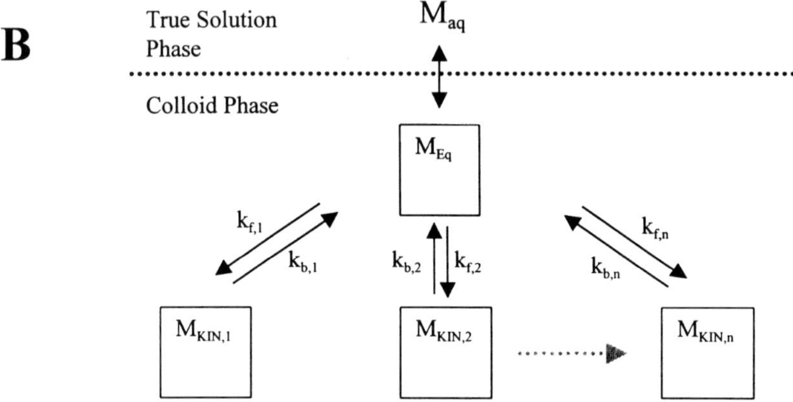

Fig. 3. Possible relationships between the kinetic fractions; (A) 100% in series; (B) 100% in parallel. M_{aq} represents free metal and each box a humic fraction; M_{Eq} is the exchangeable fraction and $M_{KIN,1}$, $M_{KIN,2...}$ $M_{KIN,n}$, the non-exchangeable fractions with progressively slower desorption rates.

could trap the metal within the structure of the aggregate between monomers, hiding the metal from solution (Fig. 4).

For the Random Coil/Penetrable Gel models, dissociation and reassociation cannot trap the metal. However, if the humic has an open structure, which is penetrated by solvent and small ions, then metal ions could migrate into the interior and become hidden. Now, the exact mechanism by which they could become trapped is unclear. Choppin (1988) talks about the metal moving from exchangeable (weaker) to non-exchangeable (stronger) sites. By contrast, Warwick et al. (2000) represent the system as shown in Fig. 5. There, the metal 'moves' from an exchangeable site on the surface to a non-exchangeable one inside the colloid. In both of these cases, the metal is required to move from one place on the humic to another. The alternative is illustrated in Fig. 6. The metal binds at an exchangeable site within the humic and, initially, is available for further reaction. However, with time, the humic rearranges its structure, and surrounds the metal ion, completely coordinating it. The metal is now surrounded by the humic organic skeleton, and a hydrophobic zone has been formed, isolating the metal from aqueous solution phase chemistry, and rendering it non-exchangeable.

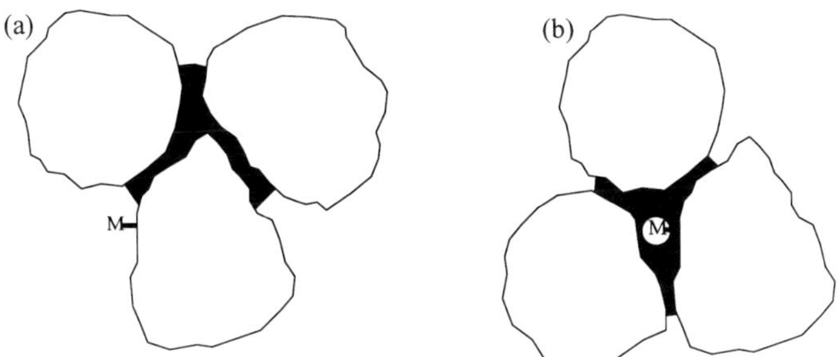

Fig. 4. Self-association explanation for kinetics; (a) initial binding to one of the small 'monomers'; (b) following rearrangement of the humic, metal becomes trapped inside the structure, in the region between monomers (shaded area).

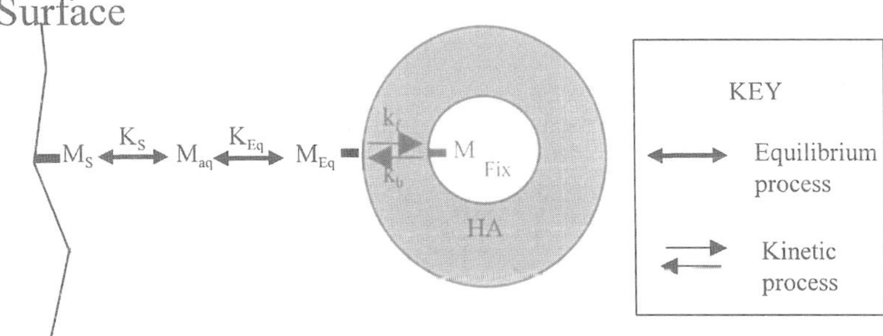

Fig. 5. Conceptual model used by Warwick et al. (2000); M_{aq} is the free metal, HA is the humic, M_{Eq} is exchangeable metal, M_{FIX} is the non-exchangeable metal.

The apparent problem is that slow desorption kinetics have been reported for fulvic acid samples, as well as humic acids (King et al., 1999). Buffle (1977) suggested a typical structure for fulvic acid (Fig. 7). This structure has a relative molecular mass of 637, and species around this mass are known to exist in fulvic acid samples. Whereas it is easy to understand how a humic acid species of mass 40,000 could shield and hide a metal ion, it is hard to see how the structure in Fig. 7 could do so. However, at least in this case, the observations are not as simple as they seem at first. Fulvic acids are generally regarded as smaller than humic acids, and have average molecular weights in the region of thousands. However, the separation of fulvic and humic acids is not made on the basis of size, but rather using differences in solubility. Now, because the smaller species tend to have higher charge densities, and hence greater solubilities, there is some degree of size separation between the two. However, the underlying picture is often highly complex. The molecular weight distribution of the fulvic acid used by King et al. (1999) has been measured by analytical ultracentrifugation and Field Flow Fractionation (Higgo et al., 1998). The number average molecular weight was found to be 2000, but the weight average weight was 4000, evidence of significant polydispersity. In fact, the molecular weight distribution was skewed to high mass, and measurable amounts of material were found up to masses of 17,000. Hence, although some of the species, like the one shown in Fig. 7, might not be

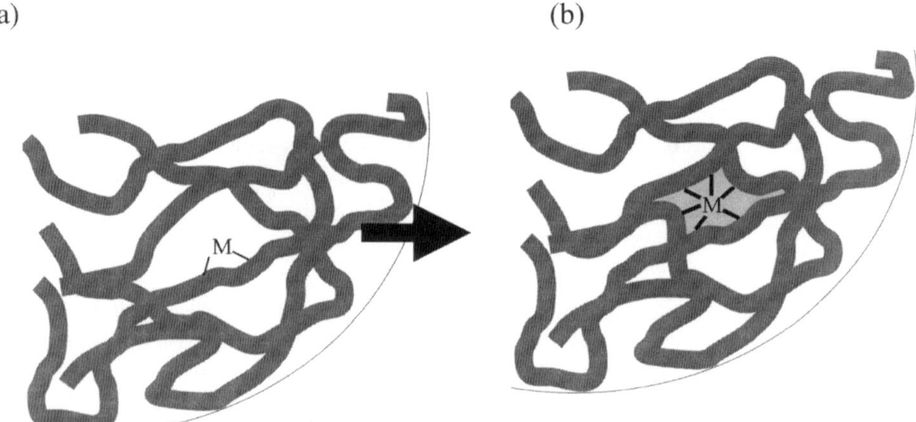

Fig. 6. Possible Random Coil/Penetrable Gel model explanation for kinetics; (a) exchangeable binding; (b) with time, non-exchangeable interaction develops and a hydrophobic zone (shaded) is formed.

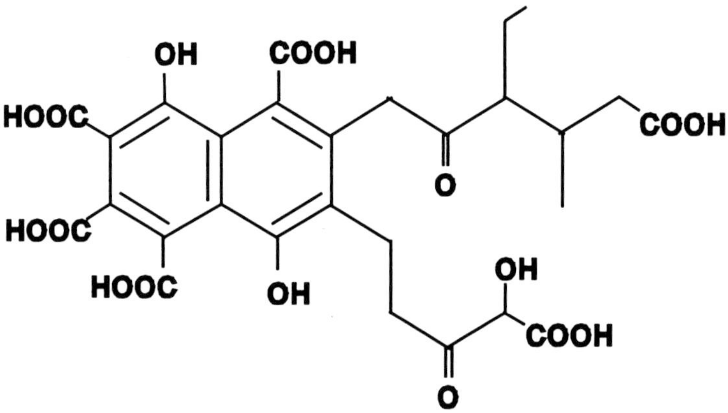

Fig. 7. Suggested fulvic acid structure (Buffle, 1977).

expected to bind metals non-exchangeably, there are much larger species present, which could. In fact, King et al. (1999) found that, although the fulvic sample did bind metal non-exchangeably, with similar rate constants to humic acid, the absolute amount of metal in this fraction was significantly lower for the fulvic acid than the humic acid. Perhaps the lower proportion of higher weight material in the fulvic sample explains this behaviour.

This example illustrates the problem with the definitions of fulvic and humic acids. The only certain difference is solubility, and to label fulvic acid as 'small' and humic acid as 'large' is far too simplistic. The same observation is borne out in other studies. For example, Bryan et al. (1998b) found that humic acid samples contained significant amounts of material of mass less than 5000. Even those samples aggregated by the addition of metal still contained some very low mass material.

Natural kinetics
Most kinetic studies to date have involved 'synthetic' metal-humate complexes. That is, metals and humics are mixed in the laboratory, and allowed to equilibrate, before the desorption kinetics are measured. However, Geckeis et al. (1999) have found that metals introduced to the humic via natural processes show different desorption kinetics. They found that metals naturally present in humic samples desorbed with rates significantly slower than for the same elements added in the lab. In fact, they have found some evidence for 'pseudo-irreversible' binding, i.e. some fraction of the natural loading would not desorb even after long times (several months). This apparent difference in behaviour is significant, since if it is repeated, it means that the results of laboratory experiments may not be directly applicable to the environment. The origin of this effect is uncertain, but it may be due to the much longer effective equilibration times available for natural complexes.

6. Modelling

It is clear that humics have a significant effect upon the mobility of metals in the environment, including toxic heavy metals and most radionuclides. For this reason, there is a great deal of interest in predicting their behaviour, and a number of mathematical models have been developed. There has been a significant improvement in the reliability and sophistication of these models over the last two decades (Jones & Bryan, 1998). The models described below are the result of that improvement.

There are two distinct conceptual approaches to this modelling; *predictive* and *descriptive*. In predictive modelling, the aim is to obtain the closest possible fit to the experimental data, and although there is often an attempt to relate the model parameters to physically realistic quantities, they are often really fitting parameters, which are varied in order to obtain the best possible fit. In descriptive modelling, the aim is to derive information regarding the physical nature of the system, rather than to produce the closest possible fit to the experimental data. Descriptive modelling is characterised by a lack of fitting parameters. The only inputs to these models are experimentally determined quantities and physical constants. Predictive and descriptive studies provide different and complementary information, and both are equally important.

As one might expect, in the same way that most experimental studies have focussed upon the initial, exchangeable interaction, so most modelling, predictive and descriptive, has also concentrated on this aspect. Indeed, very few kinetic modelling studies have been undertaken. Although the exchangeable and non-exchangeable binding modes are really part of one system, their modelling is not yet properly combined and hence we will treat them separately here.

Exchangeable modelling

Predictive
Most modelling to date has been predictive, aimed at the exchangeable interaction. These models make many different, and often contradictory, assumptions. Nevertheless, many are able to reproduce some aspect of humate chemistry. Most are, to some extent, empirical,

and rely partially upon data fitting. As a result, the fact that a model is able to reproduce the behaviour of the system is no guarantee that the assumptions are correct. Indeed, some are so contradictory, but still work, that this must be the case.

Within these predictive models there are two distinct types, discrete site and continuous distribution. The discrete site models, whilst acknowledging that humics actually have a distribution of binding sites and affinities, assume that it is possible to reproduce their binding behaviour with a fixed number of sites with the same affinities. On the other hand, continuous distribution models use a mathematical function to define a continuum of binding sites. Both types have their advantages: discrete site types are mathematically more convenient, but distribution models are theoretically more elegant.

There have been so many different modelling studies that it is not possible to discuss them all here. Therefore, a small number are described below, and give an indication of the breadth and diversity of approaches to this problem. However, the list is by no means exhaustive.

One of the simplest empirical models is the Charge Neutralisation Model of Kim et al. (Kim & Czerwinski, 1996). This model assumes that each metal ion binds to a number of deprotonated humic carboxylates equal to its charge: hence, a Cu^{2+} ion binds to, and neutralises two carboxylates, and an Am^{3+} ion binds to three. If Z_H is the number of charges on the humic, and z is the cation charge, then the metal binding reaction is represented thus,

$$M^{z+} + HA^{Z_H-} \rightarrow MHA^{(Z_H-z)-} \tag{7}$$

and a simple binding constant, β, can then be defined for the reaction. This type of model has been widely applied to the simulation of actinide data (Kim et al., 1989, 1991, 1993; Kim & Sekine, 1991; Kim & Czerwinski, 1996; Czerwinski et al., 1994, 1996; Choppin & Labonne-Wall, 1997), and so is particularly relevant to the study of environmental radioactivity.

One of the most complex of the discrete site models is the 'Model V' of Tipping & Hurley (1992). This model assumes that the humic has 8 binding site types; 4 carboxylate and 4 phenolic. Each site, numbered 1–8, has a different pK, defined in terms of the constants pK_A and pK_B for the carboxylic and phenolic sites respectively. All of the carboxylate sites are assumed to have the same equilibrium constant with any given metal. Similarly, the phenolic sites have the same constant. In addition to these simple, monodentate sites, certain combinations of these are permitted, giving a total of 12 bidentate site types. The binding constants for these sites are derived from the constants for the constituent monodentate sites. The model allows the ratio of monodentate to bidentate sites to be varied.

In common with most advanced models, Model V explicitly takes into account the effects of the double layer through the use of empirical corrections (Tipping & Hurley, 1992; Higgo et al., 1996; Warwick et al., 1996). Finally, the model also takes into account the amount of metal removed from the bulk solution by entrapment in the double layer.

Model V has been widely applied to exchangeable binding data for a wide range of metals (e.g. Tipping & Hurley, 1992; Tipping, 1993a). Like the Charge Neutralisation Model, it has been applied to the binding of the actinides (Tipping, 1993b). It has also been included in an associated geochemical speciation model, WHAM (Tipping, 1994).

Continuous distribution models

A large number of continuous distribution models have been developed (Jones & Bryan, 1998). Most are based upon early work by Posner (1966) and then Perdue & Lytle (1983), and use log-normal distributions of binding strengths. Unlike the discrete models, distribution models tend not to use classical equilibrium constants, because this would make the solution of the equations too complex. All distribution models follow the same general strategy in that they define a local adsorption isotherm, which describes the interaction of an individual site with the metal. The model then integrates over the whole population of sites to calculate the humic metal loading. This model was later developed by Dobbs et al. (1989a, b) and Allison et al. (1991). Bimodal distributions have also been used to predict binding (Manunza et al., 1992,1995).

Of all of the many continuous distribution models, there is one recent one that stands out from the rest, both in terms of its success and its complexity, and that is the model of De Wit and coworkers, which has been developed over the last decade (De Wit et al., 1990, 1993; Nederlof, 1993; Milne et al., 1995a, b; Benedetti et al., 1995; Van Riemsdijk et al., 1996; Kinniburgh et al., 1996).

The CONICA version of the model uses the Henderson-Hasselbach/Rudzinski isotherm, which defines a fractional coverage for each site in the distribution in terms of the concentration of free metal in the region of the binding site. The model defines both 'carboxylate' and 'phenolic' distributions and allows bidentate as well as mondentate binding.

The model takes account of humic electrostatic effects via a semi-empirical Donnan approach, which is consistent with a penetrable, gel-like structure for humic substances. It assumes that sufficient cations are attracted towards the humic to exactly balance its charge. The model performs very well in predicting the binding of metals, particularly Ca, Cu and Cd (De Wit et al., 1990, 1993; Nederlof, 1993; Milne et al., 1995a, b; Benedetti et al., 1995; Van Riemsdijk et al., 1996; Kinniburgh et al., 1996).

The best model?

Several studies have attempted to determine which of these models is the 'best' (e.g. Higgo et al., 1996; Warwick et al., 1996; Choppin & Labonne-Wall, 1997). Many, particularly the authors of the various models, have strong opinions about which is the best. However, to varying extents, they all fulfil their function in fitting the data. Whether that success should be taken to infer anything about the physicochemical nature of the system is open to question. A good illustration of this point is the relative performance of the discrete site and continuous distribution models. A number of comparative studies (Higgo et al., 1996; Warwick et al., 1996) have recommended Tipping's Model V discrete site model as 'the best'. However, there is clear evidence that humic acids are heterogeneous, and effectively have a continuous distribution of binding sites (Buffle et al., 1990), and the success of Model V cannot be taken to infer that humics actually have eight types of identical functional groups. In fact, despite the impressive performance of some of these models, there are a number of properties which they all ignore. In particular, they all treat the humic sample as a species with a single molecular weight whereas, in reality, humics are highly polydisperse.

Descriptive modelling

For obvious reasons, most effort has concentrated upon mathematical models aimed at obtaining the best possible fit to experimental data. However, a small number of studies have attempted to investigate the mechanisms of binding.

Random structural modelling

In the eighties, Murray and Linder (Murray & Linder, 1983, 1984; Linder & Murray, 1987) developed a random structural model, which was a computer program designed to identify the most probable metal binding sites present in fulvic acids. The model 'constructed' hypothetical strands of fulvic acid, which conformed to measurements of real samples. These strands were identical to the real samples in terms of carbon, hydrogen and oxygen composition, as well as carboxylate, phenol and alcoholic functional group concentrations. Having constructed the fragment, the model searched for patterns of functional groups, or binding sites. The procedure was repeated 1000 times with different fulvic strands, which all conformed to the same parameters. In this way, average site concentrations were built up.

The procedure was enhanced by Mountney & Williams (1992), who combined the equilibrium constants associated with the ligands equivalent to the binding site patterns, along with the concentrations from the random model, to make predictions for the amount of metal that would bind to the fulvic. They found similarities between the model and reality. Later, Bryan et al. (1997) extended the model to include nitrogen based functional groups and binding sites. They used the approach of Mountney & Williams (1992) for calculating metal binding, but attempted to include electrostatic effects by analysis of pH titration curves. The results showed that the binding strength predicted by the model was significantly less than for the real sample. This discrepancy was attributed to the fact that the model only considered functional groups as contributing to binding sites if they were on a contiguous section of the humic strand. Hence, the model underestimated the complexity of the binding sites.

Mechanistic modelling

One problem with humic-metal predictive models is that they are all able to simulate metal binding data, regardless of their underlying conceptual approaches or assumptions, and hence, they do not allow us to deduce very much about the nature of the system. In a way, it is as if the raw metal binding data themselves were too easy to fit. One approach, therefore, is to extend the scope of modelling to include other aspects of the interaction, specifically enthalpies and entropies of reaction.

By inspection of the thermodynamic data, it is possible to deduce that exchangeable binding is driven by a large charge entropy change, and Rao & Choppin (1995) have speculated that the dehydration of the metal ion could be driving the process. At present, the system is poorly understood, and hence it is not possible to model the binding of one metal, and then directly extend that to other metals. Specific data are required for a metal, before any reliable calculation can be made. More than this, it is not automatically possible to extend from one set of conditions, pH, I etc., to another, even for the same metal (Bryan et al., 2000).

In an attempt to address these problems, Bryan et al. (1998a, 2000) have developed a mechanistic model. It concentrates upon reaction entropies, and has two sources of entropy, the dehydration of the metal ion, and the relaxation of the double layer. Initially, both the humic and the metal impose a high degree of order on the system, the metal by the waters held in the hydration sphere, and the humic by the counterions trapped in the double layer. When the metal binds, and partially neutralises the humic charge, the waters around the metal and the cations trapped within the double layer are released.

There is a clear dependence of raw binding strength upon the chemical identity of the cation (Read & Falck, 1996). The double layer contribution is entirely physical in nature and will depend only on the charge number of the cation. Therefore, this behaviour is explained by the dehydration of the cation. Metals which impose higher degrees of order upon the solvent will produce greater entropies of reaction when they are dehydrated and will be bound more strongly. However, dehydration alone is insufficient to explain all observations. For example, metal binding is very ionic strength dependent. This is due to the double layer contribution.

The model uses the penetrable Poisson-Boltzmann theory, as described in equations (1)–(6), to calculate the electrical free energies of the counterions and the humic itself, plus the statistical (entropic) free energies of the counterions. By calculating the magnitude of the three components before and after the addition of a metal ion, the electrostatic contribution to metal binding is derived. Partial molar entropies are used as a measure of the contribution of dehydration.

The model is able to explain several aspects of metal-humate behaviour (Bryan et al., 1999b, 2000). For example, it is able to explain the effect of ionic strength on metal binding. Furthermore, Fig. 8 shows the correlation between the model and some experimental entropies. Finally, there is a strong correlation between the model predicted entropies, and the binding strengths of a range of metals.

These results are very encouraging, since the correlations have been obtained without fitting parameters. They offer the hope that, in the near future, we will be able to make predictions for metal-humate behaviour, with confidence, even for systems where there are few or no experimental data.

Kinetic modelling

Although the slow dissociation of metals from humic colloids was first studied over a decade ago (Choppin, 1988), it is only recently that the importance of these slow processes has been realised, and mathematical models which take kinetics into account developed. In fact, the true importance of kinetics was partly revealed as a result of model development. In the past, it had proved difficult to predict the results of column experiments (Warwick et al., 2000). In these experiments, solutions of metal and humic substance are passed through a column packed with some porous material, which can range from acid-washed sand to real, site-specific sediment. Often, in the absence of humic, the transmission of metal is negligible (Warwick et al., 2000). Modelling these experiments is inherently more complex than predicting the amount of metal bound to a humic substance, since the model not only needs to simulate the speciation of a solution, but also needs to 'move' the solution along the column.

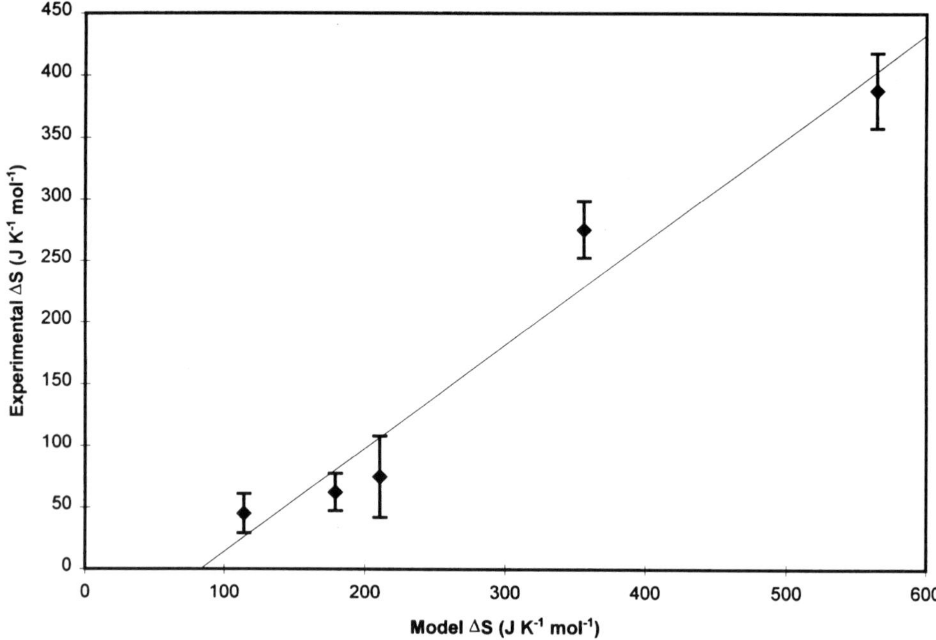

Fig. 8. Correlation between mechanistic model and experimental reaction entropies.

These transport models are often 'calibrated' with fairly simple, laboratory experiments, and the results 'up-scaled' to make predictions on the larger, field scale. Hence, mathematical simplicity is important when constructing them. One simplification is to make the assumption of local equilibrium, i.e. that all chemical processes are fast compared to the movement of the solution. In simple terms, equilibria are used to describe all interactions including all solution phase complexation, whether humic or not, and all sorption on surfaces: kinetics are assumed to be unimportant.

However, these equilibrium models have repeatedly failed to simulate column experiments (Warwick et al., 2000). Schussler et al. (1999) discovered that the behaviour in the column, and the amount of metal emerging from the end of the column were both heavily dependent upon the pre-equilibration time of the metal-humate complex prior to injection and also on the residence time in the column, i.e. typical kinetic behaviour. Despite the fact that humic kinetics dominate the behaviour of the system, equilibria are still important. Not only should the initial, exchangeable binding of metals be described as an equilibrium, but it is usual to describe the interaction of metals with surfaces using an equilibrium approach. Hence, in order to describe these systems accurately, a model that includes both chemical kinetics and equilibria is required. Due to this complexity, initial attempts to model the system did not treat the transport phenomena explicity (Schussler et al., 1998). Later, the k1D model was developed, which includes kinetics and equilibria in a rigorous transport code (Warwick et al., 2000). This model has been applied to europium (Warwick et al., 2000), cobalt (Bryan et al., 1999b) and americium (Schussler et al., 1999) data.

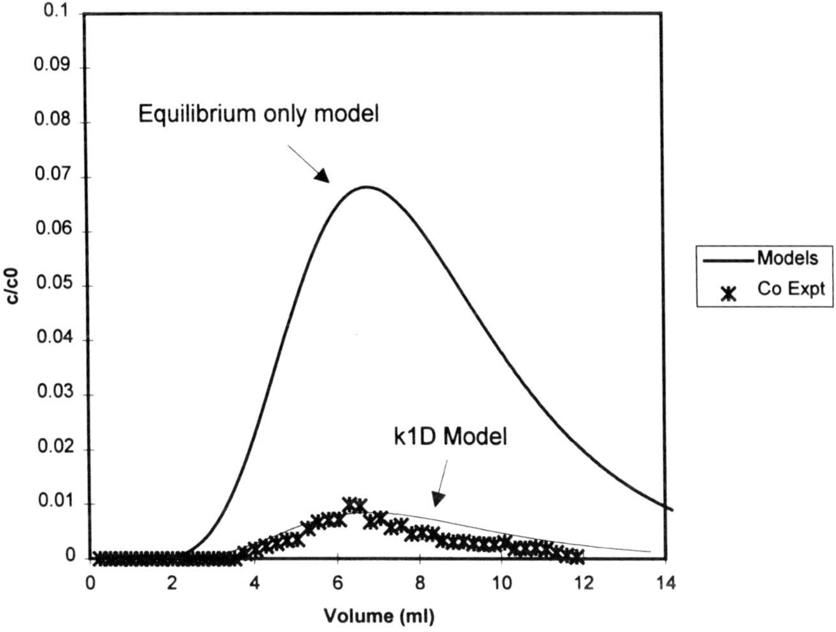

Fig. 9. Elution profile from a Co column experiment, showing the improvement in fit obtained by including chemical kinetics.

The combined equilibrium and kinetic approach has lead to a significant improvement in the fit between experiment and model. Figure 9 shows an elution profile from a Co column experiment along with the best predictions from equilibrium-only and equilibrium-plus-kinetic approaches. Clearly, the new combined model provides a much better fit to the experiment. This improvement is mirrored in studies using different metals and types of column experiment (Schussler et al., 1999, Warwick et al., 2000).

A particular feature of these new models is their relative simplicity. Metals desorb from humic complexes with a spectrum of first order rate constants, ranging from instantaneous to 1×10^{-7} s^{-1}. However, in order to treat the results of column experiments, it is only necessary to include the slowest fraction, since it is this component that dominates the transport of metals. Figure 5 shows the system used by Warwick et al. (2000), which is mathematically identical to that used by Bryan et al. (1999b) and Schussler et al. (1999). In many ways this approach is conceptually much simpler than the CONICA or Model V speciation models, but it is still able to simulate the behaviour in column experiments with a high degree of success.

Predicting the effects of humic substances on the field scale

Whether or not humics will enhance the migration of radionuclides is a matter for concern, and predicting their effect in the environment is an important aim for humic research. Generally, the scenario to be modelled is a waste repository, buried underground.

Models must predict the extent of migration of radionuclides out of the repository, into the environment, and towards the food chain and the human population. Therefore, the simulations assume that the volume surrounding the repository starts uncontaminated, and that all the activity is confined within the repository. Clearly, these calculations will be highly complex, and humic substances will be just one small part of the process. However, despite the complexity of the mathematical model, the central issue is very simple: will the radionuclide be held in solution? If so, then it will migrate, but if it precipitates from solution, or is sorbed onto solid surfaces, then it will remain contained.

It has been known for some time that, purely as strong ligands, with large equilibrium constants for exchangeable binding, humic substances had the potential to mediate transport (e.g. Choppin, 1988; Livens, 1991). However, the existence of non-exchangeable binding modes will have a significant impact upon predictions. In the past, it might have been possible to argue that the vast excess of surface sorption sites would be sufficient to strip metals from humics, regardless of their large binding constants. However, metal 'trapped' in the non-exchangeable sites will be isolated from all solution and surface chemistry and the usual mechanisms of retardation will not be effective.

Most humic substances are actually found in the solid phase, rather than solution. Therefore, it has been suggested that, even if humics have sufficiently high affinities that they compete with inorganic surfaces, the fact that the majority of the humic mass is part of the solid phase would prevent migration. However, this only holds true if the metal-humic interaction is entirely exchangeable, and metal is free to 'jump' from one humic colloid (in solution) to another (in the solid) which is clearly not the case. If a radionuclide starts its journey into the geosphere bound non-exchangeably to a particular colloid, then it does not have the opportunity to interact with the solid phase humic material. Only if the humic colloid itself is retarded will migration be prevented, but column experiments have found little evidence for this (Schussler et al., 1999; Warwick et al., 2000).

Much will depend upon two distinct factors. The first is whether the humic is able to enter the contaminated area and interact with radionuclides, and then carry them out into the environment. If humics are excluded from the system, then their effect might be lessened. The second factor is simply the rate of desorption from the non-exchangeable site. The slower the metal desorbs, the further it will migrate (Bryan et al., 1999a).

Calculations designed to test the performance of waste repositories routinely cover periods of 10,000 years, and even longer. The half-times for desorption in laboratory experiments are of the order of 80 days. At first sight, it might seem, since the reaction times are tiny compared to the total calculation time, that humic kinetics will not be significant. However, the critical time is not the total calculation time, but rather the residence time, which is the time taken for any given humic colloid to travel from the repository to the limit of the area covered by the calculation. For example, if calculations are to be performed for 100 m around the repository, and the linear groundwater flow rate is 1×10^{-6} m s^{-1} (a realistic value), then the residence time is approximately 3 years, and if the flow rate increases to 1×10^{-5} m s^{-1} then the residence time becomes 115 days. In either case, but particularly at the higher flow rate, the kinetics observed in the laboratory would have an effect. However, there is some evidence that desorption rates observed in nature may be significantly slower than for synthetic, laboratory systems (Geckeis et al., 1999). Clearly, if the natural rates are slower, then there will be a significant impact upon the

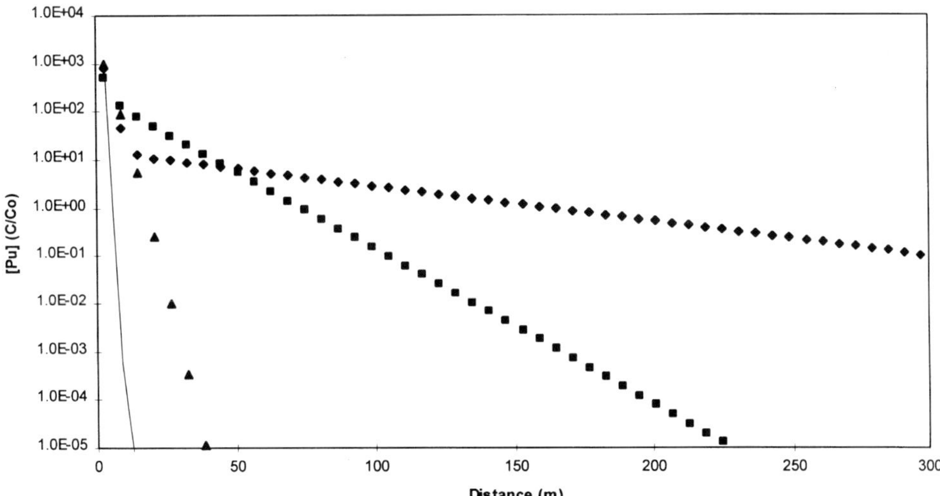

Fig. 10. Field-scale predictions of Pu migration; (i) prediction in absence of humics (line); (ii) including humics, but exchangeable interaction only (triangles); (iii) including kinetics, $k_b = 2 \times 10^{-7}$ s^{-1} (squares); (iv) including kinetics $k_b = 2 \times 10^{-8}$ s^{-1} (diamonds).

extent of migration. Further, if the observation that some metals in natural samples appear to be 'pseudo-irreversibly' bound is generally true, then the implications are potentially serious. In the absence of humic retardation, any radionuclides bound in that fraction would transport conservatively.

Using the new equilibrium plus kinetic transport models and parameters derived from laboratory studies, we can start to make some quantitative predictions about the effect of humic substances on the migration of metals and radionuclides in the environment. Figure 10 shows a field-scale prediction made for Pu migration using the k1D model of Warwick et al. (2000). Calculations were performed for a 300 m distance from a hypothetical repository, a flow rate of 1×10^{-6} m s^{-1}, a longitudinal dispersion of 10 m and a dissolved organic carbon concentration of 3 mg l^{-1}. The data are plotted as the ratio of the Pu concentration (sorbed + solution) at any distance divided by the input Pu concentration at zero distance. The results are correct for any input concentration less than 1×10^{-7} M. The results of four calculations are shown: the first shows the result expected in the absence of humic substances; the second includes humic exchangeable binding only; the third includes both exchangeable and non exchangeable binding, and uses a rate constant expected from lab experiments (Davis et al., 1999) (2×10^{-7} s^{-1}); the fourth also includes both components, but this time, has a rate constant an order of magnitude lower (2×10^{-8} s^{-1}), representing the slower rates which might be expected from a 'natural' sample.

One significant point is that there is only a small difference between the calculation with no humic and that which includes exchangeable binding only. The most significant differences are between the kinetic and non-kinetic plots. The slow desorption kinetics clearly promote migration, and a change in the constant by an order of magnitude has made a very significant difference. Therefore, one would expect that the results of migration calculations would be highly sensitive to kinetic rates. In at least one case (McCarthy et al., 1998), humic kinetics have been implicated in the migration of radionuclides away from a real repository. Hence, these calculations would seem to be realistic.

7. Conclusions

In general terms, there are currently several plausible suggestions for the transformation of complex components of plant debris into the heterogeneous macromolecules which constitute humic substances. However, there is no clear understanding of the detailed mechanisms, nor of the roles which different groups or species of microorganisms play in these processes. In the absence of such knowledge, it is difficult to develop a predictive understanding of the role of biological transformations in the formation and degradation of humic substances and, therefore, in the interactions of humic substances and radionuclides.

By contrast, in recent years there have been considerable advances in the understanding of the chemical properties of humic substances and, in particular, their interactions with metals. We now know that in the solution phase, they are highly charged, hydrophilic and penetrated by water and small ions. There are two possible explanations for their properties, the Self Association and Random Coil models. Although both are plausible, the Random Coil approach does seem to fit the observations best. That said, humic substances are unlikely to exist as a single, contiguous molecular strand. Rather, there is likely to be cross-linking between various strands to produce a structure that is flexible, can expand and contract as solution conditions change and allows access to its interior.

The exchangeable interaction between humic substances and metals has been well studied over the last few decades, and data are available for a wide range of metals, humic samples and solution conditions. As a result, there have been enormous improvements in the understanding of this part of the interaction which is best illustrated by the progress of metal binding models. Early models were very limited but now there is a range of advanced predictive models which simulate metal-humate speciation with a high degree of success. In recent years, other, descriptive models have started to explain the underlying processes and define the driving forces behind the reactions.

By comparison, the non-exchangeable interactions of metals with humic substances are poorly understood. The limited range of metals studied so far show broadly analogous behaviour, that is, a range of kinetic fractions, which range from almost exchangeable behaviour to slow desorption, with first order rate constants of the order of 1×10^{-6} to 1×10^{-7} s^{-1}. The origin of the effect is still unclear but the mechanism is likely to be complex and may reflect the colloidal properties of humic substances. It is possible that there is significantly different behaviour between metals naturally present in humic samples and those added in the laboratory. This lack of understanding is mirrored in modelling and further experimental studies are required to provide the necessary data.

Despite, these problems, we are now at the stage where we can at least begin to make transport predictions on the field-scale, albeit with fairly large uncertainties. The extent of migration will depend heavily upon the rate of desorption of the metal from the humate complex. If long equilibration times do lead to much slower desorption rates, or even pseudo-irreversible behaviour, then the effects are likely to be dramatic. Indeed, there is already some evidence of humic mediated transport at one disposal site.

Acknowledgements

The authors would like to thank Malcolm Jones (University of Manchester) for many helpful suggestions and conversations over the years; also the partners in the EC HUMICS project, and particularly Gunnar Buckau (FZK, Karlsruhe), Jenny Higgo (BGS), Sam King (University of Loughborough), Wolfram Schussler (FZK, Karlsruhe) and Peter Warwick (University of Loughborough).

References

Alexander, M. (1977). *Introduction to Soil Microbiology*, New York: John Wiley.

Allison, J. D., Brown, D. S. & Novo-Gradac, K. J. (1991). *MINTEQA2/PRODEFA2. A Geochemical Assessment Model for Environmental Systems: version 3.0.* Environmental Research Laboratory, Office of Research and Development, US Environment Protection Agency, Athens Georgia 30 613, USA.

Bartschat, B. M., Cabaniss, S. E. & Morel, F. M. M. (1992). Oligoelectrolyte model for cation binding by humic substances. *Environmental Science and Technology*, *26*, 284–294.

Benedetti, M. F., Milne, C. J., Kinniburgh, D. G., Van Riemsdijk, W. H. & Koopal, L. K. (1995). Metal ion binding to humic substances: application of the non-ideal competitive adsorption model. *Environmental Science and Technology*, *29*, 446–457.

Benedetti, M. F., Van Riemsdijk, W. H. & Koopal, L. K. (1996). Humic substances considered as a heterogeneous donnan gel phase. *Environmental Science and Technology*, *30*, 1805–1813.

Bowell, R. J., Gize, A. P. & Foster, R. P. (1993). The role of fulvic acid in the supergene migration of gold in tropical rain forest soils. *Geochimica et Cosmochimica Acta*, *57*, 4179–4190.

Boyd, S. A., Sommers, L. E. & Nelson, D. W. (1981). Copper(II) and iron(III) complexation by the carboxylate group of humic acid. *Soil Science Society of America Journal*, *45*, 1241–1242.

Bryan, N. D., Livens, F. R. & Jones, M. N. (1998b). The application of ultracentrifugation methods to ion-induced association of humic substances. *Biochemical Society Transactions*, *26*, 762–765.

Bryan, N. D., Griffin, D. & Regan, L. (1999a). Implications of humic chemical kinetics for radiological performance assessment. In G. Buckau (Ed.), *Effects of Humic Substances on the Migration of Radionuclides: Complexation and Transport of the Actinides* (pp. 339–356). Wissenschaftliche Berichte (FZKA 6324, ISSN 0947-8620), Forschungszentrum Karlsruhe Technik und Umwelt, Karlsruhe, Germany.

Bryan, N. D., Hesketh, N., Livens, F. R., Tipping, E. & Jones, M. N. (1998a). Metal ion-humic substance interaction—a thermodynamic study. *Journal of the Chemical Society—Faraday Transactions*,*94*, 95–100.

Bryan, N. D., Robinson, V. J., Livens, F. R., Hesketh, N., Jones, M. N. & Lead, J. R. (1997). Metal-humic interactions: a random structural modelling approach. *Geochimica et Cosmochimica Acta*, *61*, 805–820.

Bryan, N. D., Jones, D., Griffin, D., Regan, L., King, S., Warwick, P., Carlsen, L. & Bo, P. (1999b). Combined mechanistic and transport modelling of metal humate complexes. In G. Buckau (Ed.), *Effects of Humic Substances on the Migration of Radionuclides: Complexation and Transport of the Actinides*, (pp. 301–338). Wissenschaftliche Berichte (FZKA 6324, ISSN 0947-8620), Forschungszentrum Karlsruhe Technik und Umwelt, Karlsruhe, Germany.

Bryan, N. D., Jones, D. M., Appleton, M., Livens, F. R., Jones, M. N., Warwick, P., King, S.J. & Hall, A. (2000). A Physicochemical model of metal-humate interactions. *Physical Chemistry Chemical Physics*, 2, 1291–1300.

Buffle, J. (1977). Conference Proceedings. *de la Commission d'Hydrologie Appliquee de l'A.G.H.T.M.L.*, Universite d'Orsay, France.

Buffle, J., Altmann, R. S., Filella, M. & Tessier, A. (1990). Complexation by natural heterogeneous compounds, site occupation distribution functions, a normalised description of metal complexation. *Geochimica et Cosmochimica Acta*,*54*, 1535–1553.

Buffle, J., Wilkinson, K. J., Stoll, S., Filella, M. & Zhang, J. W., (1998). A generalized description of aquatic colloidal interactions: the three-colloidal component approach. *Environmental Science and Technology*, *32*, 2887–2899.

Buchwalter, D. B., Linder, G. & Curtis, L. R. (1996). Modulation of cupric ion activity by pH and fulvic acid as determinants of toxicity in Xenopus laevis embryos and larvae. *Environmental Toxicology and Chemistry*, *15*, 568–573.

Cacheris, W. P. & Choppin, G. R. (1998). Dissociation kinetics of thorium-humate complex. *Radiochimica Acta*, *42*, 185–190.

Carter, R. J., Hoxey, A. & Verheyen, T. V. (1992). Complexation capacity of sediment humic acids as a function of extraction technique. *Science of the Total Environment*, *125*, 25–31.

Cameron, R. S., Thornton B. K., Swift R. S. & Posner A. M. (1972a). Molecular weight and shape of humic acid from sedimentation and diffusion measurements on fractionated extracts. *Journal of Soil Science*, *23*, 394–408.

Cameron, R. S., Thornton B. K., Swift R. S. & Posner A. M. (1972b). Calibration of gel permeation chromatography materials for use with humic acid. *Journal of Soil Science*, *23*, 342–349.

Chakrabati, C. L., Lu. Y., Gregorie, D. C., Back D. C. & Schroeder W. H. (1994). Kinetic studies of metal speciation using chelex cation exchange resin: application to cadmium, copper and lead speciation in river water and snow. *Environmental Science and Technology*, *28*, 1957–1967.

Chin, W. C., Orellana, M. V. & Verdugo, P. (1998). Spontaneous assembly of marine dissolved organic matter into polymer gels. *Nature*, *391*, 568–572.

Choppin, G. R. (1988). Humics and radionuclide migration. *Radiochimica Acta*, *44/45*, 23–28.

Choppin, G. R. & Clark, S. B. (1991). The kinetic interactions of metal ions with humic acid. *Marine Chemistry*,*36*, 27–38.

Choppin, G. R. & Labonne-Wall, N. (1997). Comparison of two models for metal-humic interactions. *Journal of Radioanalytical and Nuclear Chemistry*, *221*, 67–71.

Conte, P. & Piccolo, A. (1999). Conformational arrangement of dissolved humic substances. Influence of solution composition on association of humic molecules. *Environmental Science and Technology*, *33*, 1682–1690.

Czerwinski, K. R., Buckau, G., Scherbaum, F. & Kim, J. I. (1994). Complexation of the uranyl-ion with aquatic humic-acid. *Radiochimica Acta*,*65*, 111–119.

Czerwinski, K. R., Kim, J. I., Rhee, D. S. & Buckau, G. (1996). Complexation of trivalent actinide ions (Am^{3+}, Cm^{3+}) with humic acid: the effect of ionic strength. *Radiochimica Acta*, *72*, 179–187.

Davis, J., Higgo, J., Moore, Y. & Milne, C. (1999). The characterisation of a fulvic acid and its interactions with uranium and thorium. In G. Buckau (Ed.), *Effects of Humic Substances on the Migration of Radionuclides: Complexation and Transport of the Actinides*. (pp. 301–338), Wissenschaftliche Berichte (FZKA 6324, ISSN 0947-8620), Forschungszentrum Karlsruhe Technik und Umwelt, Karlsruhe, Germany.

DeNobili, M. & Chen, Y. (1999). Size exclusion chromatography of humic substances: limits, perspectives and prospectives. *Soil Science*, *164*, 825–833.

Devitt, E. C. & Wiesner, M. R. (1998). Dialysis investigations of atrazine-organic matter interactions and the role of a divalent metal. *Environmental Science and Technology*, *32*, 232–237.

De Wit J. C. M., Van Riemsdijk, W. H. & Nederlof, N. M. (1990). Analysis of ion binding on humic substances, and the determination of intrinsic affinity distributions. *Analytica Chimica Acta*, *232*, 189–207.

De Wit, J. C. M., Van Riemsdijk, W. H. & Koopal, L. K. (1993). Proton binding to humic substances, 1: Electrostatic effects, 2: Chemical heterogeneity and adsorption models. *Environmental Science and Technology*, *27*, 2005–2022.

Dobbs, J. C., Susetyo, W., Knight, F. E., Castles, M. A., Carreira, L. A. & Azarraga L. V. (1989a). Characterization of metal-binding sites in fulvic-acids by lanthanide ion probe spectroscopy. *Analytical Chemistry*, *61*, 483–488.

Dobbs, J. C., Susetyo, W., Carreira, L. A. & Azarraga, L. V. (1989b). Competitive-binding of protons and metal-ions in humic substances by lanthanide ion probe spectroscopy. *Analytical Chemistry*, *61*, 1519–1524.

Engebretson, R. R. & Von Wandruszka, R., (1998). Kinetic aspects of cation-enhanced aggregation in aqueous humic substances. *Environmental Science and Technology, 32*, 488–493.

Fukushima, M., Nakayasu, K., Tanaka, S. & Nakamura, H. (1995). Chromium(III) binding abilities of humic acids. *Analytica Chimica Acta, 317*, 195–206.

Geckeis, H., Rabung, T. & Kim, J. I. (1999). Kinetic aspects of metal ion binding to humic substances. In G. Buckau (Ed.), *Effects of Humic Substances on the Migration of Radionuclides: Complexation and Transport of the Actinides* (pp. 47–58). Wissenschaftliche Berichte (FZKA 6324, ISSN 0947-8620), Forschungszentrum Karlsruhe Technik und Umwelt, Karlsruhe, Germany.

Gjessing, E. T., Riise, G., Peterson, R. C. & Andruchow, E. (1989). Bioavailability of aluminium in the presence of humic substances at low and moderate pH. *The Science of the Total Environment, 81/82*, 683–690.

Goodman, B. A., Cheshire, M. V. & Chadwick, J. (1991). Characterisation of the Fe(III)-fulvic acid reaction by Mossbauer spectroscopy. *Journal of Soil Science, 42*, 25–38.

Gregor, J.E. & Powell, H.K.J. (1987). Effects of extraction procedures on fulvic acid properties. *The Science of the Total Environment, 62*, 3–12.

Hatcher, P. G., Breger, I. A., Dennis, L. W. & Maciel, G. E. (1983). Solid state C-NMR of sedimentary humic substances: new revelations on their chemical composition. In R. F. Christman & E. T. Gjessing (Eds), *Aquatic and Terrestrial Humic Materials* (pp. 37–82). Michigan, USA: Ann Arbor Science.

Harvey, G. R. & Boran, D. A. (1985). Geochemistry of humic substances in seawater. In G. R. Aiken & D. M. Mcknight (Eds), *Humic Substances in Soil, Sediment, and Water: Geochemistry, Isolation and Characterization* (pp. 223–247). New York: John Wiley

Hayes, M. H. B. & Swift, R. S. (1978). The chemistry of soil organic colloids. In D. J. Greenland and M. H. B. Hayes (Eds), *The Chemistry of Soil Constituents* (pp. 179–320). London: John Wiley.

Hayes, M. H. B. & Swift, R. S. (1990). Genesis, isolation, composition and structures of soil humic substances. In M. F. de Boodt, M. H. B. Hayes & A. Herbillon, *Soil Colloids and Their Association in Aggregates* (pp. 245–305). New York: Plenum.

Hayes, M. H. B., McCarthy, P., Malcolm, R. L. & Swift, R. S. (1989). *Humic Substances II: In Search of Structure*. New York: John Wiley.

Hesketh, N. (1995). The characterisation of humic substances and their interactions with metallic ions, herbicides and pesticides. Ph.D thesis, University of Manchester, UK.

Hessen, D. O. & Tranvik L. J. (1998). *Aquatic Humic Substances: Ecology and Biogeochemistry*. Berlin: Springer-Verlag.

Higgo, J. J. W., Crawford, M. B., Milne, C. J. & Kinniburgh, D. G. (1996). *British Geological Survey Technical Report: Comparative Evaluation of Models for Ion-Binding by Humic Substances: PHREEQEV/Model V and NICA-Donnan* WE/95/51.

Higgo, J. J. W., Davis, J. R., Smith, B. & Milne, C. (1998). Extraction, purification and characterisation of fulvic acid. In G. Buckau (Ed.), *Effects of Humic Substances on the Migration of Radionuclides: Complexation and Transport of the Actinides* (pp. 2450-262). Wissenschaftliche Berichte (FZKA 6124, ISSN 0947-8620), Forschungszentrum Karlsruhe Technik und Umwelt, Karlsruhe, Germany

Jones, M. N. & Bryan, N. D. (1998). Colloidal properties of humic substances. *Advances in Colloid and Interface Science, 78*, 1–48.

Jouany, C. & Chassin, P. (1987). Wetting properties of Fe and Ca humates. *The Science of the Total Environment, 62*, 267–270.

Kaiser, K. (1998). Fractionation of dissolved organic matter affected by polyvalent metal cations. *Organic Geochemistry, 28*, 849–854.

Khan, S., Khan, N. N. & Iqbal, N. (1985). Influence of organic acids and bases on the mobility of some heavy metals in soil. *Journal of Indian Society of Soil Science, 33*, 779–784.

Kim, J. I. & Czerwinski K. R. (1996). Complexation of metal ions with humic acid: metal ion charge neutralisation model. *Radiochimica Acta, 73*, 5–10.

Kim, J. I. & Sekine, T. (1991). Complexation of neptunium(V) with humic-acid. *Radiochimica Acta, 55*, 187–192.

Kim, J. I., Wimmer, H. & Klenze, R. (1991). A study of curium(III) humate complexation by time resolved laser fluorescence spectroscopy (TRLFS). *Radiochimica Acta, 54*, 35–41.

Kim, J. I., Buckau, G., Bryant, E. & Klenze, R. (1989). Complexation of americium(iii) with humic-acid. *Radiochimica Acta, 48*, 135–143.

Kim, J. I., Rhee, D. S., Wimmer, H., Buckau, G. & Klenze R. (1993). Complexation of trivalent actinide ions (Am^{3+}, Cm^{3+}) with humic acid—a comparison of different experimental methods. *Radiochimica Acta, 62*, 35–43.

King, S., Warwick, P. & Bryan, N.D. (1999). A Study of metal complexation with humic and fulvic acid: the effect of Temperature on Association and Dissociation. In G. Buckau (Ed.), *Effects of Humic Substances on the Migration of Radionuclides: Complexation and Transport of the Actinides* (pp. 277–300). Wissenschaftliche Berichte (FZKA 6324, ISSN 0947-8620), Forschungszentrum Karlsruhe Technik und Umwelt, Karlsruhe, Germany

Kinniburgh, D. G., Milne, C. J., Benedetti, M. F., Pinheiro, J. P., Filius, J., Koopal, L. K. & Van Riemsdijk W.H. (1996). Metal ion binding by humic acid: application of the NICA-Donnan model. *Environmental Science and Technology, 30*, 1687–1698.

Kirk, T. K. (1984). The degradation of lignin. In D.T. Gibson, *Microbial Degradation of Organic Compounds*, Microbiology Series (Vol. 13) (pp. 399–437). New York: Marcel Dekker.

Kononova, M. M. (1966). *Soil Organic Matter* (2nd edn). Oxford, UK: Pergamon Press.

Lakatos, B. T., Tibai, T. & Meisel, L. (1977). EPR spectra of humic acids and their metal complexes. *Geoderma, 19*, 319–338.

Linder, P. W. & Murray, K. (1987). Statistical determination of the molecular structure and metal- binding sites of fulvic acids. *The Science of the Total Environment, 64*, 149–161.

Livens, F. R. (1991). Chemical reactions of metals with humic material. *Environmental Pollution, 70*, 183–208.

Livens, F. R. & Singleton, D. L. (1991). Plutonium and americium in soil organic matter. *Journal of Environmental Radioactivity, 13*, 323–339.

McBride, B. R. (1989). *Factors influencing the distribution of humic substances and trace metals in an upland catchment and their implications for water quality*. A thesis submitted to the University of Manchester (UK) for the degree of M.Sc.

McCarthy, J. F., Czerwinski, K. R., Sanford, W.E., Jardine, P. M. & Marsh, J. D. (1998). Mobilization of transuranic radionuclides from disposal trenches by natural organic matter. *Journal of Contaminant Hydrology, 30*, 49–77.

McKnight, D. M. & Aiken, G. R. (1998). Sources and age of aquatic humus, In D. O. Hessen & L. J. Tranvik (Eds), *Aquatic Humic Substances: Ecology and Biogeochemistry*. Springer-Verlag, Berlin.

Manahan, S. E., (1994). *Environmental Chemistry*, (6th edn.). Boca Raton, FL: CRC Press.

Manunza, B., Gessa, C., Deianna, S. & Rausa, R. (1992). A normal distribution model for the titration curves of humic acids. *Journal of Soil Science, 43*, 127–131.

Manunza, B., Deiana, S., Maddau, V., Gessa, C. & Seeba, R. (1995). Stability-constants of metal-humate complexes—titration data analyzed by bimodal gaussian distribution. *Soil Science Society of America Journal, 59*, 1570–1574.

Marinsky, J. A. & Ephraim, J. (1986). A unified physiochemical description of the protonation and metal ion complexation equilibria of natural organic acids (humic and fulvic acids) 1. Analysis of the influence of polyelectrolyte properties on protonation equilibria in ionic media: fundamental concepts. *Environmental Science and Technology, 20*, 349–354.

Martin, J. P. & Haider, K. (1971). Microbial activity in relation to humus formation. *Soil Science, 111*, 54–63.

Milne, C. J., Kinniburgh, D. G., De Wit, J.C.M., Van Riemsdijk, W. H. & Koopal, L. K. (1995a). Analysis of proton binding by a peat humic acid using a simple electrostatic model. *Geochimica et Cosmochimica Acta, 59*, 1101–1112.

Milne, C. J., Kinniburgh, D. G., De Wit, J. C. M., Van Riemsdijk, W. H. & Koopal, L .K. (1995b). Analysis of metal-ion binding by a peat humic-acid using a simple electrostatic model. *Journal of Colloid and Interface Science, 175*, 448–460.

Mountney, A. W. & Williams, D. R. (1992). Computer-simulation of metal-ion humic and fulvic-acid interactions. *Journal of Soil Science, 43*, 679–688.

Murray, K. & Linder, P. W. (1983). Fulvic acids: structure and metal binding, I a random molecular model. *Journal of Soil Science, 34*, 511–523.

Murray, K. & Linder, P. W. (1984). Fulvic acids: structure and metal binding, II: predominant metal binding sites. *Journal of Soil Science, 35*, 217–222.

Nash, K. L. & Choppin, G. R. (1980). Interaction of humic and fulvic acids with Th(IV). *Journal of Inorganic and Nuclear Chemistry, 42*, 1045–1050.

Nederlof, M. M., De Wit, J. C. M., Van Riemsdijk, W. H. & Koopal, L. K. (1993). Determination of proton affinity distributions for humic substances. *Environmental Science and Technology, 27*, 846–856.

Oden, S. (1914). Zur Kolloidchemie der Humusstoffe. *Kolloid Z., 14*, 123–130.

Petersen, R. (1982). Influence of copper and zinc on the growth of a freshwater alga, Scenedesmus quadricauda: the significance of chemical speciation. *Environmental Science and Technology, 16*, 443–447.

Piccolo, A. (1997). New insights on the conformational structure of humic substances as revealed by size exclusion chromatography. In J. Drozd, S. S. Gonet, N. Senesi & J. Weber (Eds), *The Role of Humic Substances in Ecosystems and in Environmental Protection*. Wroclaw, Poland: IHSS.

Piccolo, A., Nardi, S. & Concheri, G. (1996). Macromolecular changes of humic substances induced by interaction with organic acids. *European Journal of soil science, 47*, 319–328.

Posner, A. M. J. (1966). The humic acids extracted by various reagents from a soil. Part I yield, inorganic components and titration curves. *Journal of Soil Science, 17*, 65–78.

Perdue, E. M. & Lytle, C. R. (1983). Distribution model for binding of protons and metal ions by humic substances. *Environmental Science and Technology, 17*, 654–660.

Ramsay, J. D. F. (1988). The role of colloids in the release of radionuclides from nuclear waste. *Radiochimica Acta, 44/45*, 165–170.

Rao, L. & Choppin, G. R. (1995). Thermodynamic study of the complexation of neptunium(V) with humic acids. *Radiochimica Acta, 69*, 87–95.

Read, D. & Falck, W. E. (1996). *CHEMVAL 2: A Coordinated Research Initiative for Evaluating and Enhancing Chemical Models in Radiological Risk Assessment*. A Commission of the European Community Report, Contract Number: FI2W-CT91-0065, Report No. EUR 16648 EN, ISBN: 92-827-8988-8.

Rheinheimer, G. (1974). *Aquatic Microbiology*. London: John Wiley

Saar, R. A. & Weber, J. H. (1982). Fulvic acid: modifier of metal ion chemistry. *Environmental Science and Technology, 16*, 510–517.

Samadfam, M., Niitsu, Y., Sato, S. & Ohashi, H. (1996). Complexation thermodynamics of Sr(II) and humic acid. *Radiochimica Acta, 73*, 211–216.

Schimpf, M. E. & Petteys, M. P., (1997). Characterization of humic materials by flow field-flow fractionation. *Colloids and Surfaces A: Physicochemical and Engineering Aspects, 120*, 87–100.

Schnitzer, M., (1978). Humic substances: chemistry and reactions. In M. Schnitzer and S. U. Khan (Eds), *Soil Organic Matter* (pp. 1–64). Amsterdam: Elsevier Science.

Schnitzer, M. & Khan, S. U. (1972). *Humic Substances in the Environment*. New York: Marcel Dekker.

Schnitzer, M., Wright, J. R. & Desjardins, J. G. (1958). A comparison of the effectiveness of various extractants for rganic matter from two horizons of a podzol profile. *Canadian Journal of Soil Science, 38*, 49–53.

Schulton, H. R. & Schnitzer, M. (1993). A state of the art structural concept for humic substances. *Naturwissenschaften, 80*, 29–30.

Schussler, W., Artinger, R., Kienzler, B. & Kim, J. I. (1998). Modelling of humic colloid mediated transport of americium (III) by a kinetic approach. In G. Buckau (Ed.), *Effects of Humic Substances on the Migration of Radionuclides. Complexation and Transport of the Actinides* (pp. 91–101). Wissenschaftliche Berichte (FZKA 6124, ISSN 0947-8620), Forschungszentrum Karlsruhe Technik und Umwelt, Karlsruhe, Germany.

Schussler, W., Artinger, R., Kim, J. I., Bryan, N.D. & Griffin D, (1999). Modelling of Humic Colloid Borne Americium(III) Migration in Column Experiments Using the Transport/Speciation Code k1-D and the KICAM Model. Paper presented at MIGRATION 99, Chemistry and Migration Behaviour of Actinides and Fission Products in the Geosphere, Lake Tahoe, California, 1 October to 26 September 1999

Sedlaceck, J., Gjessing, E. T. & Kallqvist, T. (1989). Influence of different aquatic humus fractions on uptake of cadmium to alga Selenastrum capricornutum Printz. *The Science of the Total Environment, 81/82*, 711–718.

Senesi, N., Bocian, D. F. & Sposito, G. (1985a). Electron Spin Resonance investigation of copper(II) complexation by soil fulvic acid. *Soil Science Society of America Journal, 49*, 114–119.

Senesi, N., Bocian, D. F. & Sposito, G. (1985b). Electron Spin Resonance investigation of copper(II) complexation by fulvic acid extracted from sewage sludge. *Soil Science Society of America Journal, 49*, 119–126.

Steinberg, C. & Munster, U. (1985). Geochemistry and ecological role of humic substances in lake water. In G. R. Aiken, D. M. McKnight, R. L. Wershaw & P. McCarthy (Eds), *Humic Substances in Soil, Sediment and Water: Geochemistry, Isolation and Characterisation*. New York: John Wiley.

Stevenson, F. J. (1982). *Humus Chemistry: Genesis, Composition, Reactions*. New York: John Wiley.

Swift, R. S. (1989a). Molecular shape and size of soil humic substances by ultracentrifugation. In M. H. B. Hayes, P. McCarthy, R. L. Malcolm & R. S. Swift (Eds), *Humic Substances II: In Search of Structure* (pp. 467–495). Chichester, UK: John Wiley.

Swift, R. S. (1989b). Molecular weight, size, shape and charge characteristics of humic substances: some basic considerations. In M. H. B. Hayes, P. McCarthy, R. L. Malcolm & R. S. Swift (Eds), *Humic Substances II: In Search of Structure* (pp. 449–465). Chichester, UK: John Wiley.

Swift, R. S. (1996). Organic matter characterisation. In D. L. Sparks (Ed.), *Methods of Soil Analysis: Chemical Analysis* (3rd edn.) (pp. 1011–1070). Madison, WI: American Society of Agronomy. bib

Swift, R. S. (1999). Macromolecular properties of soil humic substances: fact, fiction, and opinion. *Soil Science, 164*, 790–802.

Swift, R. S., DeLisle, G. & Leonard, R. L. (1987). Biodegradation of humic acids from New-Zealand soils. *The Science of the Total Environment, 62*, 423–430.

Tanford, C. (1961). *Physical Chemistry of Macromolecules*. New York: John Wiley.

Tipping, E. (1993a). Modelling ion-binding by humic acids. *Colloids and Surfaces A: Physicochemical and Engineering Aspects, 73*, 117–131.

Tipping, E. (1993b). Modelling the binding of europium and the actinides by humic substances. *Radiochimica Acta, 62*, 141–152.

Tipping E. (1994). WHAM—A chemical equilibrium model and computer code for waters, sediments, and soils incorporating a discrete site/electrostatic model of ion-binding by humic substances. *Computers and Geosciences, 20*, 973–1023.

Tipping, E. & Hurley, M. A. (1992). A unifying model of cation binding by humic substances. *Geochimica et Cosmochimica Acta, 56*, 3627–3641.

Tombacz, E., Rice, J. A. & Ren, S. Z., (1997). Fractal structure of polydisperse humic acid particles in solution studied by scattering methods. *ACH-Models in Chemistry, 134*, 877–888.

Van Riemsdijk, W. H., De Wit, J. C. M., Mous, S. L. J., Koopal, L. K. & Kinniburgh, D. G. (1996). An analytical isotherm equation (CONICA) for nonideal mono- and bidentate competitive ion adsorption to heterogeneous surfaces. *Journal of Colloid and Interface Science, 183*, 35–50.

Von Wandruszka, R., Ragle, C. & Engebretson, R., (1997). The role of selected cations in the formation of pseudomicelles in aqueous humic acid. *Talanta, 44*, 805–809.

Warwick, P., Hall, T. & Read, D. (1996). A comparative study employing three different models to investigate the complexation properties of humic and fulvic acids. *Radiochimica Acta, 73*, 11–19.

Warwick, P., Hall, A., King, S. J., Zhu, J. & Van der Lee, J. (1998). Thermodynamic aspects of nickel humic acid interactions. *Radiochimica Acta, 81*, 215–222.

Warwick, P., Hall, A., Pashley, V., Bryan, N. D. & Griffin, D. (2000). Modelling the effect of humic substances on the transport of europium through porous media: a comparison of equilibrium and equilibrium/kinetic models. *Journal of Contaminant Hydrology, 42*, 19–34.

Wershaw, R. L. (1986). A new model for humic materials and their interactions with hydrophobic chemicals in soil-water or sediment-water systems. *Journal of Contaminant Hydrology, 1*, 29–45.

Wershaw, R. L. (1989). Sizes and shapes of humic substances by scattering techniques. In M. H. B. Hayes, P. McCarthy, R. L. Malcolm & R.S. Swift (Eds), *Humic Substances II: In Search of Structure* (pp. 545–559). Chichester, UK: John Wiley.

Wershaw, R. L. (1993). Model for humus in soils and sediments. *Environmental Science and Technology, 27*, 814–816.

Wood, M. (1995). *Environmental Soil Biology*. Blackie Academic and Professional.

Xia, K., Bleam, W. & Helmke, P. A. (1997). Studies of the nature of binding sites of first row transition elements bound to aquatic and soil humic substances using X-ray absorption spectroscopy. *Geochimica et Cosmochimica Acta, 61*, 2223–2235.

Miranda J. Keith-Roach and Francis R. Livens (Editors)
© 2002 Elsevier Science Ltd. All rights reserved

Chapter 6

Microbial interactions with metals/radionuclides: the basis of bioremediation

Geoffrey M. Gadd

Division of Environmental and Applied Biology, Biological Science Institute, School of Life Sciences, University of Dundee, Dundee DD1 4HN, Scotland, UK

1 Introduction

Contamination of the environment by radionuclides, toxic metals, metalloids and organometals is of considerable economic and environmental significance (Gadd, 1992a, b, 1997, 2000a, b; White et al., 1995; Wainwright & Gadd, 1997). There is therefore considerable interest in the ways microbiological processes can effect the behaviour of contaminants in natural and engineered environments and their potential to bioremediate the contaminants. The extent to which these processes can affect contaminants is dependent on the identity and chemical form of the radionuclide released and the physical and chemical nature of the contaminated site or substance. For example, mineral components may contain considerable quantities of contaminants which are biologically unavailable (Wainwright & Gadd, 1997; White & Gadd, 1998a). Microbial processes which solubilise radionuclides increase their bioavailability and potential toxicity, whereas those that immobilise them, reduce bioavailability. The relative balance between mobilisation and immobilisation varies depending on the radionuclide/metal, the organisms, their environment and physicochemical conditions. As well as being an integral component of biogeochemical cycles for metallic species, these processes may be exploited for the treatment of contaminated solid and liquid wastes (Gadd, 1992a, 1996, 1997, 2000a, b; Gadd & White, 1993; White et al., 1997, 1998a, b).

Microorganisms can mobilise radionuclides/metals through autotrophic and heterotrophic leaching, chelation by microbial metabolites and siderophores, and methylation, which can result in volatilisation. Conversely, immobilisation can result from sorption to cell components or exopolymers, transport into cells and intracellular sequestration or precipitation as insoluble organic and inorganic compounds, e.g. oxalates (Sayer and Gadd, 1997; Gharieb et al., 1998), sulfides or phosphates (Yong and Macaskie, 1995; White and Gadd, 1996a, b). In addition, some microorganisms can mediate the reduction of certain redox-sensitive radionuclides/metals to a lower valency, which may also aid mobilisation, e.g. Mn(IV) reduction to the more soluble Mn(II), or immobilisation, e.g.

Tc(VII) to the less soluble Tc(IV) or U(VI) to U(IV) (Lovley et al., 1991; White & Gadd, 1998a; Lloyd et al., 1998). In the context of bioremediation, solubilisation provides a route for removal from solid matrices such as soils, sediments, dumps and industrial wastes. Alternatively, immobilisation processes may enable metals to be transformed in situ into insoluble and chemically inert forms and are particularly applicable to removing metals from mobile aqueous phases. This chapter will detail certain microbiological processes which are of significance in determining radionuclide/metal mobility and which have actual or potential applications in bioremediation of metal/radionuclide and metalloid pollution. These include autotrophic and heterotrophic leaching, biosorption, reduction, precipitation and transformation.

2. Radionuclide/metal leaching

Autotrophic leaching

Radionuclides/metals can be leached from solid matrices, and thus solubilised, as a result of autotrophic metabolism. Most autotrophic leaching is carried out by chemolithotrophic, acidophilic bacteria which fix carbon dioxide and obtain energy from the oxidation of ferrous iron or reduced sulfur compounds. These metabolic processes yield Fe(III) or H_2SO_4 as the respective end products, as outlined in the equations below. The microorganisms involved in autotrotrophic leaching include sulfur-oxidising bacteria, e.g. *Thiobacillus thiooxidans*, iron- and sulfur-oxidising bacteria, e.g. *Thiobacillus ferrooxidans* and iron-oxidising bacteria, e.g. *Leptospirillum ferrooxidans* (Ewart & Hughes, 1991; Bosecker, 1997). Both the *Thiobacillus* species are able to oxidise inorganic sulfur (equation 1) and common metal sulfides such as pyrite (equation 2; Baldi et al., 1992). *T. ferrooxidans* and *L. ferrooxidans* are able to oxidise soluble ferrous iron (equation 3) producing ferric iron which can then indirectly solubilise metal sulfides (equation 4; Luther, 1987; Bosecker, 1997).

$$2S^0_{(s)} + 3O_2 + 2H_2O \rightarrow 2H_2SO_{4(aq)} \tag{1}$$

$$2FeS_{2(s)} + 7O_2 + 2H_2O \rightarrow 2FeSO_{4(aq)} + 2H_2SO_{4(aq)} \tag{2}$$

$$4FeSO_{4(aq)} + O_2 + 2H_2SO_{4(aq)} \rightarrow 2Fe_2(SO_4)_{3(aq)} + 2H_2O \tag{3}$$

$$FeS_{2(s)} + 14Fe^{3+}_{(aq)} + 8H_2O \rightarrow 15Fe^{2+}_{(aq)} + 2SO^{2-}_{4(aq)} + 16H^+ \tag{4}$$

As a result of sulfur and iron oxidation by these bacteria, metal sulfides are solubilised and the pH of their immediate environment is decreased, which enhances the solubilisation of other metal compounds. *T. thiooxidans*, *T. ferrooxidans* and *L. ferrooxidans* are all mesophilic bacteria although moderately thermophilic iron- and sulfur-oxidising bacteria with optimum temperatures for metal sulfide solubilisation of 45–50°C have also been isolated (Marsh & Norris, 1983; Wood & Kelly, 1983; Ghauri & Johnson, 1991).

The autotrophic leaching of metal sulfides by *Thiobacillus* species and other acidophilic bacteria is well established for use in industrial scale biomining processes. Low grade copper and uranium ores are leached to extract the metal and refractory gold ores are leached to remove arsenopyrite passivation layers before conventional cyanidation and extraction of the gold. The bacteria used for extracting U, principally of the genus *Thiobacillus* and *Leptospirillum*, are able to grow in highly acidic environments with high U and Th levels (Muñoz et al., 1995). Moderately thermophilic (45–50°C) microorganisms such as the genus *Sulfobacillus* and highly thermophilic bacteria (60–80°C) such as *Sulfolobus* have been detected in uranium heaps, aiding bioleaching in the centre of heaps and at the even higher temperatures reached in some industrial operations (Muñoz et al., 1995).

In a bioremediation context, autotrophic production of sulfuric acid by *Thiobacillus* species has been used to solubilise metals from sewage sludge, thus enabling separation from the sludge which can then be used as a fertiliser (Sreekrishnan & Tyagi, 1994). Autotrophic leaching with *T. ferrooxidans* required acidification of the sludge to approximately pH 4.0 before adequate growth of the bacteria could occur although sulfur-oxidising *Thiobacillus* sp. have been isolated which were able to grow at pH 7.0 making preliminary acidification unnecessary (Sreekrishnan & Tyagi, 1994). The elemental sulfur, which acts as an energy source for the process, can be of either chemical or biological origin (Tichy et al., 1994) and it has been shown that solid lumps of sulfur are as effective as powder. The use of sulfur as rods or similar may allow unused sulfur to be removed and therefore prevent wastage of sulfur and acidification of the sludge after disposal (Ravishankar et al., 1994). Simultaneous sewage sludge digestion and metal leaching under acidic conditions (pH 2.0–2.5) has an advantage over conventional aerobic and anaerobic digestion in that the acidity leads to a decrease in pathogenic microorganisms as well as the digestion of the sludge and removal of toxic metals (Blais et al., 1997).

Autotrophic leaching has been used to remediate other metal-contaminated solid materials including soil (Zagury et al., 1994) and red mud, the main waste product of Al extraction from bauxite (Vachon et al., 1994). Most of the processes and laboratory studies involving autotrophic leaching have used mesophilic chemolithotrophic bacteria, however, moderately thermophilic, chemolithotrophic bacteria may prove increasingly important in the future.

Heterotrophic leaching

Heterotrophic metabolism can also lead to leaching, and this is most important in the case of fungi. Leaching occurs as a result of several processes, including the efflux of protons from hyphae and the production of siderophores (see Section 5), but in most fungal strains, leaching occurs mainly by the production of organic acids (Burgstaller & Schinner, 1993; Sayer & Gadd, 1997; Gadd, 1999; Gadd & Sayer, 2000). Solubilisation of insoluble metal compounds results from protonation of the anion of the compound, which makes it less available to the metal cation (Hughes & Poole, 1991). The production of organic acids provides both a source of protons and a metal-chelating anion to complex the metal cation, with complexation being dependent on the relative concentrations of the anions and metals in solution, pH and the stability constants of the various complexes (Denêvre et al., 1996).

Organic acids are released into the soil by both plant roots and fungal hyphae, with citric and oxalic acids being most commonly reported (Fox & Comerford, 1990; Jones & Kochian, 1996). Citrate and oxalate anions can form stable complexes with a large number

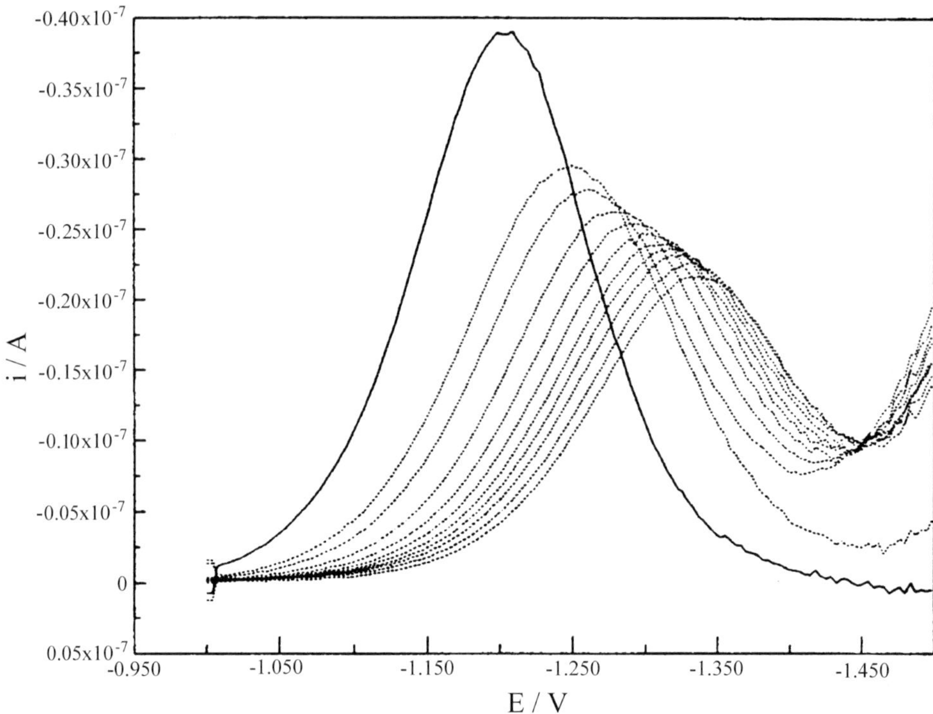

Fig. 1. Complexation of 100 μM Co^{2+} (solid line) by citric acid, shown by polarograms obtained by successive additions of citric acid (100–1000 μM, broken lines). Successive additions of citric acid result in a decrease in peak height and a shift in evolution potential towards more negative values which is indicative of complex formation. Similar decreases in peak height and potential shifts occur on addition of oxalic acid or *A. niger* culture filtrates (adapted from White et al., 1997).

of metals, for example, cobalt (Fig. 1). Figure 1 shows the complexation of Co^{2+} by citric acid, as demonstrated by polarography, a technique used to assess metal complexation by a variety of substances, including microbial metabolites. Uranium also forms very stable 1:1 and 1:2 uranium-citrate complexes with stability constants that are orders of magnitude higher than those of uranyl acetate, uranyl lactate, U-EDTA and uranyl ascorbate complexes (Borkowski et al., 1996). Most metal citrates are highly mobile and are not readily degraded and so the presence of citric acid in the terrestrial environment will leach contaminant metals from soils and enhance their solubility for a significant time (Francis et al., 1992). However, complex stability may be lower under more extreme conditions, for example, U-citrate complexes have been shown to dissociate in hypersaline conditions in the presence of phosphate, giving citric acid and a uranium precipitate, which is predominately $K(UO_2)_5(PO_4)_3(OH)_2{\cdot}nH_2O$ (Francis et al., 2000).

Citrate was found to enhance the dissolution of Fe from solid phase $Fe(OH)_3$ (Jones & Kochian, 1996) and leach Pu and Am from soil particles (Lu et al., 1998). Lu et al. (1998) found that the ability of citrate to leach Pu and Am is highly dependent on both the fraction of the soil they are held in and the E_h of the system. In the finest particles examined (<0.053

mm particle size), the Pu and Am were mostly associated with the iron-oxide coating and citrate leached only 20% of the alpha activity. Addition of a reducing agent improved this to 65% by aiding dissolution of the Fe oxide and improving the interaction between the Pu(IV) and citrate. However, in the larger particles (0.50–1.0 mm particle size), the Pu and Am were predominately held in the organic fraction and citrate alone leached 68%. In this case, reducing the E_h led to 88% removal, showing that while citrate will remove the majority of Pu and Am from the soil organic fraction, it is generally most effective in reducing conditions. Oxalic acid can act as a leaching agent for those metals that form soluble oxalate complexes, including Al and Fe (Strasser et al., 1994).

Fungi are ubiquitous in both polluted and unpolluted soils, and they dominate the biota under acidic conditions (Metting, 1992), where metals are more likely to be speciated into more mobile and available forms (Harter, 1983). However, at low pH, the toxicity of metal cations can be reduced due to protonation of metal-binding sites on the fungal cell walls which results in decreased uptake (Gadd, 1993a). Solubilisation of insoluble metal compounds is important in the natural environment for the release of anions, such as phosphate, and essential cations into biogeochemical cycles (Gadd, 1999). Most phosphate fertilisers are applied in a solid form (calcium phosphate), which has to be solubilised before it is available to plants and other organisms. In fact, increased phosphate uptake by mycorrhizal plants is believed to be due to the high phosphate-solubilising ability of the mycorrhizal partner (Lapeyrie et al., 1991). However, such solubilisation in the environment can also be hazardous, for example, bacterial activity in mining areas can leach significant levels of metal cations into water systems (Banks et al., 1994). The presence of fulvic acid in the soil, which may arise from the breakdown of organic matter by microorganisms, has also been found to increase the transport of thorium from soil to stream by more than two orders of magnitude (Tipping, 1996). It has been suggested that certain microbiological processes, including proton efflux, are included in the safety assessment of waste repositories (Arter et al., 1991; see Chapters 9 and 10).

A screening method for the incidence of solubilisation and metal tolerance of fungal strains and isolates has been developed (Sayer et al., 1995). Fungal strains, including environmental isolates, are inoculated onto media containing the desired insoluble metal compound: a halo of solubilisation around a colony indicates an active strain (Burgstaller & Schinner, 1993). The incidence of solubilising ability in natural soil isolates was found to be high, and approximately one-third of soil isolates screened (including isolates from lead and nickel contaminated soils) were able to solubilise at least one of the three test metal compounds ($Co_3(PO_4)_2$, ZnO and $Zn_3(PO_4)_2$); 11% were able to solubilise all three. Twenty percent of the isolates showed no decrease in growth rate on the metal compounds, indicating that 80% were susceptible, to varying degrees, to the metal compounds whether solubilisation occurred or not, although it should be noted that a small amount of solubilisation may not have been detected by this method (Sayer et al., 1995).

The soil fungus *Aspergillus niger* is able to solubilise a wide range of insoluble metal compounds, including phosphates, sulfides and mineral ores such as cuprite (CuS) (Sayer et al., 1995, 1997, 1999; Sayer & Gadd, 1997). This fungus, a prolific organic acid producer, will produce organic acids and acidify its surrounding medium whether a metal compound is present or not (Burgstaller & Schinner, 1993; Sayer and Gadd, 1997). Fungal growth media can be manipulated, e.g. by changing the N and P balance or pH, to maximise organic acid production. It is known, for example, that a deficiency of manganese (less than 10^{-8} M) in the growth medium leads to the production of large amounts of citric acid by *A. niger* (Meixner et al., 1985), and typical concentrations of citric acid produced industrially by this fungus can reach 600 mM (Mattey, 1992). The pH of non-regulated *A. niger* cultures can fall to values between 1.5 and 2.0 due to high citric acid production as the optimal pH for citric acid production is below 3.5 (Schrickx et al., 1994). Oxalic acid production can be manipulated to yield concentrations of up to 200 mM on low cost carbon sources with the optimum pH for oxalic acid production being around neutrality (Strasser et al., 1994).

In an industrial context, most biological leaching has been autotrophic (Rawlings & Silver, 1995) because chemoautotrophic acidophilic bacteria such as *Thiobacillus* species have the advantages of not needing a carbon source and having a high acidification capacity. However, their acidophilic nature may make them unable to tolerate the higher pH values of many industrial wastes (Burgstaller & Schinner, 1993). Many fungi can tolerate these higher pH conditions, giving heterotrophic leaching a definite advantage in these cases. There may be additional advantages in that fungi are more easily manipulated in bioreactors than thiobacilli and, by altering growth conditions, can be induced to produce high concentrations of organic acids (Mattey, 1992; Burgstaller & Schinner, 1993; Strasser et al., 1994). Furthermore, many species can tolerate high concentrations of metals (Gadd, 1993a) and can grow in both low and high pH environments (Burgstaller & Schinner, 1993).

Many species of fungi are able to leach metals from industrial wastes and byproducts, low grade ores (Burgstaller & Schinner, 1993) and metal-bearing minerals (Tzeferis et al., 1994; Drever & Stillings, 1997; Sayer et al., 1999; Gadd & Sayer, 2000). There are several examples in the literature of fungal leaching of industrial waste containing insoluble metal compounds. A strain of *Penicillium simplicissimum*, isolated from a metal contaminated site, has been used successfully to leach Zn from insoluble ZnO contained in industrial filter dust. This fungus only developed the ability to produce citric acid (>100 mM) in the presence of the filter dust (Schinner & Burgstaller, 1989; Franz et al., 1991, 1993). Cd, Zn, Cu, Pb and Al have been leached from municipal waste fly ash using *A. niger* (Bosshard et al., 1996). Culture filtrates from *A. niger* have also been used to leach Cu, Ni and Co from copper converter slag (Sukla et al., 1992). Al has been leached from red mud (the waste product of the extraction of Al from bauxite) with various fungal strains and adapted thiobacilli. The thiobacilli were best, with the most efficient of the fungal strains being *P. simplicissimum*, the fungal-derived acids (mainly citric) having a much greater ability to leach Al than pure citric acid (Vachon et al., 1994). A heterotrophic mixed culture has been employed for leaching manganiferous minerals through the reduction of MnO_2, with the process having potential for the treatment of materials not treatable by conventional processes (Veglio, 1996).

3. Biosorption

Biosorption can be defined as the microbial uptake of organic and inorganic metal species, both soluble and insoluble, by physicochemical mechanisms such as adsorption. In living cells, metabolic activity may also influence this process because of changes in pH, E_h, organic and inorganic nutrients and metabolites in the cellular microenvironment. Biosorption can also provide nucleation sites for the formation of stable minerals (Beveridge & Doyle, 1989). Almost all biological macromolecules have some affinity for metal species with cell walls and associated materials being of the greatest significance in biosorption. The affinity also depends on the oxidation state and chemical speciation of the metal since binding depends, to a large extent, on the electrostatic interaction between binding site and metal species. As well as this sorption to the cellular surface, some cationic species can be accumulated within cells, via transport systems through the cell wall of varying affinity and specificity. Once inside cells, metal species may be bound, precipitated, localised within intracellular structures or organelles, or translocated to specific structures depending on the element concerned and the organism (Gadd, 1996, 1997; White et al., 1997).

Biosorption by cell walls and associated components

Microbial exopolymers can be composed of polysaccharide, glycoproteins and lipopolysaccharide which may be associated with protein (Geesey & Jang, 1990). Many such exopolymers act as polyanions under natural conditions, and negatively charged groups can interact with cationic metal/radionuclide species although uncharged polymers are also capable of binding and entrapment of insoluble forms (Beveridge & Doyle, 1989). Peptidoglycan carboxyl groups are the main binding site for cations in Gram-positive bacterial cell walls with phosphate groups contributing significantly in Gram-negative species (Beveridge & Doyle, 1989). Chitin is an important structural component of fungal cell walls and this is an effective biosorbent for radionuclides, as are chitosan and other chitin derivatives (Tobin et al., 1994). In *Rhizopus arrhizus*, U biosorption involves coordination to the amine N of chitin, adsorption in the cell wall chitin structure and further precipitation of hydroxylated derivatives (Tsezos & Volesky, 1982). Fungal phenolic polymers and melanins possess many potential metal-binding sites with oxygen-containing groups including carboxyl, phenolic and alcoholic hydroxyl, carbonyl and methoxyl groups being particularly important (Gadd, 1993a).

Biosorption by free and immobilised biomass

Biosorption has the potential to remove radionuclides/metals from industrial waste waters. Both freely-suspended and immobilised biomass from bacterial, cyanobacterial, algal and fungal species have received attention with immobilised systems appearing to possess several advantages including higher mechanical strength and easier biomass/liquid separation (Macaskie & Dean, 1989). Biomass of all groups has been immobilised by encapsulation or cross-linking using supports which include agar, cellulose, alginates, cross linked ethyl acrylate-ethylene glycol dimethylacrylate, polyacrylamide, silica gel and cross-linking reagents such as toluene diisocyanate and glutaraldehyde (Macaskie & Dean, 1989; Brierley, 1990; Macaskie, 1991; Tobin et al., 1994). Immobilised living biomass has mainly taken the form of bacterial biofilms on inert supports and is used in a variety of bioreactor configurations including rotating biological contactors, fixed bed reactors, trickle filters, fluidised beds and air-lift bioreactors (Gadd, 1988; Macaskie &

Dean, 1989; Gadd & White, 1990, 1993). Refer to Chapter 12 of this volume for a detailed discussion of this subject.

4. Metal-reduction

Metal-reducing bacteria

A taxonomically diverse range of microorganisms is able to use oxidised species of metallic elements, e.g. Fe(III), U(VI) or Pu(IV), as terminal electron acceptors. Many of these organisms can utilise more than one terminal electron acceptor including several metals or other anions, such as nitrate or sulfate. Most of these organisms are anaerobic although a few are facultative anaerobes and oxygen may also be respired. The majority of dissimilatory metal-reducing bacteria are respiratory heterotrophic organisms possessing an electron transport chain. The range of substrates utilised, comprising organic acids, alcohols and aromatic compounds, reflects this metabolic pattern (Lovley & Phillips, 1988; Lovley et al., 1989, 1993; Shen & Wang, 1993), as does inhibition of dissimilatory metal reduction by respiratory inhibitors (Myers & Nealson, 1988). However, a small number of metal-reducing strains is able to reduce metals within a fermentative metabolic framework (Ghani et al., 1993; Rusin et al., 1993). A significant feature of Fe(III)- and Mn(IV)-reducing bacteria is the much greater affinity that these organisms have for both organic substrates and hydrogen compared to sulfate-reducing bacteria and methanogens. This can lead to the total competitive inhibition of sulfate reduction and methanogenesis when substrates are limiting (Lovley & Phillips, 1987; Caccavo et al., 1992) and produce zonation of anaerobic heterotrophic activities in groundwater (Chapelle & Lovley, 1992).

Processes using dissimilatory metal reduction

As yet, biotechnological processes using microbial metal reduction are at the stage of laboratory demonstration and there are no near commercial applications (Lovley & Coates, 1997). Fe(III) and Mn(IV) appear to be the most commonly utilised metals as terminal electron acceptors in the biosphere and metal-reducing organisms from many habitats frequently utilise both of these metals (Lovley, 1993). However, since the solubility of both Fe and Mn is increased by bacterial reduction, and neither metal is significantly toxic, other metals are targeted in waste treatment. Molybdenum(VI) was reduced to molybdenum blue by a strain of *Enterobacter cloacae* which was isolated from a molybdate-polluted aquatic environment (Ghani et al., 1993). Another strain of *E. cloacae*, also isolated from a polluted habitat, was able to reduce Cr(VI) to Cr(III) under similar conditions (Wang et al., 1989) and in a bioreactor Cr(III) was precipitated from a simulated waste water by this organism (Fujie et al., 1994). The fact that these strains were isolated from polluted waters may indicate that natural waters do not contain sufficient chromate or molybdate to support similar organisms and that this type of metabolism is an adaptation specifically exploiting oxyanions present in industrial effluent. Dissimilatory Cr(VI) reduction was also carried out by a strain of *Escherichia coli* under both anaerobic and aerobic conditions, albeit at a slower rate (Shen and Wang, 1993). Metal reduction processes may also be useful as pretreatments for other processes, e.g. the reduction of Cr(VI) compounds to Cr(III) facilitates removal by processes such as biosorption or (bio)precipitation (Aksu et al., 1991).

Perhaps the most promising potential application of dissimilatory biological metal reduction is uranium precipitation, which is performed by a number of organisms and may have potential both in waste treatment and in concentrating uranium from low-grade sources. While U(VI) compounds are readily soluble, U(IV) compounds such as the hydroxide or carbonate have low solubility and readily form precipitates at neutral pH. A strain of *Shewanella (Alteromonas) putrefaciens* which reduced Fe(III) and Mn(IV) also reduced U(VI) to U(IV) forming a black precipitate of U(IV) carbonate (Lovley et al., 1993). When the organisms were contained by dialysis tubing, the precipitate was associated with the organisms, indicating that it was the result of an enzymatic reaction (Gorby & Lovley, 1992). U(VI) was also reduced by the sulfate-reducing bacterium *Desulfovibrio desulfuricans* in the presence of sulfate, utilising the electron transport chain and producing a very pure precipitate of U(IV) carbonate thus providing a potential alternative to more conventional chemical technologies (Lovley and Phillips, 1992a, b). It was also reported that *Desulfovibrio vulgaris* carried out a similar enzymic reduction of uranium(VI) (Lovley et al., 1993). Bacterial uranium reduction has also been combined with chemical extraction to produce a potential process for soil bioremediation and has been suggested as a means of immobilising U from ground water (see Chapter 7; Phillips et al., 1995). Metal precipitation by sulfate-reducing bacteria is discussed further in Section 4.

The solubility of other radionuclides can be increased by reduction and this may favour their removal from matrices such as soils. For example, iron-reducing bacterial strains solubilised 40% of the Pu present in contaminated soils within 6–7 days through reduction of Pu(IV) to the more soluble Pu(III) (Rusin et al., 1993) and both iron- and sulfate-reducing bacteria were able to solubilise Ra from uranium mine tailings, although solubilisation occurred largely by disruption of reducible host minerals (Landa & Gray, 1995).

Metal precipitation by sulfate-reducing bacteria

The sulfate-reducing bacteria (SRB) are strictly anaerobic heterotrophic bacteria commonly found in environments where oxygen is excluded and where carbon substrates and sulfate are available. Examples of such habitats are freshwater, marine and estuarine sediments and waters with a high organic content. SRB are largely mesophilic (Postgate, 1984; Barnes et al., 1991; White & Gadd, 1996b) although thermophilic strains have been recovered from habitats such as hydrothermal vents (Prieur et al., 1995). Sulfate-reducing bacteria are almost entirely neutrophilic with maximum growth obtained in the range pH 6–8 (Postgate, 1984). However, some isolates can grow in moderately acid conditions such as mine and surface waters where the bulk phase pH is in the range 3–4. In these environments the sulfate-reducing bacteria are found in sediments and their apparent acid tolerance is derived from the existence of more neutral microenvironments which are maintained by the buffering effect resulting from the low dissociation of H_2S (Hedin & Nairn, 1991; White & Gadd, 1996a). Sulfate-reducing bacteria utilise an energy metabolism in which the oxidation of organic compounds or hydrogen is coupled to the reduction of sulfate as the terminal electron acceptor, producing sulfide. They dissimilate carbon via respiratory mechanisms which have been described in detail elsewhere (Postgate, 1984; Peck, 1993). The range of carbon/energy sources used by SRB as a group is very wide and includes alcohols, organic acids and hydrocarbons. However, individual strains are only able to metabolise a limited range of these substrates and substrate preferences have been used to divide sulfate-reducing bacteria into groups which also differ in important aspects such as growth rate (Widdel, 1988). The hydrogen-lactate group comprises mainly *Desulfovibrio* and *Desulfotomaculum* species which utilise lactic, pyruvic, succinic, fumaric and malic acids in addition to ethanol, formate and glycerol. Some organisms with this metabolic pattern can also use hydrogen as electron-donor in the presence of CO_2, acetate or another organic carbon source. This ability has been utilised in a laboratory-scale packed-bed reactor using producer gas (synthesis gas), which contains a substantial proportion of hydrogen, as a substrate for sulfate-reducing bacteria to remove sulfate from simulated industrial waste waters (Du Preez et al., 1992). Sulfate-reducing bacteria utilising this metabolic pattern show the fastest growth with doubling times of 3–4 hours on lactate (Postgate, 1984; Hansen, 1993). *Desulfobacter* species are capable of completely oxidising acetate to CO_2. Although some species can also utilise lactate or ethanol, the range of substrates is generally very limited and this group are unable to utilise H_2 as electron donor. Growth is slow, with a doubling time of approximately 20 h (Postgate, 1984). Growth on other substrates is slower still with doubling times in excess of 48 hours. Higher molecular weight fatty acids are used, e.g. by *Desulfobulbus* and some *Desulfovibrio* species and polyols are also occasionally utilised (Dwyer & Tiedje, 1986; Widdel, 1988). Some *Desulfotomaculum* species can utilise aromatic compounds such as benzoate, phenol, catechol, *p*- and *m*-cresol and benzyl alcohol (Widdel, 1988; Drzyzga et al., 1993; Kuever et al., 1993; Beller et al., 1996). Sugars are apparently not used with the exception of fructose (Daumas et al., 1988; Ollivier et al., 1988). While many or most SRB strains are able to grow phototrophically, the availability of complex nitrogen sources has been found to enhance growth in many cases (Postgate, 1984) and may also enhance the yield of sulfide per unit of carbon substrate by alleviating biosynthetic requirements (White & Gadd, 1996a). Inorganic nutrients such as phosphate or ammonium also enhanced growth of SRB in a number of studies (Okabe & Characklis, 1992; Barnes et al., 1994). The role of cationic trace nutrients is less well established but requirements for Fe and Cu have been inferred from the advantageous effect on growth of chelating agents such as citrate or EDTA (Postgate, 1984). Bacterial sulfate reduction results in the formation of sulfide, which may reach significant concentrations in sediments or chemostat cultures (White & Gadd, 1996a, b). Although low concentrations (e.g. 2–5 mM) of sulfide benefit SRB growth by ensuring a low E_h, high concentrations of sulfide are inhibitory (Postgate, 1984; McCartney & Oleszkievicz, 1991). A sulfide concentration of 16.1 mM was toxic to an SRB culture derived from an anaerobic treatment plant (Reis et al., 1992). However, such sulfide concentrations are not generally encountered due to precipitation of sulfide with metals. With the exception of the alkali and alkaline-earth metals, metal sulfides are essentially insoluble and the resultant precipitation of sulfides has been observed to protect SRB against metal toxicity (Lawrence & McCarty, 1965; Postgate, 1984): metals similarly protect the organisms against sulfide toxicity.

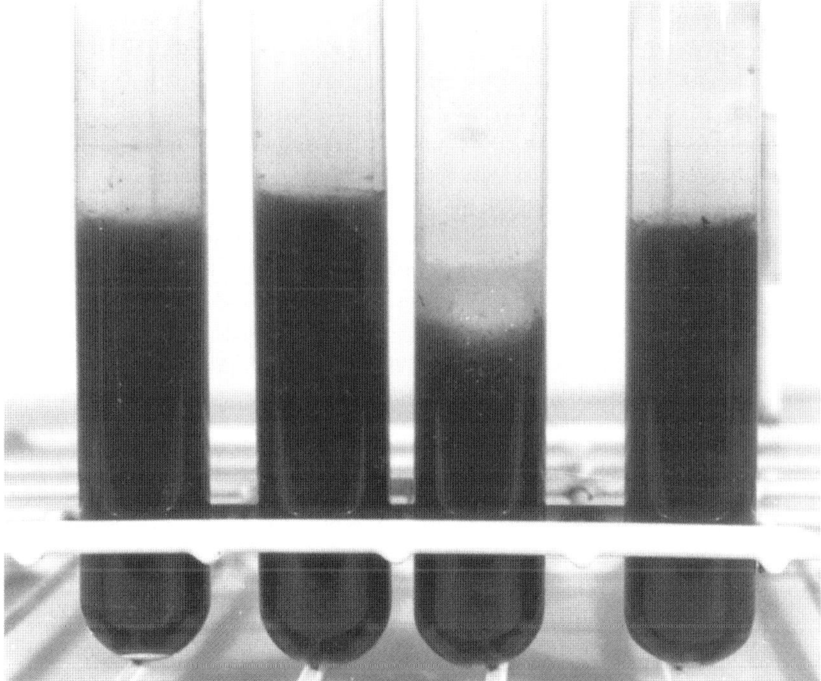

Fig. 2. Metal sulfide precipitates arising from bacterial sulfate-reduction. A mixed culture of sulfate-reducing bacteria was grown in a solid agar medium containing copper sulfate; the resulting black precipitate of CuS can be seen in the lower anaerobic regions where growth has occurred (adapted from Gadd, 1996).

The main mechanism whereby sulfate-reducing bacteria remove toxic metals from solution is via the formation of metal sulfide precipitates (Fig. 2) by reactions of the following type:

$$M^{2+} + SO_4^{2+} + 2CH_3CH_2OH \xrightarrow{BSR} 2\,CH_3COOH + 2H_2O + MS\downarrow \quad (5)$$

$$M^{2+} + SO_4^{2-} + 2CH_3CHOHCOOH \xrightarrow{BSR} 2CH_3COOH + 2CO_2 + 2H_2O + MS\downarrow \quad (6)$$

$$M^{2+} + SO_4^{2-} + CH_3COOH \xrightarrow{BSR} 2CO_2 + 2H_2O + MS\downarrow \quad (7)$$

(BSR indicates bacterial sulfate reduction).

The solubility products of most heavy metal sulfides are very low, in the range of 4.65×10^{-14} (Mn) to 6.44×10^{-53} (Hg) (Chang, 1993) so that even a moderate output of sulfide can remove metals to levels permitted in the environment (Crathorne & Dobbs, 1990; Taylor & McLean, 1992), with metal removal being directly related to sulfide production (White & Gadd, 1996a, 1998a, b, 2000; White et al., 1998). Sulfate-reducing bacteria can also create extremely reducing conditions, which can chemically reduce metals such as uranium(VI), albeit at a slower rate than enzymic reduction. In addition, sulfate reduction partially eliminates acidity from the system as a result of the shift in equilibrium when sulfate (dissociated) is converted to sulfide (largely protonated) (White & Gadd 1996a). This can result in the further precipitation of metals such as copper or aluminium as hydroxides as well as increasing the efficiency of sulfide precipitation.

Acid mine drainage occurs through the activities of sulfur- and iron-oxidising bacteria and, due to the quantities of sulfate available, sulfate reduction is an important process controlling the efflux of metals and acidity in mine effluents (Fortin et al., 1995; Ledin & Pedersen, 1996; Schippers et al., 1996). Laboratory studies indicate that sulfate reduction can provide both in situ (Uhrie et al., 1996) and ex situ metal removal from such waters (Hammack & Edenborn, 1992; Lyew et al., 1994; Christensen et al., 1996) and contribute to the removal of metals and acidity in artificial and natural wetlands (Hedin & Nairn, 1991; Perry, 1995), although other mechanisms, such as biosorption may predominate in these systems (Wieder, 1993; Karathanasis & Thompson, 1995).

Technetium reduction and precipitation by the sulfate-reducing bacterium *Desulfovibrio desulfuricans* has been successfully demonstrated by Lloyd et al. (1998, 1999; see also Chapter 11). In the absence of sulfur, TcO_4^- was enyzmatically reduced to a lower valence oxide, probably Tc_2O_5 or TcO_2, resulting in a precipitate that was associated with the bacterial cells. This was extremely rapid, with 85% of the Tc added to the culture precipitating within 1 hour when hydrogen was supplied as the electron donor. However, in the presence of sulfur, technetium cannot compete as an electron acceptor and instead reacts chemically with the enzymatically reduced H_2S. Thus, the Tc forms extracellular technetium sulfide precipitates of either Tc_2S_7 or TcS_2 (Lloyd et al., 1998). The enzymatic reduction has been tested within an experimental bioreactor (Lloyd et al., 1999), with immobilised cells precipitating large amounts of Tc. The advantages of using this kind of system to remove Tc from waste waters include the use of non-growing cells, minimising bulk waste, and lack of toxic H_2S production.

Large-scale bioreactor systems based on bacterial sulfate reduction have been developed, with the most extensive use to date in the treatment of contaminated ground water at the Budelco zinc smelting works at Budel-Dorplein in the Netherlands. A process integrating bacterial sulfate reduction with bioleaching by sulfur-oxidising bacteria has also been developed to remove contaminating toxic metals from soils (Fig. 3) (White et al., 1998). In this process sulfur- and iron-oxidising bacteria are employed to liberate metals from soils by the breakdown of sulfide minerals and production of sulfuric acid, which liberates acid labile forms such as hydroxides, carbonates (Chang, 1993; White et al., 1997, 1998) or sorbed metals. The bioreactor to treat the acid leachate contained a mixed, undefined culture of sulfate-reducing bacteria produced by combining a number of metal-tolerant enrichment cultures from different environmental origins (White & Gadd, 1996a). It was supplied with a concentrated nutrient mixture containing ethanol as carbon/energy source with additional inorganic phosphate and ammonium as well as organic nitrogen (White & Gadd, 1996b). This allowed the nutrient supply to match the sulfate concentration of the inflowing leachate and thus allowed maximisation of sulfate conversion and acidity removal (White & Gadd, 1996a, b). Metals were mainly precipitated as solid sulfides and, overall, the bioreactor removed more than 98% of the target metals with the exception of

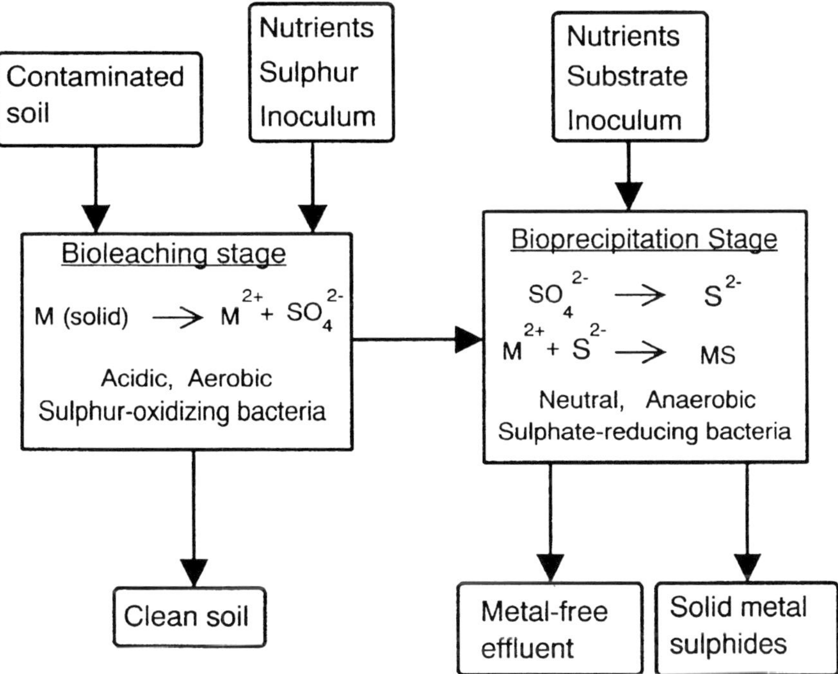

Fig. 3. Diagram showing the outline integrated process for bioremediation of metal-contaminated soils. The outline reactions and conditions for the bioleaching and bioprecipitation stages are shown in addition to the inputs and organisms utilised. Target metals are considered to be divalent cations and are indicated as M^{2+} (adapted from White et al., 1997, 1998).

Mn and, to a lesser extent, Ni and Pb. The process effectively removed the contaminating metal load from the soil and converted it to sulfides which were concentrated 100- to 200-fold in the solid phase, while the concentrations of metals in the liquid effluent were low enough to meet environmental discharge criteria and allowed recycle of the liquor to the bioleaching stage and significant water conservation during operation.

Phosphatase-mediated metal precipitation

Radionuclides and metals may also undergo precipitation reactions with phosphate. Precipitation requires the metal to be in a tri- or tetravalent, low solubility state, so this process is of particular interest when a metal does not precipitate efficiently on reduction. For example, reduction of Np(V) to Np(IV) removes only 12–15% from solution, but when reduction is coupled to Np(IV)-phosphate precipitation, 95% is removed (Lloyd et al., 2000). In this process, metal or radionuclide accumulation by bacterial biomass is mediated by a phosphatase enzyme, induced during metal-free growth, which liberates inorganic phosphate from a supplied organic phosphate donor molecule, e.g. glycerol 2-phosphate. Metal/radionuclide cations are then precipitated as phosphates on the biomass often to high levels (Macaskie & Dean, 1989; Macaskie, 1991). Most work has been carried out with a *Citrobacter* sp. and a range of bioreactor configurations, including those using immobilised biofilms, have been described (Macaskie et al., 1994; Tolley et al., 1995).

High gradient magnetic separation

Another technique for metal ion removal from solution is to use bacteria rendered susceptible to magnetic fields. 'Non-magnetic' bacteria can be made magnetic by the precipitation of metal phosphates (aerobic) or sulfides (anaerobic) on their surfaces as described previously above. For those organisms producing iron sulfide, it has been found that this compound is not only magnetic but also an effective adsorbent for metallic elements (Watson & Ellwood, 1994; Watson et al., 1995). Solutions treated by high gradient magnetic separation can have very low residual levels of metal ions remaining in solution (Watson & Ellwood, 1994; Watson et al., 1996). This method has been applied experimentally to Pu removal from water, and reduced the level to 50% of the initial activity (Bahaj et al., 1998).

5. Metal-binding proteins, polysaccharides and other biomolecules

As described earlier (Section 3), microbial cell surfaces are able to bind and immobilise metals (Gadd, 1988). However, a diverse range of specific and non-specific metal binding compounds are also produced by microorganisms, some of which are released into the environment. Unbound small, soluble molecules aid metal mobilisation, but the larger insoluble extracellular polysaccharides can lead to metal immobilisation.

Non-specific metal binding compounds are byproducts of microbial metabolism and range in size from simple organic acids and alcohols to macromolecules such as polysaccharides, humic and fulvic acids (Birch & Bachofen, 1990). Humic and fulvic acids are undefined macromolecules resulting in soil and water from the microbial degradation of cellulose, lignin and other complex organic compounds, and have been shown to bind radionuclides and toxic metals (Choppin, 1992; Spark et al., 1997) and are discussed in detail in Chapter 5. Extracellular polymeric substances (EPS), a mixture of polysaccharides, mucopolysaccarides and proteins (Zinkevich et al., 1996) are produced by bacteria, algae and fungi and also bind significant amounts of potentially toxic metals (Schreiber et al., 1990; Beech & Cheung, 1995). Extracellular polysaccharides of microbial origin are able to both bind metals and also adsorb or entrap particulate matter such as precipitated metal sulfides and oxides (Flemming, 1995; Vieira & Melo, 1995). One process has been developed which uses silage as a support for cyanobacterial growth. Floating mats were formed which removed metals from waters, the metal-binding process being due to large polysaccharides (>200,000 Da) produced by the cyanobacteria (Bender et al., 1994).

Specific metal binding compounds may be produced in response to the levels of metals present in the environment. The most well known extracellular metal-binding compounds are siderophores which are low molecular weight ligands (500–1000 Da) possessing a high affinity for Fe(III) (Neilands, 1981). Siderophores are produced extracellularly in response to low iron availability. They scavenge for Fe(III), and complex and solubilise it, making it available for microorganisms. Although primarily produced as a means of obtaining iron, siderophores are also able to bind other metals such as magnesium, manganese, chromium(III), gallium(III) and radionuclides such as plutonium(IV), U(VI, IV) and Th(IV) (Bulman, 1978; Birch & Bachofen, 1990, Brainard et al., 1992). The tetravalent actinides have extremely high binding constants with siderophores because, like Fe(III), they are hard Lewis acids. For example, the binding constants of Pu(IV) and Fe(III) with the siderophore enterobactin are estimated to be equally high at $\sim 10^{50}$ (Harris et al., 1979, Brainard et al., 1992). Experiments have shown that PuO_2 solubilisation by enterobactin is $\sim 10^3$ times more effective than the other chemical chelators tested, such as citrate and DTPA (N,N-bis[2-(bis[carboxymethyl]amino)-ethyl]glycine). Perhaps most importantly, solubilisation was most effective in the presence of Fe and occurred at environmentally realistic concentrations (Brainard et al., 1992). Other metal binding molecules have also been identified. Specific, low molecular weight (6000–10,000 Da) metal binding proteins, termed metallothioneins, are produced by animals, plants and microorganisms in response to the presence of toxic metals (Howe et al., 1997). Other metal binding proteins, phytochelatins and related peptides, all contain glutamic acid and cysteine at the amino-terminal position, and have been identified in plants, algae and several microorganisms (Rauser, 1995).

The metal binding abilities of siderophores, metallothioneins, phytochelatins and other similar molecules have potential for bioremediation. For example, a laboratory-scale process has been developed using one kind of metallothionein (ovotransferrin). Metal contaminated water passes through an affinity column containing ovotransferrin attached to CNBr-activated Sepharose 4B. The bound metal can be removed from the column using a low pH, weakly chelating buffer such as HEPES and the column reused (Spears & Vincent, 1997). Such processes may have potential for the remediation of large quantities of water, which contain only low concentrations of a metal.

6. Transformations

Microorganisms can transform certain metal, metalloid and organometallic species by oxidation, reduction, methylation and dealkylation (Gadd, 1993b; Lovley, 1995; Brady et al., 1996). Reduction of metals/radionuclides to lower oxidation states has been discussed earlier (Section 4), and this is the most important transformation reaction for radionuclides. Here, some other types of transformation reactions focusing on metalloids will be discussed. Transformation processes modify the mobility and toxicity of metalloids, have biogeochemical significance, and are also of biotechnological potential in bioremediation (Tamaki & Frankenberger, 1992; Gadd, 1993b, 1996; Lovley, 1993, 1995; Karlson & Frankenberger, 1993; Brady et al., 1996).

Microbial reduction of metalloid oxyanions

The reduction of metalloid oxyanions such as SeO_4^{2-}, SeO_3^{2-} and TeO_3^{2-} to elemental selenium or tellurium can be catalysed by numerous microbial species (Lovley, 1993; Gharieb et al., 1995). Some bacteria isolated from anoxic sediments can metabolise acetate coupled to the reduction of SeO_4^{2-}, as shown in equation 8, to support growth (Oremland et al., 1989):

$$4CH_3COO^- + 3SeO_4^{2-} \rightarrow 3Se^0 + 8CO_2 + 4H_2O + 4H^+ \tag{8}$$

A *Pseudomonas* sp. was able to respire SeO_4^{2-} to SeO_3^{2-}, with oxidation of ^{14}C-labelled acetate to $^{14}CO_2$ (equation 9) (Macy et al., 1989):

$$CH_3COO^- + H^+ + 4SeO_4^{2-} \rightarrow 2CO_2 + 4SeO_3^{2-} + 2H_2O \tag{9}$$

A novel species, *Thauera selenatis*, is also capable of respiring SeO_4^{2-} to SeO_3^{2-} anaerobically, with concomitant reduction of NO_3^- (Demoldecker and Macy, 1993). Some other anaerobic bacteria are able to reduce SeO_3^{2-} to Se^0 but cannot reduce SeO_4^{2-}. It is generally believed that SeO_4^{2-} is used as a terminal electron acceptor whereas SeO_3^{2-} reduction is a method of detoxification (Lovley, 1993). Reduction of TeO_3^{2-} to Te^0 is also apparently a means of detoxification found in bacteria (Walter & Taylor, 1992). Numerous filamentous and unicellular fungal species are also capable of reducing SeO_3^{2-} and TeO_3^{2-} to their elemental forms. Intra- and extracellular deposits of these elements result in a red coloration of colonies in the case of S^0 (Zieve et al., 1985; Gharieb et al., 1995; Morley et al., 1996), and black or grey colonies in the case of Te^0 (Smith, 1974).

Oremland et al. (1990, 1991) utilised these metabolic processes for the in situ removal of SeO_4^{2-} from agricultural drainage regions of Nevada. Exposed reservoir sediments were flooded to create anoxic conditions, in which the natural bacterial population reduced and immobilised large quantities of the selenium that was present in the sediments (Long et al., 1990).

Methylation of metalloids

Microbial methylation of metalloids to yield volatile derivatives, e.g. dimethylselenide or trimethylarsine, can be effected by a variety of bacteria, algae and fungi (Gadd, 1993b; Karlson & Frankenberger, 1993). Selenium methylation appears to involve transfer of methyl groups as carbonium (CH_3^+) ions via the S-adenosyl methionine system, and there is also evidence of dimethyltelluride and dimethylditelluride production in fungi (Karlson & Frankenberger, 1993). Several bacterial and fungal species have been shown to methylate arsenic compounds such as arsenate (As(V), AsO_4^{3-}), arsenite (As(III), AsO_2^-) and methylarsonic acid ($CH_3H_2AsO_3$) to volatile dimethyl- ($(CH_3)_2HAs$) or trimethylarsine ($(CH_3)_3As$) (see Tamaki & Frankenberger, 1992). Microbial methylation of selenium, resulting in volatilisation, has been used successfully for in situ bioremediation of selenium containing land and water at Kesterson Reservoir, California, reducing the selenium concentrations to acceptable levels (Thompson-Eagle & Frankenberger, 1992).

7. Concluding remarks

Microorganisms play important roles in the environmental fate of radionuclides, toxic metals and metalloids with a multiplicity of physicochemical and biological mechanisms effecting transformations between soluble and insoluble phases. Such mechanisms are important components of natural biogeochemical cycles, with some processes being of potential application to the treatment of contaminated materials. Although the biotechnological potential of most of these processes has only been explored at laboratory scale, some mechanisms, notably bioleaching, biosorption and precipitation, have been employed at a commercial scale. Of these, autotrophic leaching is an established major process in mineral extraction but has also been applied to the treatment of contaminated land. There have been several attempts to commercialise biosorption using microbial biomass but success has been limited, primarily due to competition with commercially produced ion exchange media. As a process for immobilising metals, precipitation of metals as sulfides has achieved large scale application, and this holds out promise of further commercial development. Exploitation of other biological processes will undoubtedly depend on a number of scientific, economic and political factors, but primarily on the availability of a market niche.

Acknowledgements

The author gratefully acknowledges financial support for his own work from NERC/AFRC (Special Topic Programme: Pollutant Transport in Soils and Rocks), BBSRC (BCE 03292, SPC 2922, SPC 02812, BSW 05375, SPC 05211), the Royal Society (London) (638072:P779 Project grant), BNFL, the Royal Society of Edinburgh (Scottish Office Education Department/RSE Support Research Fellowship 1994-1995) and NATO (ENVIR.LG.950387 Linkage grant).

References

Aksu, Z., Kutsal, T., Gun, S., Haciomanoglu, N. & Gholaminejad, M. (1991). Investigation of biosorption of Cu(II), Ni(II) and Cr(II) ions to activated sludge bacteria. *Environmental Technology, 12*, 915–921.

Arter, H.E., Hanselmann, K. W. & Bachofen, R. (1991). Modelling of microbial-degradation processes - the behaviour of micro-organisms in a waste repository. *Experientia, 47*, 542–548.

Bahaj A. S., Croudace I. W., James P. A. B., Moeschler F. D. & Warwick P. E. (1998). Continuous radionuclide recovery from wastewater using magnetotactic bacteria. *Journal of Magnetism and Magnetic Materials, 184*, 241–244.

Baldi, F., Clark, T., Pollack, S. S. & Olson, G. J. (1992). Leaching of pyrites of various reactivities by *Thiobacillus ferrooxidans*. *Applied and Environmental Microbiology, 58*, 1853–1856.

Banks, M. K., Waters, C. Y. & Schwabb, A. P. (1994). Influence of organic acids on leaching of heavy metals from contaminated mine tailings. *Journal of Environmental Science and Health, A29*, 1045–1056.

Barnes, L. J., Janssen, F. J., Sherren, J., Versteegh, J. H., Koch, R. O. & Scheeren, P. J. H. (1991). A new process for the microbial removal of sulfate and heavy metals from contaminated waters extracted by a geohydrological control system. *Transactions of the Institution of Chemical Engineers, 69*, 184–186.

Barnes, L. J., Scheeren, P. J. M. & Buisman, C. J. N. (1994). Microbial removal of heavy metals and sulfate from contaminated groundwaters. In J. L. Means & R. E. Hinchee (Eds), *Emerging Technology for Bioremediation of Metals* (pp. 38–49). Boca Raton, FL: Lewis Publishers.

Beech, I. B. & Cheung, C. W. S. (1995). Interactions of exopolymers produced by sulfate-reducing bacteria with metal ions. *International Biodeterioration and Biodegradation, 35*, 59–72.

Beller, H. R., Spormann, A. M., Sharma, P. K., Cole, J. R. & Reinhard, M. (1996). Isolation and characterization of a novel toluene-degrading sulfate-reducing bacterium. *Applied and Environmental Microbiology, 62*, 1188–1196.

Bender, J., Rodriguez-Eaton, S., Ekanemesang, U.M. & Phillips, P. (1994). Characterization of metal-binding bioflocculants produced by the cyanobacterial component of mixed microbial mats. *Applied and Environmental Microbiology, 60*, 2311–2315.

Beveridge, T. J. & Doyle, R. J. (1989). *Metal Ions and Bacteria*. New York: John Wiley.

Birch, L. & Bachofen, R. (1990). Complexing agents from micro-organisms. *Experientia, 46*, 827–834.

Blais, J. F., Meunier, N. & Tyagi, R. D. (1997). Simultaneous sewage sludge digestion and metal leaching at controlled pH. *Environmental Technology, 18*, 499–508.

Borkowski M., Lis S. & Choppin G. R. (1996). Complexation study of NpO^{2+} and UO_2^{2+} ions with several organic ligands in aqueous solutions of high ionic strength. *Radiochimica Acta, 74*, 117–121.

Bosecker, K. (1997). Bioleaching: metal solubilization by micro-organisms. *FEMS Microbiology Reviews, 20*, 591–604.

Bosshard, P. P., Bachofen, R. & Brandl, H. (1996). Metal leaching of fly-ash from municipal waste incineration by *Aspergillus niger. Environmental Science and Technology, 30*, 3066–3070.

Brady, J. M., Tobin, J. T. & Gadd, G. M. (1996). Volatilization of selenite in aqueous medium by a *Penicillium* species. *Mycological Research, 100*, 955–961.

Brainard, J. R., Strietelmeier, B. A., Smith, P. H., Langston-Unkefer, P. J., Barr M. E. & Ryan, R.,R. (1992). Actinide binding and solubilization by microbial siderophores. *Radiochimica Acta, 58/59*, 357–363.

Brierley, C. L. (1990). Bioremediation of metal-contaminated surface and groundwaters. *Geomicrobiology Journal, 8*, 201–223.

Bulman, R. A. (1978). Chemistry of plutonium and the transuranics in the biosphere. *Structure and Bonding, 34*, 39–77.

Burgstaller, W. & Schinner, F. (1993). Leaching of metals with fungi. *Journal of Biotechnology, 27*, 91–116.

Caccavo, F., Blakemore, R. & Lovley, D. R. (1992). A hydrogen-oxidizing, Fe(III)-reducing micro-organism from the Great Bay Estuary, New Hampshire. *Applied and Environmental Microbiology, 58*, 3211–3216.

Chang, J. C. (1993). Solubility product constants: In D. R. Lide (Ed.), *CRC Handbook of Chemistry and Physics* (pp. 8–39). Boca Raton, FL: CRC Press.

Chapelle, F. H. & Lovley, D. R. (1992). Competitive exclusion of sulfate reduction by Fe(III)-reducing bacteria: a mechanism for producing discrete zones of high-iron ground water. *Ground Water, 30*, 29–36.

Choppin G. R. (1992). The role of natural organics in radionuclide migration in natural aquifer systems. *Radiochimica Acta, 58/59*, 113–120.

Christensen, B., Laake, M. & Lien, T. (1996). Treatment of acid-mine water by sulfate-reducing bacteria—results from a bench-scale experiment. *Water Research, 30*, 1617–1624.

Crathorne, B. & Dobbs, A. J. (1990). Chemical pollution of the aquatic environment by priority pollutants and its control: In R. M. Harrison (Ed.), *Pollution: Causes, Effects and Control* (pp. 1–18). Cambridge, UK: The Royal Society of Chemistry.

Daumas, S., Cord-Ruwisch, R. & Garcia, J. L. (1988). *Desulfotomaculum geothermicum* sp. nov., a thermophilic, fatty acid degrading, sulfate-reducing bacterium isolated with H_2 from geothermal ground water. *Antonie van Leeuwenhoek, 54*, 165–178.

Demolldecker, H. & Macy, J. M. (1993). The periplasmic nitrite reductase of *Thaueria selenatis* may catalyze the reduction of selenite to elemental selenium. *Archives of Microbiology, 160*, 241–247.

Denêvre, O., Garbaye, J. & Botton, B. (1996). Release of complexing organic acids by rhizosphere fungi as a factor in Norway Spruce yellowing in acidic soils. *Mycological Research, 100*, 1367–1374.

Drever, J. I. & Stillings, L. L. (1997). The role of organic acids in mineral weathering. *Colloids and Surfaces, 120*, 167–181.

Drzyzga, O., Kuever, J. & Blotevogel, K.-H. (1993). Complete oxidation of benzoate and 4–hydroxybenzoate by a new sulfate-reducing bacterium resembling *Desulfoarculus. Archives of Microbiology, 159*, 109–113.

Du Preez, L. A., Odendaal, J. P., Maree, J. P. & Ponsonby M. (1992). Biological removal of sulphate from industrial effluents using producer gas as energy source. *Environmental Technology, 13*, 875–882.

Dwyer, D. F. & Tiedje, L. M. (1986). Metabolism of polyethylene glycol by two anaerobic bacteria, *Desulfovibrio desulfuricans* and a *Bacteroides* sp. *Applied and Environmental Microbiology, 52*, 852–856.

Ewart, D. K. & Hughes, M. N. (1991). The extraction of metals from ores using bacteria. *Advances in Inorganic Chemistry, 36*, 103–135.

Flemming, H-K. (1995). Sorption sites in biofilms. *Water Science and Technology, 32*, 27–33.

Fortin, D., Davis, B., Southam, G. & Beveridge, T. J. (1995). Biogeochemical phenomena induced by bacteria within sulphidic mine tailings. *Journal of Industrial Microbiology, 14*, 178–185.

Fox, T. R. & Comerford, N. B. (1990). Low-molecular weight organic acids in selected forest soils of the southeastern USA. *Soil Science Society of America Journal, 54*, 1139–1144.

Francis, A. J., Dodge, C. J. & Gillow, J. B. (1992). Biodegradation of metal citrate complexes and implications for toxic metal mobility. *Nature, 356*, 140–142.

Francis, A. J., Dodge, C. J., Gillow, J. B. & Papenguth, H. W. (2000). Biotransformation of uranium compounds in high ionic strength brine by a halophilic bacterium under denitrifying conditions. *Environmental Science and Technology, 34*, 2311–2317.

Franz, A., Burgstaller, W. & Schinner, F. (1991). Leaching with *Penicillium simplicissimum*: influence of metals and buffers on proton extrusion and citric acid production. *Applied and Environmental Microbiology, 57*, 769–774.

Franz, A., Burgstaller, W., Muller, B. & Schinner, F. (1993). Influence of medium components and metabolic inhibitors on citric acid production by *Penicillium simplicissimum. Journal of General Microbiology, 139*, 2101–2107.

Fujie, K., Tsuchida T., Urano K. & Ohtake H. (1994). Development of a bioreactor system for the treatment of chromate wastewater using *Enterobacter cloacae HO1. Water Science and Technology, 30*, 235–243.

Gadd, G. M. (1988). Accumulation of metals by micro-organisms and algae. In H-J. Rehm (Ed.), *Biotechnology, Vol 6b* (pp. 401–433). Weinheim: VCN Verlagsgesellschaft.

Gadd, G.M. (1992a). Microbial control of heavy metal pollution: In J. C. Fry, G. M. Gadd, R. A. Herbert, C. W. Jones & I. A. Watson-Craik (Eds), *Microbial Control of Pollution. Society for General Microbiology Symposium* (Vol. 48) (pp. 59–88). Cambridge, UK: Cambridge University Press.

Gadd, G. M. (1992b). Molecular biology and biotechnology of microbial interactions with organic and inorganic heavy metal compounds: In R. A. Herbert & R. J. Sharp (Eds), *Molecular Biology and Biotechnology of Extremophiles* (pp. 225–257). Glasgow: Blackie.

Gadd, G.M. (1993a). Interactions of fungi with toxic metals, *New Phytologist, 124*, 25–60.

Gadd, G. M. (1993b). Microbial formation and transformation of organometallic and organometalloid compounds. *FEMS Microbiology Reviews, 11*, 297–316.

Gadd, G. M. (1996). Influence of microorganisms on the environmental fate of radionuclides. *Endeavour, 20*, 150–156.

Gadd, G. M. (1997). Roles of microorganisms in the environmental fate of radionuclides: In G. R. Bock & G. Cardew (Eds), *CIBA Foundation 203: Health Impacts of Large Releases of Radionuclides* (pp. 94–104). Chichester, UK: John Wiley.

Gadd, G. M. (1999). Fungal production of citric and oxalic acid: importance in metal speciation, physiology and biogeochemical processes. *Advances in Microbial Physiology 41*, 47–92.

Gadd, G. M. (2000a). Bioremedial potential of microbial mechanisms of metal mobilization and immobilization. *Current Opinion in Biotechnology, 11*, 271–279.

Gadd, G. M. (2000b). Heavy metal pollutants: environmental and biotechnological Aspects. In J. Lederberg (Ed.), *The Encyclopedia of Microbiology* (2nd edn) (pp. 607–617). San Diego, CA: Academic Press.

Gadd, G. M. & Sayer, G. M. (2000). Fungal transformations of metals and metalloids. In D. R. Lovley (Ed.), *Environmental Microbe-Metal Interactions* (pp. 237–256). Washington: American Society for Microbiology, Washington.

Gadd, G. M. & White, C. (1990). Biosorption of radionuclides by yeast and fungal biomass. *Journal of Chemical Technology and Biotechnology, 49*, 331–343.

Gadd, G. M. & White, C. (1993). Microbial treatment of metal pollution—a working biotechnology? *Trends in Biotechnology, 11*, 353–359.

Geesey, G. & Jang, L. (1990). Extracellular polymers for metal binding. In H. L. Ehrlich and C. L. Brierley (Eds), *Microbial Mineral Recovery* (pp.223–275). New York: McGraw-Hill.

Ghani, B., Takai, M., Hisham, N. Z., Kishimoto, N., Ismail, A. K., Tano, T. & Sugio, T. (1993). Isolation and characterisation of a Mo^{6+} reducing bacterium. *Applied and Environmental Microbiology, 59*, 1176–1180.

Gharieb, M. M., Wilkinson, S. C. & Gadd, G. M. (1995). Reduction of selenium oxyanions by unicellular, polymorphic and filamentous fungi: cellular location of reduced selenium and implications for tolerance. *Journal of Industrial Microbiology, 14*, 300–311.

Gharieb, M.M., Sayer, J.A. & Gadd, G.M. (1998). Solubilization of natural gypsum ($CaSO_4 \cdot 2H_2O$) and the formation of calcium oxalate by *Aspergillus niger and Serpula himantioides. Mycological Research, 102*, 825–830.

Ghauri, M. A. & Johnson, D.B. (1991). Physiological diversity amongst some moderately thermophilic iron-oxidising bacteria. *FEMS Microbial Ecology, 85*, 327–334.

Gorby, Y. & Lovley, D. R. (1992). Enzymatic uranium precipitation. *Environmental Science and Technology, 26*, 205–207.

Hammack, R. W. & Edenborn, H. M. (1992). The removal of nickel from mine waters using bacterial sulphate-reduction. *Applied Microbiology and Biotechnology, 37*, 674–678.

Hansen, T. A. (1993). Carbon metabolism in sulfate-reducing bacteria. In J. M. Odom & R. Singleton (Eds), *The Sulfate-Reducing Bacteria: Contemporary Perspectives* (pp. 21–40). New York: Springer-Verlag.

Harris W. R., Carrano C. J. & Raymond K. N. (1979) Coordination chemistry of microbial iron transport compounds, 16: Isolation, characterization, and formation constants of ferric aerobactin. *Journal of the American Chemical Society, 101*, 2722–2727.

Harter, R. D. (1983). Effect of soil pH on adsorption of lead, copper, zinc and nickel. *Soil Science Society of America Journal, 47*, 47–51.

Hedin, R. S. & Nairn, R. W. (1991). Contaminant removal capabilities of wetlands constructed to treat coal mine drainage: In G. A. Moshiri (Ed.), *Proceedings of the International Symposium on Constructed Wetlands for Water-Quality Improvement* (pp. 187–195). Chelsea MI: Lewis Publishers.

Howe, R., Evans, R. L. & Ketteridge, S. W. (1997). Copper-binding proteins in ectomycorrhizal fungi. *New Phytologist, 135*, 123–131.

Hughes, M. N. & Poole, R. K. (1991). Metal speciation and microbial growth—the hard (and soft) facts. *Journal of General Microbiology, 137*, 725–734.

Jones, D. L. & Kochian, L. V. (1996). Aluminium-organic acid interactions in acid soils. *Plant and Soil, 182*, 221–228.

Karathanasis, A. D. & Thompson, Y. L. (1995). Mineralogy of iron precipitates in a constructed acid-mine drainage wetland. *Soil Science Society of America Journal, 59*, 1773–1781.

Karlson, U. & Frankenberger, W. T. (1993). Biological alkylation of selenium and tellurium, In H. Sigel & A. Sigel (Eds), *Metal Ions in Biological Systems* (pp. 185–227). New York: Marcel Dekker.

Kuever, J., Kuylmer, J., Jaansen, S., Fischer, U. & Blotevogel, K.-H. (1993). Isolation and characterisation of a new spore-forming sulfate-reducing bacterium growing by complete oxidation of catechol. *Archives of Microbiology, 159*, 282–288.

Landa, E. R. & Gray, J. R. (1995). US Geological Survey—results on the environmental fate of uranium mining and milling wastes. *Journal of Industrial Microbiology, 26*, 19–31.

Lapeyrie, F., Ranger, J. & Vairelles, D. (1991). Phosphate solubilizing activity of ectomycorrhizal fungi *in vitro. Canadian Journal of Botany, 69*, 342–346.

Lawrence, A. W. & McCarty, P. L. (1965). The role of sulphide in preventing heavy metal toxicity in anaerobic treatment. *Journal of the Water Pollution Control Federation, 37*, 392–406.

Ledin, M. & Pedersen, K. (1996). The environmental impact of mine wastes—roles of micro-organisms and their significance in the treatment of mine wastes. *Earth Science Reviews, 41*, 67–108.

Lloyd, J. R., Nolting, H.-F., Solé, V. A., Bosecker, K. & Macaskie ,L. E. (1998). Technetium reduction and precipitation by sulfate-reducing bacteria. *Geomicrobiology Journal, 15*, 45–58.

Lloyd, J. R., Ridley, J., Khizniak, T., Lyalikova, N. N., Macaskie, L. E. (1999). Reduction of technetium by *Desulfovibrio desulfuricans*: biocatalyst characterisation and use in a flowthrough bioreactor. *Applied and Environmental Microbiology, 65*, 2691–2696.

Lloyd, J. R., Yong, P. & Macaskie, L. E. (2000). Biological reduction and removal of Np(V) by two micro-organisms. *Environmental Science and Technology, 34*, 1297–1301.

Long, R. H. B., Benson, S. M., Tokunaga, T. K. & Yee, A. (1990). Selenium immobilization in a pond sediment at Kesterson Reservoir. *Journal of Environmental Quality, 19*, 302–311.

Lovley, D. R. (1993). Dissimilatory metal reduction. *Annual Review of Microbiology, 47*, 263–290.

Lovley, D. R. (1995). Bioremediation of organic and metal contaminants with dissimilatory metal reduction. *Journal of Industrial Microbiology, 14*, 85–93.

Lovley, D. R. & Coates, J. D. (1997). Bioremediaition of metal contamination. *Current Opinion in Biotechnology, 8*, 285–289.

Lovley, D. R. & Phillips, E. J. P. (1987). Competitive mechanisms for inhibition of sulfate-reduction and methane production in the zone of ferric iron reduction in sediments. *Applied and Environmental Microbiology, 53*, 2636–2641.

Lovley, D. R. & Phillips, E. J. P. (1988). Novel mode of microbial energy metabolism: organic carbon oxidation coupled to dissimilatory reduction of iron or manganese. *Applied and Environmental Microbiology, 54*, 1472–1480.

Lovley, D. R. & Phillips, E. J. P. (1992a). Bioremediation of uranium contamination with enzymatic uranium reduction. *Environmental Science and Technology, 26*, 2228–2234.

Lovley, D. R. & Phillips, E. J. P. (1992b). Reduction of uranium by *Desulfovibrio desulfuricans. Applied and Environmental Microbiology, 58*, 850–856.

Lovley, D. R., Phillips, E. F. & Lonergan, D. J. (1989). Hydrogen and formate oxidation coupled to dissimilatory reduction of iron or manganese by *Alteromonas putrefaciens. Applied and Environmental Microbiology, 55*, 700–706.

Lovley, D. R., Phillips, E. J. P., Gorby,Y. A. & Landa, E. R. (1991). Microbial reduction of uranium. *Nature, 350*, 413–416.

Lovley, D. R., Giovannoni, S. J., White, D. C., Champine, J. E., Phillips, E. J. P., Gorby, Y. A. and Goodwin, S. (1993). *Geobacter metallireducens* gen. nov. sp.nov., a micro-organism capable of coupling the complete oxidation of organic compounds to the reduction of iron and other metals. *Archives of Microbiology, 159*, 336–344.

Lu, N., Kung, K. S., Mason, C. F. V., Triay, I. R., Cotter, C. R., Pappas, A. J. & Pappas, M. E. G. (1998). Removal of plutonium-239 and americium-241 from Rocky Flats soil by leaching. *Environmental Science and Technology, 32*, 370–374.

Luther, G. W. (1987). Pyrite oxidation and reduction: molecular orbital theory considerations. *Geochimica et Cosmochimica Acta, 51*, 3193–3199.

Lyew, D., Knowles, R. & Sheppard, J. (1994). The biological treatment of acid-mine drainage under continuous-flow conditions in a reactor. *Process Safety and Environmental Protection, 72*, 42–47.

Macaskie, L. E. (1991). The application of biotechnology to the treatment of wastes produced by the nuclear fuel cycle—biodegradation and bioaccumulation as a means of treating radionuclide-containing streams. *Critical Reviews in Biotechnology, 11*, 41–112.

Macaskie, L. E. & Dean, A. C. R. (1989). Microbial metabolism, desolubilization and deposition of heavy metals: uptake by immobilized cells and application to the treatment of liquid wastes: In A. Mizrahi (Ed.), *Biological Waste Treatment* (pp. 150–201). New York: Alan R. Liss.

Macaskie, L. E., Jeong, B. C. & Tolley, M. R. (1994). Enzymically-accelerated biomineralization of heavy metals: application to the removal of americium and plutonium from aqueous flows. *FEMS Microbiology Reviews, 14*, 351–368.

Macy, J. M., Michel, T. A. & Kirsch, D. G. (1989). Selenate reduction by a new *Pseudomonas* species: a new mode of anaerobic respiration. *FEMS Microbiology Letters, 61*, 195–198.

Marsh, R. M. & Norris, P. R. (1983). The isolation of some thermophilic, autotrophic, iron- and sulphur-oxidising bacteria. *FEMS Microbiology Letters, 17*, 311–315.

Mattey, M. (1992). The production of organic acids. *Critical Reviews in Biotechnology, 12*, 87–132.

McCartney, D.M. & Oleszkievicz, J.A. (1991). Sulfide inhibition of anaerobic degradation of lactate and acetate. *Water Research, 25*, 203–209.

Meixner, O., Mischack, H., Kubicek, C. P. & Rohr, M. (1985). Effect of manganese deficiency on plasma-membrane lipid composition and glucose uptake in *Aspergillus niger. FEMS Microbiology Letters, 26*, 271–274.

Metting, F. B. (1992). Structure and physiological ecology of soil microbial communities. In F. B. Metting (Ed.), *Soil Microbial Ecology, Applications and Environmental Management* (pp. 3–25). New York: Marcel Dekker.

Morley, G. F., Sayer, J. A., Wilkinson, S. C., Gharieb, M. M. & Gadd, G. M. (1996). Fungal sequestration, mobilization and transformation of metals and metalloids: In J. C. Frankland, N. Magan & G. M. Gadd (Eds), *Fungi and Environmental Change* (pp. 235–256). Cambridge, UK: Cambridge University Press.

Muñoz, J. A., González, F., Blázquez, M. L. & Ballester A. (1995). A study of the bioleaching of a Spanish uranium ore. Part I: A review of the bacterial leaching in the treatment of uranium ores. *Hydrometallurgy, 38*, 39–57.

Myers, C. R. & Nealson, K. H. (1988). Bacterial manganese reduction and growth in manganese oxide as the sole electron acceptor. *Science, 240*, 1319–1321.

Neilands, J. B. (1981). Microbial iron compounds. *Annual Review of Biochemistry, 50*, 715–731.

Okabe, S. & Characklis, W. G. (1992). Effects of temperature and phosphorus concentration on microbial sulphate reduction by *Desulfovibrio desulfuricans. Biotechnology and Bioengineering, 39*, 1031–1042.

Ollivier, B., Cord-Ruwisch, R., Hatchikan, E. C. & Garcia, J. L. (1988). Characterisation of *Desulfovibrio fructosovorans* sp. nov. *Archives of Microbiology, 149*, 447–450.

Oremland, R. S., Hollibaugh, J. T., Maest, A. S., Presser, T. S., Miller, L. G. & Culbertson, C. W. (1989). Selenate reduction to elemental selenium by anaerobic bacteria in sediments and culture: biogeochemical significance of a novel sulfate-independent respiration. *Applied and Environmental Microbiology, 55*, 2333–2343.

Oremland, R. S., Steinberg, N. A., Maest, A. S., Miller, L. G. & Hollibaugh, J. T. (1990). Measurement of *in situ* rates of selenate removal by dissimilatory bacterial reduction in sediments. *Environmental Science and Technology, 24*, 1157–1164.

Oremland, R. S., Steinberg, N. A., Presser, T. S. & Miller, L. G. (1991). *In situ* bacterial selenate reduction in the agricultural drainage systems of Western Nevada. *Applied and Environmental Microbiology, 57*, 615–617.

Peck H. D. (1993). Bioenergetic strategies of the sulfate-reducing bacteria. In J. M. Odom & R. Singleton (Eds), *The Sulfate-Reducing Bacteria: Contemporary Perspectives* (pp. 41–75). New York: Springer-Verlag.

Perry, K. A. (1995). Sulfate-reducing bacteria and immobilization of metals. *Marine Georesearch and Geotechnology, 13*, 33–39.

Phillips, E. J. P., Landa, E. R. & Lovley, D. R. (1995). Remediation of uranium contaminated soils with bicarbonate extraction and microbial U(VI) reduction. *Journal of Industrial Microbiology, 14*, 203–207.

Postgate, J. R. (1984). *The Sulphate-Reducing Bacteria*. Cambridge, UK: Cambridge University Press.

Prieur, D., Erauso, G. & Jeanthon, C. (1995). Hyperthermophilic life at deep-sea hydrothermal vents. *Planet. Space Science, 43*, 115–122.

Rauser, W. E. (1995). Phytochelatins and related peptides. *Plant Physiology, 109*, 1141–1149.

Ravishankar, B. R., Blais, J. F., Benmoussa, H. & Tyagi, R. D. (1994). Bioleaching of metals from sewage sludge: elemental sulfur recovery. *Journal of Environmental Engineering, 120*, 462–470.

Rawlings, D. E. & Silver, S. (1995). Mining with microbes. *Biotechnology, 13*, 773–778.

Reis, M. A. M., Almeida, J. S., Lemos, P. C. & Carrondo, M. J. T. (1992). Effect of hydrogen sulfide on the growth of sulfate-reducing bacteria. *Biotechnology and Bioengineering, 40*, 593–600.

Rusin, P. A., Sharp, J. E., Oden, K. L., Arnold, R. G. & Sinclair, N. A. (1993). Isolation and physiology of a manganese-reducing *Bacillus polymyxa* from an Oligocene silver-bearing ore and sediment with reference to Precambrian biogeochemistry. *Precambrian Research, 61*, 231–240.

Sayer, J. A. & Gadd, G. M. (1997). Solubilization and transformation of insoluble inorganic metal compounds to insoluble metal oxalates by *Aspergillus niger. Mycological Research, 101*, 653–661.

Sayer, J. A., Raggett, S. L. & Gadd, G. M. (1995). Solubilization of insoluble metal compounds by soil fungi: development of a screening method for solubilizing ability and metal tolerance. *Mycological Research, 99*, 987–993.

Sayer, J. A., Kierans, M. & Gadd, G. M. (1997). Solubilization of some naturally-occurring metal-bearing minerals, limescale and lead phosphate by *Aspergillus niger. FEMS Microbiology Letters, 154*, 29–35.

Sayer, J. A., Cotter-Howells, J. D., Watson, C., Hillier, S. & Gadd, G. M. (1999). Lead mineral transformation by fungi. *Current Biology, 9*, 691–694.

Schinner, F. & Burgstaller, W. (1989). Extraction of zinc from an industrial waste by a *Penicillium* sp. *Applied and Environmental Microbiology, 55*, 1153–1156.

Schippers, A., Von Rege, H. & Sand, W. (1996). Impact of microbial diversity and sulphur chemistry on safeguarding sulphidic mine waste. *Minerals Engineering, 9*, 1069–1079.

Schreiber, D. R., Millero, F. J. & Gordon, A. S. (1990). Production of an extracellular copper-binding compound by the heterotrophic marine bacterium *Vibrio alginolyticus. Marine Chemistry, 28*, 275–284.

Schrickx, J. M., Raedts, M. J. H., Stouthamer, A. H. & Van Versveld, H. W. (1994). Organic acid production by *Aspergillus niger* in recycling culture analysed by capillary electrophoresis. *Analytical Biochemistry, 231*, 175–181.

Shen, H. & Wang, Y.-T. (1993). Characterization of enzymatic reduction of hexavalent chromium by *Escherichia coli*, ATCC 33456. *Applied and Environmental Microbiology, 59*, 3771–3777.

Smith, D. G. (1974). Tellurite reduction in *Schizosaccharomyces pombe. Journal of General Microbiology, 83*, 389–392.

Spark, K. M., Wells, J. D. & Johnson, B. B. (1997). The interaction of a humic acid with heavy metals. *Australian Journal of Soil Research, 35*, 89–101.

Spears, D. R. & Vincent, J. B. (1997). Copper binding and release by immobilized transferrin: a new approach to heavy metal removal and recovery. *Biotechnology and Bioengineering, 53*, 1–9.

Sreekrishnan, T. R. & Tyagi, R. D. (1994). Heavy metal leaching from sewage sludges: a techno-economic evaluation of the process options. *Environmental Technology, 15*, 531–543.

Strasser, H., Burgstaller, W. & Schinner, F. (1994). High yield production of oxalic acid for metal leaching purposes by *Aspergillus niger. FEMS Microbiology Letters, 119*, 365–370.

Sukla, L. B., Kar, R. N. & Panchanadikar, V. (1992). Leaching of copper converter slag with *Aspergillus niger* culture filtrate. *Biometals, 5*, 169–172.

Tamaki, S. & Frankenberger, W. T. (1992). Environmental biochemistry of arsenic. *Reviews of Environmental Contamination and Toxicology, 124*, 79–110.

Taylor, M. R. G. & McLean, R. A. N. (1992). Overview of clean-up methods for contaminated sites. *Journal of the Insitute of Water and Environmental Management, 6*, 408–417.

Thompson-Eagle, E. T. & Frankenberger, W. T. (1992). Bioremediation of soils contaminated with selenium. In R. Lal & B. A. Stewart (Eds), *Advances in Soil Science* (pp. 261–309). New York: Springer-Verlag.

Tichy, R., Janssen, A., Grotenhuis, J. T. C., Lettinga, G. & Rulkens, W. H. (1994). Possibilities for using biologically-produced sulphur for cultivation of *Thiobacilli* with respect to bioleaching processes. *Bioresource Technology, 48*, 221–227.

Tipping, E. (1996). Hydrochemical modelling of the retention and transport of metallic radionuclides in the soils of an upland catchment. *Environmental Pollution, 94*, 105–116.

Tobin, J., White, C. & Gadd, G. M. (1994). Metal accumulation by fungi: applications in environmental biotechnology. *Journal of Industrial Microbiology, 13*, 126–130.

Tolley, M. R., Strachan, L. F. & Macaskie, L. E. (1995). Lanthanum accumulation from acidic solutions using a *Citrobacter* sp. immobilized in a flow-through bioreactor. *Journal of Industrial Microbiology, 14*, 271–280.

Tsezos, M. & Volesky, B. (1982). The mechanism of uranium biosorption by *Rhizopus arrhizus. Biotechnology and Bioengineering, 24*, 385–401.

Tzeferis, P. G., Agatzini, S. & Nerantzis, E. T. (1994). Mineral leaching of non-sulphide nickel ores using heterotrophic micro-organisms. *Letters in Applied Microbiology, 18*, 209–213.

Uhrie, J. L., Drever, J. I., Colberg, P. J. S. & Nesbitt, C. C. (1996). *In situ* immobilization of heavy metals associated with uranium leach mines by bacterial sulphate reduction. *Hydrometallurgy, 43*, 231–239.

Vachon, P., Tyagi, R. D., Auclair, J. C. & Wilkinson, K. J. (1994). Chemical and biological leaching of aluminium from red mud. *Environmental Science and Technology, 28*, 26–30.

Veglio, F. (1996). The optimisation of manganese-dioxide bioleaching media by fractional factorial experiments. *Process Biochemistry, 31*, 773–785.

Vieira, M. J. & Melo, L. F. (1995). Effect of clay particles on the behaviour of biofilms formed by *Pseudomomonas fluorescens. Water Science and Technology, 32*, 45–52.

Wainwright, M. & Gadd, G.M. (1997). Industrial pollutants. In D. T. Wicklow & B. Soderstrom (Eds), *The Mycota, Vol. V: Environmental and Microbial Relationships* (pp. 85–97). Berlin: Springer-Verlag.

Walter, E. G. & Taylor, D. E. (1992). Plasmid-mediated resistance to tellurite: expressed and cryptic. *Plasmid, 27*, 52–64.

Wang, P.-C., Mori, T., Komori, K., Sasatu, M., Toda, K. & Ohtake, H. (1989). Isolation and characterisation of an *Enterobacter cloacae* strain that reduces hexavalent chromium under anaerobic conditions. *Applied and Environmental Microbiology, 55*, 1665–1669.

Watson, J. H. P. & Ellwood, D. C. (1994). Biomagnetic separation and extraction process fron heavy metals from solution. *Minerals Engineering, 7*, 1017–1028.

Watson, J. H. P., Ellwood, D. C., Deng, Q. X., Mikhalovsky, S., Hayter, C. E. & Evans, J. (1995). Heavy metal adsorption on bacterially-produced FeS. *Minerals Engineering 8*, 1097–1108.

Watson, J. H. P., Ellwood, D. C. & Duggleby, C. J. (1996). A chemostat with magnetic feedback for the growth of sulphate-reducing bacteria and its application to the removal and recovery of heavy metals from solution. *Minerals Engineering, 9*, 937–983.

White, C. & Gadd, G. M. (1996a). Mixed sulphate-reducing bacterial cultures for bioprecipitation of toxic metals: factorial and response-surface analysis of the effects of dilution rate, sulphate and substrate concentration. *Microbiology, 142*, 2197–2205.

White, C. & Gadd, G. M. (1996b). A comparison of carbon/energy and complex nitrogen sources for bacterial sulphate-reduction: potential applications to bioprecipitation of toxic metals as sulphides. *Journal of Industrial Microbiology, 17*, 116–123.

White, C and Gadd, G. M. (1997). An internal sedimentation bioreactor for laboratory-scale removal of toxic metals from soil leachates using biogenic sulphide precipitation. *Journal of Industrial Microbiology, 18*, 414–421.

White, C. & Gadd, G. M. (1998a). Reduction of metal cations and oxyanions by anaerobic and metal-resistant organisms: chemistry, physiology and potential for the control and bioremediation of toxic metal pollution. In W. D. Grant & T. Horikoshi (Eds), *Extremophiles: Physiology and Biotechnology* (pp. 233–254). New York: John Wiley.

White, C. & Gadd, G. M. (1998b). Accumulation and effects of cadmium on sulphate-reducing bacterial biofilms. *Microbiology, 144*, 1407–1415.

White, C. & Gadd, G. M. (2000). Copper accumulation by sulphate-reducing bacterial biofilms and effects on growth. *FEMS Microbiology Letters, 183*, 313–318.

White, C., Wilkinson, S. C. & Gadd, G. M. (1995). The role of microorganisms in biosorption of toxic metals and radionuclides. *International Biodeterioration and Biodegradation, 35*, 17–40.

White, C., Sayer, J. A. & Gadd, G. M. (1997). Microbial solubilization and immobilization of toxic metals: key biogeochemical processes for treatment of contamination. *FEMS Microbiology Reviews, 20*, 503–516.

White, C., Sharman, A. K. & Gadd, G. M. (1998). An integrated microbial process for the bioremediation of soil contaminated with toxic metals. *Nature Biotechnology 16*, 572–575.

Widdel F. (1988). Microbiology and ecology of sulfate- and sulfur-reducing bacteria. In A. B. Zehner (Ed), *Biology of Anaerobic Micro-organisms* (pp. 469–585). New York: John Wiley.

Wieder, R. K. (1993). Ion input/output budgets for 5 wetlands constructed for acid coal- mine drainage treatment. *Water, Air and Soil Pollution, 71*, 231–270.

Wood, A. P. & Kelly, D. P. (1983). Autotrophic and mixotrophic growth of three thermoacidophilic iron-oxidising bacteria. *FEMS Microbiology Letters, 20*, 107–112.

Yong, P. & Macaskie, L. E. (1995). Enhancement of uranium bioaccumulation by a *Citrobacter* sp. via enzymically-mediated growth of polycrystalline $NH_4UO_2PO_4$. *Journal of Chemical Technology and Biotechnology, 63*, 101–108.

Zagury, G. J., Narasasiah, K. S. & Tyagi, R. D. (1994). Adaptation of indigenous iron- oxidizing bacteria for bioleaching of heavy metals in contaminated soils. *Environmental Technology, 15*, 517–530.

Zieve, R., Ansell, P. J., Young, T. W. K. & Peterson, P. J. (1985). Selenium volatilization by *Mortierella* species. *Transactions of the British Mycological Society, 84*, 177–179.

Zinkevich, V., Bogdarina, I., Kang, H., Hill, M. A. W., Tapper, R. & Beech, I. B. (1996). Characterization of exopolymers produced by different isolates of marine sulphate-reducing bacteria. *International Biodeterioration and Biodegradation, 37*, 163–172.

© 2002 Elsevier Science Ltd. All rights reserved

Chapter 7

Microbial redox interactions with uranium: an environmental perspective

Robert T. Anderson, Derek R. Lovley

University of Massachusetts, Department of Microbiology, Morrill Science Center IVN, Amherst, MA 01003, USA

1. Introduction

Microbially catalysed processes have the potential to affect the fate of uranium profoundly in a variety of environmental settings. Redox interactions are particularly important because the mobility of uranium in the environment is largely determined by its oxidation state. Reduced uranium, U(IV), is highly insoluble and is the oxidation state most often associated with uranium-containing ores (Langmuir, 1978). Oxidised uranium, U(VI), is relatively soluble and therefore mobile in the environment. Microorganisms catalyse the oxidation and reduction of uranium and therefore influence uranium mobility in the environment. Recent interest in the use of microorganisms for metal removal from waste streams and from groundwater of metal-contaminated environments has sparked interest in microbial redox interactions with uranium and potential applications at uranium-contaminated sites (Lovley & Phillips, 1992a; Lovley, 1995a; Lovley & Coates, 1997). While microbially enhanced oxidation mobilises uranium, perhaps best exemplified during bioleaching of uranium from low grade ore (Brierley, 1978; Hutchins et al., 1986; Rawlings & Silver, 1995; Bosecker, 1997), microbially catalysed reduction processes immobilise uranium.

Until quite recently, uranium reduction was generally thought to be dominated by abiotic reactions (Jensen, 1958; Hostetler & Garrels, 1962; Langmuir, 1978; Maynard, 1983; Nakashima et al., 1984, 1999). The discovery of anaerobic microorganisms capable of coupling growth to uranium reduction (Lovley et al., 1991) demonstrates a biogeochemical cycle for uranium consistent with geochemical observations of uranium accumulation in anoxic sediments (Bertine et al., 1970; Bonatti et al., 1971; Colley & Thomson, 1985; Cochran et al., 1986; Anderson, 1987; Klinkhammer & Palmer, 1991; Barnes & Cochran, 1993) and proposed conditions of uranium ore formation (Jensen, 1958; Hostetler & Garrels, 1962; Adler, 1974; Langmuir, 1978; Mohagheghi et al., 1985). The stimulation of uranium reduction within contaminated aquifers has been proposed as a method to remove uranium from contaminated groundwater in situ and could prove widely applicable at many uranium-contaminated sites (Lovley et al., 1991; Abdelouas et al., 1998a, b, 1999b). Microbial uranium reduction also has potential applications in industry for removing uranium from waste streams (Lovley & Phillips, 1992a; Lloyd & Macaskie, 2000). In this chapter, we will review microbial interactions with uranium, particularly emphasising the biogeochemical aspects and implications of uranium reduction.

2. Microbially catalysed redox reactions of uranium

It is becoming increasingly apparent that microorganisms play a significant, and perhaps dominant, role in the biogeochemical cycling of metals in the environment (Ehrlich, 1990; Lovley, 2000). Metal-microbe redox interactions have implications for ore formation and recovery of metals from ore-containing materials. Aerobic, metal-oxidising bacteria catalyse the oxidation of reduced metals, enhancing metal recovery from ore-containing materials during bioleaching processes (Brierley, 1978). Anaerobic bacteria found in sedimentary environments catalyse metal reduction. Since the reduced forms of many elements are insoluble, anaerobic sediments frequently serve as environmental sinks. Uranium(VI) is relatively mobile in aerobic environments but once reduced to U(IV) becomes quite insoluble. Understanding the microbially catalysed redox interactions of uranium is key to understanding recovery of uranium from ore-containing materials and uranium accumulation in anaerobic sediments under naturally occurring or engineered conditions.

Aerobic interactions with uranium: bioleaching processes

Microbially enhanced oxidation of uranium from U(IV) to U(VI) is widely used in the recovery of uranium from low grade ores (Tuovinen & Kelly, 1974; Brierley, 1978; Lundgren, 1980; Hutchins et al., 1986; Rawlings & Silver, 1995; Muñoz et al., 1995; Bosecker, 1997). Many globally significant ore deposits contain uranium as insoluble oxides, usually with a high proportion of U(IV) (Plant et al., 1999). Efficient recovery of uranium from such materials depends on the oxidation of U(IV) to U(VI), thereby creating a soluble and easily recovered form of uranium. Uranium can often be chemically extracted directly from high grade ores (2.5–12.2% U), but global depletion of these deposits has resulted in the increased exploitation of lower grade deposits (Brierley, 1978; Muñoz et al., 1995). Many lower grade uranium ores (0.04–0.4% U) are extracted in an aerobic, acid leaching process enhanced by the presence of acid-tolerant, Fe(II)- and S^0-oxidising bacteria (Tuovinen & Kelly, 1974; Brierley, 1978; Rawlings & Silver, 1995; Muñoz et al., 1995; Bosecker, 1997). Fe(III) is an effective oxidant for U(IV) and, when added to uranium-containing ore under acidic conditions, solubilises uranium as U(VI) in the leachate. Fe(III) is reduced during uranium oxidation but can be regenerated by acid-tolerant, Fe(II)-oxidising bacteria such as *Thiobacillus ferrooxidans*. *T. ferrooxidans* is an appropriate microbial model for bioleaching processes as this organism thrives at low pH (1.5–2.5 optimal range) and couples growth to aerobic oxidation of Fe(II), thereby indirectly enhancing uranium recovery from ore-containing materials (Brierley, 1978; Hutchins et al., 1986; Rawlings & Silver, 1995; Bosecker, 1997). Additionally, many uranium-containing ores also contain pyrite (FeS_2). Representatives of the *Thiobacillus* and *Leptospirillum* families have been identified as active aerobic bacterial species within leach heaps containing uranium and other metals (Brierley, 1978; Rawlings & Silver, 1995; Schippers et al., 1995). The presence of acid-tolerant Fe(II)- and S^0-oxidising bacteria of the *Thiobacillus* and *Leptospirillum* families enhances pyrite oxidation (Bruynesteyn, 1989; Schippers & Sand, 1999; Fowler et al., 1999) generating the Fe(III) and sulfuric acid necessary for uranium mobilisation during bioleaching.

Metal-oxidising bacteria can also directly oxidise U(IV) to U(VI) (DiSpirito & Tuovinen, 1981, 1982a, b). While organisms such as *T. ferrooxidans* derive energy from the aerobic oxidation of Fe(II) and S^0, free energy calculations also indicate a potential net energy gain from aerobic U(IV) oxidation (DiSpirito & Tuovinen, 1982b). However, U(VI) is toxic to many organisms and inhibits Fe(II)-oxidation by *Thiobacillus* species in laboratory cultures at concentrations approaching 1 mM (Tuovinen & Kelly, 1974). The isolation of active organisms from leaching operations containing much higher U(VI) concentrations shows that the organisms within leach piles adapt to the much higher uranium concentrations. In the laboratory, strains of *Thiobacillus* can be adapted to tolerate higher U(VI) concentrations, thus more accurately modelling the organisms found within leach piles (Tuovinen & Kelly, 1974). Strains of *Thiobacillus* cultured in the presence of U(VI) have been tested for the ability to oxidise U(IV) directly. Carbon fixation and oxygen uptake associated with the oxidation of U(IV) compounds by adapted cultures of *T. ferrooxidans* and *T. acidophilus* indicates these organisms couple metabolic processes, but not growth, to U(IV) oxidation (DiSpirito & Tuovinen, 1981, 1982a, b). These observations raise the possibility that uranium solubilisation during acidic bioleaching processes results from both direct and indirect microbial oxidation (Francis, 1990; Rawlings & Silver, 1995; Bosecker, 1997).

Mobilisation of uranium arising from the activity of metal- and sulfur-oxidising bacteria such as *Thiobacillus* is a potential mechanism for uranium contamination of groundwater. Metal- and sulfur-oxidising bacteria are ubiquitous in the environment particularly in settings where reduced minerals contact atmospheric oxygen. Uranium-contaminated groundwater is of concern in areas where past uranium milling operations have left large tailings piles (Abdelouas et al., 1998a, b). Low level leaching of uranium within these piles provides a continued source of U(VI) to the local aquifer, and groundwater uranium concentrations at many of these sites present a threat to down-gradient water resources. Remediation of uranium contaminated aquifers has tended to focus on 'pump and treat' systems which have had little success in producing sustained lowering of groundwater U(VI) concentrations (Abdelouas et al., 1999b), prompting investigation of alternative strategies.

Anaerobic interactions with uranium

Anaerobic microbial processes have the potential to remove U(VI) effectively from contaminated groundwater. A relatively novel process for the treatment of uranium contaminated aquifers is the stimulation of anaerobic processes to precipitate uranium as U(IV) within narrow zones across groundwater flow paths (Lovley et al., 1991). The idea is similar to other permeable reactive barriers (Cantrell et al., 1995; Gu et al., 1998) except that, in this case, the indigenous anaerobic bacteria are stimulated within the subsurface by the addition of a suitable electron donor. Groundwater at many uranium-contaminated sites is aerobic and uranium is therefore mobile. The creation of anaerobic conditions, as would occur upon the addition of organics, creates conditions favourable for reduction and therefore precipitation of uranium in situ. A variety of anaerobic organisms are known to reduce U(VI) and the production of anaerobic conditions within aquifers results in a predictable succession of microbially catalysed redox processes that can influence uranium immobilisation.

Succession of anaerobic microbial processes in sedimentary environments

Uranium and other heavy metals often accumulate in anaerobic sedimentary environments as a direct consequence of microbial metabolism. Anaerobic conditions develop in sediments due to the depletion of oxygen by aerobic bacteria which couple the oxidation of organic matter to the reduction of dissolved oxygen as the terminal electron acceptor. Oxygen is therefore most rapidly depleted in sediments containing large amounts of organic matter (Chapelle, 1993). Diffusion of oxygen into sediments is slow and often limits aerobic processes to fringe areas along the oxic/anoxic boundary in organics-contaminated aquifers or at the sediment/water interface of aquatic or marine sediments (Klinkhammer & Palmer, 1991; Anderson & Lovley, 1997). On the depletion of oxygen, microbial respiratory processes shift to alternative terminal electron acceptors such as NO_3^-, Mn(IV), Fe(III), SO_4^{2-} and CO_2 (see Chapter 3, this volume).

Anaerobic metabolism in aquatic and marine sediments exploits a succession of terminal electron accepting processes (Ponnamperuma, 1972; Reeburgh, 1983). In an ideal system, on depletion of oxygen, microbial processes coupled to NO_3^- reduction become dominant. Deeper in the sediment, as NO_3^- concentrations are depleted, microbial processes shift to Mn(IV) reduction, followed successively by Fe(III) reduction, SO_4^{2-} reduction and finally CO_2 reduction (Fig. 1) (Froelich et al., 1979; Reeburgh, 1983). The segregation of anaerobic processes into distinct zones is not absolute and sediment heterogeneity or departure from steady state can lead to microsites where several terminal electron accepting processes exist together. Under steady state conditions, there is generally one predominant terminal electron accepting process (Froelich et al., 1979; Reeburgh, 1983).

While the distribution of terminal electron accepting processes within sediments under steady state conditions correlates with the thermodynamic energy yield of each reaction, a more accurate model of terminal electron accepting process distribution can be obtained by considering the physiological constraints on microbial metabolism and the effects of microbial competition (Lovley & Chapelle, 1995). Terminal electron accepting process distribution in sediments is often explained on a thermodynamic basis where organisms preferentially utilise the terminal electron acceptor yielding the most available energy (McCarty, 1972; Stumm & Morgan, 1981; Bouwer, 1992). However, reactions leading to less energy should also occur to some extent if they are thermodynamically favourable. Therefore, thermodynamics alone cannot adequately predict terminal electron accepting process distribution in sediments, as they cannot explain the general absence of reactions, such as CO_2 reduction (methanogenesis), with a lower energy yield in zones dominated by other, more energetically favourable terminal electron accepting processes. Organic matter degradation under anaerobic conditions initially results in the generation of H_2 and low

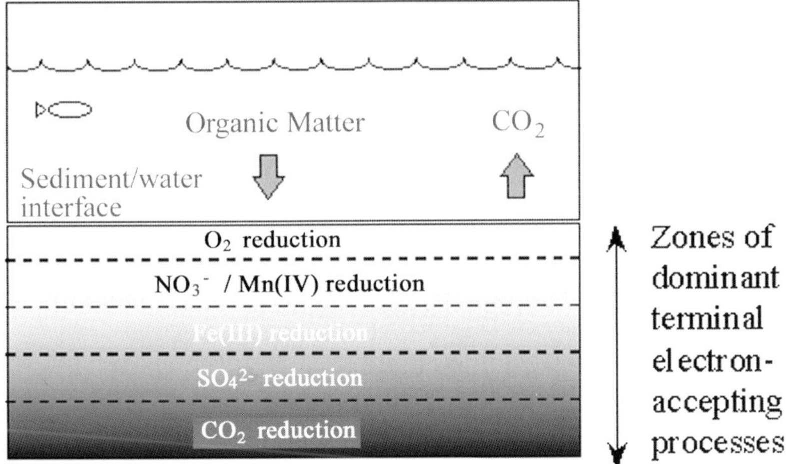

Fig. 1. The distribution of terminal electron-accepting processes (TEAPs) found with depth in aquatic and marine sediments.

molecular weight organic acids such as acetate via the successive activities of hydrolytic and fermentative organisms (Fig. 2) (Lovley & Chapelle, 1995). Anaerobic respiratory bacteria compete for hydrogen and organic acids as substrates, and different electron accepting processes have different threshold substrate concentrations (Lovley & Goodwin, 1988; Lovley et al., 1994). For example, in the presence of Fe(III), Fe(III) reducers will outcompete SO_4^{2-} reducers because Fe(III)-reducing bacteria can metabolise substrates to levels too low to support SO_4^{2-} reduction (Lovley and Phillips, 1987). Therefore, Fe(III) reduction will predominate in sediments where NO_3^- is depleted and Fe(III) is available. Similarly, SO_4^{2-} reducers can metabolise substrates to levels too low to support methanogenesis (Lovley & Klug, 1983; Lovley and Goodwin, 1988). In the presence of SO_4^{2-} and in the absence of Fe(III) and NO_3^-, SO_4^{2-} reduction will predominate. Threshold substrate concentrations form the physiological basis for microbial competition and explain terminal electron accepting process distributions in anaerobic sediments (Lovley & Goodwin, 1988; Lovley et al., 1994).

The succession of microbial processes observed in aquatic and marine anoxic sediments is also found within aquifers contaminated with organic compounds (Lyngkilde & Christensen, 1992; Baedecker et al., 1993; Patterson et al., 1993; Vroblesky & Chapelle, 1994; Lovley et al., 1994; Bjerg et al., 1995; Borden et al., 1995; Rugge et al., 1995). Pristine aquifers generally contain low concentrations of organics and are therefore generally dominated by aerobic processes (Chapelle, 1993). In other words, microbial processes in pristine, aerobic aquifers are donor limited. The amount of electron acceptor in the form of oxygen exceeds the amount of electron donor (organic carbon). When aquifers become contaminated with organic materials, such as petroleum hydrocarbons or landfill leachate, the amount of electron donor (organic carbon) far exceeds the amount of dissolved oxygen and microbial metabolism shifts to anaerobic processes due to consumption of dissolved oxygen by aerobic degradation processes (Chapelle, 1993; Anderson & Lovley, 1997).

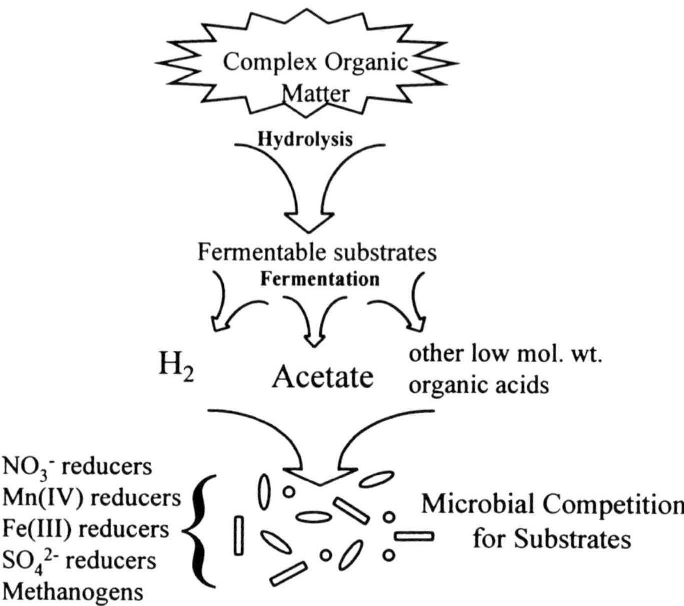

Fig. 2. Organic matter degradation in anaerobic environments.

In heavily contaminated aquifers, large areas dominated by anaerobic processes develop downgradient of source areas as soluble contaminants migrate with the groundwater to produce a contaminant plume (Lyngkilde & Christensen, 1992; Baedecker et al., 1993; Bjerg et al., 1995; Borden et al., 1995; Rugge et al., 1995). In the anaerobic portions of a contamination plume, a succession of terminal electron accepting processes develops, similar to that observed with depth in anoxic aquatic or marine sediments. Areas dominated by methanogenic processes tend to be found closest to the source, where contamination has existed for the longest period of time and where all potential electron acceptors other than CO_2 have been exhausted. Methanogenic conditions are followed by successive downgradient zones dominated by SO_4^{2-} reduction, Fe(III) reduction, Mn(IV) reduction, NO_3^- reduction and aerobic conditions once again at the contaminant plume boundary (Fig. 3) (Chapelle, 1993; Lovley et al., 1994; Lovley, 1997; Anderson & Lovley, 1997). The composition of terminal electron accepting process zones within contaminated aquifers varies, based on the availability of potential electron acceptors and, for most aquifers, is much more complex spatially than Fig. 3 implies. Nonetheless, this model serves as a useful basis for understanding microbial processes in contaminated groundwater systems and for the application of some in situ bioremediation techniques (Reinhard et al., 1997; Anderson et al., 1998; Hutchins et al., 1998; Anderson & Lovley, 2000).

The terminal electron accepting process distribution observed in organic-contaminated aquifers is a useful model for understanding in situ microbial immobilisation of uranium. Many metal-contaminated aquifers, including sites contaminated with uranium, are not usually associated with high concentrations of organic materials and are therefore aerobic (Abdelouas et al., 1999b) but in situ immobilisation of uranium can be accomplished by

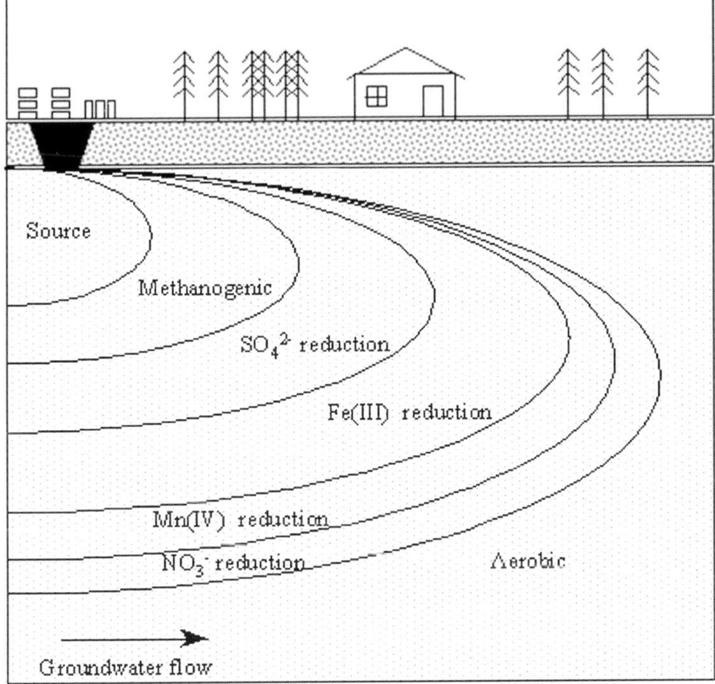

Fig. 3. The distribution of terminal electron-accepting processes (TEAPs) observed within anaerobic portions of aquifers contaminated with organic compounds.

creating anaerobic conditions within the subsurface. Uranium immobilisation will depend on the anaerobic process(es) stimulated in situ and the way in which microorganisms found in each terminal electron accepting process zone interact with uranium.

Uranium reduction under Fe(III)-reducing conditions

Microbial processes coupled to Fe(III) reduction are predicted to be important during stimulated in situ uranium reduction. Groundwater at uranium contaminated sites often contains high concentrations of NO_3^- and SO_4^{2} derived from past acidic extractions and ongoing bioleach processes within uranium mill tailings piles (Abdelouas et al., 1999b). In the presence of NO_3^-, stimulated anaerobic processes within the subsurface are likely to be dominated initially by NO_3^- reduction for the reasons mentioned above. Uranium reduction is not a favoured process under these conditions and removal of NO_3^- is a prerequisite for in situ uranium reduction. Uranium may in fact be mobilised under NO_3^- reducing conditions in some brine environments where U(VI), present as insoluble hydroxides and phosphates, solubilises in the presence of bicarbonate produced by microbial oxidation of organics (Francis et al., 2000). Additionally, NO_3^- reduction could, in principle, be coupled to the microbial oxidation of U(IV). Both these mechanisms have the potential to increase the mobility of uranium in groundwater.

Although, in the absence of NO_3^-, metabolic processes coupled to Mn(IV) reduction are expected to be the most favourable, Mn(IV) abundance in sediments is generally low so that Mn(IV) is not likely to be a major electron acceptor in comparison to other metals such as Fe(III) (Lovley, 1995b). Fe(III), on the other hand, is widely available in sediments and metabolic processes coupled to Fe(III) reduction are likely to be dominant in many sediments upon the depletion of NO_3^- (Lovley, 1995a). Fe(III)-reducing bacteria are known to reduce U(VI) (Lovley et al., 1991; Gorby & Lovley, 1992) and are predicted to be the dominant U(VI)-reducing microorganisms during in situ U(VI) reduction.

A variety of bacteria are capable of interacting with U(VI) by sorption, accumulation or reduction (Lovley & Phillips, 1992b; Suzuki & Banfield, 1999; Lloyd & Macaskie, 2000). However, Fe(III)-reducing bacteria are the only organisms known to have the ability to couple growth to the reduction of U(VI) (Lovley et al., 1991; Gorby & Lovley, 1992;). Microbial U(VI) reduction by Fe(III) reducers has been investigated largely with organisms in the genera *Geobacter* and *Shewanella* (Lovley et al., 1991; Gorby & Lovley, 1992; Ganesh et al., 1997; Truex et al., 1997). These organisms utilise U(VI) as a terminal electron acceptor, oxidising acetate (*Geobacter*) or hydrogen (*Shewanella*) and deriving energy for growth (Lovley et al., 1991). Uranium is precipitated as insoluble uraninite (UO_2) during these processes (Gorby & Lovley, 1992). A recent estuarine isolate, *Desulfotomaculum*, which is capable of utilising U(VI) as an electron acceptor, couples growth with the oxidation of short chain fatty acids such as butyrate (Tebo & Obraztsova, 1998). This organism is also capable of utilising Fe(III) or SO_4^{2-} as an electron acceptor during metabolism. However, its distribution in sediments is as yet unclear. By contrast, members of the *Geobacter* family are of particular note as they have been found in a wide variety of anaerobic aquifer sediments and molecular studies have suggested that they are the dominant members of the Fe(III)-reducing microbial community in subsurface environments (Rooney-Varga et al., 1999; Snoeyenbos-West et al., 2000). Uranium reduction in *Geobacter metallireducens* is thought to proceed via electron transport from a c-type cytochrome to U(VI) (Lovley et al., 1993a). The observation that *Geobacter* species utilise acetate as an electron donor during U(VI) reduction is important because acetate is one of the most common degradation intermediates in anaerobic environments and the addition of acetate has been shown to stimulate the enrichment of *Geobacter* species in anaerobic aquifer sediments (Snoeyenbos-West et al., 2000). Therefore, acetate addition to sediments contaminated with uranium is likely to stimulate the growth and activity of known U(VI)-reducing *Geobacters*. Furthermore, Fe(III)-reducing organisms degrade a wide variety of organic contaminants such as aromatic hydrocarbons (Lovley et al., 1993a). Stimulation of *Geobacter* organisms within aquifers contaminated with both uranium and organic contaminants could potentially result in the immobilisation of uranium coupled to the oxidation of contaminant organics (Lovley et al., 1991).

Uranium reduction under thermophilic conditions

Recent isolations of thermophilic Fe(III)-reducing organisms capable of U(VI) reduction imply that many higher temperature uranium deposits may have biological origins (Kieft et al., 1999; Kashefi & Lovley, 2000). A deep subsurface *Thermus* isolate obtained from a South African gold mine grows optimally at 60°C by Fe(III) reduction and will reduce U(VI) in cell suspension with lactate as the electron donor (Kieft et al., 1999). The hyperthermophilic organism *Pyrobaculum islandicum* grows at 100°C and is also capable of U(VI) reduction, using hydrogen as the electron donor (Kashefi & Lovley, 2000). Neither of these organisms couples growth to U(VI) reduction but the potential they demonstrate for uranium immobilisation at high temperature is consistent with conditions proposed for the formation of some sandstone-type uranium ore deposits (Hostetler & Garrels, 1962) and other well-known deposits such as the Oklo nuclear reactor (Brookins, 1990).

Respiration of humic substances: implications for U(VI) reduction

Fe(III)-reducing bacteria utilise humic acids as electron acceptors, for example by reducing quinone groups (Scott et al., 1998) and these are then a potential source of reductant for uranium immobilisation (Lovley et al., 1996). Some types of uranium ore deposits are associated with humic-related materials such as lignites or black shales (Plant et al., 1999). Uranium immobilisation in these deposits is thought to occur upon complexation and reduction by organic matter (Nakashima et al., 1984, 1999; Landais et al., 1987).

Microbially reduced humics are known to reduce metals such as Fe(III) abiotically. On metal reduction, the reoxidised humics are once again available for use as bacterial electron acceptors. This process, known as 'electron shuttling', greatly accelerates the rate of Fe(III) reduction in sediments (Lovley et al., 1998; Nevin & Lovley, 2000) and suggests a biological source for the reduced organic matter observed within some organic-rich uranium deposits (Plant et al., 1999).

Uranium reduction under SO_4^{2-}-reducing conditions

For some time, geochemical studies have shown that the accumulation of uranium in anoxic marine sediments is a globally important sink for uranium in the environment (Bertine et al., 1970; Bonatti et al., 1971; Colley & Thomson, 1985; Anderson et al., 1989a, b; Klinkhammer & Palmer, 1991; Barnes & Cochran, 1993). On a geological time scale, mobile U(VI) produced from the oxidation of U(IV)-containing materials is transported into the marine environment and becomes immobilised by reduction to U(IV) in anoxic sediments (Ferguson, 1987; Klinkhammer & Palmer, 1991). Sulfate reduction is a dominant microbial process within anoxic marine sediments, so abiotic reduction of U(VI) by sulfide, hydrogen or organic matter was previously thought to account for uranium accumulation in these environments (Jensen, 1958; Hostetler & Garrels, 1962; Langmuir, 1978; Maynard, 1983, Landais et al., 1987). Similarly, investigations of terrestrial uranium ore deposits identified an accumulation of uranium in anoxic zones along steep redox gradients in association with organic matter and pyrite (FeS_2) (Adler, 1974; Maynard, 1983). The close association of sulfide minerals with uranium mineral deposits suggested sulfide reduction of U(VI) as a potential mechanism for uranium precipitation in anoxic environments (Jensen, 1958; Hostetler & Garrels, 1962; Adler, 1974). Abiotic reduction of U(VI) by sulfide has been demonstrated at relatively high concentrations of U(VI) (>3 mg l^{-1}, pH 7, 35°C) (Mohagheghi et al., 1985). However, the persistence of environmentally relevant concentrations of U(VI) (<0.8 mg l^{-1}) in the presence of sulfide in the environment (Anderson et al., 1989a) and in laboratory studies (Lovley et al., 1991) indicates little potential for abiotic reduction of U(VI) in natural waters.

SO_4^{2-}-reducing bacteria previously associated with uranium precipitation (Mohagheghi et al., 1985) are now known to reduce U(VI) to U(IV) enzymatically, thereby providing a mechanism for rapid accumulation of uranium in many sedimentary environments (Lovley & Phillips, 1992b; Lovley et al., 1993b, c; Tebo & Obraztsova, 1998). SO_4^{2-} reducers couple the oxidation of hydrogen or lactate to the reduction of U(VI) by electron transport via a c_3-type cytochrome (*Desulfovibrio vulgaris*) but cannot obtain energy for growth by this mechanism (Lovley & Phillips, 1992b; Lovley et al., 1993c). The SO_4^{2-}-reducing *Desulfotomaculum* described earlier is reportedly capable of growing using U(VI) as an electron acceptor (Tebo & Obraztsova, 1998). However, detailed investigations of the potential for enzymatic uranium reduction have focused on organisms of the *Desulfovibrio* family (Mohagheghi et al., 1985; Lovley & Phillips, 1992b; Lovley et al., 1993b, c; Phillips et al., 1995; Tucker et al., 1996, 1998; Ganesh et al., 1997, 1999; Panak et al., 1998; Spear et al., 1999;). These organisms tolerate high concentrations of uranium in solution (up to 24 mM), are relatively easy to grow and freeze-dried cell preparations exposed to oxygen retain activity upon reconstitution in media, a practical advantage for use in biotreatment applications (Lovley & Phillips, 1992a, b). *Desulfovibrio*-catalysed U(VI) reduction is relatively insensitive to a wide variety of potentially co-contaminating metals (<100 mM Zn(II), Cu(II), Ni(II), Co(II)) (Lovley & Phillips, 1992a) and will precipitate uranium from U(VI)-containing groundwaters and simulated waste streams (Lovley & Phillips, 1992a; Ganesh et al., 1999).

Potential industrial applications of *Desulfovibrio*-catalysed U(VI) reduction have been investigated in immobilised cell systems (Lovley & Phillips, 1992a; Tucker et al., 1998), batch cultures (Ganesh et al., 1997, 1999) and flow-through bioreactors (Tucker et al., 1996; Spear et al., 1999) for the treatment of uranium-containing waste streams. Industrial effluent or contaminated groundwater commonly contains elevated concentrations of NO_3^- and SO_4^{2-}, and/or U(VI) complexed with organic ligands (Lloyd & Macaskie, 2000). Elevated NO_3^- and SO_4^{2-} concentrations can interfere with *Desulfovibrio*-based bioreduction processes. For example, results obtained from immobilised cell systems indicate a decrease in U(VI) removal efficiency in the presence of as little as 50 mg l^{-1} NO_3^-, presumably due to NO_2^- toxicity (Tucker et al., 1998) although SO_4^{2-} concentrations as high as 2000 mg l^{-1} did not affect U(VI) removal. Batch cultures reduced U(VI) in the presence of higher anion concentrations (up to 5000 mg l^{-1} NO_3^- and SO_4^{2-}), typical of those observed in uranium-containing waters (Ganesh et al., 1999). Additional batch culture data show that reduction rates for U(VI) complexed with organic ligands vary depending on the denticity of the ligands and the identity of the bacterial species used (Ganesh et al., 1997), so the treatment of mixed uranium-organic waste streams may be difficult (Macaskie, 1991).

In sediments, SO_4^{2-} reduction generally takes place in zones distinct from Fe(III) reduction because Fe(III)-reducing bacteria can out-compete SO_4^{2-} reducers for substrates (Lovley & Phillips, 1987; Chapelle & Lovley, 1992). However, SO_4^{2-} reducers can reduce Fe(III), but cannot grow by this metabolism (Coleman et al., 1993; Lovley et al., 1993b), and it is possible for SO_4^{2-}-reducing organisms to compete effectively for Fe(III) at the Fe(III)/SO_4^{2-} reduction zone interface or in sediments under non-steady state conditions (Lovley et al., 1993b). Minimum threshold hydrogen concentrations are lower for SO_4^{2-}-reducing organisms when reducing U(VI) rather than Fe(III), suggesting SO_4^{2-} reducers may also reduce U(VI) within the zone of Fe(III) reduction zone prior to significant sulfide accumulation (Lovley et al., 1993b), consistent with the observed accumulation of uranium within anoxic marine sediments (Bertine et al., 1970; Bonatti et al., 1971; Colley & Thomson, 1985; Cochran et al., 1986; Anderson et al., 1989a, b; Klinkhammer & Palmer, 1991; Barnes & Cochran, 1993).

3. Engineered removal of uranium from groundwater and waste streams by microbial reduction

As mentioned earlier, in situ immobilisation of uranium by stimulating Fe(III)- or SO_4^{2-}-reducing organisms within the contaminated subsurface potentially provides a more effective method of uranium removal from groundwater. Additionally, biotechniques based on microbial U(VI) reduction have been developed to remove uranium from industrial waste streams to prevent further entry of uranium into the environment.

Reduction-based bioremediation of uranium contaminated aquifers

Uranium mill tailings piles are notable sources of aquifer contamination and reduction-based techniques provide effective methods of uranium removal. Uranium in natural waters and uranium-contaminated aquifers is generally found in low concentrations (0.1–1 mg l^{-1}) and is most likely to exist as a carbonate complex (Langmuir, 1978; Mohagheghi et al., 1985; Abdelouas et al., 1998a). U(VI) forms stable carbonate complexes and the addition of 30–100 mM bicarbonate solutions to contaminated soils and sediments removes U(VI) in quantities comparable to extraction with 1 M HNO_3^- (Phillips et al., 1995). In comparison to acid extraction, bicarbonate extraction is relatively specific for U(VI) and the leachate is easily amenable to further biotreatment (Phillips et al., 1995). *Desulfovibrio*, *Geobacter* and *Shewanella* species have been shown to precipitate uranium as insoluble uraninite (UO_2) from carbonate complexes in sediment extracts and contaminated groundwater (Lovley et al., 1991; Lovley & Phillips, 1992a; Phillips et al., 1995). This technique, based on microbial reduction of U(VI), is one of many which are potentially useful for removing uranium from groundwater, contaminated soils, sediments and waste streams (Macaskie, 1991; Thomas & Macaskie, 1996; Francis & Dodge, 1998). A comprehensive review of other biotechniques (biosorption, biomineralisation, bioprecipitation) pertinent to radionuclide removal from various sources has recently been published (Lloyd & Macaskie, 2000) and the reader is referred to this work and Chapters 11 and 12 of this book. Reduction of U(VI) is discussed here because it has the potential to immobilise uranium in situ providing an alternative to pump and treat-based techniques for aquifer remediation.

Current pump and treat remediation practices for removing uranium from groundwater generally do not provide sustained, decreased U(VI) concentrations within aquifers (Abdelouas et al., 1999b). Stimulated U(VI) reduction can be accomplished in situ upon the addition of a suitable source of organic carbon (Fig. 4). A zone of stimulated anaerobic activity positioned perpendicular to groundwater flow paths could serve as a zone of uranium immobilisation, preventing further migration within the subsurface (Lovley & Phillips, 1992a). The potential for stimulated anaerobic conditions to remove U(VI) from

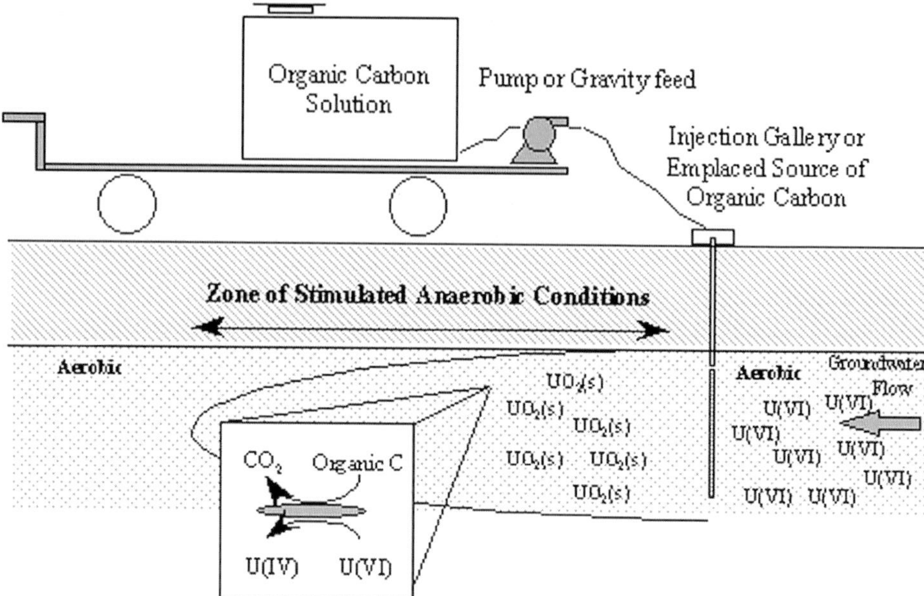

Fig. 4. Conceptualised bioremediation scheme for stimulated U(VI) reduction in situ upon bulk addition of a suitable electron donor.

groundwater has been evaluated in aquifer sediments collected from a uranium mill tailings pile near Tuba City, Arizona (USA) (Abdelouas et al., 1998a, b; Abdelouas et al., 1999a, b). Ethanol added to aquifer sediments during batch and column studies promoted anaerobic conditions and resulted in uranium removal from groundwater even in the presence of relatively high concentrations of NO_3^- and SO_4^{2-} (Abdelouas et al., 1998b). Microbial processes coupled to NO_3^- reduction were initially stimulated as predicted in our earlier discussion of terminal electron accepting processes. U(VI) reduction commenced after the depletion of NO_3^-, resulting in the precipitation of uraninite. The reduction of Fe(III) was not monitored over time, precluding an evaluation of U(VI) reduction under Fe(III)-reducing conditions in these sediments. However, stimulated SO_4^{2-} reduction resulted in the removal of uranium from incubations of sediment and groundwater, indicating that the aquifer contains indigenous SO_4^{2-} reducers capable of precipitating uranium (Abdelouas et al., 1998b). Uraninite precipitated with the reduced sulfide mineral mackinawite ($FeS_{0.9}$) is potentially shielded from oxidative dissolution on return to aerobic conditions (Abdelouas et al., 1999a). These observations imply that uranium immobilised under SO_4^{2-}-reducing conditions at this site could remain immobile for extended periods of time even in the presence of dissolved oxygen (Abdelouas et al., 1999b).

Stimulation of SO_4^{2-}-reducing conditions may not be appropriate at some sites due to the potential for sulfide production. Bulk electron donor addition to the subsurface is likely to trigger SO_4^{2-} reduction because of the overwhelming supply of electron donor relative to potential electron acceptors in the subsurface. However, controlled addition of low concentrations of electron donor may be an effective method for stimulating U(VI) reduction under Fe(III)-reducing conditions without at the same time inducing substantial SO_4^{2-} reduction. Fe(III)-reducing organisms of the *Geobacter* family, known to reduce U(VI), have been detected in diverse aquifer sediments from very different geographical locations, indicating these organisms are widespread in groundwater environments (Coates et al., 1996; Rooney-Varga et al., 1999; Snoeyenbos-West et al., 2000). Addition of various electron donors, particularly acetate but also glucose, lactate, benzoate and formate, to previously aerobic aquifer sediment stimulates the growth of potentially U(VI)-reducing *Geobacter* species (Snoeyenbos-West et al., 2000). This has been tested experimentally in sediments collected from a uranium-contaminated aquifer near Shiprock, New Mexico (USA) (Finneran et al., 2002). Acetate addition (2 mM) caused faster removal of U(VI) from batch sediment incubations compared to other potential electron donors such as lactate. Uranium removal was coincident with Fe(III) reduction but not with SO_4^{2-} reduction, despite the presence of 10–40 mM SO_4^{2-} in the groundwater used in the incubation experiments. The results indicate that rapid uranium immobilisation can be achieved with the addition of acetate and SO_4^{2-} reduction need not be induced if low concentrations of electron donor are supplied, sufficient only to deplete NO_3^- and stimulate Fe(III)-reducing conditions (Finneran et al., 2002) (Fig. 5). Coprecipitation of uranium with sulfide minerals, which could interfere with future efforts to remove the immobilised, and concentrated, uranium can therefore be prevented (Abdelouas et al., 1999b).

Bioreduction of uranium in waste streams

Removal of uranium from waste streams prevents potential discharges of high concentrations of uranium into the environment. Addition of U(VI)-reducing organisms to uranium-containing mixtures is potentially a unit process for the removal of uranium from solution by precipitation of uraninite (Gorby & Lovley, 1992; Tucker et al., 1996, 1998; Ganesh et al., 1997; Truex et al., 1997; Spear et al., 1999;). On a per cell basis, enzymatic reduction of uranium has a greater potential to remove uranium than biosorption techniques because of the limited availability of sorption sites at the cell surface (Lovley & Phillips, 1992a). Amounts of precipitated uranium reported for *Desulfovibrio* (Lovley & Phillips, 1992a) are comparable (11 g U(VI) g^{-1} dry cells) to results obtained during phosphatase-mediated uranium precipitation (9 g U(VI) g^{-1} dry cells) (Macaskie, 1991) and are likely to be greater in flow-through systems (Lovley & Phillips, 1992a). Information regarding the kinetics of U(VI) reduction in the presence of *Desulfovibrio* and *Shewanella* species is necessary for bioreduction process design in flow-through systems (Tucker et al., 1996; Truex et al., 1997; Spear et al., 1999). Half saturation constants (K_S) for U(VI) reduction average about 0.5 mM for *Desulfovibrio* species (Tucker et al., 1996; Spear et al., 1999) and 0.13 mM for *Shewanella* species (Truex et al., 1997) with maximum specific reduction rates (k) of 1.38 mmol U(VI) mg^{-1} h^{-1} and 0.24 mmol U(VI) mg^{-1} h^{-1}, respectively. The specific rate data suggest *Desulfovibrio*-mediated U(VI) reduction may be kinetically favoured compared to *Shewanella* mediated reduction (Spear et al., 1999). U(VI) reduction kinetics for *Geobacter* species have not yet been reported.

Bioreactor performance for U(VI) reduction will depend on the bacterial species selected. U(VI) in waste streams is often complexed with various organic ligands that can affect rates of bioreduction and potentially inhibit uraninite precipitation (Robinson

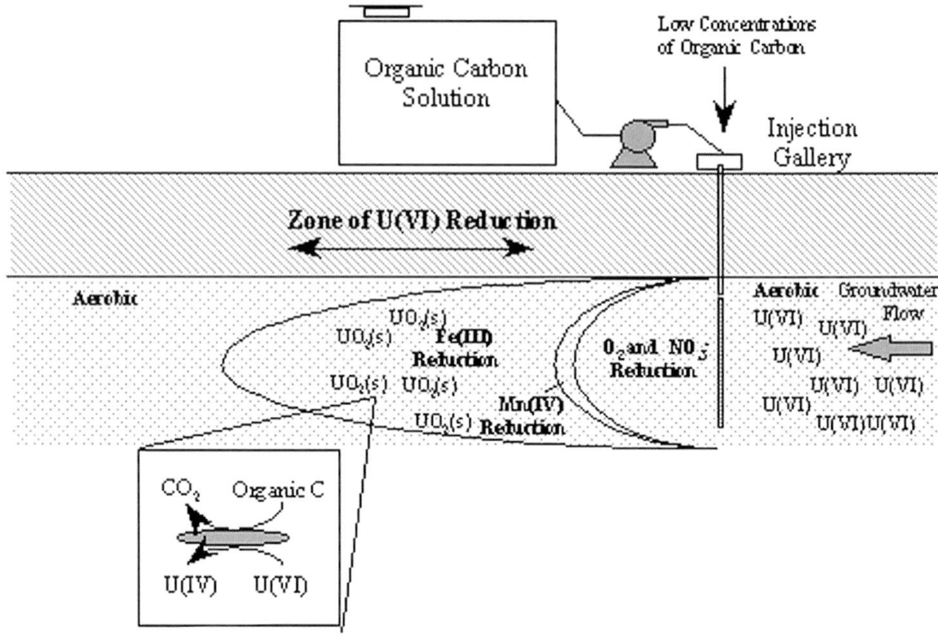

Fig. 5. Stimulated U(VI) reduction in aquifers upon the steady addition of low concentrations of a suitable electron donor such as acetate. Fe(III) and U(VI) reduction are stimulated upon the depletion of O_2, NO_3^- and Mn(IV) as electron acceptors.

et al., 1998). Increased rates of *Desulfovibrio*-based U(VI) reduction have been reported for uranium complexed with monodentate ligands, such as acetate, relative to polydentate ligands such as malonate, oxalate or citrate (Ganesh et al., 1997). In contrast, using a *Shewanella*-based bioreduction, U(VI) complexed with polydentate ligands was reduced faster than that complexed by monodentate ligands (Ganesh et al., 1997). The results imply that uranium removal rates, and therefore process design, will depend on the choice of U(VI)-reducing bacteria employed for a given biotreatment process.

4. Summary

Uranium contamination of the environment has received increased attention in recent post-Cold War years. Examples of microbial interactions with uranium, particularly U(VI) reduction, has prompted new hypotheses concerning the origin of uranium deposits, microbially-enhanced ore recovery and potential bioremediation applications for uranium removal from contaminated media. While a wide variety of techniques is available to remove uranium from waste streams, removal of uranium from contaminated aquifers remains problematic. U(VI) is quite soluble and therefore mobile in aerobic groundwater. Indigenous Fe(III)- and SO_4^{2-}-reducing bacteria found within aquifer environments are capable of reducing U(VI) to insoluble U(IV) thereby precipitating uranium from groundwater when anaerobic conditions are stimulated. Microbial U(VI) reduction has potential applications in the bioremediation of many uranium-contaminated sites such as mill tailings piles. Understanding the microbial ecology of anaerobic aquifers can aid the stimulation and control of an in situ process for uranium immobilisation. However, uranium reduction in situ has yet to be attempted at a uranium-contaminated site.

References

Abdelouas, A., Lutze, W. & Nuttall, H. E. (1998a). Chemical reactions of uranium in groundwater at a mill tailings site. *Journal of Contaminant Hydrology, 34*, 343–361.

Abdelouas, A., Yongming, L., Lutze, W. & Nuttall, H. E. (1998b). Reduction of U(VI) to U(IV) by indigenous bacteria in contaminated groundwater. *Journal of Contaminant Hydrology, 35*, 217–233.

Abdelouas, A., Lutze, W. & Nuttall, H. E. (1999a). Oxidative dissolution of uraninite precipitated on Navajo sandstone. *Journal of Contaminant Hydrology, 36*, 353–375.

Abdelouas, A., Lutze, W. & Nuttall, H. E. (1999b). Uranium contamination in the subsurface: characterization and remediation. In P. C. Burns & R. Finch (Eds), *Uranium: Mineralogy, Geochemistry and the Environment* (pp. 433–473). Washington, DC: Mineralogical Society of America.

Adler, H. H. (1974). Concepts of uranium-ore formation in reducing environments in sandstones and other sediments. *Formation Of Uranium Ore Deposits: Proceedings of a Symposium, Athens* (pp. 141–168).

Anderson, R. F. (1987). Redox behavior of uranium in an anoxic marine basin. *Uranium, 3*, 145–164.

Anderson, R. T. & Lovley, D. R. (1997). Ecology and biogeochemistry of in situ groundwater bioremediation. In J. G. Jones (Ed), *Advances in Microbial Ecology* (pp. 289–350). New York: Plenum Press.

Anderson, R. T. & Lovley, D. R. (2000). Anaerobic bioremediation of benzene under sulfate reducing conditions in a petroleum-contaminated aquifer. *Environmental Science and Technology, 34*, 2261–2266.

Anderson, R. F., Fleisher, M. Q. & LeHuray, A. P. (1989a). Concentration, oxidation state, and particulate flux of uranium in the Black Sea. *Geochimica et Cosmochimica Acta, 53*, 2215–2224.

Anderson, R. F., LeHuray, A. P., Fleisher, M. Q. & Murray, J. W. (1989b). Uranium deposition in Saanich Inlet sediments, Vancouver Island. *Geochimica et Cosmochimica Acta, 53*, 2205–2213.

Anderson, R. T., Rooney-Varga, J. N., Gaw, C. V. & Lovley, D. R. (1998). Anaerobic benzene oxidation in the Fe(III) reduction zone of petroleum-contaminated aquifers. *Environmental Science and Technology, 32*, 1222–1229.

Baedecker, M. J., Cozzarelli, I. M., Siegel, D. I., Bennett, P. C. & Eganhouse, R. P. (1993). Crude oil in a shallow sand and gravel aquifer: 3. Biogeochemical reactions and mass balance modeling in anoxic ground water. *Applied Geochemistry, 8*, 569–586.

Barnes, C. E. & Cochran, J. K. (1993). Uranium geochemistry in estuarine sediments: controls on removal and release processes. *Geochimica et Cosmochimica Acta, 57*, 555–569.

Bertine, K. K., Chan, L. H. & Turekian, K. K. (1970). Uranium determinations in deep-sea sediments and natural waters using fission tracks. *Geochimica et Cosmochimica Acta, 34*, 641–648.

Bjerg, P. L., Rugge, K., Pedersen, J. K. & Christensen, T. H. (1995). Distribution of redox-sensitive groundwater quality parameters downgradient of a landfill (Grindsted, Denmark). *Environmental Science and Technology, 29*, 1387–1394.

Bonatti, E., Fisher, D. E., Joensuu, O. & Rydell, H. S. (1971). Postdepositional mobility of some transition elements, phosphorus, uranium and thorium in deep sea sediments. *Geochimica et Cosmochimica Acta, 35*, 189–201.

Borden, R. C., Gomez, C. A. & Becker, M. T. (1995). Geochemical indicators of intrinsic bioremediation. *Ground Water, 33*, 180–189.

Bosecker, K. (1997). Bioleaching: metal solubilization by micro-organisms. *FEMS Microbiology Reviews, 20*, 591–604.

Bouwer, E. J. (1992). Bioremediation of organic contaminants in the subsurface. In R. Mitchell (Ed.), *Environmental Microbiology* (pp. 287–318). New York: John Wiley.

Brierley, C. L. (1978). Bacterial leaching. *Critical Reviews in Microbiology, 6*, 207–262.

Brookins, D. G. (1990). Radionuclide behavior at the Oklo nuclear reactor, Gabon. *Waste Management, 10*, 285–296.

Bruynesteyn, A. (1989). Mineral biotechnology. *Journal of Biotechnology, 11*, 1–10.

Cantrell, K. J., Kaplan, D. I. & Wietsma, T. W. (1995). Zero-valent iron for the in situ remediation of selected metals in groundwater. *Journal of Hazardous Materials, 42*, 201–212.

Chapelle, F. H. (1993). *Ground-water Microbiology and Geochemistry*. New York: John Wiley.

Chapelle, F. H. & Lovley, D. R. (1992). Competitive exclusion of sulfate reduction by Fe(III)-reducing bacteria: a mechanism for producing discrete zones of high-iron ground water. *Ground Water, 30*, 29–36.

Coates, J. D., Lonergan, D. J., Jenter, H. & Lovley, D. R. (1996). Isolation of *Geobacter* species from a variety of sedimentary environments. *Applied and Environmental Microbiology, 62*, 1531–1536.

Cochran, J. K., Carey, A. E., Sholkovitz, E. R. & Surprenant, L. D. (1986). The geochemistry of uranium and thorium in coastal marine sediments and sediment pore waters. *Geochimica et Cosmochimica Acta, 50*, 663–680.

Coleman, M. L., Hedrick, D. B., Lovley, D. R., White, D. C. & Pye, K. (1993). Reduction of Fe(III) in sediments by sulphate-reducing bacteria. *Nature, 361*, 436–438.

Colley, S. & Thomson, J. (1985). Recurrent uranium relocations in distal turbidites emplaced in pelagic conditions. *Geochimica et Cosmochimica Acta, 49*, 2339–2348.

DiSpirito, A. A. & Tuovinen, O. H. (1981). Oxygen uptake coupled with uranous sulfate oxidation by *Thiobacillus ferrooxidans* and *T. acidophilus. Geomicrobiology Journal, 2*, 275–291.

DiSpirito, A. A. & Tuovinen, O. H. (1982a). Kinetics of uranous ion and ferrous iron oxidation by *Thiobacillus ferrooxidans. Archives of Microbiology, 133*, 33–37.

DiSpirito, A. A. & Tuovinen, O. H. (1982b). Uranous ion oxidation and carbon dioxide fixation by *Thiobacillus ferrooxidans. Archives of Microbiology, 133*, 28–32.

Ehrlich, H. L. (1990). *Geomicrobiology*. New York: Marcel Dekker, Inc.

Ferguson, J. (1987). The distribution of uranium in space and time. *Uranium, 3*, 131–144.

Finneran, K. T., Nevin, K. P., Anderson, R. T. & Lovley, D. R. (2002). Potential for uranium bioremediation by microbial U(VI) reduction. *Journal of Soil and Sediment Contamination* (in press).

Fowler, T. A., Holmes, P. R. & Crundwell, F. K. (1999). Mechanism of pyrite dissolution in the presence of *Thiobacillus ferrooxidans. Applied and Environmental Microbiology, 65*, 2987–2993.

Francis, A. J. (1990). Microbial dissolution and stabilization of toxic metals and radionuclides in mixed wastes. *Experientia, 46*, 840–851.

Francis, A. J. & Dodge, C. J. (1998). Remediation of soils and wastes contaminated with uranium and toxic metals. *Environmental Science and Technology, 32*, 3993–3998.

Francis, A. J., Dodge, C. J., Gillow, J. B. & Papenguth, H. W. (2000). Biotransformation of uranium compounds in high ionic strength brine by a halophilic bacterium under denitrifying conditions. *Environmental Science and Technology, 34*, 2311–2317.

Froelich, P. N., Klinkhammer, G. P., Bender, M. L., Luedtke, N. A., Heath, G. R., Cullen, D., Dauphin, P., Hammond, D., Hartman, B. & Maynard, V. (1979). Early oxidation of organic matter in pelagic sediments of the eastern equatorial Atlantic: suboxic diagenesis. *Geochimica et Cosmochimica Acta., 43*, 1075–1090.

Ganesh, R., Robinson, K. G., Reed, G. D. & Sayler, G. S. (1997). Reduction of hexavalent uranium from organic complexes by sulfate and iron-reducing bacteria. *Applied and Environmental Microbiology, 63*, 4385–4391.

Ganesh, R., Robinson, K. G., Chu, L., Kucsmas, D. & Reed, G. D. (1999). Reductive precipitation of uranium by *Desulfovibrio desulfuricans*: evaluation of cocontaminant effects and selective removal. *Water Research, 33*, 3447–3458.

Gorby, Y. A. & Lovley, D. R. (1992). Enzymatic uranium precipitation. *Environmental Science and Technology, 26*, 205–207.

Gu, B., Liang, L., Dickey, M. J., Yin, X. & Dai, S. (1998). Reductive precipitation of uranium(VI) by zero-valent iron. *Environmental Science and Technology, 32*, 3366–3373.

Hostetler, P. B. & Garrels, R. M. (1962). Transportation and precipitation of uranium and vanadium at low temperatures with special reference to sandstone-type uranium. *Economic Geology, 57*, 137–167.

Hutchins, S. R., Davidson, M. S., Brierley, J. A. & Brierley, C. L. (1986). Micro-organisms in reclamation of metals. *Annual Reviews of Microbiology, 40*, 311–336.

Hutchins, S. R., Miller, D. E. & Thomas, A. (1998). Combined laboratory/field study on the use of nitrate for in situ bioremediation of a fuel-contaminated aquifer. *Environmental Science and Technology, 32*, 1832–1840.

Jensen, M. L. (1958). Sulfur isotopes and the origin of sandstone-type uranium deposits. *Economic Geology, 53*, 598–616.

Kashefi, K. & Lovley, D. R. (2000). Reduction of Fe(III), Mn(IV), and toxic metals at 100°C by *Pyrobaculum islandicum. Applied and Environmental Microbiology, 66*, 1050–1056.

Kieft, T. L., Frederickson, J. K., Onstott, T. C., Gorby, Y. A., Kostandarithes, H. M., Bailey, T. J., Kennedy, D. W., Li, S. W., Plymale, A. E., Spadoni, C. M. & Gray, M. S. (1999). Dissimilatory reduction of Fe(III) and other electron acceptors by a *Thermus* isolate. *Applied and Environmental Microbiology, 65*, 1214–1221.

Klinkhammer, G. P. & Palmer, M. R. (1991). Uranium in the oceans: where it goes and why. *Geochimica et Cosmochimica Acta, 55*, 1799–1806.

Landais, P., Connan, J., Dereppe, J. M., George, E., Meunier, J. D., Monthioux, M., Pagel, M., Pironon, J. & Poty, B. (1987). Alterations of organic matter: a clue for uranium ore genesis. *Uranium, 3*, 307–342.

Langmuir, D. (1978). Uranium solution-mineral equilibria at low temperatures with applications to sedimentary ore deposits. *Geochimica et Cosmochimica Acta, 42*, 547–569.

Lloyd, J. R. & Macaskie, L. E. (2000). Bioremediation of radionuclide-containing wastewaters. In D. R. Lovley (Ed.), *Environmental Microbe-Metal Interactions* (pp. 277–327). Washington, DC: ASM Press.

Lovley, D. R. (1995a). Bioremediation of organic and metal contaminants with dissimilatory metal reduction. *Journal of Industrial Microbiology, 14*, 85–93.

Lovley, D. R. (1995b). Microbial reduction of iron, manganese, and other metals. *Advances in Agronomy, 54*, 175–231.

Lovley, D. R. (1997). Potential for anaerobic bioremediation of BTEX in petroleum-contaminated aquifers. *Journal of Industrial Microbiology, 18*, 75–81.

Lovley, D. R. (2000). *Environmental Microbe-Metal Interactions* (405pp). Washington, DC: ASM Press.

Lovley, D. R. & Chapelle, F. H. (1995). Deep subsurface microbial processes. *Reviews of Geophysics, 33*, 365–381.

Lovley, D. R. & Coates, J. D. (1997). Bioremediation of metal contamination. *Current Opinion in Biotechnology, 8*, 285–289.

Lovley, D. R. & Goodwin, S. (1988). Hydrogen concentrations as an indicator of the predominant terminal electron-accepting reactions in aquatic sediments. *Geochimica et Cosmochimica Acta, 52*, 2993–3003.

Lovley, D. R. & Klug, M. J. (1983). Sulfate reducers can out compete methanogens at freshwater sulfate concentrations. *Applied and Environmental Microbiology, 45*, 187–192.

Lovley, D. R. & Phillips, E. J. P. (1987). Competitive mechanisms for inhibition of sulfate reduction and methane production in the zone of ferric iron reduction in sediments. *Applied and Environmental Microbiology, 53*, 2636–2641.

Lovley, D. R. & Phillips, E. J. P. (1992a). Bioremediation of uranium contamination with enzymatic uranium reduction. *Environmental Science and Technology, 26*, 2228–2234.

Lovley, D. R. & Phillips, E. J. P. (1992b). Reduction of uranium by *Desulfovibrio desulfuricans. Applied and Environmental Microbiology, 58*, 850–856.

Lovley, D. R., Phillips, E. J. P., Gorby, Y. A. & Landa, E. R. (1991). Microbial reduction of uranium. *Nature, 350*, 413–416.

Lovley, D. R., Giovannoni, S. J., White, D. C., Champine, J. E., Phillips, E. J. P., Gorby, Y. A. & Goodwin, S. (1993a). *Geobacter metallireducens* gen. nov. sp. nov., a micro-organism capable of coupling the complete oxidation of organic compounds to the reduction of iron and other metals. *Archives of Microbiology, 159*, 336–344.

Lovley, D. R., Roden, E. E., Phillips, E. J. P. & Woodward, J. C. (1993b). Enzymatic iron and uranium reduction by sulfate reducing bacteria. *Marine Geology, 113*, 41–53.

Lovley, D. R., Widman, P. K., Woodward, J. C. & Phillips, E. J. P. (1993c). Reduction of uranium by cytochrome c_3 of *Desulfovibrio vulgaris. Applied and Environmental Microbiology, 59*, 3572–3576.

Lovley, D. R., Chapelle, F. H. & Woodward, J. C. (1994). Use of dissolved H_2 concentrations to determine the distribution of microbially catalyzed redox reactions in anoxic ground water. *Environmental Science and Technology, 28*, 1205–1210.

Lovley, D. R., Coates, J. D., Blunt-Harris, E. L., Phillips, E. J. P. & Woodward, J. C. (1996). Humic substances as electron acceptors for microbial respiration. *Nature, 382*, 445–448.

Lovley, D. R., Fraga, J. L., Blunt-Harris, E. L., Hayes, L. A., Phillips, E. J. P. & Coates, J. D. (1998). Humic substances as a mediator for microbially catalyzed metal reduction. *Acta Hydrochimica et Hydrobiologica, 26*, 152–157.

Lundgren, D. G. (1980). Ore leaching by bacteria. *Annual Reviews of Microbiology, 34*, 263–283.

Lyngkilde, J. & Christensen, T. H. (1992). Redox zones of a landfill leachate pollution plume (Vejen, Denmark). *Journal of Contaminant Hydrology, 10*, 273–289.

Macaskie, L. E. (1991). The application of biotechnology to the treatment of wastes produced from the nuclear fuel cycle: biodegradation and bioaccumulation as a means of treating radionuclide-containing streams. *Critical Reviews in Biotechnology, 11*, 41–112.

Maynard, J. B. (1983). *Geochemistry of Sedimentary Ore Deposits*. New York: Springer-Verlag.

McCarty, P. L. (1972). Energetics of organic matter degradation. In R. Mitchell (Ed.), *Water Pollution Microbiology* (pp. 91–118). New York: John Wiley.

Mohagheghi, A., Updegraff, D. M. & Goldhaber, M. B. (1985). The role of sulfate reducing bacteria in the deposition of sedimentary uranium ores. *Geomicrobiology Journal, 4*, 153–173.

Muñoz, J. A., Gonzalez, F., Blazquez, M. L. & Ballester, A. (1995). A study of the bioleaching of a Spanish uranium ore. Part I: a review of the bacterial leaching in the treatment of uranium ores. *Hydrometallurgy, 38*, 39–57.

Nakashima, S., Disnar, J. R., Perruchot, A. & Trichet, J. (1984). Experimental study of mechanisms of fixation and reduction of uranium by sedimentary organic matter under diagenetic or hydrothermal conditions. *Geochimica et Cosmochimica Acta, 48*, 2321–2329.

Nakashima, S., Disnar, J.-R. & Perruchot, A. (1999). Precipitation kinetics of uranium by sedimentary organic matter under diagenetic and hydrothermal conditions. *Economic Geology, 94*, 993–1006.

Nevin, K. P. & Lovley, D. R. (2000). Potential for nonenzymatic reduction of Fe(III) during microbial oxidation of organic matter coupled to Fe(III) reduction. *Environmental Science and Technology, 34*, 2472–2478.

Panak, P., Hard, B. C., Pietzsch, K., Kutschke, S., Roske, K., Selenska-Pobell, S., Bernhard, G. & Nitsche, H. (1998). Bacteria from uranium mining waste pile: interactions with U(VI). *Journal of Alloys and Compounds, 271-273*, 262–266.

Patterson, B. M., Pribac, F., Barber, C., Davis, G. B. & Gibbs, R. (1993). Biodegradation and retardation of PCE and BTEX compounds in aquifer material from Western Australia using large-scale columns. *Journal of Contaminant Hydrology, 14*, 261–278.

Phillips, E. J. P., Lovley, D. R. & Landa, E. R. (1995). Remediation of uranium contaminated soils with bicarbonate extraction and microbial U(VI) reduction. *Journal of Industrial Microbiology, 14*, 203–207.

Plant, J. A., Simpson, P. R., Smith, B. & Windley, B. F. (1999). Uranium ore deposits–products of the radioactive Earth. In P. C. Burns & R. Finch (Eds), *Uranium: Mineralogy, Geochemistry and the Environment* (pp. 255–319). Washington, DC: Mineralogical Society of America.

Ponnamperuma, F. N. (1972). The chemistry of submerged soils. *Advances in Agronomy, 24*, 29–96.

Rawlings, D. E. & Silver, S. (1995). Mining with microbes. *Biotechnology, 13*, 773–778.

Reeburgh, W. S. (1983). Rates of biogeochemical processes in anoxic sediments. *Annual Reviews of Earth and Planetary Sciences, 11*, 269–298.

Reinhard, M., Shang, S., Kitanidis, P. K., Orwin, E., Hopkins, G. D. & Lebron, C. A. (1997). In situ BTEX biotransformation under enhanced nitrate- and sulfate reducing conditions. *Environmental Science and Technology, 31*, 28–36.

Robinson, K. G., Ganesh, R. & Reed, G. D. (1998). Impact of organic ligands on uranium removal during anaerobic biological treatment. *Water Science and Technology, 37*, 73–80.

Rooney-Varga, J., Anderson, R. T., Fraga, J. L., Ringleberg, D. & Lovley, D. R. (1999). Microbial communities associated with anaerobic benzene degradation in a petroleum- contaminated aquifer. *Applied and Environmental Microbiology, 65*, 3056–3063.

Rugge, K., Bjerg, P. L. & Christensen, T. H. (1995). Distribution of organic compounds from municipal solid waste in the groundwater downgradient of a landfill (Grindsted, Denmark). *Environmental Science and Technology, 29*, 1395–1400.

Schippers, A. & Sand, W. (1999). Bacterial leaching of metal sulfides proceeds by two indirect mechanisms *via* thiosulfate or *via* polysulfides or sulfur. *Applied and Environmental Microbiology, 65*, 319–321.

Schippers, A., Hallmann, R., Wentzien, S. & Sand, W. (1995). Microbial diversity in uranium mine waste heaps. *Applied and Environmental Microbiology, 61*, 2930–2935.

Scott, D. T., McKnight, D. M., Blunt-Harris, E. L., Kolesar, S. E. & Lovley, D. R. (1998). Quinone moieties act as electron acceptors in the reduction of humic substances by humics-reducing micro-organisms. *Environmental Science and Technology, 32*, 2984–2989.

Snoeyenbos-West, O. L., Nevin, K. P., Anderson, R. T. & Lovley, D. R. (2000). Enrichment of Geobacters species in response to stimulation of Fe(III) reduction in sandy aquifer sediments. *Microbial Ecology, 39*, 153–167.

Spear, J. R., Figueroa, L. A. & Honeyman, B. D. (1999). Modeling the removal of uranium U(VI) from aqueous solutions in the presence of sulfate reducing bacteria. *Environmental Science and Technology, 33*, 2667–2675.

Stumm, W. & Morgan, J. J. (1981). *Aquatic Chemistry*. New York: John Wiley.

Suzuki, Y. & Banfield, J. F. (1999). Geomicrobiology of uranium. In P. C. Burns & R. Finch (Eds), *Uranium: Mineralogy, Geochemistry and the Environment* (pp. 393–432). Washington, DC: Mineralogical Society of America.

Tebo, B. M. & Obraztsova, A. Y. (1998). Sulfate reducing bacterium grows with Cr(VI), U(VI), Mn(VI), and Fe(III) as electron acceptors. *FEMS Microbiology Letters, 162*, 193–198.

Thomas, R. A. P. & Macaskie, L. E. (1996). Biodegradation of tributyl phosphate by naturally occurring microbial isolates and coupling to the removal of uranium from aqueous solution. *Environmental Science and Technology, 30*, 2371–2375.

Truex, M. J., Peyton, B. M., Valentine, N. B. & Gorby, Y. A. (1997). Kinetics of U(VI) reduction by a dissimilatory Fe(III)-reducing bacterium under non-growth conditions. *Biotechnology and Bioengineering, 55*, 490–496.

Tucker, M. D., Barton, L. L. & Thomson, B. M. (1996). Kinetic coefficients for simultaneous reduction of sulfate and uranium by *Desulfovibrio desulfuricans. Applied Microbiology and Biotechnology, 46*, 74–77.

Tucker, M. D., Barton, L. L. & Thomson, B. M. (1998). Removal of U and Mo from water by immobilized *Desulfovibrio desulfuricans* in column reactors. *Biotechnology and Bioengineering, 60*, 88–96.

Tuovinen, O. H. & Kelly, D. P. (1974). Studies on the growth of *Thiobacillus ferrooxidans* II. Toxicity of uranium to growing cultures and tolerance conferred by mutation, other metal cations and EDTA. *Archives of Microbiology, 95*, 153–164.

Vroblesky, D., A. & Chapelle, F., H. (1994). Temporal and spatial changes of terminal electron-accepting processes in a petroleum hydrocarbon-contaminated aquifer and the significance for contaminant biodegradation. *Water Resources Research, 30*, 1561–1570.

Miranda J. Keith-Roach and Francis R. Livens (Editors)
© 2002 Elsevier Science Ltd. All rights reserved

Chapter 8

Diversity and activity of bacteria in uranium waste piles

Sonja Selenska-Pobell

Institute of Radiochemistry, Forschungszentrum Rossendorf, Bautzner Landstraße 128, 01314 Dresden, Germany

1. Introduction

The pollution of the environment with toxic metals is one of the most severe problems of our industrial age. Uranium mining waste piles are a subject of particular attention, because in the soils, sediments and drainage waters of these environments, significant amounts of many hazardous metals are present, such as uranium, selenium, molybdenum, arsenic, cadmium, chromium, mercury, lead, copper, nickel and zinc (Francis, 1990). In addition, significant amounts of thorium, radium, polonium and other decay products may also be present in the so-called uranium 'mill tailings' where the extraction of uranium from the ores was performed.

Even in the most heavily polluted uranium wastes, large numbers of bacteria are present (Cerda et al., 1993; Goebel & Stackebrandt, 1994; Shippers et al., 1994). Moreover, different groups of bacteria can interact in different ways with metals and radionuclides (see Chapters 3, 6 and 7, this volume). Some of the most important mechanisms by which bacteria can biotransform and influence the mobilisation and/or immobilisation of metals are listed below:

(i) Direct oxidation and/or reduction of metals, which affect their solubility (DiSpirito & Tuovinen, 1982; Lovley, 1993; Nelson et al., 1999; Sharma et al., 2000; Wildung et al., 2000);
(ii) Direct or indirect oxidation of metal sulfides and the associated solubilisation of certain elements (Bosecker, 1997; Krebs & Brandl, 1997);
(iii) Indirect alteration of metal speciation caused by microbially-induced pH and Eh changes in the medium (Bosecker, 1997; Bacelar-Nicolau & Johnson, 1999);
(iv) Bioaccumulation [biosorption by cell surface polymers (DiSpirito et al., 1983; Macaskie et al., 1992; Valentive et al., 1996; Douglas & Beveridge, 1998; Panak et al., 1999; Selenska-Pobell et al., 1999) and/or uptake of metals inside the cells (Marques et al., 1991; Purchase et al., 1997; Klaus et al., 1999)];

(v) biomineralisation, including induction of metal precipitation by specific metabolic functions and the consequent generation of minerals (Francis, 1998; Brown & Beveridge, 1998; Douglas & Beveridge, 1998); and
(vi) Release of biosorbed metals by chelation, alkylation, or decomposition (Francis, 1990; Bosecker, 1997; Francis et al., 1998).

It is clear that the bacterial activities described above strongly influence the fate and migration of toxic metals in and outside the sites where uranium mining has been performed. In addition to the living cells, significant amounts of different bacterial metabolites are present in the wastes and these also interact with the heavy metals and influence their behaviour. For these reasons, knowledge of the diversity and activity of the indigenous bacteria in the uranium waste piles is of fundamental importance for understanding the biogeochemical processes occurring in these environments and especially for modelling the migration of the heavy metals and radionuclides.

Until recently, information about bacterial diversity in uranium waste piles was limited to several studies dealing with cultivable bacteria (Francis et al., 1991; Cerda et al., 1993; Shippers et al., 1994; Goebel & Stackebrandt, 1994; Berthelot et al., 1997). This, however, is not sufficient because only a few percent of natural bacterial populations can be cultured and studied in the laboratory (see Chapter 1, this volume) due to our limited knowledge of the nutrient requirements and other life necessities for most bacterial species in nature (Ward et al., 1990; Pace, 1997; Service, 1997). The development and application of molecular approaches in bacterial ecology during the last decade has revealed a tremendous prokaryotic diversity which was overlooked by traditional culture enrichment techniques (Pace, 1997; Service, 1997; Chandler et al., 1997; Byers et al., 1998; Dojka et al., 2000).

One of these approaches, 16S rDNA retrieval, has been applied to analysis of bacterial diversity in deep granitic rocks and in an aquifer around a nuclear fuel repository in Sweden (see Chapter 10, this volume; Pedersen, 1997; Kotelnikova & Pedersen, 1998). The geological, chemical and biological composition of such environments are, however, completely different from those of the uranium mining wastes which will be discussed in this chapter. In addition, in contrast to uranium wastes, which are open and may easily exchange material with the wider environment, nuclear fuel repositories are very well protected from contact with the outside environment (Pedersen, 1997; Stroes-Gascoyne & West, 1997; Kieft et al., 1997; Kotelnikova & Pedersen, 1998).

We have recently published some results on direct molecular analysis of bacterial diversity in soil and water samples drawn from several uranium wastes in Germany and in the USA (Satschanska et al., 1999; Radeva & Selenska-Pobell, 2000; Radeva et al., 2000; Selenska-Pobell et al., 2000, 2001).

In the following chapter we will present:

(1) An analysis of the diversity of the natural bacterial communities in several uranium wastes of different geographical and geological origin in which the following molecular approaches have been applied: (a) Repetitive primer amplified polymorphic DNA (Rep-APD) using bacterial specific primers; (b) 16S rDNA retrieval; and (c) Ribosomal intergenic spacer amplification (RISA) retrieval.

(2) Measures of bacterial diversity in the uranium wastes studied, based on classical *culture-dependent* methods. A comparison of the problems and advantages of the *culture-dependent* and the *culture-independent*, direct molecular approaches mentioned above, will

demonstrate that both techniques should be used in parallel to gain an understanding of the composition and diversity of bacterial communities in uranium wastes highly polluted with radionuclides and other metals.

(3) Studies on the interactions of several bacterial isolates from the uranium mining wastes with uranium and other metals.

2. Description of the soil and water samples analysed

Forty-four soil and 21 water samples were analysed, drawn from the following wastes: two uranium mining depository waste piles, Haberlandhalde near the town of Johanngeorgenstadt and Deponie B1 in Germany; and three uranium mill tailings, two of them, Gittersee/Coschütz and Schlema/Alberoda, in Germany, and one, Shiprock, New Mexico, in the USA.

The soil samples were drawn from different sites and depths (up to 5 m below the surface) of the wastes, whereas all water samples were taken from an equal depth of 30 m below the surface of the uranium wastes studied. The soil samples were taken from the corer under sterile conditions and stored frozen at −20°C before analysis. The water samples were transported to the laboratory in sterile flasks packed in ice and immediately filtered through a glass-fibre pre-filter and a 0.2 μm nitrocellulose membrane filter. The material on both filters was stored at −20°C for further analysis.

The concentrations of uranium in the soil samples varied between 18 and 178 mg kg^{-1}. In addition, the concentrations of other metals were found to be: Al – from 5 to 17 g kg^{-1}; Cr – 10 to 20 mg kg^{-1}; Fe – 19 to 225 g kg^{-1}; Mn – 0.5 to 1.25 g kg^{-1}; Co – 10 to 60 mg kg^{-1}; Ni – 15 to 125 mg kg^{-1}; Cu – 20 to 165 mg kg^{-1}; Zn – 90 to 455 mg kg^{-1}; Sr – 3 to 19 mg kg^{-1}; Pb – 15 to 82 mg kg^{-1}; Th – 5 to 10 mg kg^{-1}. The water samples contained up to 5 mg l^{-1} U accompanied by widely varying levels of nitrate and sulfate.

3. Direct analysis of bacterial diversity in the uranium waste piles through the use of Rep-APD

Studies of natural bacterial populations, especially from extreme environments such as uranium mining wastes, which are heavily polluted with toxic metals and radionuclides, are difficult to carry out due to inadequate knowledge of most of the bacterial species present there, and of methods to culture and handle them. The inability to culture bacteria from different natural environments has been overcome to a large extent by direct, culture-independent analyses of nucleic acids (see Chapter 2, this volume; Ward et al., 1990; Selenska & Klingmüller, 1991; Pace, 1997; Service, 1997; Borneman & Triplet, 1997; Chandler et al., 1997; Byers et al., 1998; Acinas et al., 1999; Yanagibayashi et al., 1999; Selenska-Pobell et al., 2000, 2001).

One of the easiest molecular methods for monitoring complex microbial communities is random amplified polymorphic DNA PCR (Polymerase chain reaction) fingerprinting (RAPD), which uses arbitrary primers (Wikström et al., 1999; Franklin et al., 2000; Yang et al., 2000). These primers, however, cannot discriminate between bacterial and

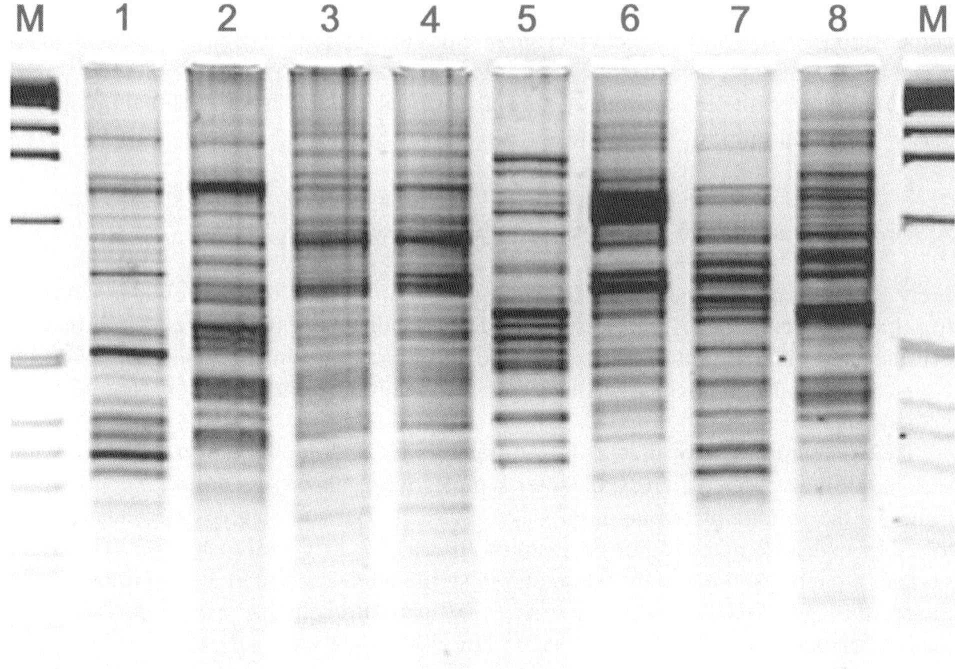

Fig. 1. ERIC-APD fingerprinting of the water DNA samples recovered from: (1) Schlema/Alberoda, well 296.I; (2) Schlema/Alberoda, well 296.II; (3) and (4), parallel DNA extractions from Gittersee/Coschütz, well 33; (5) Gittersee/Coschütz, well 10; (6) Deponie B1, well GB1; (7) Shiprock New Mexico, well UMS1-853; (8) Shiprock New Mexico, well UMS1-819. M – Marker, BRL-ladder kbl plus.

the eukaryotic or archaeal parts of the natural microbial communities. In order to monitor bacteria more precisely, primers such as BOX, eubacterial repetitive consensus sequences (ERIC), or other bacteria-specific primers as described by Versalovic et al. (1994) and Selenska-Pobell et al. (1998, 2000) may be used, instead of random primers.

One example of such Rep-APD fingerprinting is presented in Fig. 1. This analysis was performed using an ERIC primer (see Table 1) for PCR amplification of total DNA which was recovered by direct lysis (Kampf & Selenska-Pobell, 1998) from the drain waters of uranium wastes. As shown in Fig. 1, the ERIC patterns are sample-specific and indicate significant differences in the composition of the bacterial communities. There are very few DNA fragments in the ERIC patterns which are shared by the samples of different geographic origins. The very few common bands may possibly represent some bacterial groups common to the samples studied. However, from the results presented in Fig. 1 it is clear that the ERIC fingerprints of the samples drawn from different sites of the same environment do not show much similarity. This may reflect the varying degree of contamination with heavy metals which could strongly influence the structure of the bacterial communities. For example, sample WP33 of the Gittersee/Coschütz mill tailing (Fig. 1,

Table 1
Primers used in the PCR amplifications and sequence analysis

Rep-APD primer	Sequence	Target site*
ERIC	5′-AAGTAAGTGACTGGGGTGAGCG-3′	
RISA primer	**Sequence**	**Target site***
RISA1-$16S_{968F}$	5′-ACGCGNARAACCTTAC-3′	16S $rDNA_{968-983}$
RISA2-$23S_{130R}$	5′-GGGTTNCCCCATTCGG-3′	23S $rDNA_{130-115}$
16S rDNA primer	**Sequence**	**Target site***
$16S_{7F}$**	5′-AAGAGTTTGATYMTGGCTCAG-3′	16S $rDNA_{8-27}$
$16S_{43F}$**	5′-CAGGCCTAACAAATGCAAGTC-3′	16S $rDNA_{43-63}$
$16S_{968F}$**	5′-ACGCGNARAACCTTAC-3′	16S $rDNA_{968-983}$
$16S_{1404R}$**	5′-GGGCGGWGTGTACAAGGC-3′	16S $rDNA_{1404-1387}$
$16S_{1513R}$**	5′-TACGG(CT)TACCTTGTTACGACTT-3′	16S $rDNA_{1513-1492}$

* Target sites correspond to the *E. coli* numbering system (Brosius et al., 1981).
** Used also as sequencing primers.

lanes 3 and 4) was one of the most polluted with U and other metals, whereas the sample WP10 (Fig. 1, Lane 5) was almost two times less polluted.

Interestingly, even the ERIC patterns which were generated by amplification of DNA samples recovered in two parallel extractions from the same drain water portion show small differences (see lanes 3 and 4 in Fig. 1). A slightly higher degree of diversity was found in parallel soil samples drawn from the uranium mining waste pile near the town of Johanngeorgenstadt (Selenska-Pobell et al., 2001). These results are consistent with the general observation that bacteria are very inhomogeneously distributed in nature, especially in soil environments (Pace, 1997; Service, 1997; Borneman & Triplet, 1997).

The Rep-APD analysis has also been successfully applied to monitor seasonal changes in the bacterial communities of uranium mining waste piles (Selenska-Pobell et al., 2001). More generally, results from the Rep-APD analysis of soil and water samples of the uranium wastes demonstrate that in these extreme environments, which are highly polluted with toxic metals and radionuclides, bacterial communities are extremely diverse and dynamic. This is in contrast to the previous suggestion that only a limited number of particular bacterial groups predominate in those environments (Cerda et al., 1993; Goebel & Stackebrandt, 1994; Shippers et al., 1994). The Rep-APD analysis is an easily performed and informative method because it derives information from many different parts of the bacterial genomes (Versalovic et al., 1994). However, it is very difficult on the basis of this analysis to identify bacteria present in the community at species level, because of our limited knowledge of the sequences surrounding the clusters of the repetitive sequences. At present, this analysis can therefore be used mainly as an indicator for the heterogeneity and dynamics of natural bacterial communities, but without any exact phylogenetic affiliation.

4. Direct analysis of bacterial diversity in the uranium wastes using rDNA retrieval

Retrieval of 16S rDNA has been the main and most powerful tool for microbial phylogenetic studies in environmental samples during the last decade. The method is based on PCR amplification of 16S rDNA fragments in total DNA recovered from solid or liquid environmental samples by direct lysis. The PCR amplification products obtained are then usually characterised by cloning and comparative sequence analysis (Chandler et al., 1997; Byers et al., 1998; Yanagibayashi et al., 1999).

Using this approach, an unexpected complexity and diversity of natural bacterial communities has been found, not only in agricultural fields and other benign environments, but also in all the extreme natural environments studied so far, for instance in deep terrestrial samples (Boivin-Jahns et al., 1996), in deep sea sediments and aquifers (Pedersen, 1997; Kato et al. 1998; Yanagibayashi et al., 1999), in hot springs (Hugenholtz et al., 1999) and in Antarctic ice (Bowman et al., 1996). Using 16S rDNA retrieval, a large variety of relatives to known bacterial groups were also identified in environments which are severely polluted with organic and inorganic toxic matter as a result of industrial activities (Dojka et al., 1999; Sandaa et al., 1999; Roane & Pepper, 2000). Another elegant rDNA approach for direct molecular studies of environmental biodiversity is Ribosomal Intergenic Spacer Amplification (RISA) retrieval, which involves about one-third of the 16S rDNA, a short part of the 23S rDNA, and the intergenic spacer between the two genes (Acinas et al., 1999).

Current information about the composition of the natural bacterial communities in the uranium wastes, which has been obtained in our studies using both the 16S rDNA and the RISA approaches, is summarised in Table 2. To our knowledge, there is no other information published on molecular analysis of phylogenetic bacterial diversity in these extreme environments. The first results of this analysis were presented by Satchanska et al. (1999) and recently published by Selenska-Pobell et al. (2001). However, most of the data of Table 2 are not yet available in the literature.

In our initial work, several hundred full length clones of a 16S $\text{rDNA}_{7F-1492R}$ clone library (referred to as 16S rDNA I in the Table 2), corresponding to different soil samples of the uranium mining waste pile Johanngeorgenstadt and of the uranium mill tailing Gittersee/Coschütz, were categorised using restriction fragment length polymorphism (RFLP) analysis which demonstrated the presence of a large variety of 16S rDNA RFLP types. A sequence analysis of several of the clones which were found to be common for the two wastes has revealed that the *Pseudomonas* group of the γ-Proteobacteria, *Aeromonas*, *Holophaga*, some green sulfur and green non-sulfur bacteria, as well as one particular eco-type of *Acidithiobacillus ferrooxidans* are present in the two uranium wastes studied (see '16S rDNA I' in Table 2). However, the number of the individual RFLP types for each sample was very large, suggesting an extremely high phylogenetic diversity and an inhomogeneous distribution of different bacterial species in the soils of the Johanngeorgenstadt and the Gittersee/Coschütz uranium mining wastes (Satchanska et al., 1998; Selenska-Pobell et al., 2001).

Our recent results, also presented in Table 2, from the analysis of two alternative clone libraries constructed for a large number of soil and water samples drawn from

Table 2
Affiliation of the 16S rDNA and RISA clones of soil and water samples from the uranium waste piles studied

Clone name	Accession No. EMBL	No. of clones	Closest phylogenetic relative (EMBL No.)	BLAST % of similarity*	Clone library	Geographic origin
			α-Proteobacteria			
GR-WP33-45	AJ296550	2	*Rhodobacter azotoformans* SA16 (D70847)	94	16S II	Gittersee, Germany **W*****
GR-Sh1-107	AJ296551	8	*Rhodobacter sphaeroides* 2.4.1 (X53855)	95	16S II	Shiprock, NM, USA **W**
GR-296.II. 89	AJ296549	5	Petroleum-degrading bact. HD-1 (D45202)	88	16S II	Schlema, Germany **W**
			Uncult. bact. WCHB1-55 (AF050531)	88	16S II	
GR-WP33-3	AJ296548	2	Uncult. bact., clone DA122 (Y12598)	92	16S II	Gittersee, Germany **W**
			Bradyrhizobium. sp. ApB16 (AF178434)	91		
GR-296.II.111	AJ296547	7	Uncult. bact. clone DA122 (Y12598)	94	16S II	Schlema, Germany **W**
			Bradyrhizobium sp. Des.B1 (AF178436)	91		
KF-JG30-B3	AJ295650	2	Uncult. bact. clone BSV56 (AJ229208)	91	16S II	Johanngeorgenstadt, Germany **S**
			Rasbo bacterium (AF 007948)	91		
KF-JG30-B10	AJ295646	3	Uncult. bact. DA111 (Y12596)	95	16S II	Johanngeorgenstadt, Germany **S**
				88		
KF-JG30-B12	AJ295647	1	Uncul . bact. DA111 (Y12596)	95	16S II	Johanngeorgenstadt, Germany **S**
				88		
KF-JG30-C23	AJ295648	1	Uncult. bact. DA111 (Y12596)	95	16S II	Johanngeorgenstadt, Germany **S**
				88		
			β-Proteobacteria			
GR-Sh2-23	AJ296605	60	*Nitrosomonas* sp. isolate JL21 (AB000700)	97	RISA	Shiprock, NM, USA **W**
GR-Sh2-24	AJ296606	27	*Nitrosomonas* sp. isolate AL212 (AB000699)	97	RISA	Shiprock, NM, USA **W**
GR-WP33-36	AJ296553	3	Uncult. Antarctic bact. LB3-81 (AF173823)	92	16S II	Gittersee, Germany **W**
			Burkholderia caryophylli ATCC25418 (X67039)	90		
KF-Gitt2-40	AJ295643	41	Uncult. bact. 5Z-C1 (AJ224618)	99	16S II	Gittersee, Germany **S**
			Thiobacillus denitrificans NCIMB 9548 (AJ243144)	98		
GR-296.II.77	AJ296552	7	Uncult. bact. Rhodocyclus gr. HMD444 (AB015328)	88	16S II	Schlema, Germany **W**
			Zooglea ramigera ATCC 19544 (X74913)	87		

Table 2
(continued)

Clone name	Accession No. EMBL	No. of clones	Closest phylogenetic relative (EMBL No.)	BLAST % of similarity*	Clone library	Geographic origin
			γ-Proteobacteria			
GR-WP33-28	AJ296558	8	*Pseudomonas stutzeri* DSM50238 (U26416)	98	16S II	Gittersee, Germany **W**
GR-Sh2-34	AJ296559	74	*Pseudomonas stutzeri* DNSP21 (U26414)	95	16S II	Shiprock, NM, USA **W**
GR-Sh1-136	AJ296561	36	*Pseudomonas migulae* CIP105470 (AF074383)	96	16S II	Shiprock, NM, USA **W**
GR-296.II.35	AJ296562	4	*Pseudomonas gessardii* CIP105469 (AF074384)	99	16S II	Schlema, Germany **W**
KF/GS-Gitt2-53	AJ295645	2	*Pseudomonas anguilliseptica* 1123/5 (X99541)	97	16S I	Gittersee, Germany **S**
		3	*Pseudomonas anguilliseptica* 1123/5 (X99541)**			Johanngeorgenstadt, Germany **S**
GR-Sh1-163	AJ296560	4	*Pseudomonas flavescens* B62 (U01916)	96	16S II	Shiprock, NM, USA **W**
KF/GS-Gitt2-41	AJ295644	2	*Pseudomonas* sp. clone NB0.1-H (AB013843)	99	16S I	Gittersee, Germany **S**
KF/GS-JG36-13	AJ295653	2	*Pseudomonas* sp. clone NB0.1-H (AB013843)	99	16S I	Gittersee, Germany **S**
		4	*Pseudomonas* sp. clone NB0.1-H (AB013843)**			Schlema, Germany **W**
KF-JG30-B15	AJ295651	4	Uncult. soil bact. SC-I-16 (AJ252619)	92	16S II	Johanngeorgenstadt, Germany **S**
GR-Sh1-26	AJ296564	31	*Frauteuria aurantia* DSM6220 (AJ010481)	93	16S II	Shiprock, NM, USA **W**
GR-Sh1-148	AJ296563	32	*Frauteuria aurantia* DSM6220 (AJ010481	93	16S II	Shiprock, NM, USA **W**
GR-B1 -2-33	AJ296557	15	*Acinetobacter lwoffii* DSM 2403 (X81665)	98	16S II	Deponie B1, Germany **W**
GR-B1-2-35	AJ296556	1	*Acinetobacter lwoffii* DSM 2403 (X81665)	96	16S II	Deponie B1, Germany **W**
GR-B1 -2-4	AJ296554	11	*Acinetobacter lwoffii* DSM 2403 (X81665)	97	16S II	Deponie B1, Germany **W**
GR-B1 -2-39	AJ296555	2	*Acinetobacter lwoffii* DSM 2403 (X81665)	97	16S II	Deponie B1, Germany **W**
GR-Sh1-128	AJ296567	6	*Marinobacter* sp. NK-1 (AB026946)	97	16S II	Shiprock, NM, USA **W**
GR-WP33-34	AJ296566	3	*Shewanella* sp. ML-S3 (AF140017)	97	16S II	Gittersee, Germany **W**
GR-WP33-14	AJ296565	2	*Aeromonas salmonicida* (AF200392)	97	16S II	Gittersee, Germany **W**
KF/GS-JG36-20	AJ295654	3	*Aeromonas bestiarum* CIP 4730 (X60406)	99	16S I	Johanngeorgenstadt, Germany **S**
KF/GS-JG36-22	AJ295655	5	*Acidithiobacillus ferrooxidans* ATCC23270^T (AJ278718)	99	16S I	Johanngeorgenstadt, Germany **S**
		3	*Acidithiobacillus ferrooxidans* ATCC23270^T (AJ278718)**			Gittersee, Germany **S**
GR-WP59-2	AJ296607	2	*Acidithiobacillus ferrooxidans* NASF-1 (AB039820)	95	RISA	Gittersee, Germany **W**
GR-371.III.80	AJ296608	1	*Sarcobium lyticum* PCM 2298 (X66835)	92	RISA	Schlema, Germany **W**
KF-JG30-C25	AJ295652	6	Uncult. soil bact. SC-I-87 (AJ252663)	89	16S II	Johanngeorgenstadt, Germany **S**
			Nitrosococcus sp. C113 (AF153343)	88		

Table 2
(continued)

Clone name	Accession No. EMBL	No. of clones	Closest phylogenetic relative (EMBL No.)	BLAST % of similarity*	Clone library	Geographic origin
			δ-Proteobacteria			
GR-B1-2-6	AJ296609	1	*Desulfobacter postagei* (M26633)	99	RISA	Deponie B1, Germany, **W**
GR-WP54-9	AJ296610	1	*Desulfobacca acetooxicans* (AF002671)	93	RISA	Gittersee, Germany **W**
GR-Sh2-74	AJ296611	2	Uncult. bact. Sva1014 (AJ240984)	92	RISA	Shiprock, NM, USA **W**
GR-Sh2-18	AJ296571	7	Uncult. SRB 368 (AJ389629)	90	16S II	Shiprock, NM, USA **W**
			Desulfocapsa thiozymogenes DSM7269 (X95181)	88		
GR-WP33-30	AJ296569	3	*Desulfuromonas acetexigens* (U23140)	89	16S II	Gittersee, Germany **W**
GR-WP59-1	AJ296612	1	Uncult. bact. MT63 (AF211303)	91	RISA	Gittersee, Germany **W**
			Desulfuromonas thiophila DSM8987 (Y11560)	88		
GR-296.II.5	AJ296613	1	*Geobacter sulfurreducens* PCA (U13928)	87	RISA	Schlema, Germany **W**
GR-WP33-58	AJ296570	3	*Geobacter sulfurreducens* PCA (U13928)	89	16S II	Gittersee, Germany **W**
GR-296.I.8	AJ296614	1	*Nitrospina gracilis* Nb-211 (L35504)	94	RISA	Schlema, Germany **W**
GR-296.I.52	AJ296568	29	*Nitrospina gracilis* Nb-211 (L35504)	91	16S II	Schlema, Germany **W**
GR-296.I.86		53	*Nitrospina gracilis* Nb-211 (L35504)	92	16S II	Schlema, Germany **W**
GR-296.II.3	AJ296615	1	*Desulfobacter amnigenus* (X83274)	93	RISA	Schlema, Germany **W**
				90		
GR-WP54-1	AJ296616	1	*Pelobacter carbinolicus* (U23141)	94	RISA	Gittersee, Germany **W**
GR-296.II.2	AJ296624	1	*Enthoteonella palauensis* (AF130847)	92	RISA	Schlema/Alberoda, Germany **W**
GR-296.II.7	AJ296625	1	*Enthoteonella palauensis* (AF130847)	92	RISA	Schlema/Alberoda, Germany **W**
			Cytophaga/Flavobacterium/Bacteroides			
GR-Sh1-5	AJ296578	1	Cytophaga-like bact.QSSC9-14 (AF170754)	96	16S II	Schlema, Germany **W**
GR-WP33-44	AJ301579	1	Uncult. bact. mixed isolate Koll2 (AJ224942)	97	RISA	Gittersee, Germany **W**
GR-Sh1-209	AJ296579	3	Uncult. bact. mixed isolate Koll2 (AJ224942)	92	16S II	Schlema, Germany **W**
			Flexibacter aggregans catalica ATCC23190 (M58791)	90		
KF-Gitt2-47	AJ295642	8	Uncult. bact. BB23 (AF129860)	96	16S II	Gittersee, Germany **W**
			Bacteroides sp. sp4(AB003389)	92		

Table 2
(continued)

Clone name	Accession No. EMBL	No. of clones	Closest phylogenetic relative (EMBL No.)	BLAST % of similarity*	Clone library	Geographic origin
			Cytophaga/Flavobacterium/Bacteroides (continued)			
KF-Gitt2-34	AJ295641	6	*Flavobacterium heparinum* (M11657)	92	16S II	Gittersee, Germany **S**
GR-WP33-68	AJ296576	5	Uncult. bact.VC2.1.Bac.22 (AF068798)	93	16S II	Gittersee, Germany **W**
GR-WP33-K3	AJ296617	1	Uncult. bact.BURTON-6 (AF142826)	92	RISA	Gittersee, Germany **W**
GR-Sh2-12	AJ296577	21	*Sphingobacterium* sp. OM-E81 (AB020206)	93	16S II	Shiprock, NM, USA **W**
			Plactomycetales			
GR-WP33-41	AJ296629	3	Anammox KOLL2a (AJ250882)	98	RISA	Gittersee, Germany **W**
GR-WP33-37	AJ301578	2	Anammox KOLL2a (AJ250882)	99	RISA	Gittersee, Germany **W**
GR-WP33-59	AJ296618	1	Anammox KOLL2a (AJ250882)	99	RISA	Gittersee, Germany **W**
GR-WP33-66	AJ296619	1	Anammox KOLL2a (AJ250882)	97	RISA	Gittersee, Germany **W**
GR-WP54-11	AJ296620	1	Anammox KOLL2a (AJ250882)	99	RISA	Gittersee, Germany **W**
			Candidate division OP1			
GR-296.II.45	AJ301571	11	Clone OPB14 (AF027045)	87	16S II	Schlema, Germany **W**
			Candidate division OP6			
GR-296.II.10	AJ296621	1	Clone OPS 152 (AF027079)	90	RISA	Schlema, Germany **W**
			Green sulfur bacteria			
GR-296.II.1	AJ296622	3	Uncult. bact. BSV 19 (AJ229184)	91	RISA	Schlema, Germany **W**
GR-296.II.4	AJ296623	2	Uncult. bact. BSV 19 (AJ229184)	93	RISA	Schlema, Germany **W**
GR-296.II.73	AJ301570	8	Uncult. bact. BSV 19 (AJ229184)	91	16S I	Schlema/Alberoda, Germany **W**
GR-296.II.11	AJ296575	7	Uncult. bact. BSV 26 (AJ229188)	89	16S II	Schlema/Alberoda, Germany **W**
			Green non-sulfur bacteria			
GR-296.II.265	AJ296574	1	Uncult. hydrocarbon seep bact. GCA004 (AF154104)	91	16S I	Schlema/Alberoda, Germany **W**
GR-296.II.57	AJ296572	7	Uncult. bact. clone 56 (AJ225340)	94	16S II	Schlema/Alberoda, Germany **W**
GR-296.II.271	AJ296573	2	Uncult. bact. SHA-21 (AJ249103)	87	16S I	Schlema/Alberoda, Germany **W**

Table 2
(continued)

Clone name	Accession No. EMBL	No. of clones	Closest phylogenetic relative (EMBL No.)	BLAST % of similarity*	Clone library	Geographic origin
			Holophaga			
KF-JG30-C18	AJ295656	4	Uncult. bact. TM1 (X97097)	93	16S II	Johanngeorgenstadt, Germany **S**
			Acidobacterium capsulatum (D26171)	90		
KF/GS-JG36-31	AJ295657	2	Uncult. bact. M26 (U68612)	92	16S I	Johanngeorgenstadt, Germany **S**
			Uncult. bact. DA052 (Y07646)	90		Gittersee, Germany **S**
			Nitrospira group			
GR-WP54-3	AJ302943	2	*Magnetobacterim bavaricum* (X71838)	94	RISA	Gittersee, Germany **W**
			Actinobacteria			
GR-WP54-8	AJ295626	2	Uncult. bact.WCHB1-81 (AF050572)	93	RISA	Gittersee, Germany **W**
KF-JG30-B11	AJ295649	3	Uncult. bact. #0319-6E22 (AF234130)	88	16S II	Johanngeorgenstadt, Germany **S**
			Gram+, Low G+C, Bacilli			
GR-371.III.83	AJ296627	1	*Bacillus sphaericus* IAM13420 (D16280)	98	RISA	Schlema/Alberoda, Germany **W**
KF-Gitt2-16	AJ295640	3	*Paenibacillus illinoisensis* NRRL NRS-1356T (D85397)	89	16S II	Gittersee, Germany **S**
			Verrucomicrobia			
GR-WP39-22	AJ296628	1	*Verrucomicrobium* sp. VeG1c2 (X99390)	95	RISA	Gittersee, Germany **W**

* Taking those parts of the gene which were considered by the BLAST search.
** Affiliated using ARDREA.
*** W – water; S – soil.

several uranium wastes, confirm the extremely high bacterial diversity in such environments. One of these libraries, the 16S rDNA$_{43F-1404}$, referred to as 16S II in Table 2, was constructed using an alternative 16S rDNA PCR primer pair (Marchesi et al., 1998) which is supposed to be more effective and to amplify a larger variety of environmental bacterial 16S rRNA genes. The RISA library was constructed using the primers 16SrDNA$_{968F}$ and 23S rDNA$_{130F}$ (see Table 1). The choice of the upstream 23S rDNA$_{130F}$ RISA primer was due to the observation that in most α-Proteobacteria extremely variable intervening sequences are present immediately after position 135 of the 23S rRNA gene (*E. coli* numbering system; Brosius, 1981) (Selenska-Pobell and Evguenieva-Hackenberg, 1995).

From the results presented in Table 2 one can see that the sequences from the 16S rDNA and the RISA clone libraries could be assigned to the bacterial taxa already described. Almost 85% of the clones studied (549 of 652) were affiliated to bacterial groups based on a sequence similarity ($\geq$92% which, according to Schlötelburg et al. (2000), indicates a close taxonomic relationship. The rest of the clones possibly represent novel bacterial lineages.

It is clear from Table 2 that, when using 16S rDNA II retrieval, the highest number of the sequences in the highly polluted soil and water samples were affiliated to the γ-subdivision of Proteobacteria, mainly to the *Pseudomonas* group. In some samples, however, RISA retrieval identified large communities of β-Proteobacteria which were not detected by the 16S rDNA primers. As an example, see the 16S rDNA II clone GR-Sh2-34, representing a large cluster related to *Pseudomonas stutzeri* strains, and the RISA clones GR-Sh2-23 and GR-Sh2-24, representing two large microdiverse groups of ammonia oxidisers (*Nitrosomonas* sp.) identified in the same water sample. This indicates the necessity of using more than one PCR primer pair in the construction of environmental 16S rDNA libraries because, as mentioned by others (Hansen et al., 1998), in such complex mixtures of DNA templates the regions flanking the PCR amplification products strongly influence the effectiveness and the preferences of the PCR reaction. As a result of these biases, enrichment with the particular fragments preferred by the process occurs.

In Fig. 2(a) the RFLP profiles obtained using endonuclease *Msp*I of several clones representing the two *Nitrosomonas* groups mentioned above are presented. When applying this straightforward, preliminary typing of the RISA and the 16S rDNA libraries we have found groups of closely related RFLP types in many of the samples studied. The sequenced representatives of these groups demonstrated almost identical sequences, indicating interspecies microdiversity in the population. Microdiversity is a newly recognised form of bacterial diversity between strains of the same species. The 16S rDNA genes of such strains possess specific short sequence stretches which represent genetically distinct populations, adapted for optimal growth under different environmental conditions (Moore et al., 1998). Several large microdiverse groups, belonging to γ-Proteobacteria (see GR-Sh1-26 and GR-Sh1-148 which represent *Frateuria*, and GR-B1-2-4, GR-B1-2-35, GR-B1-2-39, GR-B1-2-33 representing *Acinetobacter*), and to δ-Proteobacteria (see GR-296.I.86 and GR-296.I.52, which represent *Nitrospina*-like bacteria) were found in some of the samples. The profiles of the 16S rDNA *Msp*I types of *Nitrospina* are presented in Fig. 2(b). Interestingly, the presence of *Nitrospina* was also demonstrated in the same sample by the use of the RISA approach (see clone GR-296.I.8). The latter is an indication that these ammonia

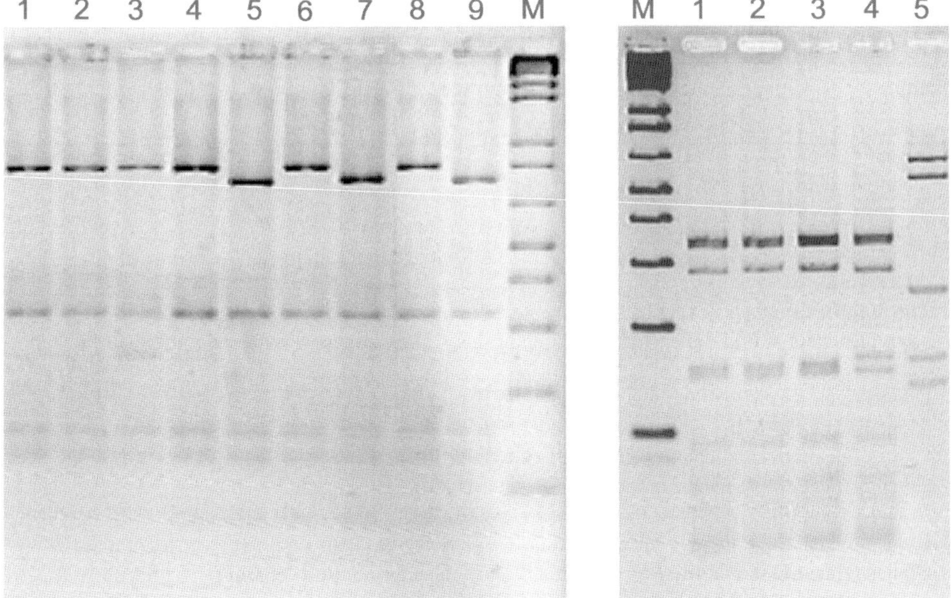

Fig. 2. *Msp*I-RFLP profiles of: (a) RISA clones related to *Nitrosomonas* group GR-Sh2-23 (lanes 1, 2, 3, 4, 6, 8) and to *Nitrosomonas* group GR-Sh2-24 (lanes 5, 7, 9); and (b) 16S rDNA II clones related to *Nitrospina* group GR-296.I.52 (lanes 1, 2, 3), to *Nitrospina* group GR-296.I.86 (lane 4), and to *Acinetobacter lwöffii* GR-B1-2-4 (lane 5). M – Marker, BRL-ladder kbl plus.

and nitrite oxidisers belonging to the δ-subdivision of Proteobacteria are predominant in the sample from well 296 of the uranium mill tailing dump at Schlema/Alberoda.

It is important to note that the method of storing samples before analysis strongly influences the structure of the bacterial communities. An example of a significant shift in the bacterial composition induced by aeration is the analysis of two parallel samples drawn from well 296 of the uranium mill tailings Schlema/Alberoda. The first sample, 296.I, was stored for several weeks at 34°C in anaerobic conditions, corresponding to the conditions at the original sampling site. The second sample, 296.II, collected from the same well of Schlema/Alberoda, was kept in sterile aerobic conditions at 34°C. As shown in Table 3, the storage conditions of sample 296.II stimulated the propagation of different bacterial groups which in the original samples were below the limit of detection by the 16S rDNA approaches. As a result of this shifting, the predominant *Nitrospina* community of the original sample 296.I was replaced by different groups of α-, β-, γ-Proteobacteria, green sulfur and green non-sulfur bacteria, see clones GR-296.II.89, GR 296.II.111 (α), GR-296.II.77 (β), GR-296.II.35 (γ), GR 296.II.73 and GR-296.II.11 (green sulfur), and GR-296.II.57 (green non-sulfur) as well as clones GR-296.II.265 and GR.II.271, which were identified by the use of 16S rDNA I also as green non-sulfur.

Interestingly, the predominant bacterial group detected in the 'shifted' sample possesses relatedness to the recently defined novel bacterial division called 'candidate division OP1' (clone GR-296.II.45, Hugenholtz et al., 1998), found in another extreme environment, the

Table 3

RFLP-typing of the 16S rDNA clones of Schlema/Alberoda, well 296

Sample	Clones obtained	Clones analysed	RFLP groups and individual RFLP types*
296.I**	106	50	(58)[1], (29)[2], (2), and 7 individual
296.II**	142	132	(11)[3], (8)[4], (7)[5], (7)[6], (7)[7], (7)[8], (5)[9], (4)[10], (4), (4), (4), (2), (2), (2), (2), and 20 individual

* The number of clones per RFLP-type are given in brackets.

** Samples were stored under anaerobic conditions. *** Samples were stored under aerobic conditions.

[1] Clones related to *Nitrospina gracilis* (GR-296.I.86 in Table 2).

[2] Clones related to *Nitrospina gracilis* (GR-296.I.52).

[3] Clones related to candidate division OP1 (GR-296.II.45).

[4] Clones related to green sulfur bacteria (GR-296.II.73).

[5] Clones related to green sulfur bacteria with less similarity (GR-296.II.11).

[6] Clones related to green non-sulfur bacteria (GR-296.II.57).

[7] Clones related to the family *Rhizobiaceae*, α-Proteobacteria (GR.296.II.111).

[8] Clones related to β-Proteobacteria (GR-296.II.77).

[9] Clones related to γ-Proteobacteria (GR-296.II.89).

[10] Clones related to δ-Proteobacteria (GR-296.II.35).

hot spring Obsidian Pool (OP) surrounded by the Yellowstone caldera in the USA. The latter is rich in reduced iron and metal sulfides and was described as a fertile ground for the discovery of novel microbial diversity in communities based on lithotrophy (Hugenholtz et al., 1998). In addition, using the RISA approach for the analysis of the sample 296.II, indications were obtained of the presence of representatives from the other candidate division OP6 and from a newly described group of green sulfur bacteria both of which are found in the Obsidian Pool (see clones GR-296.II.10, GR-296.II.1 and GR-296.II.4). Because the database matches of the sequences from the 'shifted' sample in most cases were relatively low, it is plausible that the bacterial groups identified in sample 296.II represent novel species and possibly even genera, which have not been described in the literature. The shifting of the bacterial populations in sample 296.II is confirmed by the results presented in Fig. 1 (see lanes 1 and 2) where completely different Rep-APD profiles were obtained for the two parallel samples 296.I and 296.II.

The RISA retrieval was very effective in the identification of a large community of *Planctomycetales* in drainage waters of Gittersee/Coschütz (see Table 2). Moreover, the matches of 99, 98 and 97 % (clones GR-WP33-41, GR-WP33-37, GR-WP33-59, GR-WP33-66, GR-WP54-11) of 16S rDNA identity with a newly described group of anaerobic ammonia-oxidisers, *Anammox* (Strous et al., 1999), suggests that in this case we have identified a large population of microdiverse strains of this unusual bacterial group. Anaerobic ammonia-oxidising bacteria were considered for a long time to be 'missing lithotrophs', because they were never isolable from nature. However, recently Strous et al. (1999) succeeded in purifying such bacteria from an enrichment biofilm culture using Percoll density gradient centrifugation. The purified organisms, named *Anammox*, were classified as a deep-branching Planctomycete and are capable of combining ammonia and nitrite directly into dinitrogen gas, a process of great importance for the removal of ammonia nitrogen from waste waters. With this discovery, Strous et al. (1999) have found a missing link in the biogeochemical nitrogen cycle. The presence of *Anammox* bacteria in the drainage waters of the uranium wastes is an indication that anaerobic ammonia oxidation processes occur in these environments. To our knowledge, this is the first reported direct molecular identification of closely related *Anammox planctomycetes* in nature.

No sequences related to *Anammox* were found in the 16S rDNA II clone library of the same sample. Instead, the following bacteria were identified: phototrophic and nitrogen fixing α-Proteobacteria (see clones GR-WP33-45 and GR-WP33-3), *Pseudomonas stutzeri* (GR-WP33-28) and other γ-Proteobacteria related to *Aeromonas* (clone GR-WP33-14) and to the metal-reducing *Shewanella* (GR-WP33-34), as well as metal reducers belonging to δ-Proteobacteria, *Desulfuromonas* (GR-WP33-30) and *Geobacter* (GR-WP33-58). In the same samples both RISA and the 16S rDNA retrievals indicated the presence of *Cytophaga*-like bacteria (clones GR-WP33-K3, GR-WP33-44 and GR-WP33-68).

Using the 16S rDNA approach Sandaa et al. (1999) have studied bacterial diversity in samples of soils amended with sewage sludge from industrial areas, where significant amounts of metals such as Cu, Ni, Cd, Zn and Cr are present. The authors have observed that the predominant clones of those samples represented Gram-positive bacteria with a high DNA G+C content (45%) and α-Proteobacteria (24%). In soils with a high metal content, an increase of the sequences representing α-Proteobacteria was found.

By contrast, using the same molecular approach, we have found that the soil samples from the uranium wastes are dominated by γ-Proteobacteria, especially by *Pseudomonas, Aeromonas*, and the iron and sulfur-oxidising chemolithoautotrophs *Acidithiobacillus ferrooxidans* (see clones KF/GS-Gitt2-53, KF/GS-Gitt2-41, KF/GS-JG36-13, KF-JG30-B15, KF/GS-JG36-20, KF/GS-JG36-22 and KF-JG30-C25; samples JG-36 and Gitt2 correspond to the most heavily polluted sites in the Johanngeorgenstadt and Gittersee/Coschütz wastes). The second predominant group of bacteria identified was affiliated to the genus *Thiobacillus* (KF-Gitt2-40). α-Proteobacteria were also identified in the soil samples, but mainly in the less contaminated ones such as JG30 (see Table 2). In addition, representatives of *Cytophaga* and *Holophaga* were identified in the soil samples. The number of the sequences representing the Gram-positive bacteria with a high DNA G+C content was limited to only 3, and these were not closely related to *Paenibacillus*.

The presence of different bacterial groups in these two heavy metal polluted environments, the industrial sewage sludge and the uranium mining waste piles, is indicative of the different biogeochemical processes which occur due to their different geographic and mineralogical characteristics and to the differences in the metal composition of both environments (see above).

As described by Sandaa et al. (1999), in the case of industrial sewage sludge high concentrations of heavy metals usually reduce the total amount of indigenous bacteria and the only parts of the natural bacterial populations that survive are those which can adapt to the toxic metals. Roane & Pepper (2000) identified the presence of cadmium-resistant isolates of *Bacillus*, *Arthrobacter* and *Pseudomonas* in both Cd polluted and unpolluted soils. They suggested that some representatives of the natural populations of these bacteria possess resistance mechanisms which switch on under stressful conditions. Interestingly, the same authors have found a correlation between enhanced cadmium and antibiotic resistance and considered this to be an indication that increasing Cd levels induces not only metal resistance but also resistance mechanisms to other stress factors.

In the case of uranium wastes, it seems that different bacterial groups involved in various interactions with metals are present. Some of these bacterial groups, such as *Acidithiobacillus ferrooxidans*, *Leptospirillum ferrooxidans*, *Geobacter* and *Desulfobacter*, are even gaining energy for growth from metal biotransformations (DiSpirito & Tuovinen, 1982; Lovley 1993; Rawlings et al., 1999; Magnuson et al., 2000). Since, in many cases, the more heavily polluted soil samples possessed higher densities of very diverse bacterial populations than the less polluted samples, it seems that the indigenous bacteria in uranium wastes are generally tolerant to the range of metals and radionuclides at the concentrations found in the wastes (see below).

5. Analysis of the diversity of the bacteria cultured from the uranium mining waste piles

As mentioned above, the main problem with direct molecular approaches to studying environmental samples is preferential PCR amplification, which can mask the presence of some DNA templates. It is especially difficult to use these methods to detect templates which are present in low concentrations in the samples. The detection of some sporulating bacteria may also be difficult, because of the lower recovery of DNA from spores by some direct lysis procedures.

For the identification of bacteria which are present in very low numbers or as spores in these extreme environments, the classical approach of enrichment cultures may be helpful. Of course, with this there is the limitation that only the parts of the community that are culturable can be analysed. As a compromise, enriched biofilm cultures or mixed cultures can be analysed instead of pure cultures, because most of the bacteria in the extreme environments are living in consortia which are believed to be symbiotic (Strous et al., 1999; Roane & Pepper, 2000; Schlottenburg et al., 2000). The identification of the members of such mixed cultures using 16S rDNA retrieval may provide further and very important information in addition to those derived by the direct methods.

As discussed above, direct molecular methods identified only a limited number of Bacilli in the uranium wastes. In contrast, by applying classical methods for spore isolation and culturing of spore-forming bacteria we have previously demonstrated the presence of a large number of Gram-positive spore-forming Bacilli in the soil samples of the uranium waste mining pile near the town of Johanngeorgenstadt (Selenska-Pobell et al., 1999). The predominant *Bacillus* species were *B. cereus*, *B. thuringensis*, *B. sphaericus*, *B. subtilis* and *B. megaterium*. Interestingly, only one sequence matching with *B. sphaericus* was identified, in a water sample from Schlema/Alberoda (see GR-371.III.89). The most probable explanation of this result is that Bacilli are present in low numbers in the natural populations in the uranium wastes. Another reason might be that most Bacilli are present in the wastes as spores and that the direct lysis method for the extraction of DNA was ineffective for these. The latter, however, is not the case, because the direct recovery of DNA from the soil samples was performed using a method of proven effectiveness for extraction of DNA from spores of Gram-positive bacteria (Selenska-Pobell, 1995). It is, however, possible that the disagreement between the culture-dependent and culture-independent methods can be attributed to preferential PCR amplification, discussed earlier, which in this case leads to masking of the Bacilli.

The information we have recently obtained regarding the diversity of bacteria cultured as pure and/or mixed cultures from the uranium wastes is presented in Table 4. Interestingly, we succeeded in culturing two bacterial isolates from samples from Gittersee/Coschütz and Johanngeorgenstadt, IrT-RS2 and TzT-JG-1-2, which are closely related to uncultured and not yet classified representatives of the genus Bacillus (see Table 4) recovered from other environments, specifically a hydrocarbon seep and an anoxic bulk rice paddy soil (Chin et al., 1999; Hengstmann et al., 1999). Because these isolates most probably represent novel species in the genus Bacillus which have not yet been described in the literature, their further characterisation is a subject of particular interest.

From the results in Table 4, it is clear that some of the γ-Proteobacteria, (*Pseudomonas stutzeri, Pseudomonas migulae, Acidithiobacillus ferrooxidans*) which were identified by direct rDNA retrieval were also successfully cultured from the waste samples. This may be an indication that these species are ubiquitous in the environments studied. Other isolates, for example those of *Desulfovibrio*, share relatively closely related 16S rDNA sequences with other sulfate and metal-reducing δ-Proteobacteria which were revealed by the direct approach (*Desulfobacter*, *Desulfobacca*, *Desulfobulbus*, *Geobacter* and *Pelobacter*, see Table 2). Members of the family Rhizobiaceae (*Agrobacterium* and *Rhizobium*) were also cultured, whereas the related *Bradyrhizobia*, members of the same family of α-Proteobacteria were identified by 16S rDNA retrieval.

'*Leptospirillum*-like' species rather than *Acidithiobacillus ferrooxidans* are the dominant iron-oxidisers in the process of biooxidation of pyrite and related ores in nature (Rawlings et al., 1999). However, some bacteria which might therefore be expected to be important in the uranium wastes, such as *Leptospirillum ferrooxidans*, were found mainly by the enrichment culture method and only one representative of the *Nitrospira/Leptospirillum* group, which is not very closely related to *Leptospirillum ferrooxidans*, was identified by the RISA retrieval in the Gittersee/Coschütz mill tailing (see clone GR-WP54-3 in Table 2).

Analysing a large number of *Leptospirillum ferrooxidans* individual isolates all representing a species of Leptospirillum group II (Bond et al., 2000, 2001) using the 16S rDNA RFLP (see Fig. 3(a), lanes 1 and 2) and sequence analysis we found that two predominant RFLP types of this species occur in the uranium wastes (Tzvetkova et al., 2002). Moreover, sequence analysis of the variable region III of the 16S rDNA of the isolates TzT-JG-7 and TzT B1-K3 revealed novel signatures distinguishing the 16S rRNA of these strains from all *L. ferrooxidans* 16S rRNA genes deposited at the EMBL (Tzvetkova et al., 2000, 2002).

The 16S rDNA sequence analysis of *Acidithiobacillus ferrooxidans* isolates recovered from soil samples represents another example of microdiversity (Flemming et al., 2000; Selenska-Pobell et al., 2000, 2001). In Fig. 3(b) (lanes 1, 2 and 3), the three recently described eco-types of *A. ferrooxidans* which were found in the uranium wastes are presented. An extensive analysis of a large number of soil samples drawn from different sites of the uranium mining waste pile near the town of Johanngeorgenstadt revealed that the

Table 4

Bacteria cultured from the uranium mining waste pile samples

Name	Accession No. EMBL	Closest phylogenetic relative (EMBL No.)	BLAST % of similarity*	Geographic origin
Leptospirillum ferrooxidans isolate TzT-JG-24** Type I	AJ295687	*Leptospirillum ferrooxidans* (AJ237903)		Johanngeorgenstadt, Germany
Leptospirillum ferrooxidans isolate TzT-JG-3** Type I	AJ295688	*Leptospirillum ferrooxidans* (AJ237903)		Johanngeorgenstadt, Germany
Leptospirillum ferrooxidans isolate TzT-JG-7** Type II	AJ295693	*Leptospirillum ferrooxidans* (AJ237903)		Johanngeorgenstadt, Germany
Leptospirillum ferrooxidans isolate TzT-B1-K3 Type II	AJ295685	*Leptospirillum ferrooxidans* (AJ237903)	98	Deponie B1, Germany
Leptospirillum ferrooxidans isolate TzT-B1-K4 Type I	AJ295686	*Leptospirillum ferrooxidans* (X72852)	98	Deponie B1, Germany
		α-Proteobacteria		
Clone IrT-JG14-33, mixed culture JG14	AJ295676	*Agrobacterium tumefaciens* IAM14141 (D13294)	98	Johanngeorgenstadt, Germany
Clone IrT-JG14-14, mixed culture JG14	AJ295675	*Agrobacterium tumefaciens* DSM 30105 (M11223)	97	Johanngeorgenstadt, Germany
Clone IrT-JG14-24, mixed culture JG14	AJ295674	*Agrobacterium tumefaciens* DSM 30105 (M11223)	97	Johanngeorgenstadt, Germany
Agrobacterium isolate IrT-JG-6	AJ295683	*Agrobacterium tumefaciens* DSM 30105 (M11223)	98	Johanngeorgenstadt, Germany
Clone IrT-JG15-65, mixed culture JG15	AJ295691	*Rhizobium* sp. USDA1920 (U89823)	92	Johanngeorgenstadt, Germany
Sphingomonas isolate TzT-JG-2-6	AJ295690	*Sphingomonas parapaucimobilis* (D84525)	99	Johanngeorgenstadt, Germany
		γ-Proteobacteria		
Pseudomonas stutzeri isolate IrT-JG-2	AJ295681	*Pseudomonas stutzeri* (U65012)	99	Johanngeorgenstadt, Germany
Pseudomonas stutzeri isolate IrT-JG-7	AJ295682	*Pseudomonas stutzeri* (AF063219)	99	Johanngeorgenstadt, Germany
Pseudomonas migulae isolate TzT-JG-2-3	AJ296580	*Pseudomonas migulae* (AF074383)	98	Johanngeorgenstadt, Germany
Pseudomonas migulae isolate TzT-JG-1-1**		*Pseudomonas migulae*		Johanngeorgenstadt, Germany
Erwinia herbicola isolate TzT-JG-2-8	AJ295692	*Erwinia herbicola* (Y13249)	98	Johanngeorgenstadt, Germany
Erwinia herbicola isolate TzT-JG-2-7**		*Erwinia herbicola*		Johanngeorgenstadt, Germany
Acidithiobacillus ferrooxidans ATCC 23270^{T}, Type I	AJ278718			Coal mine, USA
Acidithiobacillus ferrooxidans ATCC 33020, Type II	AJ278719			Uranium mine, Japan
Acidithiobacillus ferrooxidans isolate TzT-B1-K2, Type III	AJ278722	*Acidithiobacillus ferrooxidans* DSM 9465 (Y11595)	99	Deponie B1, Germany
Acidithiobacillus ferrooxidans isolate SS4** Type I	AJ278720			Johanngeorgenstadt, Germany
Acidithiobacillus ferrooxidans isolate SS6** Type II	AJ278721			Johanngeorgenstadt, Germany

Table 4
(continued)

Name	Accession No. EMBL	Closest phylogenetic relative (EMBL No.)	BLAST % of similarity*	Geographic origin
		δ-Proteobacteria		
Desulfovibrio intestinalis isolate JG-GI 2	AJ295680	*Desulfovibrio intestinalis* KMS2 (Y12254)	99	Johanngeorgenstadt, Germany
Clone IrT-JG1-58, mixed culture JG1	AJ295670	*Desulfovibrio vulgaris* (oxamicus) (AJ295677)	97	Johanngeorgenstadt, Germany
Clone IrT-JG1-71, mixed culture JG1	AJ295669	*Desulfovibrio vulgaris* (oxamicus) (AJ295677)	98	Johanngeorgenstadt, Germany
Desulfovibrio isolate JG1	AJ295678	*Desulfovibrio vulgaris* (oxamicus) (AJ295677)	98	Johanngeorgenstadt, Germany
Desulfovibrio isolate JG5	AJ295679	*Desulfovibrio vulgaris* (oxamicus) (AJ295677)	98	Johanngeorgenstadt, Germany
Clone IrT-JG14-32, mixed culture JG14	AJ295671	*Stenotrophomonas maltopilia* (AJ1293464)	99	Johanngeorgenstadt, Germany
Clone IrT-JG14-28, mixed culture JG14	AJ295672	*Stenotrophomonas maltopilia* (AJ293464)	99	Johanngeorgenstadt, Germany
Clone IrT-JG14-12, mixed culture JG14	AJ295673	*Stenotrophomonas maltopilia* LMG11087 (X95924)	99	Johanngeorgenstadt, Germany
		Bacillus/Clostridium group		
Clone IrT-JG1-9, mixed culture JG1	AJ295663	*Clostridium celerecrescens* (X71848)	92	Johanngeorgenstadt, Germany
Clone IrT-JG1-54, mixed culture JG1	AJ295662	*Clostridium* sp. 45 (AF191251)	99	Johanngeorgenstadt, Germany
Clone IrT-JG1-73, mixed culture JG1	AJ295661	*Clostridium* sp. 6 (AF191249)	99	Johanngeorgenstadt, Germany
Clone IrT-JG1-53, mixed culture JG1	AJ295664	*Clostridium sporosphaeroides* (M59116)	93	Johanngeorgenstadt, Germany
Clone IrT-JG1-59, mixed culture JG1	AJ295667	*Clostridium sporosphaeroides* (M59116)	90	Johanngeorgenstadt, Germany
Clone IrT-JG1-68, mixed culture JG1	AJ295665	*Clostridium sporosphaeroides* (M59116)	92	Johanngeorgenstadt, Germany
Clone IrT-JG1-52, mixed culture JG1	AJ295668	*Clostridium putrefaciens* (AF127024)	95	Johanngeorgenstadt, Germany
Clone IrT-JG1- 64, mixed culture JG1	AJ295666	*Clostridium sporosphaeroides* (M 59116)	94	Johanngeorgenstadt, Germany
Clone IrT-JG1-67, mixed culture JG1	AJ295658	*Clostridium celerecrescens* (X71848)	99	Johanngeorgenstadt, Germany
Clone IrT-JG1-22, mixed culture JG1	AJ295660	*Clostridium celerecrescens* (X71848)	98	Johanngeorgenstadt, Germany
Clone IrT-JG1-12, mixed culture JG1	AJ295659	*Clostridium celerecrescens* (X71848)	99	Johanngeorgenstadt, Germany
Bacterial isolate R52	AJ295684	Uncult. hydrocarbon seep bacterium (AF154082)	97	Gittersee/Coschütz, Germany
		Bacillus sp. SB45 (AJ229238)	97	
Isolate TzT-JG1-2	AJ295689	Unidentif. bacterium, anoxic bulk soil (AJ229200)	96	Johanngeorgenstadt, Germany
		Bacillus niacini (AJ021194.1)	95	

* Taking those parts of the gene which were considered by the BLAST search.
** Affiliated using ARDREA.

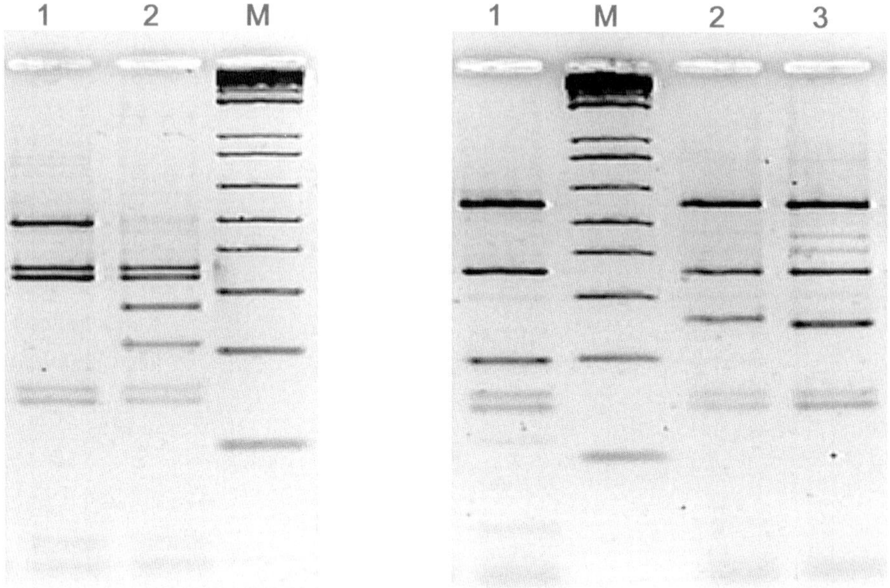

Fig. 3. 16S rDNA *Rsa*I-RFLP profiles of: (a) *Leptospirillum ferrooxidans* natural isolates of type I (lane 1) and type II (lane 2); and (b) *Acidithiobacillus ferrooxidans* natural isolates of type I (lane 1), type II (lane 2), and type III (lane 3). M – Marker, BRL-ladder kbl plus.

representatives of type I (see lane 1 in Fig. 3(b)) usually occur in the most contaminated samples from greater depths, 3–5 m below the surface (Selenska-Pobell et al., 2000, 2001). However, from sediment samples from the uranium waste site Deponie B1, which were even more contaminated with uranium, *A. ferrooxidans* of type III was mainly recovered. The representatives of type II were usually found in samples close to the surface which were less polluted. In many samples with an intermediate grade of pollution, mixed populations of two or three *A. ferrooxidans* types were identified (Selenska-Pobell et al., 2001). Interestingly, the *A. ferrooxidans* isolates of types I and III demonstrated higher tolerance to uranium in laboratory conditions. The minimal inhibitory concentration of uranium for the growth of the isolates of type I was found to be 9 mM and for type III, 10 mM, whereas type II completely stopped its growth at 4 mM of uranium (Merroun & Selenska-Pobell, 2000, 2001).

The 16S rRNA genes of the three eco-types of *A. ferrooxidans* share over 99% identity, but there are three specific stretches in their genes, also called signatures, which clearly discriminate them (Flemming et al., 2000; Selenska-Pobell et al., 2001). One of these signatures is located in helix 18 of the variable region III, the second one is between helices 21 and 22, and the third one is in helix 27 of the variable region V of the 16S rRNA gene. It is possible that these 16S rDNA signatures reflect the genetic adaptation of each of the three distinct *A. ferrooxidans* types to different concentrations of heavy metals, oxygen, and/or probably other compounds in their natural environments. It seems that these three eco-types are present in different ratios in the natural *A. ferrooxidans* populations and one or another of them is predominant, depending on conditions. The three types of *A. ferrooxidans* were also found when analysing a series of 9 isolates from Canadian uranium wastes, which were kindly provided by Leo Leduc (unpublished data). Similar cases of microdiversity have also been observed recently for different bacterial species in other environments (Moore et al., 1998; Prüß et al., 1999).

It seems that in such extreme environments as the uranium wastes, which are inhomogeneously polluted with different heavy metals and radionuclides, the phenomenon of microdiversity is widespread. The fact that microdiversity was detected even by the direct, culture-independent methods (see above) demonstrates that the bacterial populations in uranium wastes possesses extremely high plasticity and adaptivity.

In addition to the individual bacterial isolates described above, several enrichment mixed cultures were recovered from the uranium mining waste pile near Johanngeorgenstadt and characterised. One of them, called initially 'JG1', from which the sulfate-reducing *Desulfovibrio* isolate JG1 was purified, consisted of 11 diverse representatives of *Clostridium* (see the series of clones IrT-JG1 affiliated to *Bacillus/Clostridium* group in Table 4), and two representatives (clones IrT-JG1-58 and IrT-JG1-71) of the genus *Desulfovibrio* which are closely related to the isolate *Desulfovibrio* JG1 and represent another example of microdiversity.

Previously, we reported that the mixed culture of 'JG1' is able to reduce and precipitate about 1.5 g of U(VI) per g of dry weight bacterial biomass from a liquid medium between pH values of 2.8 and 6.0 (Panak et al., 1997). Our recent results demonstrate, however, that the U(VI)-reducing capacity of the pure culture of *Desulfovibrio* sp. JG1 does not differ significantly from those of other uranium-reducing bacteria, *Desulfovibrio vulgaris*, *D. desulfuricans*, *Geobacter metalireducens* and *Shewanella putrefaciens*, which is about 5 times lower than the mixed culture and in addition is pH dependent with an optimum of pH 6.8 (Gorby & Lovley, 1992; Lovley, 1993; Lovley et al., 1993). This result indicates that the greater ability of the mixed culture 'JG1' to precipitate uranium under a wider range of environmental conditions (pH for example) is a result of the combined function of the different members. Preliminary results from EXAFS spectroscopy demonstrated that only about 30% of the uranium precipitated by the mixed culture was converted into U(IV) (Reich, 1999), showing that, in addition to U(VI) reduction by the natural strains of *Desulfovibrio* sp. mentioned above (JG1, IrT-JG1-58 and JrT-JG1-71), some other processes as bioprecipitation or/and biomineralisation occur, stimulated by the metabolic functions of the Clostridia and possibly also of some other presently unidentified strains present in the mixed culture.

Another mixed culture 'JG14' consisted of different α-Proteobacteria (*Agrobacterium* and *Rhizobium*) and a microdiverse population of environmental strains of *Stenotrophomonas maltopilia* (δ-Proteobacteria). These strains of *S. maltopilia* as well as the agrobacteria and rhizobia represent different kinds of rhizospheric bacteria which interact with plants. Their presence in the sample, which was collected from a depth of 1 m below the surface, may indicate bacteria-plant symbiotic interactions, perhaps providing an opportunity for natural bioremediation. It is certainly known that some *Rhizobium* strains possess mechanisms of resistance to high concentrations of toxic metals, which do not inhibit their ability to establish symbiotic and beneficial partnerships with the host plants (Purchase et al., 1997).

6. Some activities of bacterial strains recovered from uranium mining waste piles

The role of *A. ferrooxidans* in the direct and indirect oxidation of U(IV) to U(VI), mobilising this radionuclide in uranium ores, has been known for a long time (DiSpirito et al., 1981; Cerda et al., 1993; Schippers et al., 1994; De Siloniz et al., 1995). As discussed above, we have demonstrated that the natural populations of this bacterial species in uranium wastes are inhomogeneous and consist of three different types, which tolerate different concentrations of dissolved uranium (VI) at their physiological pH optimum for growth which is between 2.0 and 2.5.

An analysis of the interactions of these three *A. ferrooxidans* types with uranium revealed that all of them accumulate U(VI) on their surfaces (Merroun & Selenska-Pobell, 2000, 2001). Surprisingly, types I and III, which are more tolerant to uranium, accumulate smaller quantities of uranium (see Fig. 4) than the type II strains, which possess lower tolerance. This is in agreement with our earlier results from the analysis of two reference strains of *A. ferrooxidans*, ATCC23270^{T} and ATCC33020, belonging to types I and II respectively (Panak et al., 1998, 1999). In the second of these papers we also demonstrated, by the use of time resolved laser fluorescence spectroscopy (TRLFS), that the complexes on the surfaces of the strain 33020, representing type II, are stronger than those built on the surface of the strain 23270 (type I). These results suggest that the more tolerant types I and III *A. ferrooxidans* strains possess a mechanism which limits uranium accumulation on their surfaces to less than lethal amounts, in contrast to the less tolerant type II strains. In Fig. 4 the accumulation of uranium by *A. ferrooxidans* at pH 4, which corresponds to the pH of the uranium waste site from which the strains were cultured, is presented. The three types of *A. ferrooxidans* are able, however, to accumulate uranium also at lower pH values down to 1.5 (Merroun and Selenska-Pobell 2001). The amount of uranium accumulated at pH 1.5 was in the same range as at pH 4, and the behaviour of the three types was the same as described above. EXAFS spectroscopic analysis of the bacterium-uranium complexes formed at pH 1.5 by *A. ferrooxidans* indicates that U complexes with organic phosphorous residues (Merroun et al., 2002a, b).

As mentioned above, a large number of bacilli was cultured from the uranium mining wastes. Three particular strains of this genus, which were classified as *B. sphaericus*, *B. cereus* and *B. megaterium*, demonstrated an ability to accumulate large amounts of U, Pb, Cd, Cu and Al selectively and reversibly from the drain waters of the uranium mining waste pile Johanngeorgenstadt (Selenska-Pobell et al., 1999). By studying one of these strains, *B. sphaericus* JG-A12, in detail, we have found that it possesses a surface layer (S-layer) protein of molecular weight 135 kDA which has a novel structure (Raff et al., 1999, 2000a, b). S-layers are highly ordered protein layers on the surface of the cell walls. They occur often in bacteria which occupy extreme environments and are believed to play an important protective role under different stress conditions (Sara & Egelseer, 1996). These S-layers interact with different metal ions by forming metal clusters and biominerals (Douglas & Beveridge, 1998; Brown at al., 1998).

We have found that the S-layer of the uranium waste pile isolate *B. sphaericus* JG-A12 interacts very effectively not only with uranium but also with Pd and Pt, forming regular metal nanoclusters with a size of 1.9 nm (Raff, 2000b; Wahl, 2000). In contrast to the published formation of gold nanoclusters on the S-layer of another strain of *B.*

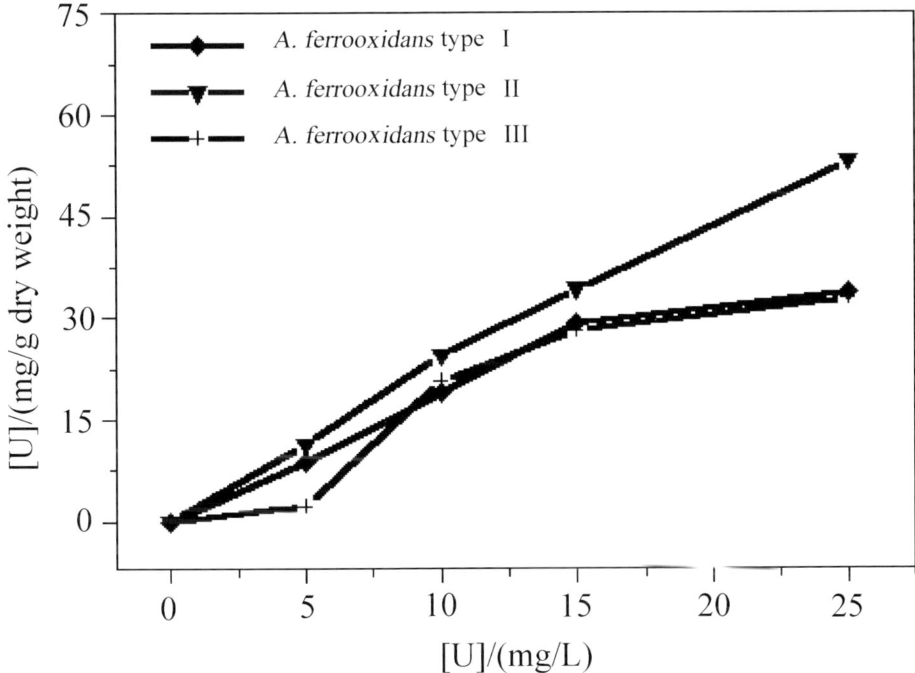

Fig. 4. Biosorption of U(VI) by the three types of *A. ferrooxidans* at pH 4.

sphaericus CCM 2177 (Dieluweit et al., 1998), no chemical modification of the protein was necessary. One possible explanation for this different behaviour of the S-layer protein from the uranium waste isolate may be the unusual structure of this protein which is described below. Possibly this S-layer protein provides the uranium mining waste pile isolate with a selection advantage in the heavy metal polluted environments from which it was recovered, because binding of radionuclides and heavy metals on the surface of the cells is one of the protective mechanisms of bacteria against these toxic compounds.

Sequence analysis of the gene of the S-layer of the uranium mining isolate JG-A12 has demonstrated that it differs significantly from those described for the *B. sphaericus* strains WHO 2362, P-1, and CCM 2177 which are isolated from other environments, not polluted with heavy metals. The S-layers of the latter three strains share about 80% homology, whereas the sequence homology of the S-layer of the strain JG-A12 to them is 35% or less. In Fig. 5, alignments of two regions located close to the N terminus of the known S-layers of *B. sphaericus* are presented. These regions possess highly conserved amino acids (shadowed) which are characteristic for the so-called S-layer homology domains (Engelhardt & Peters, 1999). As shown in Fig. 5, the S-layer of the uranium pile isolate JG-A12 possesses three unique amino acid sequences which are missing in those of the other *B. sphaericus* strains. These sequences were also not present in the rest of the approximately 45 S-layers deposited to date in the EMBL protein gene-bank. Recently, several short metal binding peptides displayed on the cell surfaces and possessing enhanced ability

```
                         41
Bsp JG-A12(AJ292965)     INRGQVVKLL GRYLEAQGQE IPADWNSKQ- --RFNDLPVT
Bsp 2362      (M28361)   ISRAEAATIF TNALE----- --LEAEGDV- --NFKDVK--
Bsp P-1       (A45814)   VTRAQAAEIF TKALE----- --LEANGDV- --NFKDVK--
Bsp 2177  (AF211170)     VTRAQAAEIF TKALE----- --LEADGDV- --NFSDVK--

                         141
Bsp JG-A12(AJ292965)     YKKANFVTEI GDLDKAYSAE QRTAIVALEY AGITNVAH--
Bsp 2362      (M28361)   FADAS---TV KPWAKSY--- ----LEIAVA NGVIKGSEAN
Bsp P-1       (A45814)   FADAS---QV KPWAKKY--- ----LEIAVA NGIFEGTDAN
Bsp 2177  (AF211170)     FADAS---QV KGWAKSA--- ----LETAVA NGIFTGSEEN
```

Fig. 5. Alignment of two parts of the known S-layer proteins of *B. sphaericus* starting from amino acid positions 41 and 141. Bsp JG-A12 – *B. sphaericus* JG-A12 isolate from the uranium mining waste pile near the town of Johanngeorgenstadt; Bsp 2362 – *B. sphaericus* WHO2362; Bsp P-1 – *B. sphaericus* P-1, Bsp2177 – *B. sphaericus* CCM2177. The highly conserved amino acids in the S-layer homology domain consensus are shaded.

to bind particular metals have been used for development of bacterium-based heavy metal biosorbents (Pazirandeh et al., 1998; Kotrba et al., 1999). It is an objective of our future work to clarify the function of the unique *B. sphaericus* JG-A12 S-layer short peptide motifs, in particular defining their role in interactions with metals and radionuclides.

Recent EXAFS measurements have shown that most of the uranium accumulated by *B. sphaericus* JG-A12 is deposited at the end of the bioaccumulation process, as phosphate complexes (Hennig et al., 2001). However, in the initial steps of the interaction with metals and radionuclides, the outermost surface layer of the bacterial cells, and especially those areas in the pores of the S-layer through which the stream of environmental compounds is passing, play an important role in defining the selectivity of binding for different metals. Thus, the ability of the uranium mining waste pile isolate JG-A12 to accumulate U and other toxic metals selectively and reversibly from the drain waters of uranium wastes makes it a good candidate for entrapping in SiO_2 ceramics using the so called sol-gel process (Al-Saraj et al., 1999) and for construction of bioceramic filters for cleaning of the polluted drain waters from uranium wastes.

7. Conclusions

The main conclusions drawn from the work presented in this chapter are the following. A large number of diverse bacterial communities is present in high densities in the soil, water, and sediments of the uranium mining waste piles. Despite the fact that a large number of species and genera which have not yet been defined is identified in these environments, the bacterial groups which predominate there belong mainly to those known to biotransform metals.

In many of the contaminated sites, closely related strains belonging to the same species coexist, perhaps indicating the high adaptivity and plasticity of the bacterial communities. The latter is important for the survival of bacterial species in these extremely complex and unstable environments where the acidity, concentrations of solubilised and bioavailable hazardous metals and other toxic compounds are continuously changing as a result of weathering and other natural processes. At the moment there is no one method that will provide complete information on bacterial diversity in these uranium wastes. The same is true for other natural environments. For this reason, it is necessary to apply a variety of methods in these analyses.

It is important to stress that the uranium mining waste piles are an unusual reservoir of bacterial diversity. Bacterial groups possessing unusual properties, especially in their interactions with heavy metals and radionuclides, are present there. Some of these bacteria may be potentially usable in development of in situ bioremediation procedures, because they are well adapted to the complex and harsh conditions in the wastes.

On the other hand, the rapid precipitation of dissolved U(VI) by the mixed bacterial consortium 'JG1' recovered from the uranium wastes suggests that, in some cases, natural consortia instead of individual bacterial strains could be preferred for bioremediation.

The different interactions of the three microdiverse types of *A. ferrooxidans* with U demonstrate that a mixed population of type I and III will be more useful for commercial bioleaching. They also suggest that the presence of type II in the uranium waste samples may enhance the formation of strong complexes with U, which will influence the mobilisation and migration of this radionuclide.

Acknowledgements

This work was supported by grants 7531.50-03-FZR/607 and 7531-50-03-FZR/708 from the Sächsisches Staatsministerium für Wissenschaft und Kunst, Dresden, Germany and GRP/9816 from the European Science Foundation. The author would like to thank her colleagues Katrin Flemming, Galina Radeva, Irena Tzvetkova, Tzvetelina Tzvetkova, Mohamed Merroun, and Johannes Raff for the use of their as yet unpublished data, and S. Kutschke and C. Puers (Forschungszentrum Rossendorf), C. Helling (TU Freiberg), and D. Balkwill (Florida State University, USA) for providing some of the water and soil samples and for measuring the metal concentrations in them.

References

Acinas, S. G., Anton, J. & Rodriguez-Valera, F. (1999). Diversity of free-living and attached bacteria in offshore Western Mediterranean waters as described by analysis of genes encoding 16S rRNA. *Applied and Environmental Microbiology, 65*, 514–522.

Al-Saraj, M., Abdel-Latif, M. S., El-Nahal, I. & Baraka, R. (1999). Bioaccumulation of some hazardous metals by sol-gel entrapped microorganisms. *Journal of Non-Crystalline Solids, 248*, 137–140.

Bacelar-Nicolau, P. & Johnson, D. B. (1999). Leaching of pyrite by acidophilic heterotrophic iron-oxidizing bacteria in pure and mixed cultures. *Applied and Environmental Microbiology, 65*, 585–590.

Berthelot, D., Leduc, L. G. & Ferroni, G. D. (1997). Iron-oxidizing autotrophs and acidophilicheterotrophs from uranium mine environments. *Journal of Geomicrobiology, 14*, 317–324.

Bosecker, K. (1997). Bioleaching: metal solubilization by microorganisms. *FEMS Microbiology Reviews, 20*, 591–604.

Borneman, J. & Triplett, E. W. (1997). Molecular microbial diversity in soils from Eastern Amazonia: evidence for unusual microorganisms and microbial population shifts associated with deforestation. *Applied and Environmental Microbiology, 63*, 2647–2653.

Boivin-Jahns, V., Rumy, R., Bianchi, A., Daumas, S. & Christen, R. (1996). Bacterial diversity in a deep-subsurface clay environment. *Applied and Environmental Microbiology, 62*, 3405–3412.

Bond, P. L. & Banfield, J. F. (2001). Design and performance of rRNA targeted oligonucleotide probes for *in situ* detection and phylogenetic identification of microorganisms inhabiting acid mine drainage environments. *Microbial Ecology, 41*, 149–161.

Bond, P. L., Druschel, G. G. & Banfield, J. F. (2000). Comparison of acidic mine drainage microbial communities in physically and geochemically distinct ecosystems. *Applied and Environmental Microbiology, 66*, 4962–4971.

Bowman, J. P., McCammon, S. A., Brown, M. V., Nichols, D. S. & McMeekin, T. A. (1997). Diversity and association of psychrophilic bacteria in antarctic sea ice. *Applied and Environmental Microbiology, 63*, 3068–3078.

Brosius, J., Dull, T. & Sleeter, D. D. (1981). Gene organization and primary structure of ribosomal RNA operon from *Escherichia coli. Journal of Molecular Biology, 148*, 107–127.

Brown, D. A., Beveridge, T. J., Keevil, C. W. & Sherriff, B. L. (1998). Evaluation of microscopic techniques to observe iron precipitation in a natural microbial biofilm. *FEMS Microbiology Ecology, 26*, 297–310.

Byers, H. K., Stackebrandt, E., Hayward, C. & Blackall, L. L. (1998). Molecular investigation of a microbial mat associated with the great artesian basin. *FEMS Microbiology Ecology, 25*, 391–403.

Cerdá, J., Gonzalez, S., Ríos, J. M. & Quintana, T. (1993). Uranium concentrates bioproduction in Spain: a case study. *FEMS Microbiology Reviews, 11*, 253–260.

Chandler, D. P., Frederickson, J. K. & Brockman, J. K. (1997). Use of 16S rDNA clone libraries to study changes in a microbial community resulting from ex situ perturbation of a surface sediment. *FEMS Microbiology Ecology, 20*, 217–230.

Chin, K. J., Hahn, D., Hengstmann, H., Liesack, W. & Janssen, P. H. (1999). Characterization and identification of numerically abundant culturable bacteria from anoxic bulk soil of rice paddy microcosms. *Applied and Environmental Microbiology, 65*, 5042–5049.

Dieluweit, S., Pum, D. & Sleytr, U. B. (1998). Formation of a gold superlattice on an S-layer with square lattice symmetry. *Supramolecular Science, 5*, 15–19.

De Silóniz, M. I., Lorenzo, P., Murúa, M. & Perera, J. (1993). Characterization of a new metal-mobilizing *Thiobacillus* isolate. *Archives of Microbiology, 159*, 237–243.

De Wulf-Durand, P., Bryant, L. J. & Sly, L. I. (1997). PCR-mediated detection of acidophilic bioleaching-associated bacteria. *Applied and Environmental Microbiology, 63*, 2944–2948.

DiSpirito, A. A. & Tuovinen, O. H. (1982). Kinetics of uranous ion and ferrous iron oxidation by *Thiobacillus ferrooxidans. Archives of Microbiology, 133*, 33–37.

DiSpirito, A. A., Talnagi, J. W. & Tuovinen, O. H. (1983). Accumulation and cellular distribution of uranium in *Thiobacillus ferrooxidans. Archives of Microbiology, 135*, 250–253.

Dojka, M. A., Hugenholtz, P., Haack, S. K. & Pace, N. R. (1998). Microbial diversity in a hydrocarbon- and chlorinated-solvent-contaminated aquifer undergoing intrinsic bioremediation. *Applied and Environmental Microbiology, 64*, 3869–3877.

Dojka, M. A., Harris, J. K. & Pace, N. R. (2000). Expanding the known diversity and environmental distribution of an uncultured phylogenetic division of bacteria. *Applied and Environmental Microbiology, 66*, 1617–1621.

Douglas, S. & Beveridge, T. J. (1998). Mineral formation by bacteria in natural microbial communities. *FEMS Microbiology Ecology, 26*, 79–88.

Flemming, K., Kutschke, S., Tzwetkova, T. & Selenska-Pobell, S. (2000). Intraspecies diversity of *Thiobacillus ferrooxidans* strains from uranium wastes. *FZR Report 285*, p. 51.

Francis, A. J. (1990). Microbial dissolution and stabilisation of toxic metals and radionuclides in mixed wastes. *Experientia, 46*, 840–851.

Francis, A. J. (1998). Biotransformation of uranium and other actinides in radioactive wastes. *Journal of Alloys and Compounds, 271–273*, 787–784.

Francis, A. J., Dodge, C. J., Gillow, J. B. & Cline, J. E. (1991). Microbial transformations of uranium wastes. *Radiochimica Acta, 52*, 311–316.

Francis, A. J., Gillow, J. B., Dodge, C. J., Dunn, M., Mantione, K., Streitelmeier, B. A., Pansoy-Hjelvik & Papenguth, H. W. (1998). Role of bacteria as biocolloids in the transport of actinides from a deep underground radioactive waste repository. *Radiochimica Acta, 84*, 347–354.

Franklin, R. B., Taylor, D. R. & Mills, A. L. (2000). The distribution of microbial communities in anaerobic and aerobic zones of a shallow coastal plain aquifer. *Microbial Ecology, 38*, 377–386.

Goebel, B. M. & Stackebrandt, E. (1994). Cultural and phylogenetic analysis of mixed microbial populations found in natural and commercial bioleaching environments. *Applied and Environmental Microbiology, 60*, 1614–1621.

Gorby, Y. A. & Lovley, D. R. (1992). Enzymatic uranium precipitation. *Environmental Science and Technology, 26*, 205–207.

Hansen, M. C., Tolker-Nielsen, T., Giskov, M. & Molin, S. (1998). Biased 16S rDNA PCR amplification caused by interference from DNA flanking the template region. *FEMS Microbiology Ecology, 26*, 141–149.

Hengstmann, U., Chin K. J., Jansen P. H. & Liesack, W. (1999). Comparative phylogenetic assignment of environmental sequences of genes encoding 16S rRNA and numerically abundant culturable bacteria from an anoxic rice paddy soil. *Applied and Environmental Microbiology, 65*, 5050–5058.

Hennig, C., Panak, P., Reich, T., Roßberg, A., Raff, J., Selenska-Pobell, S., Matz, W., Bucher, J., Bernhard, G. & Nitsche, H. (2000). XAFS investigation of uranium(VI) complexes formed at *Bacillus cereus* and *Bacillus sphaericus* surfaces. *Radiochimica Acta, 89*, 625–631.

Hugenholtz, P., Pitulle, C., Hershberger, K. L. & Pace, N. R. (1998). Novel division level bacterial diversity in a Yellowstone hot spring. *Journal of Bacteriology, 180*, 366–376.

Kampf, G. & Selenska-Pobell, S. (1998). Studies of the variability of natural bacterial communities in uranium contaminated soils and drain waters. *FZR Report 218*, pp. 59–60.

Kato, C., Li, L., Nogi, Y., Nakamura, Y., Tamaoka, J. & Horikoshi, K. (1998). Extremely barophilic bacteria isolated from the Mariana Trench, Challenger Deep at a depth of 11,000 meters. *Applied and Environmental Microbiology, 64*, 1510–1513.

Kotelnikova, S. & Pedersen, K. (1998). Distribution and activity of methanogens and homoacetogens in deep granitic aquifers at Äspö Hard Rock Laboratory, Sweden. *FEMS Microbiology Ecology, 2*, 121–134.

Kieft, T. L., Kovacik, W. P., Ringelberg, D. B., White, D. C., Haldeman, D. N., Amy, P. S. & Hersman, L. E. (1997). Factors limiting microbial growth and activity at a proposed high-level nuclear repository, Yucca mountain, Nevada. *Applied and Environmental Microbiology, 63*, 3128–3133.

Klaus, T., Joerger, R., Olsson, E. & Granqvist, C. -G. (1999). Silver-based crystalline nanoparticles, microbially fabricated. *Proceedings of the National Academy of Sciences, 96*, 13611–13614.

Kotrba, P., Doleckova, L., De Lorenzo, V. & Ruml, T. (1999). Enhanced bioaccumulation of heavy metal ions by bacterial cells due to surface display of short metal binding peptides. *Applied and Environmental Microbiology, 65*, 1092–1098.

Krebs, W., Brombacher, C., Bosshard, P. P., Bachofen, R. & Brandl, H. (1997). Microbial recovery of metals from solids. *FEMS Microbiology Reviews, 20*, 605–617.

Lovley, D. R. (1993). Dissimilatory metal reduction. *Annual Review of Microbiology, 47*, 263–290.

Lovley, D. R., Widman, P. K., Woodward, J. C. & Phillips, E. J. P. (1993). Reduction of uranium by cytochrome C_3 of *Desulfovibrio vulgaris*. *Applied and Environmental Microbiology, 59*, 3572–3576.

Magnuson, T. S., Hidges-Myerson, A. L. & Lovley, D. (2000). Characterization of a membrane-bound NADH-dependent Fe^{3+} reductase from the dissimilatory Fe^{3+}-reducing bacterium *Geobacter sulfurreducens*. *FEMS Microbiology Letters, 185*, 205–211.

Macaskie, L. E., Empson, R. M., Cheetham, A. K., Grey, C. P. & Scarnulis, A. J. (1992). Uranium bioaccumulation by *Citrobacter* sp. as a result of enzymatically mediated growth of polycrystalline HUO_2PO_4. *Science, 257*, 782–784.

Marchesi, J. R., Sato, T., Weightman, A. J., Martin, T. A., Fry, J. C., Hiom, S. J. & Wade, W. G. (1998). Design and evaluation of useful bacterium-specific PCR primers that amplify genes coding for 16S rRNA. *Applied and Environmental Microbiology, 64*, 795–799.

Marques, A. M., Roca, X., Simon-Pujol, M. D., Fuste, M. C. & Congregado, F. (1991). Uranium accumulation by *Pseudomonas* sp. ESP–5028. *Applied Microbiology and Biotechnology, 35*, 406–410.

Merroun, M. L. & Selenska-Pobell, S. (2000). Microdiverse types of *T. ferrooxidans* and their interactions with uranium. *FZR Report 285*, p. 53.
Merroun, M. & Selenska-Pobell, S. (2001). Interaction of three eco-types of *Acidithiobacillus ferrooxidans* with U(VI). *BioMetals, 14(2)*, 171–179.
Merroun, M., Geipel, G., Nicolai, R., Heise, K-H. & Selenska-Pobell, S. (2002). Complexation of uranium (VI) by three eco-types of *Acidithiobacillus ferrooxidans* studied using time-resolved laser-induced fluorescence spectroscopy and infrared spectroscopy. *BioMetals* (submitted).
Merroun, M., Hennig, C., Rossberg, A., Reich, T. & Selenska-Pobell, S. (2002). Characterization of U(VI)-*Acidithiobacillus ferrooxidans* complexes by EXAFS, electron microscopy and energy-dispersive X-ray analysis. *Radiochemica Acta* (submitted).
Moore, L. R., Rocap, G. & Chisholm, S. W. (1998). Physiology and molecular phylogeny of coexisting *Prochlorococcus ecotypes. Nature, 393*, 464–467.
Nelson, Y. M., Lion, L. W., Ghiorse, W. C. & Shuler, M. L. (1999). Production of biogenic Mn oxides by *Leptothrix discophora* SS-1 in a chemically defined growth medium and evaluation their Pb adsorption characteristics. *Applied and Environmental Microbiology, 65*, 175–180.
Pace, N. R. (1997). A molecular view of microbial diversity and the biosphere. *Science, 276*, 734–740.
Panak, P., Hard, B., Pietsch, K., Selenska-Pobell, S., Bernhard, H. & Nitsche, H. (1998). Uranium reduction by a natural *Desulfovibrio* isolate JG 1. *FZR Report 218*, pp. 63–64.
Panak, P., Selenska-Pobell, S., Kutschke, S., Geipel, G., Bernhard, G. & Nitsche, H. (1999). Complexation of uranium with the cells of *Thiobacillus ferrooxidans* and *Thiomonas cuprina. Radiochimica Acta, 84*, 183–186.
Panak, P., Raff, J., Selenska-Pobell S., Geipel, G., Bernhard G. & Nitsche H. (2000). Complex formation of U(VI) with *Bacillus*-isolates from a uranium mining waste pile. *Radiochimica Acta, 88*, 71–76.
Pazirandeh, M., Wells, B. & Ryan, R. L. (1998). Development of bacterium-based heavy metal biosorbents: enhanced uptake of cadmium and mercury by *Escherichia coli* expressing a metal binding motif. *Applied and Environmental Microbiology, 64*, 4068–4072.
Pedersen, K. (1997). Microbial life in deep granitic rock. *FEMS Microbiology Reviews, 20*, 399–414.
Prüß, B. M., Francis, K. P., Von Stetten, F. & Scherer, S. (1999). Correlation of 16S ribosomal DNA signature sequence with temperature-dependent growth rates of mesophilic psychrotolerant strains of the *Bacillus cereus* group. *Journal of Bacteriology, 181*, 2624–2630.
Purchase, D., Miles, R. J. & Young, T. W. K. (1997). Cadmium uptake and nitrogen fixing ability in heavy-metal resistant laboratory and field strains of *Rhizobium leguminosarum biovar. trifolii. FEMS Microbiology Ecology, 22*, 85–93.
Radeva, G. & Selenska-Pobell, S. (2000). Bacterial diversity in drain waters of several uranium waste piles. *FZR Report 285*, p. 56.
Radeva, G., Flemming, K. & Selenska-Pobell S. (2000). Molecular analysis of bacterial populations in ground water polluted with heavy metals. *FZR Report 285*, p. 57.
Raff, J., Kirsch, R., Kutschke, S., Mertig, M., Selenska-Pobell, S. & Pompe, W. (1999). The surface layer protein of *Bacillus sphaericus* Isolate JG-A12 from a uranium waste pile. 99th General Meeting of American Society for Microbiology, 30 May to 3 June 1999. Chicago, USA, p. 398.
Raff, J., Mertig, M., Pompe, W. & Selenska-Pobell, S. (2000a). Proteolytical analysis of the S-layer proteins of the uranium waste pile isolate *Bacillus sphaericus* and the reference strain *B. sphaericus* NCTC 9602. *FZR Report 285*, p. 55.
Raff, J., Selenska-Pobell, S., Wahl, R., Mertig, M. & Pompe, W. (2000b). A novel S-layer protein from a uranium mining waste pile isolate *B. sphaericus* JG-A12 and its interactions with uranium and other metals. Bacterial-metal/Radionuclide interactions: basic research and bioremediation. Abstract Book, p. 26.
Rawlings, D. E., Tributsch, H. & Hansford, G. S. (1999). Reasons why 'Leptospirillum'-like species rather that *Thiobacillus ferrooxidans* are the dominant iron-oxidizing bacteria in many commercial processes for biooxidation of pyrite and related ores. *Microbiology, 145*, 5–13.
Reich, T. Institute of Radiochemistry, FZR (personal communication).
Roane, T. M. & Pepper, I. L. (2000). Microbial responses to environmentally toxic cadmium. *Microbial Ecology, 38*, 358–364.

Sára, M. & Egelseer, E. M. (1996). Functional aspects of S-layer. In U. B. Sleytr, P. Messner, D. Pum & M. R. G. Sára (Eds), *Crystalline Bacterial Cell Surface Proteins* (pp. 103–131). New York: Academic Press.

Satchanska, G., Kampf, G., Flemming, K. & Selenska-Pobell, S. (1999). Molecular bacterial diversity in soils and waters of two East German uranium mining waste piles. Bacterial-metal/radionuclide interactions: basic research and bioremediation. *FZR Report 252*, pp. 96–98.

Schlötenburg, C., Von Wintzigerode, F., Hauck, R., Hagemann, W. & Göbel, U. B. (2000). Bacteria of an anaerobic 1,2-dichloropropane dechlorinating mixed culture are phylogenetically related to those of other anaerobic dechlorinating consortia. *International Journal of Systematic Environmental Microbiology, 50*, 1505–1511.

Selenska, S. & Klingmüller, W. (1991). Direct detection of *nif*-gene sequences of *Enterobacter agglomerans* in soil. *FEMS Microbiology Letters, 80*, 243–246.

Selenska-Pobell, S. (1995). Direct simultaneous extraction of DNA and RNA from soil. In A. D. L. Akkermans, J. D. Van Elsas & F. J. De Bruijn (Eds), *Molecular Microbial Ecology Manual 1.5.1* (pp. 1–17). Dordrecht, the Netherlands: Kluwer Academic Publishers.

Selenska-Pobell, S. & Evguenieva-Hackenberg, E. (1995). Fragmentations of the large-subunit rRNA in the family *Rhizobiaceae. Journal of Bacteriology, 177*, 6993–6998.

Selenska-Pobell, S., Otto, A. & Kutschke, S. (1998). Identification and discrimination of Thiobacilli using ARDREA, RAPD and Rep-APD. *Journal of Applied Bacteriology, 84*, 1085–1091.

Selenska-Pobell, S., Panak, P., Miteva, V., Boudakov, I., Bernhard, G. & Nitsche, H. (1999). Selective accumulation of heavy metals by three indigenous *Bacillus* strains, *B. cereus*, *B. megaterium* and *B. sphaericus*, from drain waters of a uranium waste pile. *FEMS Microbiology Ecology, 29*, 59–67.

Selenska-Pobell, S., Flemming, K. & Radeva, G. (2000). Direct detection and discrimination of different *Thiobacillus ferrooxidans* types in soil samples of a uranium mining waste pile. *FZR Report 285*, p. 52.

Selenska-Pobell, S., Flemming, K., Kampf, G., Radeva, G. & Satchanska, G. (2001). Bacterial diversity in soil samples from two uranium waste piles as determined by Rep-APD, RISA and the 16S rDNA retrieval. *Antonie van Leewenhuek, 79*, 149–161.

Service, R. F. (1997). Microbiologists explore life's rich, hidden kingdoms. *Science, 275*, 1740–1742.

Sharma, P. K., Balkwill, D. L., Frenkel, A. & Vairavamurthy, M. A. (2000). A new *Klebsiella planticola* strain (Cd-1) grows anaerobically at high cadmium concentrations and precipitates cadmium sufide. *Applied and Environmental Microbiology, 66*, 3083–3087.

Shippers, A., Hallmann, R., Wentzien, S. & Sand, W. (1995). Microbial diversity in uranium mine waste heaps. *Applied and Environmental Microbiology, 61*, 2930–2935.

Stroes-Gascoyne, S. & West, J. M. (1997). Microbial studies in the Canadian nuclear fuel waste management program. *FEMS Microbiology Reviews, 20*, 573–590.

Strous, M., Fuest, J. A., Kramer, E. H. M., Logemann, S., Muyzer, G., Van de Pas-Schoonen, K., Webb, R., Kuenen, J. G. & Jetten, K. S. M. (1999). Missing lithotroph identified as new planctomycete. *Nature, 400*, 446–449.

Tzvetkova, T., Flemming, K., Groudeva, V. & Selenska-Pobell, S. (2000). Recovery and characterization of *Leptospirillum ferrooxidans* in soil samples of two uranium mining waste piles. *FZR Report 285*, p. 58.

Tzvetkova, T., Groudeva, V. & Selenska-Pobell, S. (2002). Microdiversity of *Leptospirillum ferrooxidans* isolates recovered from uranium wastes and their interactions with U(VI). *FEMS Microbiology Ecology* (submitted).

Valentine, N. B., Bolton, H., Kingsley, M. T., Drake, G. R., Balkwill, D. L. & Plymale, A. E. (1996). Biosorption of cadmium, cobalt, nickel, and strontium by a *Bacillus simplex* strain isolated from the vadose zone. *Journal of Industrial Microbiology, 16*, 189–196.

Versalovic, J., Schneider, M., De Brujin, F. J. & Lupski, J. R. (1994). Genomic fingerprinting of bacteria using repetitive sequence-based polymerase chain reaction. *Methods in Molecular Cell Biology, 5*, 25–40.

Wahl, R. (personal communication).

Ward, D. M., Weller, R. & Bateson, M. M. (1990). 16S rRNA sequences reveal numerous unculturated microorganisms in a natural community. *Nature, 345*, 63–65.

Wikström, P. Anderson, A. C. & Forsman, M. (1999). Biomonitoring complex microbial communities using random amplified polymorphic DNA and principal component analysis. *FEMS Microbiology Ecology, 28*, 131–139.

Willdung, R. E., Gorby, Y. A., Krupka, K. M., Hess, N. J., Li, S. W., Plymale, A. E., McKinley, J. P. & Frederickson, J. K. (2000). Effect of electron donor and solution chemistry on products of dissimilatory reduction of technetium by *Shewanella putrefaciens. Applied and Environmental Microbiology, 66*, 2451–2460.

Yanagibayashi, M., Nogi, Y., Li, L. & Kato, C. (1999). Changes in the microbial community in Japan Trench sediment from a depth of 6. 292 m during cultivation without decompensation. *FEMS Microbiology Letters, 170*, 271–279.

Yang, Y. H., Yao, J., Hu, S. & Qi, Y. (2000). Effects of agricultural chemicals on DNA sequences diversity of soil microbial community: a study with RAPD marker. *Microbial Ecology, 39*, 72–79.

Miranda J. Keith-Roach and Francis R. Livens (Editors)
© 2002 National Environment Research Council. Published by Elsevier Science Ltd. All rights reserved

Chapter 9

Microbial effects on waste repository materials

Julia M. West[a], Ian G. McKinley[b], Simcha Stroes-Gascoyne[c]

[a]*British Geological Survey, Keyworth, Nottingham, NG12 5GG, UK*
[b] *NAGRA, Hardstrasse 73, CH-5430 Wettingen, Switzerland*
[c] *AECL, Whiteshell Laboratories, Pinawa, Manitoba, R0E 1L0, Canada*

1. Introduction

This chapter considers the potential influence of microbial activity on repositories for radioactive waste. It starts with a summary of general radioactive waste concepts and then moves on to give a brief account of the microbiology of host rocks and the repository environment. The significance of microbial activity on the various materials present in the repository is then summarised, with a focus on high-level waste (HLW) and long-lived intermediate waste. Thereafter, the total performance of the series of 'engineered barriers' is reviewed and models that have been developed for the quantification of microbial perturbations are described. Finally, approaches to testing the conclusions of such models, using laboratory, field and natural analogue studies are overviewed and areas for future work are highlighted.

The chapter explicitly considers only deep geological repositories in fully saturated host rocks. Thus, there is no discussion of particular attributes of alternative disposal methods (e.g. liquid waste injection into permeable strata, rock melters, subseabed disposal), dry host rocks (e.g. salt, anhydrite) and unsaturated sites (e.g. Yucca Mountain in the USA). Nevertheless, many of the general points considered would be equally applicable to such systems.

2. Radioactive waste concepts

Deep geological disposal has been implemented, or is being planned, in many countries for many types of waste (e.g. IAEA, 1997). As yet, however, no repository for the most radioactive HLW is operational, although significant steps in this direction have been taken in Finland and Sweden (see Chapter 10, this volume), and extensive site characterisation efforts are being carried out in the USA. An up-to-date overview of progress in national (and international) nuclear waste disposal programmes can be accessed at http://www.radwaste.org, while the background principles of radioactive waste disposal can be found in a number of textbooks (e.g. Chapman & McKinley, 1987, Savage, 1995).

Geological repositories are based on the 'multiple barrier principle', in which long-term safety is assured by a series of engineered and natural barriers. Waste is generally solidified in some kind of stable matrix, encapsulated in a container or some other type of package and placed into underground tunnels, caverns or silos. This packaged waste, along with any backfilling used to seal void spaces and other structural materials such as tunnel linings or plugs, is termed the 'engineered barrier system' (EBS). Surrounding these engineered barriers is the rock itself, termed the 'geosphere'. The 'near field' consists of the waste itself, the EBS and disturbed rock surrounding the repository (disturbed mechanically, thermally, hydraulically and chemically during repository construction and EBS emplacement). The 'far field' comprises the undisturbed geosphere and extends to the surface biosphere.

Many different combinations of engineered and geological barriers have been proposed to provide safe repository concepts for particular combinations of waste type(s) and geographical/geological setting. Repositories may contain a great diversity of materials, which can be subdivided in terms of their function as waste matrices, canisters and over packs, buffers and backfills and assorted structural elements. In simple terms, the roles played by these materials are:

- Waste matrix – immobilisation of contained radionuclides in a corrosion/leach-resistant solid;
- Canister/over pack – physical isolation and protection of the waste matrix;
- Buffer/backfill – physical, chemical and hydrological protection (buffering) of the canister/over pack and restriction of radionuclide transport;
- Structural elements – ensuring stability of the operational repository and sealing of the repository thereafter.

Depending on the wastes considered, the host rock and other practical considerations, these functions may be achieved with different materials (Table 1). High Level Waste (HLW) usually consists of spent fuel – if this is considered as a waste for 'direct disposal' – or the solidified, high-activity wastes from the reprocessing of such fuel. In the latter case, the solidification matrix is usually a borosilicate glass or some kind of ceramic such as 'SYNROC' (Lutze & Ewing, 1988). A range of corrosion resistant canisters has been developed for HLW, often using steel, copper, titanium or special alloys. The buffer used in many concepts is either purified clay (bentonite) or clay/sand mixtures and backfills may be clays, or mixtures of clays and crushed host rock (Fig. 1). The waste matrix is generally extremely stable under expected geochemical conditions, with calculated dissolution times in the order of 10^5 to 10^6 years. Canisters are generally designed to provide complete containment of the waste during an initial period, typically around 10^3 years, during which high activity, short-lived radionuclides decay. After failure of containment, canister material may continue to play an important role as a redox buffer or radionuclide sorbent. In the case of spent fuel, however, some key radionuclides (such as ^{135}Cs and ^{129}I) are present in an easily leachable form, which can lead to designs with even longer-lived canisters or over packs (e.g. Cu or Ti). In most concepts, some kind of clay-based buffer or backfill is envisaged (e.g. bentonite or a bentonite/sand mixture). Swelling clays have a range of important barrier properties, particularly:

Table 1
Materials in a nuclear waste repository

	Waste matrix	Canister/ overpack	Backfill/ buffer	Others
HLW	Glass Spent fuel (UO_2/MOX) SYNROC	**Steel** **Copper** **Titanium**	***Clay, crushed rock***	Liners – *concrete*, **steel** Plugs/seals – *concrete*, ***clay*** Grouts – cement, ***clay***
TRU/L/ILW	**Metals** *Cement/concrete* Resins Bitumen Organic waste components e.g. cellulosic materials)	**Steel** *Concrete*	***Clay*** ***Crushed rock*** *Cement* *MgO_2 (WIPP)*	

Key: Groupings: Stable waste matrices, **Metals**, *Cement and concrete*, ***Clays and crushed rock***, Organics and miscellaneous materials.
HLW = high level waste; TRU = wastes containing long lived radionuclides; ILW = intermediate level waste; LLW = low level waste; WIPP = waste isolation pilot plant; MOX = mixed oxide fuel; SYNROC = waste isolation material.

- Plasticity/self-healing (physical protection of the canister/over pack);
- Low hydraulic conductivity (solute transport predominantly by diffusion);
- Chemical buffering (especially pH, redox);
- High radionuclide sorption capacity;
- Colloid filtration.

For other wastes containing high concentrations of long-lived radionuclides (sometimes termed 'TRU' – transuranic element containing wastes – even though transuranic radionuclides may not be their most problematic components), a similar series of barriers may be considered. Given the lower toxicity, higher physical volume and heterogeneity of such wastes, however, lower barrier performance may be acceptable, leading to extensive use of cement and concrete. Other low level and intermediate level (L/ILW) waste types may also be disposed of in deep repositories, but the requirement on the engineered barriers for these is generally rather minimal. Low and intermediate level waste may be produced in a wide range of solidification matrices (e.g. cement, bitumen, resins, metals), but is usually contained in steel drums or concrete packages and disposed in caverns backfilled with cementitious grout (Fig. 2).

Despite the great variety in the details of individual repository systems, a common concern is the quantification of their behaviour over very long periods of time. The formalised procedures used to carry out such a 'performance assessment' (PA) utilise a database of the characteristics of the waste, the EBS, the surrounding geological barrier and the surface 'biosphere'. This provides input and boundary conditions for a series of

Safety barrier system for high-level waste

Glass matrix (in steel mould)

- Low corrosion rate of glass
- High resistance to radiation damage
- Homogeneous radionuclide distribution

or

Spent fuel elements

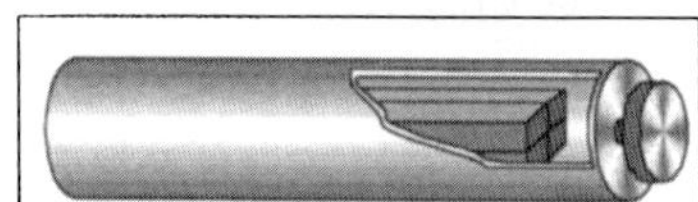

- Low UO_2 dissolution rate
- High radiological / thermal stability of UO_2-matrix

Steel canister

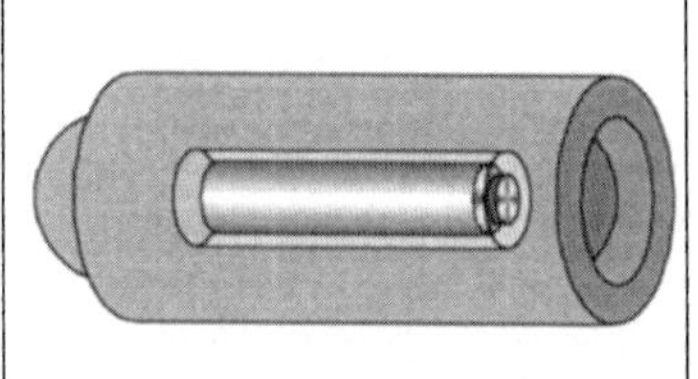

- Completely isolates waste for > 1000 years
- Corrosion products act as a chemical buffer
- Corrosion products take up radionuclides

Bentonite backfill

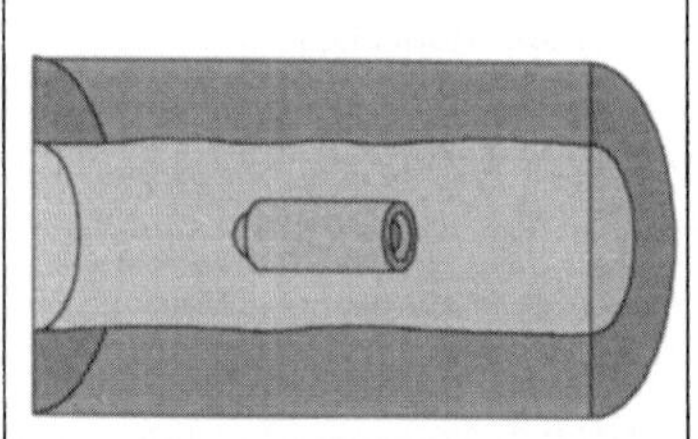

- Long resaturation time
- Low solute transfer rates (diffusion)
- Retardation of radionuclide transport (sorption)
- Chemical buffer
- Low radionuclide solubility in leachate
- Colloid filter
- Plasticity (self-healing following physical disturbance)

Geological barriers

Repository zone:
- Low water flux
- Favourable geochemistry
- Mechanical stability

Geosphere:
- Retardation of radionuclides (sorption, matrix diffusion)
- Reduction of radionuclide concentration (dilution, radioactive decay)
- Physical protection of the engineered barriers (e.g. from glacial erosion)

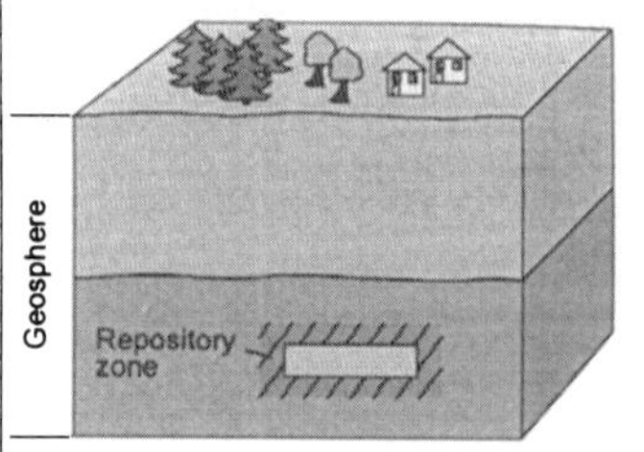

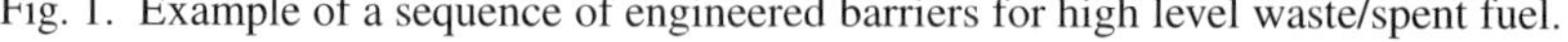

Fig. 1. Example of a sequence of engineered barriers for high level waste/spent fuel.

Safety barrier system for short-lived waste

Solidification matrix
(cement, bitumen, resins)

- Low radionuclide release rate

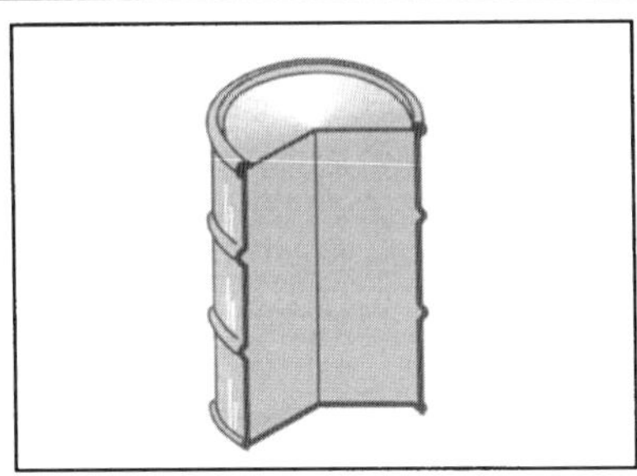

Container, container infill, waste drums
(concrete / cement grout / steel)

- Low solute transport rates
- Radionuclide sorption
- Chemical buffer

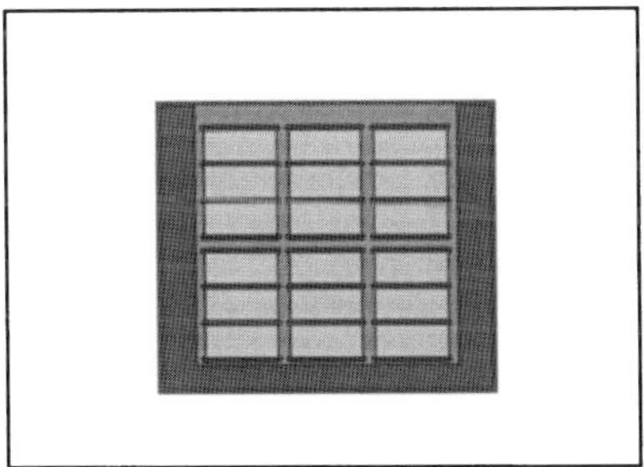

Emplacement cavern, lining and backfill
(concrete / special mortar)

- Limited water access
- Delay start of release
- Low solute transport rates
- Chemical buffer
- Radionuclide sorption
- Allows gas escape

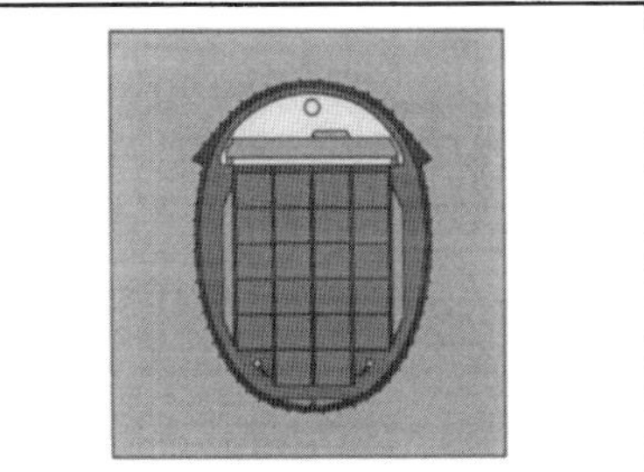

Geological barriers

Repository zone:
- Low water flux
- Favourable hydrochemistry
- Mechanical stability

Geosphere:
- Retardation of radionuclides (sorption, matrix diffusion)
- Reduction of radionuclide concentration (dilution, radioactive decay)
- Physical protection of the engineered barriers (e.g. from glacial erosion)

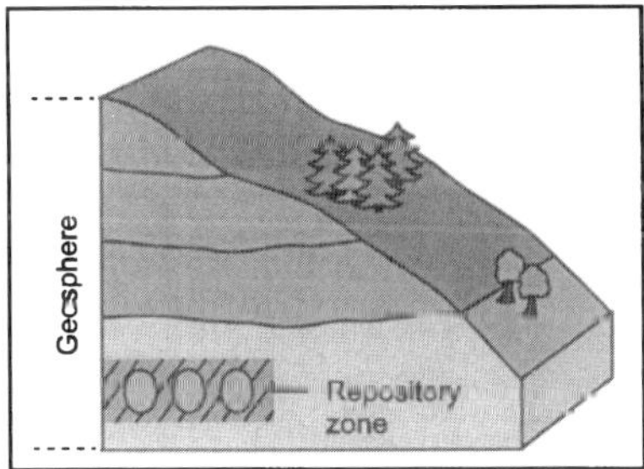

Fig. 2. Example of a sequence of engineered barriers for L/ILW.

models to represent slow EBS degradation, the release of radionuclides from the waste and their transport through the geosphere to the accessible environment. As it is not possible to predict the future, such analyses are carried out for a range of possible futures, termed scenarios. Given the requirement for PA, it is important to consider the many potential effects of microbial activity in a waste repository. The following sections consider, in turn, the presence of microorganisms within the repository system, their possible influence on the performance of the engineered and natural barriers and, finally, the more challenging questions of how such influences can be quantified – especially in view of the enormously long timescales involved.

3. Microbiology of relevant geological formations

Following initial literature searches, an early part of several national geomicrobiology programmes was to establish the presence of microbes in the geological formations being considered as host rocks for repositories, the rationale being that if no indigenous populations are present then microbes may not be viable in the environment selected and hence may not be a problem. Studies have taken place in a variety of rock types such as granites, sedimentary rocks, evaporites and volcanic tuffs. Details are given elsewhere (e.g. West et al., 1982, 1991a; West & McKinley, 1984; Christofi & Philp, 1991; West, 1995) and are summarised in Table 2. An example of work on microbial characterisation of igneous rocks in Sweden is described in Chapter 10 of this volume.

The composition of microbial populations varies with site (West, 1995). It is, however, increasingly recognised that the analysis of ambient microbial populations is part of site characterisation and is as critical in understanding the hydrogeochemistry of the

Table 2

Bacterial populations in deep groundwater environments (after West et al., 1991; West, 1995)

Location	Geology	Depth (mbgl)	Bacterial count
Canada	Granite	350–400	Total count 10^3–10^5 cells ml^{-1}
Finland	Granite (various)	200–950	Total counts 10^5–10^6 cells ml^{-1} (includes fresh, brackish and saline waters)
Japan	Granite	Approx. 400–790	Total counts 10^2–10^7 bacteria ml^{-1}
Stripa, Sweden	Granite	799–1240	Total counts 2.0×10^1 to 1.3×10^5 cells ml^{-1}
Aspo, Sweden	Granite	129–1078	Total counts 1.5×10^4 to 1.8×10^6 cells ml^{-1}
Grimsel, Switzerland	Granite	Approx. 350 m	9.5×10^1 to 9.0×10^4 CFU ml^{-1*}
Altnabreac, UK	Granite	10–281	9.4×10^5 CFU ml^{-1*}
Mol, Belgium	Boom clay	190–223	1.2×10^3 CFU ml^{-1*}
Harwell, UK	Oxford clay	165–331	8.6×10^3 to 3.5×10^5 CFU ml^{-1*}
Asse, Germany	Salt	750	ND
Yucca Mountain, USA	Volcanic tuff	60	10^2–10^3 bacteria g^{-1} dry weight (NB above water table)

Key: CFU = colony forming units; * = aerobic heterotrophs; ND = not detected; mbgl = metres below ground level.

site as in assessing the role of microbes on repository performance. For example, the oxidising/reducing (redox) conditions of a site are of fundamental importance in assessing site suitability. However, many of the redox couples that collectively determine the value of the 'Eh' are multielectron transfers between S, N and C species, which are kinetically slow at relevant temperatures and will proceed only with microbial catalysis (see Chapters 3 and 4, this volume). Indeed, in experiments under differing Eh conditions with rock and groundwater from a granitic environment, sulfate and iron-reducing bacteria appeared to catalyse secondary smectite formation (Bateman et al., 1999). It is important, therefore, to focus on the in situ activity levels of relevant physiological microbial groups, such as sulfate-reducing bacteria (SRB), sulfur oxidisers, iron reducers and methanogens, rather than counting numbers of cells or determining microbial families or species.

There are a number of different ways in which microbial activity can affect the behaviour of radionuclides and other contaminants in the near and far field. Microbes can produce extracellular polymers that can act as ligands and form complexes with dissolved radionuclides and render them less likely to sorb (Birch & Bachofen, 1990). Trace amounts of radionuclides or non-radioactive contaminants can be sorbed on, and subsequently desorbed from, the outer surfaces of microbes (Gadd, 1988). From a contaminant transport perspective, these contaminated microbes can then act as 'living' radiocolloids, which have the potential to be transported in flowing groundwater through the far field. In fact, because radiocolloids are larger than ions, they are less likely to diffuse into the interconnected pore space of the rock fabric adjacent to fractures than dissolved species. This inaccessibility of the rock mass may reduce retardation which, in turn, results in an enhanced transport of colloids compared to dissolved species (Vilks & Bachinski, 1994). Microbes can also incorporate trace amounts of radionuclides into their cell structure and these isotopes are released when the microbes subsequently decompose (Gadd, 1988). It is also possible that, under the oligotrophic conditions that exist in most geological host formations, microbes will form biofilms that strongly adhere to the geological surfaces lining the flow paths in the far field (West & McKinley, 1984; Costerton et al., 1986; Brown & Hamon, 1994). These biofilms can affect the transport of radionuclides and other contaminants by acting as a barrier to radionuclide sorption on geological surfaces or, alternatively, by providing new sorption sites (potentially important for normally poorly or non-sorbing radionuclides). It is also possible that the biofilms have little or no effect on sorption if they are porous and do not inhibit the access of radionuclides to the underlying geological surfaces. Finally, it is also conceivable that fragments of biofilm containing radionuclides could be sloughed off and transported as particulates (Vandergraaf et al., 1997).

4. The microbiology of the near field

The near field environment of a HLW repository is often perceived as being too extreme for life because it is hot and dry, with few nutrients to sustain life and, in some designs, highly radioactive.

However, conditions in a repository will vary with waste type and with materials used in its construction (Table 3). For HLW, conditions will be very radioactive (hundreds to thousands of Sieverts at the surface of the waste) and hot (in most cases ~50–100°C). In

Table 3

An extract from the Swiss National Inventory providing some properties for HLW (WA-1) and various types of L/ILW – average values for single containers (Alder & McGinnes, 1995)

Waste sort	Raw waste	Material of conditioning	α activity (Bq/container)	β-γ activity (Bq/container)	Surface dose rate* (Sv/h)	Heat output (W)
WA (reprocessing waste)						
WA-1	High-level vitrified residues	Glass	1.1×10^{14}	2.8×10^{16}	3.4×10^{3}	2.8×10^{3}
WA-2	Precipitates and sludges or $BaCO_3$ and MEB crud	Bitumen or cement	3.6×10^{8} to 1.8×10^{10}	3.5×10^{11} to 4.3×10^{12}	1.2×10^{-1} to 5.2×10^{-1}	8.2×10^{-2} to 4.5×10^{-1}
WA-4	Hulls and ends	Cement	1.4×10^{11} to 2.8×10^{11}	1.4×10^{14} to 8.2×10^{14}	2.4×10^{1} to 3.9×10^{1}	2.0×10^{1} to 1.2×10^{2}
WA-5	Technological low-level waste	Compacted	2.6×10^{6} to 1.2×10^{8}	8.0×10^{7} to 3.8×10^{9}	2.1×10^{-5} to 1.0×10^{-4}	4.5×10^{-6} to 2.1×10^{-4}
WA-6	α-emitting technological waste	Cement	7.4×10^{10}	7.3×10^{11}	7.0×10^{-3}	8.0×10^{-2}
WA-7	Centrifuge cake slurry	Cement	3.8×10^{11}	4.0×10^{13}	2.0	3.5
MIF (waste from medicine, industry and research)						
MIF-1	β, γ-emitting waste	Cement	0.0	3.7×10^{10}	9.5×10^{-3}	3.1×10^{-3}
MIF-2	Tritium-bearing waste	Cement	0.0	1.2×10^{13}	3.0×10^{-5}	1.1×10^{-2}
MIF-3	'α' waste	Cement	1.2×10^{8} to 1.3×10^{10}	0.0 to 1.1×10^{9}	3.0×10^{-9} to 2.5×10^{-6}	1.0×10^{-4} to 1.2×10^{-2}
MIF-4	Radium-bearing waste	Cement	2.1×10^{9}	0.0	5.4×10^{-6}	1.6×10^{-3}
MIF-5	'α, β, γ' waste	Cement	7.6×10^{8} to 1.9×10^{9}	5.9×10^{10} to 1.7×10^{11}	1.7×10^{-2} to 3.2×10^{-2}	7.7×10^{-3} to 1.5×10^{-2}

* Dose is dominated by γ-radiation therefore absorbed dose (in Gray-Gy) is approximately equal to individual dose equivalent (in Sieverts –Sv).

Table 4
Tolerance of microbes to extreme environments (updated from West, 1995)

Condition	Example of organism	Limit of growth
High temperature	'Black smoker' bacteria	250°C (at 26.9 MPa)
Low temperature	*Sporotrichum carnis*	−20°C
High pH	Nitrifying bacteria	12
Low pH	*Thiobacillus ferrooxidans*	0
High salinity	*Halobacterium halobium*	50% salt by wt
Low salinity	*Salmonella oranienburg*	70 ppb dissolved salts
High pressure	*Desulfovibrio desulfuricans*	180 MPa
Radiation	*Deinococcus radiodurans*	Single dose 5000 Gy

the early phases of some concepts for the proposed Yucca Mountain site in Nevada, USA, the temperatures could be in excess of 200°C. There could also be considerable pressure generated from the overlying water column and rock burden (perhaps 10–25 MPa) and, in some concepts, high salinity occurs. Heat and radiation will be much less in a L/ILW repository, but the concretes used will generate a hyperalkaline environment.

Such extreme conditions for HLW were initially thought to preclude life and hence most early safety or performance assessments implicitly assumed that a repository would be an abiological environment. However, there are many examples of individual microbial species tolerating specific extremes of environment (West et al., 1982; West, 1995). Some of those relevant to a repository are given in Table 4. However, it is difficult to find reported work on microbial activity in natural environments where all the conditions that would be generated in a repository were present in that environment at the same time (West et al., 1991a).

The ability of microbes to tolerate high radiation doses and high temperatures simultaneously is of particular interest to HLW programmes. Studies in Canada suggest that radiation and desiccation effects could create a zone of depleted or reduced microbial activity extending a few tens of cm into buffer material surrounding waste containers (Stroes-Gascoyne & West, 1997). Work on indigenous microbes from the Yucca Mountain and Nevada test site demonstrate that they are capable of surviving gamma radiation up to 10 kGy (at 1.63 Gy min^{-1}) in a viable but non-culturable form and can be resuscitated to a culturable form (Pitonzo et al., 1999a, b). A very recent study (Billi et al., 2000) on ionising radiation (X ray) resistance in desiccation tolerant cyanobacteria showed recovery of viable cells after 15 kGy but not after exposure to 20 kGy. The bacteria studied were isolated from desert and hypersaline environments and the radiation resistance observed was thought to reflect the ability of these extreme-environment bacteria to survive prolonged desiccation through efficient DNA damage repair. These findings could be of significance for unsaturated and saline disposal environments. Other radiation work with sulfate-reducing bacteria found a tolerance of up to 10^3 Gy over 40 hours (West, 1995). Observations of the reactor core at Three Mile Island showed that microbes were present in spite of the regular use of H_2O_2 as a biocide and receiving doses of 10 Gy h^{-1} (Booth, 1987). All these conditions are comparable with dose estimates for HLW disposal.

The effects of moisture content on microbial presence were investigated further in an in situ experiment with a full-scale nuclear fuel waste disposal container (with a heater mimicking the heat output of a nuclear fuel waste container), which was surrounded by compacted buffer (50% bentonite and 50% sand) (Stroes-Gascoyne et al., 1997). Microbes could only be cultured from the buffer material where the water activity (a_w) was greater than 0.96. This suggests that buffer material will be populated by active microorganisms only where a_w is such that free water is available for active life.

The effects of temperature on microbe survival and migration were studied for the Yucca Mountain project. Here a block of tuff was heated to a maximum temperature of 142°C in some locations. Two test isolates were found to tolerate the conditions and to migrate through the tuff itself to a distance of 1.5 m from their injection point (Chen et al., 1999).

Tolerance to alkaline conditions has been shown in a study of alkaline groundwaters in Jordan where a range of microbes tolerated pH 12 and above (West et al., 1995). Work in France has shown also that microbes will grow when in contact with cement and will reduce the solution pH (Perfettini et al., 1991). In experiments where sulfate-reducing bacteria have been grown over a range of pH and Eh values, activity at pH 8–10 was found to be enhanced by decreasing Eh (Fukunaga et al., 1995).

Microbial tolerances to a whole range of extreme environmental conditions demonstrate that a repository, even for HLW, or a repository backfilled with cement, cannot be assumed to be sterile for its entire lifetime. Given this fact, it then becomes clear that the only certain controls on life will be the availability of water, nutrients and energy sources. Chapter 1 of this volume discusses microbial lifestyles in more detail. Thus, an assessment must be made of the likely impacts of microbial activity on the waste itself, on the containment materials and on subsequent radionuclide transport.

During repository excavation, construction and waste emplacement, a range of microbes, together with exploitable energy and nutrient sources will be added to the subsurface environment, making it less oligotrophic – at least for some period of time. Work on disposal of spent fuel in granitic environments, for example, has shown that a significant amount of nutrients may be introduced from explosive residues associated with excavated rock reused in backfill material. Also, organic matter in backfill clays has been shown to increase microbial viability when treated with heat and radiation (Stroes-Gascoyne & West, 1997). For assessing the microbial effects on repository materials, measurements on microbial populations in materials are probably more relevant than studies of the host rock. Alternatively, studies of natural alkaline environments could be relevant where cement and concrete are of interest in the waste disposal concept (Bath et al., 1987; West, 1995).

5. Biodegradation of repository materials

The various groups of materials listed in Table 1 are considered in terms of the various roles they may have in a repository. It should be noted that the specific examples discussed concentrate on the main materials involved and do not consider the exotic components that inevitably arise in any national waste programme (especially waste from research centres – e.g. Bi/Pb eutectics used as beam-dumps in particle accelerators).

Stable HLW waste matrices

Little, if any, work has been reported on the biodegradation of HLW glass, SYNROC or spent fuel. Experimental studies are, however, carried out in conditions that are certainly not sterile, and hence empirical degradation rates may contain a contribution from microbially catalysed processes. Direct microbially induced corrosion of materials such as borosilicate glass, uranium dioxide or SYNROC is not to be expected; rather microbial byproducts could be of significance. For example, etching of glass by fungal byproducts such as organic acids is well known (West et al., 1982) but is related to contamination of glass surfaces by organic material.

Metals

Microbially influenced corrosion (MIC) may affect the integrity of metals and alloys used as encapsulation (container) materials, tunnel liners or even included as waste. For metals acting as containers/over packs, an important factor is their lifetime before physical failure, so that localised corrosion is particularly important. However, for metals which are wasteforms (e.g. activated steel or zircalloy) only the total corrosion rate, which will govern the rate of release of contained radionuclides, is important.

In addition to corrosion rate, the products of microbially catalysed corrosion can be important if they differ from those resulting from 'inorganic' corrosion. The radionuclide sorption and redox buffering properties of solid phase products are of particular relevance. Gas may be also generated by such corrosion and this could exert pressure on containment materials, again affecting integrity.

MIC can be either direct, with microbes using the metals as energy sources, or indirect, where microbes change the local conditions allowing chemical corrosion to take place. MIC of metallic materials is an electrochemical process and the physical presence of microbial cells, in addition to their metabolic activities, modifies this process. Adsorbed cells grow, reproduce and form colonies that develop into unevenly distributed biofilms, occluding some areas and leaving others exposed to the bulk environment, resulting in the formation of zones of differential aeration and leading to enhanced, uneven corrosion (Geesey, 1993; Little & Wagner, 1996). In all cases, such effects will be controlled by the amounts of nutrients and energy sources available to the organisms. In laboratory experiments simulating a L/ILW disposal regime, sulfate-reducing bacteria enhanced carbon steel corrosion by three times the rate of a control (Philp et al., 1991). Other experiments in realistic conditions have also shown more important localised and deep pitting of steels, which was directly attributable to microbial action (West et al., 1998). In Canada, Ogundele & Jain (1999) determined the effect of a host of microbial metabolites on the corrosion of copper in saline solution. They found that, within the concentration ranges examined, the corrosion rate of copper was slightly enhanced by microbial metabolites compared to the average abiological rate.

Titanium and copper alloys are candidate encapsulation materials in several HLW disposal concepts. The use of such expensive material is justified by their corrosion resistance over very long periods of time so the possible role of microbes in affecting these materials has been a significant area of study. Within a HLW repository, the environmental conditions will change with time. Upon emplacement of the waste, the conditions will be warm and oxidising. As the radionuclides contained in the waste decay and the initially trapped oxygen is consumed (by corrosion of the container, by reduced components in the rocks and backfill and by microbial activity), conditions will change to be cool and anoxic. Consequently, the nature of the chemical corrosion reactions will also change with time. In general, localised corrosion and fast uniform corrosion are only expected to occur in the oxidising phase (Stroes-Gascoyne & West, 1996). For HLW, these conditions would occur when radiation and thermal effects are most hostile to life and where lack of moisture might inhibit microbial activity near containers (Stroes-Gascoyne et al., 1997) and, therefore, MIC effects. Repopulation of the area near the containers by microbes later in the life of a repository will depend on whether organisms can move through the pore space in the compacted buffer material. Experiments to investigate buffer material repopulation with viable *Pseudomonas stutzeri* after a 'sterilisation' period suggested that movement could not take place within a compacted bentonite matrix (Stroes-Gascoyne & West, 1997). Swedish studies by Motomedi et al. (1996) and Pusch (1999) have also demonstrated the lack of movement (and even survival) in compacted backfills consisting mostly of bentonite. However, Stroes-Gascoyne et al. (1997) showed that movement could occur at interfaces between the backfill and other experimental materials. This suggests that fractures or discontinuities, for example cracks in the buffer resulting from initial desiccation, may be a preferred pathway for microbial migration. Resealing will occur as groundwater resaturates the buffer, but resealing may be slower than microbial movement. If a zone of depleted microbial activity is created around the waste containers during the high heat and desiccation period and repopulation is limited by the pore size of the buffer material then microbial activity would be limited to regions outside this depleted zone. Only anaerobic corrosion, probably involving sulfate reducing bacteria, could then occur, as the repository would no longer be in the oxidising phase. In this case, the only microbial impact on the encapsulation materials will result from the diffusion of microbially reduced sulfur species to the container surface. Modelling studies have predicted the extent of sulfate reduction in such a situation and the consequent effects on copper corrosion are expected to be minimal (King et al., 1999).

There are no reports on MIC of titanium (Wagner & Little, 1993) and no literature has been found on direct studies of biodegradation of other possible metallic waste components (e.g. zircalloy, control rod materials) under repository conditions although the extensive studies of MIC in reactor systems may contain relevant information.

Cement and concrete

Cement and concrete may be present in tunnel liners, seals and grouts used in HLW repositories but, in many cases, it is planned to keep such material away from the waste packages because of the potentially detrimental interaction of cement leachates with clay-based backfills. In such roles, the main features of interest are the structural integrity of the cementitious material (mechanical strength) and its hydraulic properties (sealing groundwater movement).

Cementitious materials are generally used much more widely in repositories for other waste types – performing other structural roles and acting as backfills, buffers and containers. In addition to the desirable properties noted above, radionuclide sorption, chemical buffering and colloid filtration may all be important. Activated concrete or concrete/cement solidification matrices are often major waste components and here a further property of importance is the leach rate of radionuclides from the waste matrix.

Degradation of cements and concretes is commonly observed under aerobic conditions (Philp et al., 1991) and could occur in the early phase of a repository. Sulfur-oxidising bacteria such as *Thiobacillus* sp. oxidise sulfur, sulfides and thiosulfates under aerobic conditions, producing sulfuric acid. Nitrifying organisms use ammonia and produce nitric acid in the same conditions. These acids can then attack the concrete matrix by dissolving calcium silicate hydrate gel and $Ca(OH)_2$. Direct anaerobic corrosion of concretes is not known, although organic acids produced from microbial attack on organic materials could be a significant factor (Perfettini et al., 1991). Biofilms can also develop on surfaces (Colasanti et al., 1991; Rogers 1995) although the organisms may be utilising the organic plasticisers added to concrete to increase their workability (Haveman et al., 1996). The locally alkaline conditions produced by the concretes seem unlikely to prevent potential microbial growth.

Clays

Clay-based backfills/buffers, often involving compacted bentonite or bentonite/sand mixtures, are included in the EBS designs for most HLW concepts and some designs for other waste types – particularly if a barrier to advective water flow is required. The very low permeability of such clays also leads to their use in plugs, seals and grouts. As noted previously, the compacted clays have a wide range of favourable properties due to their microporous nature and the presence of large areas of reactive surfaces. The properties of such clays are determined under conditions that are generally not sterile and hence ambient microbial populations are not likely to cause degradation of performance. There is an increasing awareness that microbial activity can contribute to mineral alteration but this would not be expected to be critical in the very low energy environments of interest. In general, it may be reasonable to conclude that direct microbial degradation would not be expected to be a problem for such material.

Organics and other miscellaneous waste materials

A very wide range of materials can be included as waste/waste matrices but, in terms of both mass and potential for biodegradation, organic materials including resins, bitumen, polymers and cellulosic materials are particularly important. Other materials such as ceramics, ashes, precipitation and sludges may be important in particular waste streams but, in most repositories, do not comprise a large component of the total waste inventory and are not likely to be greatly affected by microbial activity. A possible exception is iodine filters in spent-fuel reprocessing waste – these may dominate safety assessments due to the mobility of contained ^{129}I. Direct microbial degradation may not be critical, but I speciation is often influenced by microbial activity and hence the release rate and mobility of this critical nuclide may depend on microbial activity levels.

Much work has been undertaken on bitumen degradation as it is extensively used in L/ILW nuclear waste solidification, e.g. for encapsulation of ion-exchange resins. Laboratory studies have confirmed that biodegradation both of the bitumen and of the resins themselves can be expected (Roffey & Norqvist, 1991; Wolf & Bachofen, 1991). In an operational L/ILW repository, biodegradation is likely to be more rapid as consortia of microbes will be present which will work together to degrade such materials more efficiently than under laboratory conditions. However, recalcitrant organic fractions are likely to remain in the waste (Stotzky, 1986), which are unavailable for microbial use.

Much work has been performed on cellulose degradation as this is a major component of some LLW (Colasanti et al., 1991). A variety of organic acids is produced, with isosaccharinic acid being particularly significant. The impact of this acid production in relation to an alkaline repository is being intensively studied and modelling approaches have been developed (Van Loon & Glaus, 1998).

6. Microbial influence on EBS performance

In the multibarrier concept, the various EBS components provide 'defence in depth' – ensuring that no significant release of radionuclides occurs. As discussed in the previous section, the presence of microbes could potentially degrade the functioning of some of these barriers and the challenge is thus to assess if such degradation is significant when compared to the considerable safety reserves included.

Focusing first on the relatively simple HLW case, it was concluded above that microbial activity was unlikely to affect the performance of the clay backfill. The corrosion rate and failure time of the canister could potentially be influenced, however, although for vitrified HLW this is not usually a factor that critically affects the safety case (e.g. NAGRA, 1994). For spent fuel, however, safety cases, which assume very long canister lifetimes, need to be carefully assessed to ensure that microbial effects have been fully accounted for.

When the canister fails, corrosion of the waste form commences. As indicated above, this corrosion rate is not expected to be influenced directly by microbial activity. The corrosion of spent fuel is, however, very dependent on redox conditions and the potential exists for radiolysis of water to lead to locally oxidising conditions and thus accelerate dissolution. Although the significance of this mechanism is actively debated at present, microbial activity would be likely to have a beneficial effect through consumption of oxidants and extraction of energy from their reaction with kinetically hindered reductants present in the system (e.g. H_2, metallic Fe).

The release of many key radionuclides is constrained not by the corrosion rate of the matrix but by their very low solubility under reducing neutral alkaline conditions. Microbial activity could play an important role – either enhancing solubility via byproducts (organic complexants, carbonate, acids) or reducing it by catalysis of redox reactions (especially for radionuclides which are released in a kinetically stable oxidised form (e.g. UO_2^{2+}, TcO_4^-)).

It is known that common groups of microbes often 'carry' trace nuclides in their catalysis of reactions of major elements, e.g. sulfate-reducing bacteria producing selenosulfides (West et al., 1995). Another example is the iron-reducing bacteria which have been shown to reduce U(VI) to U(IV), hence, making U less soluble and mobile (Gorby & Lovley, 1992). It is possible that these organisms may also reduce other actinides such as Pu.

The production of organic byproducts which can complex and therefore mobilise trace elements is well known for a wide range of relevant organisms. The most extreme examples are organic molecules which are directly utilised for this purpose (e.g. siderophores) which can complex extremely strongly (effectively irreversibly) with a range of relevant elements – notably actinides. Nevertheless, a very wide range of byproducts ranging from simple organics (e.g. formate, acetate, oxalate), larger biodegradation products (e.g. isosaccharinic acid from breakdown of cellulose) and large macromolecules (fulvic and humic acids) can be of importance. This has been reviewed in detail elsewhere (Birch & Bachofen, 1990).

The final key factor constraining the release rates of nuclides from the EBS is retardation during their slow transport through the different EBS components. Microbes could potentially decrease sorption by covering active mineral surfaces, or increase sorption if immobile organisms (i.e. in biofilms) actively take up radionuclides.

The net effect of such sorption depends on the extent of its reversibility and the mobility of the organisms involved. Irreversible uptake is of most significance but its net effect is detrimental only if the organisms are mobile. Biofilms, which are expected to form extensively in oligotrophic rock matrices, play a crucial role in controlling both the mobility of organisms and the sorption of radionuclides in and around organisms in the biofilm. Uptake into microorganisms may, superficially, be difficult to distinguish from sorption on to outer membranes. The difference becomes particularly significant for the cases in which internal mineralisation occurs, which is well known for many microbial groups (Fortin et al., 1997). In the extreme case, this can result in the immobilisation of radionuclides in mineral forms, as occurs in the formation of ores of elements such as uranium. The same process can be detrimental, however, if the microbes containing the concentrated radionuclide can migrate through the engineered or natural barriers and release it when they die or if the mineralised form is released as a mobile colloid – both processes being most unlikely in a compacted bentonite buffer.

The same basic appraisal can be carried out for the other waste streams, but it is very much more difficult to scope the likely magnitude of possible effects due to greater system heterogeneity. Thus biodegradation may have little direct influence on the performance of cement or zircalloy waste forms but be significant for steels and bitumen and very important for cellulosic materials. Similarly, the effect of microbes on radionuclide solubility or mobility is much harder to assess in the hyperalkaline environment found in repositories containing large quantities of cementitious materials. Concrete may be a very effective buffer when intact, but fracturing due either to non-biological processes or microbial byproducts (e.g. gas), could greatly reduce performance, enhancing transport through, for example, advective flow or colloid transport.

Fracturing of the backfill is an example of a perturbation that could potentially short-circuit a key barrier. Microbial processes which could cause such a short-circuit (such as biogenic gas production) or which could increase the consequences of such a short circuit (movement of microbes as 'organic colloids') are of particular concern.

Gas production is of considerable concern in repositories where the waste has a high organic content (L/ILW) or where it contains large quantities of metals (e.g. HLW containers). This is because of both the potential for gas pressurisation, which may compromise the repository by creating transport pathways, and the inherent properties of gas (radio- or chemotoxicity, flammability and chemical reactivity). Gases can be formed by microbial activity in three ways:

1. Direct biodegradation of organic materials which act as a nutrient source for biogenic gas formation;
2. Direct catalysis of anaerobic corrosion of metals; and
3. Indirectly, by producing chemical environments which cause production of gases, e.g. acid conditions which enhance corrosion of metals.

Microbes can also act as consumers of gases (Bachofen, 1991). For example, carbon dioxide is utilised as an electron acceptor by methanogens and acetogens while hydrogen can be used as an electron donor (methane oxidisers). This is discussed in more detail in Chapters 3 and 10 of this volume.

Under the aerobic conditions encountered in a repository soon after waste emplacement carbon dioxide is likely to be the main gas produced by direct microbial action. As the redox conditions change, a range of other gases may be produced. Sulfate-reducing bacteria generate sulfide, which may form toxic hydrogen sulfide gas, and, under very reducing conditions, methane may be formed by methanogens. The net effect of microbial activity is difficult to quantify as gas may also be produced by inorganic mechanisms. Studies of degradation of cellulosic waste under alkaline conditions have shown that carbon dioxide is the major gas produced although it is quickly consumed by carbonation reactions. However, as the redox potential drops methane is also generated and becomes the principal component of the produced gas (Colasanti et al., 1991). Little hydrogen was produced overall, probably because it is being used as a substrate by a number of species. For HLW, biogenic gas production is not expected to be very significant, as the amount of organic material is generally much lower. Studies of gas production from laboratory systems containing backfill and granitic groundwater have shown that the backfill may suppress methane production, probably due to competing sulfate reduction processes (Stroes-Gascoyne & West, 1997). This appears to be supported by a recent study by Stroes-Gascoyne et al. (2000). Here changes in gas composition in compacted 50% bentonite buffer, which had been sealed in a borehole for 6.5 years at in situ (13–17°C) temperatures, were examined and compared to the original gas composition. There was no significant difference between the gas compositions despite the presence of a large population of viable microbes. This suggests that the population was inactive, that the organics associated with the bentonite were not easily degradable and that the resulting lack of oxygen reduction suppressed methanogenesis. Some sulfate-reducing bacterial activity was indicated, not by the presence of hydrogen sulfide but by a small but measurable increase in sulfide in the buffer.

In terms of a perturbed EBS where significant radionuclide transport could occur, microbial influences on such migration need to be assessed. The radionuclide transport properties of any medium depend on both the processes moving solute in solution (advection, diffusion, dispersion) and those transferring solute from dissolved to a solid phase (sorption, precipitation).

The physical processes of advection, diffusion and dispersion can be influenced by microbes only if they can increase (due to dissolution of solid phase) or close (clogging, formation of biofilms) fluid filled pore space. If very intense microbial activity occurred in an EBS in which porosity was very small, an influence might well be expected. Significant effects for macroporous EBS systems, or for the geosphere would require either very high concentrations of biomass (or biofilms), or secondary alteration products such as production of fines. Laboratory experiments have, rather unexpectedly, observed blocking of flow cells within 2 days due to intense microbial growth/alteration of crushed rock (Bateman et al., 1999), but the extent to which this is relevant to real systems is unknown. Lucht et al. (1997) found no evidence for pore clogging in backfill (a mixture of 25% bentonite and 75% crushed granite rock) by bacterial activity over a 180-day period. Biofilms on fracture surfaces may limit access to the matrix and thus greatly reduce the retardation of many key radionuclides. To complicate the situation, radionuclides could also be directly sorbed by the biofilms themselves, which may immobilise them.

Sorption is a generic term encompassing a range of processes which lead to uptake of solute on to surfaces. Such sorption is generally assessed in laboratory uptake experiments and, in cases where microbiology is considered at all, it is often assumed that the experimental systems already include microbial effects as no attempts are made to limit microbial growth. In some cases, the agreement between the predicted rate of retardation based on laboratory sorption measurements with that observed in in situ tracer tests argues that this assumption is justifiable. Experiments showing that microbial activity plays a role (West et al., 1991b) do, however, suggest that care must be taken when attempting to develop mechanistic sorption models based entirely on physicochemical mechanisms.

The effects of microbes on radionuclide migration can be advantageous (if sorbed on to attached organisms or onto biofilms) or detrimental (if sorbed on to mobile or motile organisms). Biofilms could potentially form on any wet surface in the near and far fields. They have been observed forming on surfaces in deep subsurface environments (Brown & Hamon, 1994) and some have been shown to include *Gallionella* spp. This organism produces iron oxyhydroxide which can sorb radionuclides. Although such biofilms are not necessarily representative of those expected in the host rock or in the EBS, these studies demonstrate the changed physicochemical regimes possible in biofilms and the potential role that they may play in radionuclide retardation. Extensive biofilms are more unlikely in compacted buffer because of the small pore space present but they are more likely in the backfill and host rock. Laboratory experiments investigating ^{75}Se, ^{113}Sn and ^{137}Cs migration through artificial granite fractures containing biofilms indicated that the presence of biofilms had no effect on retardation (Vandergraaf et al., 1997). However, some studies of similar low nutrient environments also suggest that some microbes become starved – becoming small and mobile and capable of penetrating deeply into geological formations (Lappin-Scott & Costerton, 1990). This effect also has implications for radionuclide retardation. Experiments where microbes are in the solution phase suggest that they have varying effects in conventional radionuclide sorption experiments. For example, in experiments with ^{137}Cs microbes compete for sites on rock materials with the radionuclide (West et al.,

1991b). Indeed, at the Yucca Mountain site there is reported to be a potential risk of the subsurface microbial population promoting plutonium transport, based on the high potential for biogenic chelation and CO_2 production which could accelerate transport (Hersman, 1997).

Finally, the effects of perturbations in the geological barrier, which result from the presence of the repository, need to be assessed. To date, one hydrological and two chemical perturbations of this type have been considered from the viewpoint of microbiology. Due to excavation practices in hard rock, a so called excavation damage zone (EDZ), with higher permeability than the bulk rock, may exist, extending some distance (<1 m) into the rock. Microbial transport could be enhanced in this zone. The EDZ could be extensively altered by microbially catalysed processes during the operation of the repository – which could be especially significant for host rocks containing reactive minerals such as pyrite (McKinley & Bradbury, 1989). The chemical perturbations include an oxidising redox front arising from radiolysis of water (especially important for directly disposed spent fuel) and a plume of high pH, hyperalkaline leachates from cement containing repositories. The extent to which these reaction fronts can be locations of enhanced microbial activity and the consequences of such activity have been reviewed (McKinley et al., 1997). It was concluded that microbial processes probably play a critical role in the development and movement of redox fronts as commonly observed in nature (e.g. roll-front ore deposits). The role microbes can play in the oxidation and reduction of Fe(II) and Fe(III) respectively, could be of particular relevance to repository concepts where C-steel is included (King & Stroes-Gascoyne, 2000). A microbial role in the high pH front was shown to be possible in principle, but was identified as an area where relevant experimental data and field observations were lacking.

7. Modelling and model testing

The significance of the effects discussed above in any repository concept must be considered within a performance assessment. However, until recently there have been few studies aimed at predicting the extent of microbial activity and fewer still that enable the consequences of this microbial activity to be determined in a manner useful for performance assessment purposes. For example, several models have been developed that predict the size of the microbial population within a disposal vault (Stroes-Gascoyne, 1989; Baker et al., 1998). However, this information, by itself, is of little use if it cannot be translated into a measurable effect such as the extent of container corrosion. Some microbiological modelling of HLW corrosion effects (McKinley et al., 1985) and of L/ILW waste degradation (McKinley & Grogan, 1991) has been performed, using mass balance and thermodynamic approaches for Swiss waste inventories and rock and groundwater compositions. Although models such as these are based on a number of assumptions, carefully designed laboratory experiments can test these assumptions to check the validity of the model predictions. For example, West et al. (1998) carried out experiments, which confirmed microbial survival in realistic Swiss L/ILW conditions but in lower numbers than predicted by the microbiological models. The experiments also provided kinetic information for future models and confirmed that iron was a major energy source, with localised corrosion of steels being observed. Modelling of microbiological activity at a hyperalkaline water site in Jordan was also confirmed by observations in the field (West et al., 1995). Humphreys et al. (1995) have developed a model for L/ILW that predicts in detail the extent of microbial activity, the associated changes in repository conditions such as pH and the concentrations of various metabolic byproducts, such as isosaccharinic acid, which can affect radionuclide mobility. King et al. (1999) developed a mathematical model to predict the extent of sulfate reduction by sulfate-reducing bacteria in a nuclear fuel waste repository and its effect on corrosion of copper containers. The model is based on a series of mass balance equations which describe the kinetics of sulfate reduction by sulfate-reducing bacteria, their growth and death, the effects of radiation and desiccation, the supply and consumption of nutrients and sulfate and the fate of the produced sulfide (i.e. precipitation by Fe(II) or transport by diffusion to the copper container). Simulations with this model showed that the amount of sulfate limited the sulfate reducing bacterial activity and that sulfide precipitated as FeS rather than causing corrosion. The latter was confirmed recently by results from an isothermal test at AECL's underground research laboratory. Buffer (50% bentonite/50% silica sand) was buried for 6 years in a granite borehole. Upon excavation, viable sulfate-reducing bacteria were found and the concentration of sulfide in the buffer had increased slightly over the 6-year period. However, the rate of sulfate reduction derived from this test suggested a much lower in situ sulfate reducing bacterial activity than assumed in the model by King et al. (1999).

During development of these models, attempts have been made to represent the heterogeneities within a repository by examining the interfaces between different materials. McKinley et al. (1997) have examined high pH fronts and redox fronts but the predictions still require testing against observations from laboratory experiments and natural systems. Observations from natural systems or analogues have already been used to substantiate some laboratory experiments, for example the determination of microbial activity in clay-based buffer and backfill environments. Considerations were that pore throats in compacted or high clay content environments are smaller than average microbial dimensions. Laboratory studies showed that, depending on clay content, microbes could not migrate into compacted clay and in some instances could not even survive. Observations from the natural environment supported these considerations and findings: clay is not an optimum environment for microbes and their diversity and numbers are low when clay content is high in sediments (Chapelle 1993). Very few viable bacteria were, for example, found deep in the Boom clay deposit (Merceron, 1994). In a different context, field observations at Poços de Caldas in Brazil have shown that the movement of uranium across a redox front was influenced by sulfur cycle organisms (West et al., 1992).

8. Conclusions and key requirements for future work

Over the last decade the study of geomicrobiology in the context of deep disposal of nuclear waste has moved from a rather exotic sideline to a significant component of many integrated R&D and performance assessment programmes. The present consensus is that major perturbations from microbial activity are not to be expected in repositories for HLW but, because of the extremely long periods over which the safety of such waste has to be assured, further work to build confidence in such a conclusion is needed. Here natural analogue studies could be of particular value.

For other waste forms, hard conclusions are difficult to draw at present. Significant microbial activity is to be expected, at least for some waste streams that contain abundant nutrients. Potential perturbations of the performance of individual barriers could either be very detrimental (production of gas or complexant byproducts) or very helpful (catalysis of redox reactions, sorption of radionuclides by biofilms). Further efforts to produce experimental data, under relevant chemical conditions, develop models and test conclusions through field studies are needed, especially for the most problematic transuranium waste types.

Acknowledgements

This chapter is published with the permission of the Director of the British Geological Survey (NERC). We would like to thank Keith Bateman (BGS) and Peter Hogarth for their constructive reviews. Thanks also to Aline Playfair (NAGRA) for assistance in preparation of the text.

References

Alder, J. C. and McGinnes, D. F. (1995). Model radioactive waste inventory for Swiss waste disposal projects (Vol 1). Main Report. Vol. 2: databases. *NAGRA Technical Report 93-21*, Baden, Switzerland.

Bachofen, R. (1991). Gas metabolism of microorganisms. *Experientia, 47*, 508–513.

Baker, S. J., West, J. M., Metcalfe, R., Noy, D. J., Yoshida, H. & Aoki, K. (1998). A biogeochemical assessment of the Tono site, Japan. *Journal of Contaminant Hydrology 35*, 331–340.

Bateman, K., Coombs, P., Hama, K., Hards. V. L., Milodowski, A. E., West. J. M. & Yoshida. H. (1999). Laboratory examination of microbial perturbations in a geological disposal site for radioactive waste. *Proceedings of the 9th Annual V. M. Goldschmidt Conference*. Abstract No. 7276. LPI Contribution No. 971, Lunar and Planetary Institute, Houston TX.

Bath, A. H., Christofi, N., Neal, C., Philp, J. C., Cave, M. R. & McKinley, I. G. (1987). Trace elements and microbiological studies of alkaline groundwater in Oman, Arabian Gulf: a natural analogue for cement pore waters. *FLPU 87-2*, British Geological Survey, Nottingham, UK.

Billi, D., Friedmann E. I., Hofer, K. G., Grilli Caiola, M. & Ocampo-Friedmann, R. (2000). Ionizing radiation resistance in the desiccation-tolerant *Cyanobacterium chroococcidiopsis*. *Applied and Environmental Microbiology, 66*, 1489–1492.

Birch, L. & Bachofen, R. (1990). Complexing agents from microorganisms. *Experientia, 47*, 827–833.

Booth, W. (1987). Postmortem on Three Mile island. *Science, 238*, 1342–1345.

Brown, D. A. and Hamon C. J. (1994). Initial investigation of groundwater microbiology at AECL's Underground Research Laboratory. *Atomic Energy of Canada Limited Technical Record* TR608/COG-93-171. Whiteshell, Canada.

Chapelle, F. H. (1993). *Ground-Water Microbiology and Geochemistry*. New York: John Wiley.

Chapman, N. A. & McKinley, I. G. (1987). *The Geological Disposal of Nuclear Waste*. London: John Wiley.

Chen, C. I., Meike, A., Chuu, Y-J., Sawvel, A. & Lin. W. (1999). Investigation of bacterial transport in the large- block test, a thermally perturbed block of Topopah Spring Tuff. *Materials Research Society Symposium Proceedings, 556*, 1151–1158.

Christofi, N. & Philp, J. (1991) . Microbiology of subterranean waste sites. *Experientia, 47*, 524–527.

Colasanti, R., Coutts, D., Pugh, S.Y.R. & Rosevear, A. (1991). The microbiology programme for UK NIREX. *Experientia, 47*, 560–572.

Costerton, J. W., Nickel, J. C. & Ladd, T. I. (1986). Suitable methods for the comparative study of free-living and surface-associated bacterial populations. In J. S. Pointdexter & R. Leadbetter (Eds), *Bacteria in Nature* (pp. 49-84). New York: Plenum Press.

Fortin, D., Ferris, F. G. & Beveridge. T. J. (1997). Surface-mediated mineral development by bacteria. *Reviews in Mineralogy, 35*, 161–180.

Fukunaga, S., Yoshikawa, H., Fujiki, K. & Asano, H. (1995). Experimental investigation on the active range of Sulphate-reducing bacteria for geological disposal. *Materials Research Society Symposium Proceedings, 353*, 173–180.

Gadd, G. M. 1988. Accumulation of metals by microorganisms and algae. In H-J. Rehm & G. Reed (Eds), *Biotechnology* (pp. 402–433, 6b). Weinheim: VCH verlagsgesellsshaft.

Geesey, G. (1993). A review of the potential for microbially influenced corrosion of high-level nuclear waste containers. *Center for Nuclear Waste Regulatory Analyses*, Report CNWRA 93-014, San Antonio, TX.

Gorby, Y. A. & Lovley, D. R. (1992). Enzymatic uranium precipitation. *Environmental Science and Technology, 26*, 205–207.

Haveman, S. A., Stroes-Gascoyne, S. & Hamon, C. J. (1996). Biodegradation of a sodium sulphonated naphthalene formaldehyde condensate by bacteria naturally present in granitic groundwater. *Atomic Energy of Canada Limited Technical Record* TR-721, COG-95-547, Whiteshell, Canada

Hersman, L. E. (1997). Subsurface Microorganisms: effects on the transport of radioactive wastes. In P. Amy & D. L. Haldeman (Eds), *Microbiology of the Terrestrial Subsurface.* Boca Raton, FL: CRC Press.

Humphreys, P., Johnstone, T., Trivedi, D. & Hoffmann, A. (1995). The biogeochemical transport code DRINK: a mechanistic description. *Materials Research Society Symposium Proceedings, 353*, 211–218.

IAEA (1997). Planning and operation of low level waste disposal Facilities. *Proceedings of an International Symposium on Experience in the Planning and Operation of Low Level Waste Disposal Facilities.* Vienna, 17-21 June 1996. *Proceedings Series. International Atomic Energy Agency*, Wien, Austria.

King, F. & Stroes-Gascoyne, S. (2000). An assessment of the long-term corrosion behaviour of C-steel and the impact on the redox conditions inside a nuclear fuel waste disposal container. Prepared by Atomic Energy of Canada for Ontario Power Generation. Ontario Power Generation, Nuclear Waste Management Division Report 06819-REP-01200-10028-R00. Toronto, Ontario.

King, F., Kolar, M., Stroes-Gascoyne, S., Bellingham, P., Chu, J. & Dawe, P. V. (1999). Modelling the activity of sulphate-reducing bacteria and the effects on container corrosion in an underground nuclear waste disposal vault. *Materials Research Society Symposium Proceedings, 556*, 1167–1174.

Lappin-Scott, H. M. & Costerton, J. W. (1990). Starvation and penetration of bacteria in soils and rocks. *Experientia, 46*, 807–812.

Little, B. & Wagner, P. (1996). An overview of microbiologically influenced corrosion of metals and alloys used in the storage of nuclear wastes. *Canadian Journal of Microbiology, 42*, 367–374.

Lucht, L. M., Stroes-Gascoyne, S., Miller, S. H., Hamon, C. J. & Dixon, D. A. (1997). Colonization of compacted backfill materials by microorganisms. *Atomic Energy of Canada Limited Report* AECL-11832, COG-97-321-I, Whiteshell, Canada.

Lutze, W. & Ewing, R. C. (1988). *Radioactive Waste Forms for the Future.* Amsterdam: Elsevier Science.

McKinley, I. G. & Bradbury, M. (1989). Near-field geochemistry of vitrified HLW in sedimentary host rock. *Materials Research Society Symposium Proceedings, 127*, 645–651.

McKinley, I. G. & Grogan, H. A. (1991). Consideration of microbiology in modelling the near- field of a L/ILW repository. *Experientia, 47*, 573–577.

McKinley, I. G., West, J. M. & Grogan, H. (1985). An analytical overview of the consequences of microbial activity in a Swiss HLW repository. *NAGRA Technical Report* NTB 85-43, Baden, Switzerland.

McKinley, I. G., Hagenlocher, I., Alexander. W. R. & Schwyn, B. (1997). Microbiology in nuclear waste disposal: interfaces and reaction fronts. *FEMS Microbiology Reviews, 20*, 545–556.

Merceron, Th. (1994). Archimede-clay project: acquisition and regulation of water chemistry in clay. Project Archimede – Argile, 13-014 Janvier 1994, CEA Cadarache, France.

Motamedi. M., Karland, O. & Pedersen, K. (1996). Survival of sulfate-reducing bacteria at different water activities in compacted bentonite. *FEMS Microbiology Letters, 141*, 83–87.

NAGRA (1994). Kristallin-I Safety Assessment report. *NAGRA Technical Report* 93-22, Baden, Switzerland.

Ogundele, G. I. and Jain, D. K. (1999). The effect of microbial metabolites on corrosion of Finnish OFP copper in Standard Canadian Shield Saline Solution (SCSSS). *Prepared by Ontario Power technologies for Ontario Power Generation, Nuclear Waste Management Division* Report 06819-REP-01200-10010-R00. Toronto, Ontario.

Perfettini, J. V., Revertegat, E., & Langomazino, N. (1991). Evaluation of cement degradation induced by the metabolic products of two fungal strains. *Experientia, 47*, 527–533.

Philp, J. C., Taylor, K. J. & Christofi, N. (1991). Consequences of sulphate-reducing bacterial growth in a lab-simulated waste disposal regime. *Experientia, 47*, 553–559.

Pitonzo, B., Amy, P. S. & Rudin, M. (1999a). Effect of gamma radiation on native endolithic microorganisms from a radioactive waste deposit site. *Radiation Research, 152*, 64–70.

Pitonzo, B. J. Amy, P. S. & Rudin, M. (1999b). Resuscitation of microorganism after gamma irradation. *Radiation Research, 152*, 71–75.

Pusch, R. (1999). Mobility and survival of sulphate-reducing bacteria in compacted and fully water saturated bentonite – microstructural aspects. *SKB technical Report* TR-99-30, Stockholm, Sweden.

Roffey, R. & Norqvist, A. (1991). Biodegradation of bitumen used for nuclear waste disposal. *Experientia, 47*, 539–542.

Rogers, R. D. (1995). Assessment of the effects of microbially influenced degradation on a massive concrete structure. Final report UCRL-CR-122068, Lawrence Livermore National Laboratory, Livermore, CA.

Savage, D. (1995) *The Scientific and Regulatory Basis for the Geological Disposal of Radioactive Waste.* Chichester, UK: John Wiley.

Stotzky, G. (1986). Influence of soil mineral colloids on metabolic processes, growth, adhesion, and ecology of microbes and viruses. In P. M. Huang & M. Schnitzer (Eds), *Interactions of Soil Minerals with Natural Organics and Microbes*. Madison, WI: Soil Science Society of America.

Stroes-Gascoyne, S. (1989). The potential for microbial life in a Canadian high level nuclear fuel waste disposal vault: a nutrient and energy source analysis. Atomic Energy of Canada Limited Report, AECL-9574, Whiteshell, Canada.

Stroes-Gascoyne, S. & West, J. M. (1996). An overview of microbial research related to high-level nuclear waste disposal with emphasis on the Canadian concept for the disposal of nuclear fuel waste. *Canadian Journal of Microbiology, 42*, 349–366.

Stroes-Gascoyne, S. & West, J. M. (1997). Microbial studies in the Canadian nuclear fuel waste management program. *FEMS Microbiology Reviews, 20*, 573–590.

Stroes-Gascoyne, S., Pedersen, K., Haveman, S. A., Dekeyser, K., Arlinger, J., Daumas, S., Ekendahl, S., Hallbeck, L., Hamon, C. J., Jahromi, N. & Delaney, T. L. (1997). Occurrence and identification of microorganisms in compacted clay-based buffer material designed for use in a nuclear fuel waste disposal vault. *Canadian Journal of Microbiology, 4*, 1133–1146.

Stroes-Gascoyne, S., Hamon, C. J. & Vilks, P. (2000). Microbial analysis of the isothermal test at AECL's Underground Research Laboratory. Prepared by Atomic Energy of Canada Limited for Ontario Power Generation. Ontario Power Generation, Nuclear Waste Management Division Report 06819-REP-01200-10023-R00. Toronto, Ontario.

Vandergraaf, T. T., Miller, H. G., Jain, D. K., Hamon, C. J. & Stroes-Gascoyne, S. (1997). The effect of biofilms on radionuclide transport in the geosphere: results from an initial investigation. Atomic Energy of Canada Limited Technical Record TR-774, COG-96-635-I, Whiteshell, Canada.

Van Loon, L. R. and Glaus, M. A. (1998). Experimental and theoretical studies on alkaline degradation of cellulose and its impact on the sorption of radionuclides. *NAGRA Technical Report 97-04*, Baden, Switzerland.

Vilks, P. and Bachinski, D. B. (1994). Colloid and suspended particle migration experiments in a granite fracture. In *Proceedings of the Fourth International Conference on the Chemistry and Migration Beha-*

viour of Actinides and Fission Products in the Geosphere, Charleston SC, December 12-17 1993, pp. 229–234.

Wagner, P. & Little, B. (1993). Impact of alloying on microbiologically influenced corrosion – a review. *Materials Performance, 32*, 65–68.

West, J. M. (1995). A review of progress in the geomicrobiology of radioactive waste disposal. *Radioactive Waste Management and Environmental Restoration, 19*, 263–283.

West, J. M. and McKinley, I. G. (1984) The geomicrobiology of nuclear waste disposal. *Materials Research Society Symposium Proceedings, 25*, 487–494.

West, J. M., McKinley, I. G. & Chapman, N. A. (1982) . Microbes in deep geological systems and their possible influence on radioactive waste disposal. *Radioactive Waste Management and the Nuclear Fuel Cycle, 3*, 1–15.

West, J. M., Grogan, H. A. & McKinley, I. G. (1991a). The role of microbiological activity in the geological containment of radioactive wastes. In J. Berthelin (Ed.), *Diversity of Environmental Biogeochemistry*. New York: Elsevier.

West, J. M., Haigh., D. G., Hooker. P. J. & Rowe, E. J. (1991b). Microbial influence on the sorption of ^{137}Cs onto material relevant to the geological disposal of radioactive waste. *Experientia, 47*, 549–552.

West, J. M., McKinley, I. G. & Vialta, A. (1992). Microbiological analysis at the Pocos de Caldas natural analogue study sites. *Journal of Geochemical Exploration, 45*, 439–449.

West, J. M. Coombs, P., Gardner, S. J. & Rochelle, C. A. (1995). The microbiology of the Maqarin site, Jordan – a natural analogue for cementitious radioactive waste repositories. *Materials Research Society Symposium Proceedings, 353*, 181–188.

West, J. M. Cave, M., Coombs, P., Milodowski, A. E. & Rochelle, C. A. (1998). Alteration of repository structural materials within the first few years. *Materials Research Society Symposium Proceedings, 506*, 503–510.

Wolf, M. & Bachofen, R. (1991) Microbial alteration of bitumen. *Experientia, 47*, 542–548.

Miranda J. Keith-Roach and Francis R. Livens (Editors)
© 2002 Elsevier Science Ltd. All rights reserved

Chapter 10

Microbial processes in the disposal of high level radioactive waste 500 m underground in Fennoscandian Shield rocks

Karsten Pedersen

Göteborg University, Department of Cell and Molecular Biology, Microbiology Section Box 462, SE-405 30 Göteborg, Sweden

1. Introduction

Radioactive waste in Sweden arises mainly from the production of nuclear power. Some waste also comes from research, hospitals and industry. The bulk of the radionuclides produced in a nuclear power reactor remains in the spent fuel elements, characterised as HLW. Radioactivity will decay with time, but some long-lived radionuclides will make the HLW hazardous for a very long time. The spent fuel elements will be encapsulated in copper-steel canisters and placed in deposition holes in tunnels at an expected depth of about 500 m (Fig. 1). The amount of spent fuel in a canister and the distances between the canisters in the repository are chosen so that the peak temperature is about 80–90°C at the canister surface. The restriction in temperature is mainly there to guarantee the long-term performance of the bentonite buffer that will surround the canisters. The low solubility of the spent fuel matrix, the copper canister, the bentonite buffer and the depth of emplacement in stable host rock are the main barriers to protect man from the radionuclides.

Independent scientific work has unambiguously demonstrated that life is present in most deep geological formations investigated, down to depths of several kilometres (Pedersen, 1993, 2000a). Sedimentary rocks (Fredrickson & Onstott, 1996), igneous rocks (Pedersen, 1997a) and subsea floor environments (Wellsbury et al. 1997; Fisk et al., 1998) all harbour life. The distribution of underground life is conceptually restricted only by temperature and, at present, the known temperature limit for life is 113°C (Stetter, 1996). This temperature is reached at very different depths around our planet, from the seafloor surface at marine hot springs to 10 km or deeper in massive sedimentary rock formations. In the Fennoscandian Shield, the temperature typically increases by 1–2°C per 100 m, which suggests that microbial life may extend as far down as 6–7 km at any igneous rock site chosen for a future Swedish high level radioactive waste (HLW) repository. Microbial processes

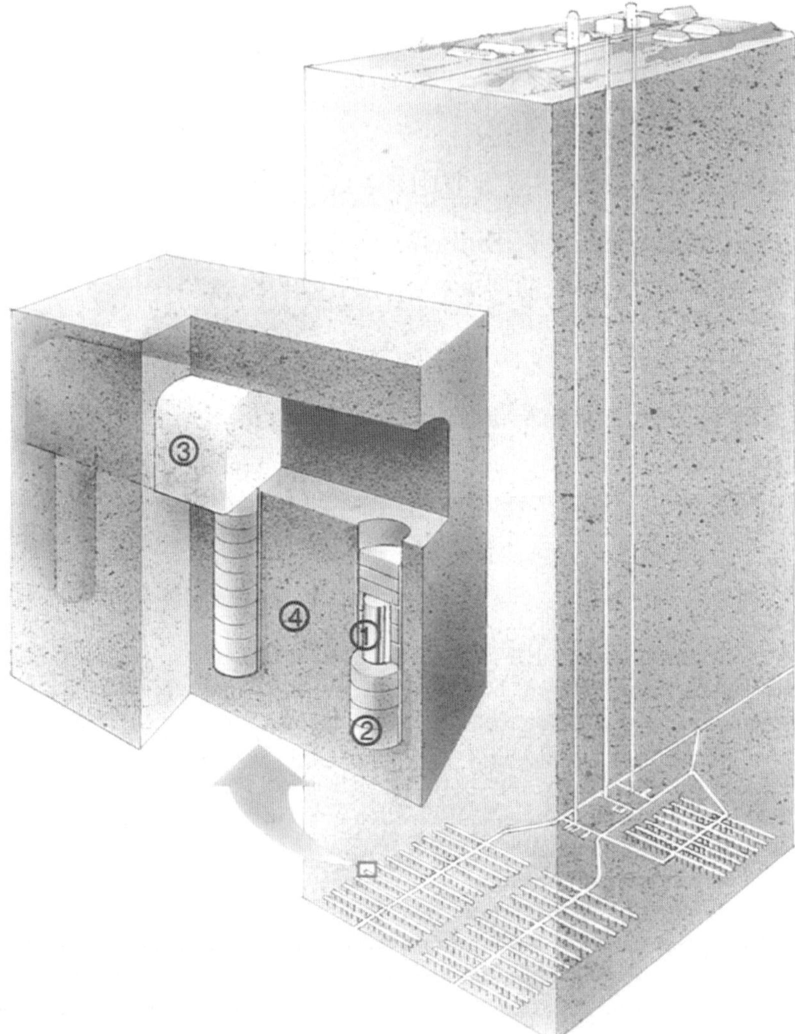

Fig. 1. Schematic drawing of a deep repository. A system of tunnels with vertical deposition holes is built at a depth of about 500 m. The spent fuel assemblies are encapsulated in copper canisters. The canisters are placed in the holes, where they are embedded in bentonite clay. Multiple barriers will protect the spent fuel in the deep repository. (1) *Copper canister*. The canister isolates the fuel from groundwater contact. The fuel itself is in solid form and has very low solubility. (2) *Blocks of bentonite clay*. The clay prevents groundwater flow around the canister and protects it against minor movements in the rock. The diffusivity of radionuclides is very low in bentonite, which will prevent radionuclide migration in case of a failing canister. (3) *A tunnel backfill*. A mixture of sand and bentonite, or crushed rock, fills up the tunnels. (4) *A rock barrier*. The rock offers a durable environment, both mechanically and chemically. It also acts as a filter for radionuclides possibly released to the groundwater.

have consequently become an integral part of the performance safety assessment of the Swedish HLW repository. This chapter summarises the last decade's work in microbiology research within the HLW disposal programme and gives current perspectives on microbial processes in HLW disposal, with the Swedish repository as a template.

In 1987, microbiology became an operative part of the Swedish Nuclear Fuel and Waste Management Co (SKB) scientific programme for the safe disposal of HLW in igneous rock and the research is still ongoing. The goal of the microbiology programme is to understand how microbes would affect the performance of a future HLW repository. In the early stages of the research programme, previously unknown microbial ecosystems were revealed in igneous rock aquifers at depths exceeding 1000 m (Pedersen & Ekendahl, 1990). This discovery triggered a thorough exploration of the subterranean biosphere in the aquifers of the Fennoscandian Shield (Pedersen, 1997b). Similarly, the Canadian radioactive waste disposal programme has stimulated investigations of microorganisms in deep igneous rock aquifers of the Canadian Shield (Stroes-Gascoyne & Sargent, 1998). Early investigations examined the potential risk of radionuclide migration caused by microorganisms able to survive in the deep groundwater systems (Birch & Bachofen, 1990; Pedersen & Albinsson, 1991, 1992). It soon became apparent that microbial communities exist in most, if not all, deep aquifers (Pedersen & Ekendahl, 1990). Attention then shifted to examining the potential effects of these microorganisms using radiotracer methods (Pedersen & Ekendahl, 1992a, b; Ekendahl & Pedersen, 1994; Kotelnikova & Pedersen, 1998). The results suggested remarkable metabolic and species diversity, which led to DNA extraction and 16/18S rDNA cloning and sequencing for assessment of subterranean microbial diversity (Ekendahl et al., 1994; Pedersen et al., 1996a, 1997c). The work revealed several previously unknown microbial species adapted to life in igneous rock aquifers (Motamedi & Pedersen, 1998; Kotelnikova et al., 1998; Kalyuzhnaya et al., 1999). The presence of hydrogen, methane and acetogenic bacteria and methanogens suggest that autotrophic activity may be an important process at repository depths (Pedersen, 1999).

Sulfate-reducing bacteria (SRB) produce sulfide and have been commonly observed in groundwater environments typical of a Swedish HLW repository (Pedersen, 1999; Haveman & Pedersen, 2000). The potential for sulfide corrosion of the copper canisters used in HLW storage must consequently be considered. The bentonite buffer around the copper canisters will be a hostile environment for most microbes, owing to the combination of radiation, temperature and low availability of water. Discrete microbial species can overcome each of these constraints and it is theoretically possible that sulfide producing microbes may be active inside a buffer, although the experiments conducted thus far have indicated the opposite (Motamedi et al., 1996; Pedersen et al., 2000a, b). A special concern is the interface between the copper canister and the buffer. Nowhere are the environmental constraints for life as strong as in this area. Still, it has been suggested that sulfate-reducing bacteria could survive and locally produce sulfide in concentrations large enough to cause damage to a canister.

The importance of the microbial processes listed above for the disposal of HLW 500 m underground in Fennoscandian Shield rock is described in more detail below, and the results from the 14-year scientific programme are reviewed.

2. Microbial processes in closed and open systems

Batch culture

The common way to culture microorganisms in the laboratory is by using a batch culture. A culture vessel is supplied with all constituents necessary for growth, and inoculated with the microbe of interest. A typical batch growth curve is shown in Fig. 2. First, there is an adaptation phase during which the cells adjust to the conditions in the culture vessel. Then the cells start to divide and grow exponentially to high counts, doubling their number at even time intervals. Finally, growth is arrested when some limiting component is used up, or when a toxic component forms at too high a concentration (e.g. alcohol in fermentation cultures). Figure 2 shows that the cells in a batch culture are, basically, active only during the exponential growth phase. The batch culture represents a closed system with no input or output of components from the system. It is a superb tool for many research purposes in the laboratory but it does not mimic the life of microbes in natural environments. The natural environment generally consists of a large number of open systems linked by continuous input and output of material. Models of microbial processes in the repository should therefore be based on continuous culture situations, as described below, rather than on batch culture situations.

Continuous culture

Hard rock aquifers can be considered as open systems. A particular fracture will have a water composition that reflects the origin of the water and various reactions between solid and liquid phases occur along the flow path. A new composition may be the result of two fractures meeting and of their water mixing. These processes may be slow but there is a continuum of varying geochemical conditions in hard rock aquifers at repository depth, and the repository with all its alien construction components will add variance to these conditions. Microbes are experts on utilising any energy in the environment that becomes thermodynamically available for biochemical reactions. A slow but steady flow of organic carbon from the surface or a flow of reduced gases such as hydrogen and methane from the interior of our planet will ultimately be the driving force for the activity of deep aquifer microbes.

Continuous growth of microbes can be studied in the laboratory using a chemostat. The culture vessel is continuously supplied with energy by a slow inflow of nutrients. The inflow is balanced by an outflow that removes waste products and some cells. The number of cells will therefore remain constant in the chemostat. The microbes will, however, be active (Fig. 3). A chemostat system is, unlike the batch system, open as it has an influx and outflow of matter. The continuous culture situation of the chemostat is directly relevant to any hard rock aquifer experiencing a flux of matter through a continuous mixing of groundwater of varying compositions. The flows may be very slow but, over geological timescales, they will be significant.

The open, continuous culture system concept can be used for the interpretation of microbiology data from groundwater, such as the number of cells in arbitrarily chosen groundwater measured at various times (see e.g. Pedersen et al., 1996a). Generally, the total

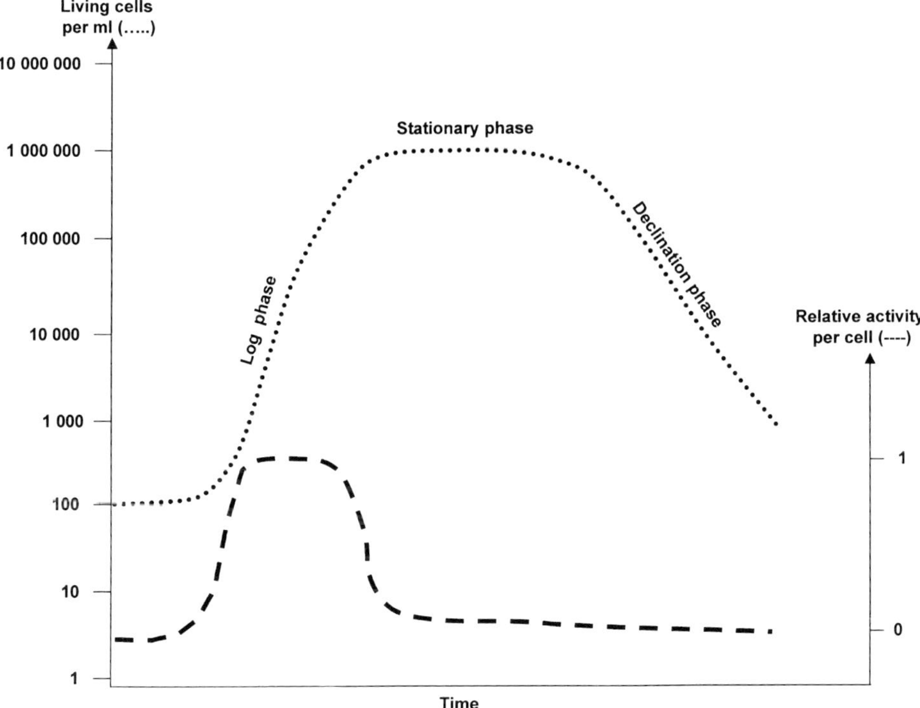

Fig. 2. A schematic representation of microbial growth in a closed batch culture. The microbes are basically active only during the exponential growth phase, when they double their number during a specific time period. The doubling time can be as short as 15 minutes for some easily culturable microbes and it may be many hours for microbes more difficult to work with.

number of cells observed lies at a specific, stable level in a borehole over time. If we apply the batch concept (Fig. 2), we would conclude that the microbes are not growing and are inactive because we do not register any increase in cell numbers over time. In contrast, with the continuous culture concept, it can be predicted that the microbes are active and grow slowly in constant environmental conditions over the time period studied. This prediction requires processes that balance an increase in cell numbers due to growth. There are several possibilities for balancing growth of cell populations in deep aquifers. Phages (i.e. viruses that attack microbes) may balance cell growth. Their activity results in lysis of infected cells and in the production of new phages. This process has not been demonstrated as yet in deep aquifers, but occurs in most surface environments. Many unicellular animals graze on other microbes and their possible presence and activity in deep aquifers would also counterbalance cell growth.

A special case is the possible occurrence of microbes that grow attached to aquifer surfaces, a phenomenon which has been repeatedly observed in groundwater from deep hard rock aquifers (Ekendahl & Pedersen, 1994; Pedersen et al., 1996a). Such biofilms will increase their cell numbers until they reach steady state, as previously described for continuous growth of unattached microbes.

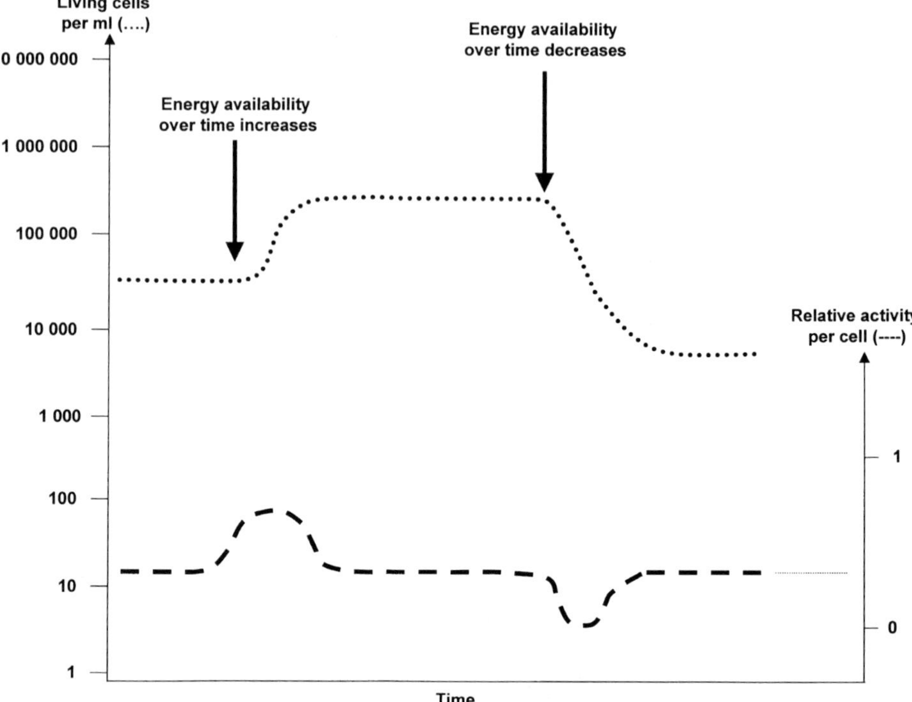

Fig. 3. The graph is a schematic representation of microbial growth in an open, continuous culture system. The microbes are continuously active except for periods when there is a decrease in availability of energy over time. The doubling time of the population can be very long. More than a month has been registered for deep groundwater biofilms (Ekendahl & Pedersen, 1994).

The microbial enigma – death or survival

In periods of inactivity due to lack of energy and necessary nutrients, or other environmental constraints such as desiccation or slowly decreasing water activity, microbes can do one of two things. They die or enter one of many different possible states of survival. Different species have different ways of addressing the problem of unfavourable conditions for active life. The most resistant form of survival is the endospore formed by certain Gram-positive bacteria and SRB. There is no measurable sign of life in an endospore, yet after many years of inactivity, it can germinate to an actively growing cell within hours. It resists desiccation, radiation, heat and aggressive chemicals far better than does the living cell.

The endospore is the most resistant state of survival of any known life form but there are many more survival strategies among the microbes, which are more or less resistant to environmental constraints. Transforming into morphologically specific survival states is an advantage when the environment changes. However, in response to mere nutrient and energy deficiency, many microbes just shut down their metabolism to an absolute minimum level, at which they may survive for many years. Most such responses result in shrinkage of the cell to a fraction of its volume at optimal growth conditions. All these survival strategies have in common that the cell is active at an absolutely minimal level, or shows no activity at all. It is consequently possible that certain microbes may survive initially harsh conditions in a repository, radiation, desiccation, heat, high pH, and so on, until the conditions for growth again become favourable. However, if the conditions are so difficult that all survival forms die off, and if the pore size of the environment does not allow for transport of microbes, as in high density bentonite, then it is possible that specific environments in the repository stay free of microbes once the original microbe population has disappeared. It is at present uncertain whether this will be the case.

3. Subterranean microorganisms in a HLW repository environment

The Fennoscandian Shield rock of a future HLW repository will be fractured and groundwater from these fractures, the aquifers, will fill up the repository after closure. Sampling for microorganisms in such groundwater at repository depth, and also deeper, requires penetration of the rock to reach the aquifers in target. This requirement is commonly fulfilled via drilling of holes or sometimes by construction of tunnels. Investigations of subterranean microorganisms in Fennoscandian Shield rocks have been performed since 1987, and studies have so far been performed at 75 levels in 52 boreholes, reaching up to 1700 m depth (Table 1). The total numbers of microorganisms have been determined by routine analytical techniques and a plot of numbers versus depth demonstrates that microbes are present down to the deepest level counted, which was about 1500 m (Fig. 4). Igneous rocks are too hot when formed to host life of any kind. Therefore, life observed in hard rock must have entered after cooling and fracturing of the rock mass. The Fennoscandian rocks are very old, commonly in the range 1500–1800 million years (Gàal & Gorbatschev, 1987). The processes of fracturing, cycles of cooling and heating, and possible colonisation with microbes have been ongoing for very long time.

Drilling and excavation to access microbial ecosystems in hard rock are vigorous operations. The risk of microbial contamination of the aquifers by the water used to cool the drill bit and transport the drill cuttings out of the boreholes is obvious. Investigations have been undertaken to study this risk (Pedersen et al., 1997c). It was found, using molecular, culturing and counting methods, that although large numbers of contaminating bacteria were introduced into the boreholes during drilling, they did not become established in the aquifers at detectable levels. The presence of microbial fossils in calcite precipitates in an aquifer from 200 m underground attest the presence of microorganisms in the rock prior to drilling (Pedersen et al., 1997b). Furthermore, isolation and description of three new species with characteristics making them well adapted to the deep aquifer environment (Motamedi & Pedersen, 1998; Kotelnikova et al., 1998, Kalyuzhnaya et al., 1999) also suggest that microorganisms occur naturally in deep hard rock aquifers. Based on these listed investigations and observations, it seems reasonably safe to propose that the microorganisms observed in hard rock aquifers at repository depth are indigenous to that environment.

Table 1

Site information for all boreholes in igneous rock investigated for microbiology since 1987. Detailed information about the boreholes and the rock formation studied can be obtained elsewhere (Pedersen, 2000b)

Site	Year	Borehole	Depths (m)	
Hålö (S)	1992–1996	HBH01, HBH02	10–45	(2 levels)
Hästholmen (F)	1997–1998	HH-KR1, HH-KR2, HH-KR3, HH-KR4, HH-KR5, HH-KR6	65–943	(6 levels)
Kivetty (F)	1997–1998	KI-KR5, KI-KR13	497–721	(2 levels)
Laxemar (S)	1988–2000	KLX01, KLX02	100–1700	(7 levels)
Olkiluoto (F)	1998–1999	OL-KR3, OL-KR4, OL-KR8, OL-KR9, OL-KR-10	248–863	(7 levels)
Palmottu (F)	1998–1999	R302, R337, R387	32–309	(5 levels)
Romuvaara (F)	1998–1999	RO-KR10, RO-KR11	543–564	(2 levels)
Stripa (S)	1987–1991	V1, V2	799–1240	(4 levels)
Ävrö (S)	1987	KAV01	420–924	(4 levels)
Äspö (S)	1988–1996	KAS02, KAS03, KAS04	129–1002	(11 levels)
Äspö HRL tunnel (S)	1992–2000	KR0012, KR0013, KR0015, SA813B, SA923A, SA1062A, HA1327B, SA1420A, KA2511A, KA2512A, KA2858A, KA2862A, KA3005A, KA3010A, KA3067A, KA3105A, KA3110A, HD0025A, KA3385A, KA3539G, KA3548A01, KA3600F, KJ0050F01, KJ0052F02, KJ0052F03	68–450	(25 levels)
11 sites	**1987–1999**	**53 boreholes**	**10–1700**	**(75 levels)**

F = boreholes in Finland; S = boreholes in Sweden.

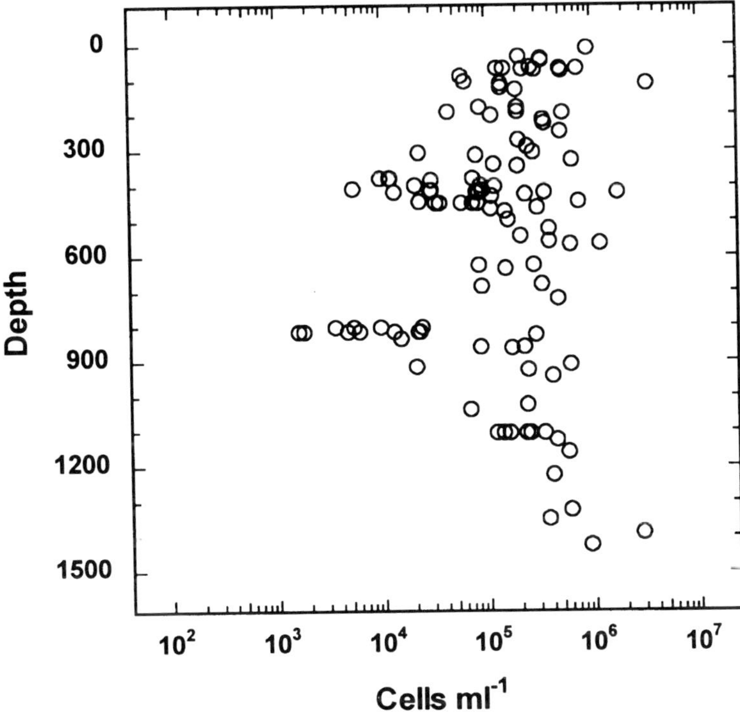

Fig. 4. Total number of cells observed at the sites presented in Table 1. At least one determination was done on each borehole level in each borehole. The number of observations depicted in the figure is 112, from Kotelnikova & Pedersen (1998), Ekendahl and Pedersen (1994), Pedersen & Ekendahl (1992a, b), Pedersen et al. (1997), Pedersen (1997), Pedersen & Ekendahl (1990), Pedersen et al. (1996), Haveman et al. (1999), Haveman & Pedersen (2002) and from investigations during 1999, not yet published.

Once a borehole or a tunnel has been constructed, sampling must be performed in such a way as to ensure non-contaminated samples. This can be achieved in several different ways as described in the review by Pedersen (2000b). Briefly, access to aquifer material and groundwater occurs via drilling of boreholes from ground surface or tunnels. After retrieval of drill core material, the boreholes are packed off in one or several sections which each isolate one or more specific aquifers. Down-hole pumps force groundwater from the aquifer to the ground surface for subsampling. Borehole samplers that can be opened and closed from the ground surface have been successfully operated (Torstensson, 1984; Haveman et al., 1999). One or more sample vessels can be used simultaneously. Boreholes can also be drilled from tunnels and they do not require pumps when the tunnel is under the groundwater table. Aquifers can be packed off and connected to sampling devices in the tunnel with pressure-resistant tubes. Finally, open fractures in tunnels can be sampled directly and represent groundwater with minimal disturbance, except for the pressure decrease due to entering the tunnel.

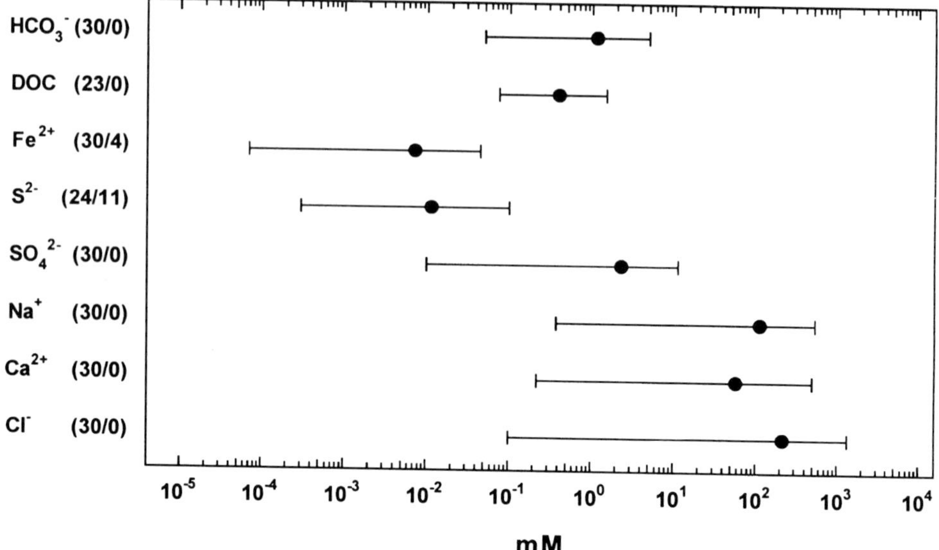

Fig. 5. Chemical data observed at the sites investigated. The data are extracted from Haveman et al. (1999), Haveman & Pedersen (2002) and Pedersen (1997a). The figure shows the mean concentration of the chemical species measured (filled circles); the bar gives the range. The numbers in parenthesis for each chemical species are the total number of observations followed by the number of observations below detection limit. DOC = dissolved organic carbon.

The groundwater environment

The groundwater environment in hard rock aquifers is generally anaerobic with a pH normally between 7 and 8. It is seldom below 7, but can approach pH 10 in some waters (Pedersen & Ekendahl, 1990). The presence of ferrous iron and sulfide indicate reduced and anoxic conditions (Fig. 5). The content of organic carbon is generally low, in the range of 0.1–10 mM. The content of dissolved solids is usually dominated by chloride, sodium and calcium (Fig. 5), but these vary by 4 orders of magnitude in concentration. This variability in groundwater composition makes it difficult to decode chemical information of the type presented in Fig. 5. Multivariate mixing and mass balance calculations (so-called M3 modelling) can be used to trace the effect from present and past groundwater flow with good accuracy (Laaksoharju et al., 1999). This technique has identified five end members, so-called reference waters, that mix to form most types of groundwaters found in the Fennoscandian Shield aquifers. These waters have meteoric (precipitation), glacial, marine (Baltic sea), altered marine or deep brine characteristics (Fig. 6).

Several gases occur in the groundwaters studied (Fig. 7). Nitrogen dominates mostly, followed by the noble gases helium and argon. Traces of carbon monoxide are found, while carbon dioxide occurs in much higher concentrations. Hydrogen is almost always present, albeit varying 5 orders of magnitude in concentration. Methane is always found and is the dominant gas in some of the deep Finnish Olkiluoto boreholes (Haveman et al.,

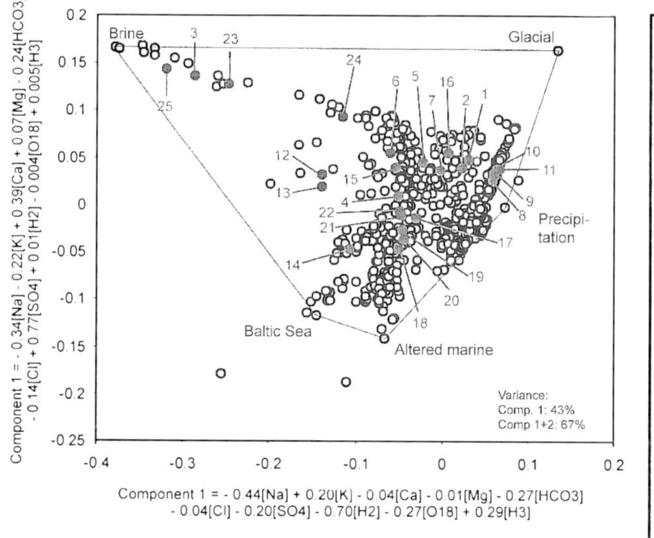

1 OL-KR3, 248 m
2 OL-KR3, 441 m
3 OL-KR4, 863 m
4 OL-KR8, 306 m
5 OL-KR9, 473 m
6 OL-KR9, 567 m
7 OL-KR10, 328 m
8 KI-KR13, 497 m
9 KI-KR5, 721 m
10 RO-KR11, 543 m
11 RO-KR10, 564 m
12 HH-KR1, 943 m
13 HH-KR2, 910 m
14 HH-KR3, 214 m
15 HH-KR4, 686 m
16 HH-KR5, 385 m
17 HH-KR6, 65 m
18 KA3566GO2:2, 450 m
19 KA3573A:1, 450 m
20 KA3573A:2, 450 m
21 KA3600F:2, 450 m
22 KA3600F:1, 450 m
23 KLX02, 1350 m
24 KLX02, 1160 m
25 KLX02, 1389 m

Fig. 6. Principal component plot calculated with groundwater chemistry data of 25 samples analysed for microbiology (closed circles) and with other Fennoscandian shield groundwater samples (open circles). Samples 1–7 are from Olkiluoto, 8–9 from Kivetty, 10–11 from Romuvaara, 12–17 from Hästholmen, 18–22 from Äspö HRL and 23–25 from Laxemar. Data used to construct the plot came from a cooperative Hydrochemical Stability project between the Swedish Nuclear Fuel and Waste Management Company and Posiva OY (Haveman & Pedersen, 2002). Note the end members of which the groundwaters are composed.

1999). Ethane and propane occur, suggesting that some or most of the methane is of mantle origin, as discussed further below.

Attempts to correlate microbial culturing numbers and total numbers with individual groundwater characteristics have not been very informative. No correlation could be found, apart from a relationship between total numbers and dissolved organic carbon (DOC). However, taking all groundwater characteristics into consideration in a M3 plot, it was found that cultivable sulfate and iron-reducing bacteria disappear as the water composition tends towards brine. However, other physiological types such as heterotrophic acetogens could be cultured from almost any groundwater type in the M3 plot (Haveman & Pedersen, 2002). The complexity in deep groundwater hydrogeochemistry suggests that many different ecological niches may exist, as each point in the M3 plot in Fig. 6 represents a more or less unique mix of different groundwaters. If this is correct, then a significant microbiological diversity should be expected, with many different species, each adapted to one or some of all the observed different hydrogeochemical conditions.

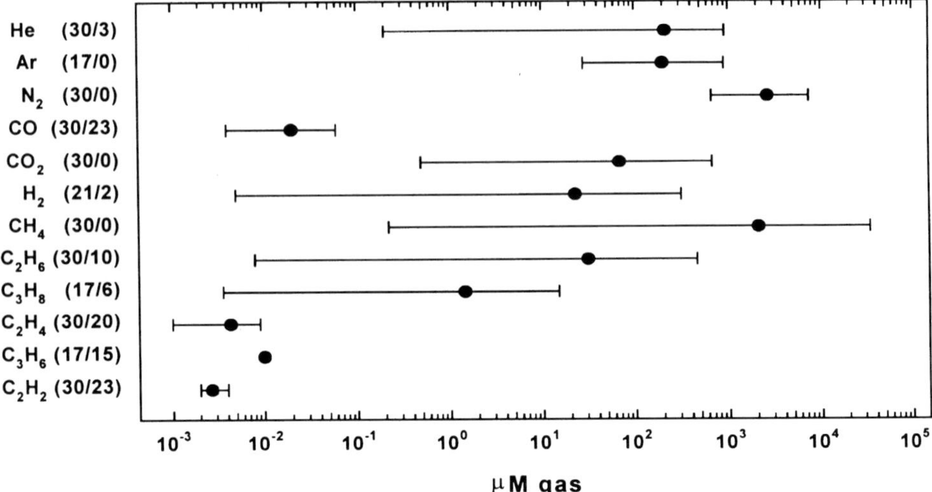

Fig. 7. The composition of the gas observed at the sites investigated. The data are extracted from Haveman et al. (1999), Haveman & Pedersen (2002) and Pedersen (1997a). The figure shows the mean concentration of the chemical species measured (filled circles); the bar gives the range. The numbers in parenthesis for each chemical species are the total number of observations followed by the number of observations below detection limit.

Microbial diversity

The cloning and sequencing of ribosomal 16/18S rDNA from microbes living in their natural environments has revealed a genetic diversity beyond any earlier estimates (Pace, 1997). This methodology has been applied to microbes from Fennoscandian Shield rock aquifers with a similar result. A total of 385 clones from two sites, the Stripa research mine and the Äspö hard rock laboratory (HRL), were sequenced (Ekendahl et al., 1994; Pedersen et al., 1996a, 1997c). The number of unique sequences found was 122, each representing a possible species that was not recorded in international databases at the time the comparisons were done. On average then, approximately one-third of the sequenced clones represented unique species. These studies clearly have not exhausted the sequences because new sequences were found in nearly every additional sample repetition. Similar results were obtained with DNA sequences from the alkaline springs of Maqarin in Jordan (Pedersen et al., 1997a), the natural nuclear reactors in Oklo, Gabon (Pedersen et al., 1996b), and in nuclear waste buffer material (Stroes-Gascoyne et al., 1997). This molecular work indicates that the deep groundwater environments studied are inhabited by diverse microbial populations, consistent with the large variability of the hydrogeochemical conditions.

The molecular work described above has provided good insight into the phylogenetic diversity of igneous rock aquifer microorganisms, but it does not reveal species-specific information unless 100% identity of the 16S rDNA gene with a known and described microorganism is obtained. The huge diversity of the microbial world makes the probability of such a hit very small. None of the 122 specific sequences mentioned above had 100% identities with described species. Still, if a 100% identity is obtained, there may yet be strain-specific differences in some characteristics that are unraveled by the 16S rDNA information (Fuhrman & Campbell, 1998). If species information is required, time-consuming methods in systematic microbiology must be applied to a pure culture. Known genera or species can be identified through these methods. Several isolates from the Laxemar, Äspö and Äspö HRL sites have been identified as *Shewanella putrefaciens*, *Pseudomonas vesicularis* and *Pseudomonas fluorescens* (Pedersen & Ekendahl, 1990; Pedersen et al., 1996a). An isolate that does not match a known genus or species obviously provides the opportunity to describe a new species.

Three new subterranean species from deep igneous rock aquifers have been described and reported. Sulfate-reducing bacteria are common in the deep aquifers studied and three species, based on their different 16S rDNA sequences, were repeatedly isolated from different boreholes in the Äspö HRL tunnel (Pedersen et al., 1996a). One of them had a 16S rDNA identity of above 99% with *Desulfomicrobium baculatum*. This genus seems to be very common at Äspö since its 16S rDNA sequence was repeatedly retrieved from tunnel boreholes at this location (Pedersen et al., 1997c). The isolate Aspo-2 was characterised in detail and described as a new species, *Desulfovibrio aespoeensis* (Motamedi & Pedersen, 1998). It is a mesophilic species with growth characteristics which appear well adapted to life in the aquifers, from where it was isolated.

Three autotrophic, methane-producing strains of Archaea were isolated from the Äspö HRL tunnel boreholes at depths of 68, 409 and 420 m. These organisms were non-motile small, thin rods, 0.1–0.15 μm in diameter, and able to utilise $H_2 + CO_2$ or formate as substrates for growth and methanogenesis. One of the isolates, denoted A8p, was studied in detail. Phylogenetic characterisation based upon 16S rRNA gene sequence comparisons placed this isolate in the genus *Methanobacterium*. Phenotypic and phylogenetic characters indicate that the alkaliphilic, halotolerant strain A8p represented a new species and we proposed the name *Methanobacterium subterraneum*. It grew with a doubling time of 2.5 hours under optimal conditions (20–40°C, pH 7.8–8.8, and 0.2–1.2 M NaCl). *Methanobacterium subterraneum* is eurythermic since it can grow at a wide range of temperatures, from 3.6 to 45°C.

Methane is common in most of the groundwaters studied (Fig. 7) and there has been growing interest in methanotrophs. Their consumption of oxygen, with methane as electron donor, is beneficial for HLW repositories and their activities have therefore been studied in detail (Kotelnikova & Pedersen, 1999). During the investigations of microbial methane oxidation in the Äspö HRL tunnel, several oxygen dependent methanotrophic isolates were obtained and one, SR5, was successfully described in close collaboration with Russian experts on methylotrophic bacteria (Kalyuzhnaya et al., 1999). Methane utilising bacteria were first enriched from deep granitic rock environments and affiliated by amplification of the functional and phylogenetic gene probes. Type I methanotrophs belonging to the genera *Methylomonas* and *Methylobacter* dominated in the enrichment cultures from depths below 400 m. A pure culture of strain SR5, an obligate methanotroph, was isolated and characterised. Based on phenotypical and genotypic characteristics, we proposed strain SR5 as a new species, *Methylomonas scandinavica*. It grows at temperatures of 5–30°C, with an optimum of 15°C, close to the in situ temperature. Whole cell protein, and enzyme and physiological analyses of *M. scandinavica* revealed significant differences between this and the other Type I methanotrophs. The prospect of anaerobic methane oxidation is an intriguing possibility which has been approached in different environments (Hindrichs et al., 1999). However, absolute evidence in the form of a laboratory culture of an anaerobic methane-consuming species or consortium is still lacking.

Microbial activity

Radioactive compounds for the estimation of microbial activity have been in use in microbial ecology for decades (e.g. Grigorova & Norris, 1990). With this technique, samples are incubated with the radiotracer of interest and then examined. Cells, or products, can be separated and examined for radioactivity using standard liquid scintillation techniques. This method will give a measure of average biological activity for the whole sample. The activity of individual cells can be examined using a microautoradiography (MARG) technique (Tabor & Neihof, 1982). Both the liquid scintillation and the MARG technique were applied to microorganisms from the Laxemar, Stripa and Äspö HRL sites, with varying radiotracers and incubation times (Pedersen & Ekendahl 1990, 1992a, b; Ekendahl & Pedersen, 1994; Kotelnikova & Pedersen, 1998;). The populations studied transformed all added compounds at varying rates. Carbon dioxide was assimilated at relatively low rates, as were formate, acetate and glucose. The fastest uptake was obtained with lactate. Significant formation rates of methane and acetate were also found. Heterotrophic methane formation and acetate formation followed the trends observed with culturing methods (Kotelnikova & Pedersen, 1998). These techniques are excellent tools for understanding the present situation but will not reveal activity in the past. That task can, however, be analysed using stable isotopes.

Carbon stable isotope analysis of methane in groundwater offers information about sources of methane. It is indicative of present and past autotrophic methanogenesis (Des Marais, 1999) and of thermocatalytic methane formation (Apps & Van de Kemp, 1993). When $\delta^{13}CH_4$ was measured for Finnish groundwaters, the values plotted in either biogenic or thermogenic areas (Haveman et al., 1999). This indicates that the methane is of thermogenic origin, with some in-mixing of biogenic methane. A weak biogenic signal can be explained by biogenic methane production elsewhere followed by groundwater transport and mixing in the aquifers sampled or by autotrophic methanogenesis at low levels at the sampled depths, as suggested by the M3 modelling (Fig. 6). The main component of the methane in Fig. 7 seems to be thermogenic, however, diffusing up from deep in the rock. The stable isotope composition of carbon in these groundwaters is similar to those found by Sherwood Lollar et al. (1993) for groundwater from Canada and elsewhere in Finland. Based on a model, their explanation for the intermediate values is mixing of different end members, with a bacterial end member consisting of anywhere between 20 and 94% of the groundwater. Anaerobic methane oxidation would be a complicating factor in interpreting these carbon stable isotope data (Hindrichs et al., 1999) because methane oxidisers would preferentially use $^{12}CH_4$, making the $\delta^{13}CH_4$ values less negative. If this is the case, the methane reported in Fig. 7 may have more of a biogenic origin than is indicated by the $\delta^{13}CH_4$ values.

A deep, hydrogen-driven biosphere

Culturing efforts have revealed iron and sulfate-reducing bacteria, acetogens and methanogens to be common in Fennoscandian Shield groundwaters (Pedersen, 2000b). Hydrogen may be an important electron and energy source, and carbon dioxide an important carbon source in deep subsurface ecosystems. Hydrogen, methane and carbon dioxide have been found in μM concentrations at all sites that have been tested for these gases (Fig. 7). Methane is a major product of autotrophic methanogens. Therefore, a model has been proposed of a hydrogen-driven biosphere in deep igneous rock aquifers of the Fennoscandian Shield (Pedersen, 2000a). The organisms at the base of this ecosystem are assumed to be autotrophic acetogens capable of reacting hydrogen with carbon dioxide to produce acetate, autotrophic methanogens that produce methane from hydrogen and carbon dioxide, and acetoclastic methanogens that produce methane from the acetate product of the autotrophic acetogens. All components needed for this life cycle have been found in the deep igneous rock aquifers and the microbial activities expected have been demonstrated (Pedersen, 2000b), so the model is supported by the qualitative data obtained so far. Ideally, the next step will be to obtain quantitative data, which would require very sensitive experimental conditions, because of the very slow metabolic rates that are expected under non-disturbed conditions. The central question for a HLW repository is whether in situ hydrogen-driven microbial chemolithoautotrophic activities may support reduced conditions, especially during times of glaciation (Fig. 8).

4. Microbial oxygen reduction in a HLW repository

The optimum performance of a HLW repository requires an oxygen-free, reduced environment. Oxygen is corrosive for the copper canisters and some radionuclides, such as Np, Pu, Tc and U, are more soluble and mobile under oxygenated conditions. Oxygen will be introduced with air to the repository during the open phase. At closure, some of this air will be captured in voids of the repository. Other routes of oxygen intrusion have also been suggested, such as transport with groundwater from the ground surface, penetration of oxygen into the rock in the tunnels which will create an oxidised rock environment and, finally, radiolysis of water to oxygen and hydrogen if radionuclides escape owing to a canister failure. Periods of glaciation provide a special scenario, when the transport of surface water from melting ice deep into the ground can be significant (Fig. 8).

The possibility that microorganisms may be able to buffer against an oxidising disturbance in bentonite, backfill and the deep host rock environment had previously been overlooked. Microbial decomposition and the production of organic material depend on the sources of energy and on the electron acceptors present. Hydrogen, organic carbon, methane and reduced inorganic molecules are possible energy sources in subterranean environments. During microbial oxidation of these energy sources, the microbes may use several different electron acceptors (see also Chapters 3 and 7, this volume). First, oxygen is used, followed by nitrate, manganese, iron, sulfate, sulfur and carbon dioxide. Simultaneously, fermentative processes supply the respiring microbes with hydrogen and short chain organic acids. The solubility of oxygen in water is low and for many microbes,

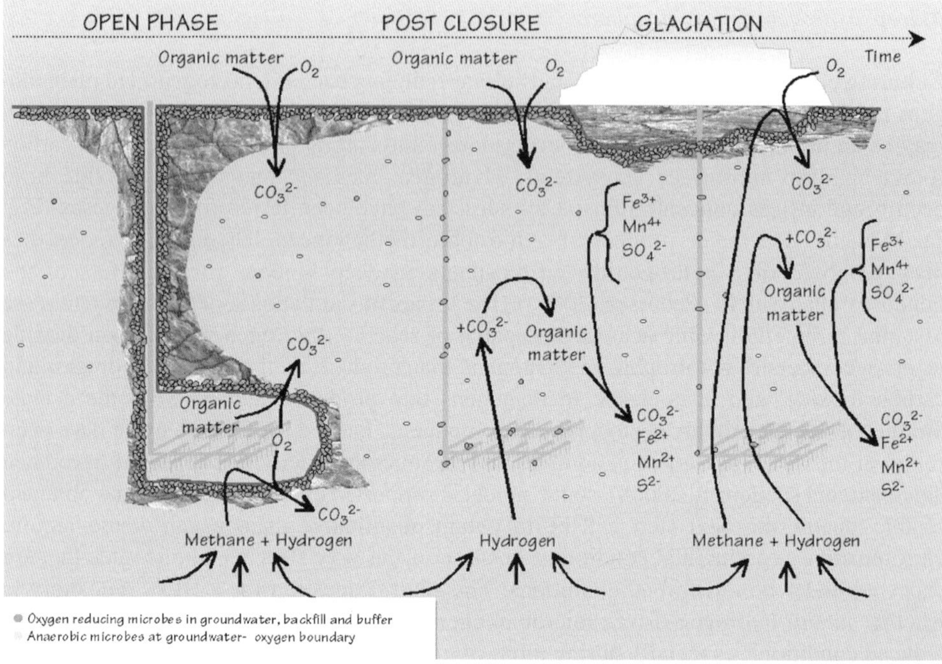

Fig. 8. A schematic model illustrating how microbes in the geosphere would stop oxygen from reaching a HLW repository and keep the groundwater redox potential at low levels. See text for details.

oxygen is the preferred electron acceptor because the use of oxygen yields much more energy per organic molecule than other electron acceptors. The presence of an active and diversified microbiota at repository depths is well documented in this chapter, as is the reducing capacity of microorganisms in any environment. The major oxygen buffers that can be used are methane and organic carbon. Hydrogen, sulfide and ferrous iron can also contribute but these chemical species generally appear in much lower concentrations than do methane and organic carbon (see Figs 5 and 7). However, locally, they may have a significant effect.

During the past decade, a series of different projects have been launched, aiming at understanding the fate of oxygen in a repository as well as the redox buffer capacity of rock and groundwater. The general conclusion from these projects is that microbes will dominate the oxygen removal and redox control processes. The projects are reviewed below.

Redox investigations in block scale

The first project, 'REDOX' studied the induction of organic carbon oxidation during large-scale shallow (0–70 m) water intrusion into a vertical fracture zone at the Äspö HRL (Banwart et al., 1996). Initial models suggested that oxygenated groundwater should reach the tunnel at 70 m depth, 3–4 weeks after the fracture zone was intersected by the tunnel

(Banwart, 1995). Oxygen never appeared. Instead, an increased ferrous iron concentration and an increase in alkalinity were observed. The conclusion was that microbes degrading organic carbon rapidly consume intruding oxygen. The degradation continues after all oxygen has been consumed, but with ferric iron replacing oxygen as the dominating electron acceptor.

Microbial oxygen reduction in a tunnel environment

A variety of bacteria, the methanotrophs, readily oxidise methane with oxygen. They utilise oxygen as an electron donor for energy generation and methane as a sole source of carbon. Most of these bacteria are aerobes and they are widespread in natural soils and waters. They also present a morphological diversity, and are apparently related solely through their ability to oxidise methane. Methanotrophs are found wherever stable sources of methane are present. There is some evidence that although methane oxidisers are obligate aerobes, they are sensitive to oxygen and prefer microaerophilic habitats for development. Recently published data, however, indicate that methane oxidation can occur in some anaerobic environments (Hindrichs et al., 1999). The methane oxidisers are often concentrated in a narrow band between anaerobic and aerobic zones where methane meets an oxygenated system. Such environments will be common in future repositories during the open phase and for some time after closure (Fig. 8). Once established, this group of bacteria will be active for as long as oxygen is present for the oxidation of methane. After closure, they will most probably react all available methane with the remaining oxygen.

The project 'Microbe-REX' established that methane oxidisers are very common in the Äspö HRL tunnel (Kotelnikova & Pedersen, 1999), so a deep repository will rapidly become anoxic after closure if methane is in excess. One molecule of methane contains eight electrons which can be used to reduce two molecules of oxygen. In the worst case scenario, there will be approximately 250 μM dissolved oxygen in groundwater close to the repository, which can be balanced by 125 μM methane. The concentration of methane in many deep groundwaters is therefore usually large enough (Fig. 7) for microbial removal of all oxygen. All that is needed is one to five volumes of methane-containing groundwater to mix with one volume of oxygen-containing groundwater. The time required for this process depends on the bacterial activity, but it will probably take much less than a year, as most microbes work very fast when given the chance to proliferate. The 'Microbe-REX' project also documented significant microbial oxygen consumption with the organic carbon naturally present in groundwater. A model was developed to predict the microbial oxygen reduction in a repository based on chemical groundwater data, microbial groups present, and specific kinetic properties of these microbes, the Km and V_{max} (Kotelnikova & Pedersen, 1999).

Redox experiment on a detailed scale

The main objective of the redox project on a detailed scale, REX, was to investigate dissolved molecular oxygen consumption by creating a controlled oxidising perturbation in a deep rock environment at the Äspö HRL. The fieldwork for this project was completed in the summer of 1999 and the publication of the results is currently in progress. Briefly, the results demonstrated that introduction of oxygen to an anaerobic hard rock aquifer induces growth of aerobic attached and unattached microbial populations. During the experiments, the proportion of culturable aerobic microbes increased continuously, while the number of culturable strict anaerobic microbes decreased. The aerobic microbes were demonstrated to have a dominating role in the oxygen reduction during a series of oxygen injection experiments. Laboratory work with a replica rock core confirmed the dominating influence of microbes on oxygen reduction in a HLW host rock environment.

Microbial reduction of oxygen with ferrous iron, sulfide or manganese

Tunnels in hard rock commonly develop brown, black or white precipitates on walls, in ponds and in ditches (Pedersen & Karlsson, 1995). The majority of these masses are built up of iron oxides and sulfur grains, mixed with very large numbers of microorganisms. The microbes take advantage of the energy available in ferrous iron, manganese and sulfur where the tunnel wall interface provides an aerobic environment. They use oxygen to oxidise the metals and the sulfur, and this gives them energy for carbon dioxide fixation in organic molecules. Such microbes are so-called 'gradient organisms', living in redox gradients between anoxic and oxic environments. Their activity stops oxygen from migrating into the rock and they produce organic carbon from carbon dioxide.

Model for interaction of microbial activity and groundwater geochemistry

The three projects described above, with experimentally independent methods, all concluded that microbes in hard rock aquifers and tunnels are capable of reducing oxygen. The results indicate that a major benefit of the presence of geosphere microbes for repository performance is their massive capacity to protect the host rock and repository from oxygen, and their production of groundwater components that lower the redox potential. Figure 8 illustrates a possible geosphere scenario. Oxygen will move with recharging groundwater into the basement rock and will diffuse from the tunnel air into the rock matrix. However, the recharging groundwater will contain organic matter and microbes will continuously reduce this oxygen by oxidising organic carbon. Anaerobic microbes in the hard rock aquifers in the host rock are known to reduce ferric iron, manganese(IV) and sulfate to ferrous iron, manganese(II) and sulfide with organic carbon. These metals and the sulfur will react with oxygen when the water reaches a tunnel. Mats of microbes develop on the tunnel walls where groundwater seeps out and produce organic carbon with the energy derived from these groundwater components. Other microbes can later use the organic matter for additional oxygen reduction. Thus the microbes close biogeochemical cycles (Pedersen & Karlsson, 1995).

Periods of glaciation present a special case (Fig. 8). During such events, the input of organic carbon with recharging groundwater will be low because during a glaciation, photosynthetic production of organic carbon will cease. The REX projects demonstrated a significant activity of methane-oxidising bacteria. Methane is produced in deep magmatic rocks and migrates upwards (Apps & Van de Kamp, 1993). The continuous flow of methane from deep mantle rocks will not depend on glaciation events. Hydrogen is an even better oxygen reducer for microbes than is methane, but this gas appears in lower concentrations in deep groundwater (Fig. 7).

The current model of microbial activity in the backfill environment

Thus far, there have been few results from backfill research. A full-scale backfill and plug test has been started at the Äspö HRL (SKB AB, 1999). Laboratory cultures of various bacteria were introduced at specific positions in the middle of the backfill and the microbes in the backfill were analysed. The results showed a significant diversity of culturable bacteria in the backfill material at the outset. We found sulfate-reducing bacteria that were culturable at 30°C and 60°C. *Pseudomonas stuzeri* and *Stenotrophomonas maltophilia* appeared in the rock/bentonite (70/30%) backfill. *Aeromonas encheleia* appeared in 100% crushed rock. *Bacillus thermophilica* (cultured at 60°C) was found in both backfill mixtures. Other species also appeared, but were not identified. Obviously, many different microbes are present in the backfill, and more will travel via the groundwater into the backfill pores. There will be a relatively dense population of microbes in the backfill, which may be beneficial for the repository, as outlined below.

As mentioned above, backfill has a large diversity of microbes in relatively abundant numbers. A main concern about backfill is its oxygen content at the outset (e.g. Wersin et al., 1994). This oxygen has a corrosive effect on the copper canisters. The REDOX, Microbe REX, and REX experiments discussed above all indicated that microbes would be very efficient in removing oxygen from groundwater, if introduced. It can be hypothesised that this will also occur in the backfill as depicted in Fig. 9. Also, active sulfate and iron-reducing bacteria would produce sulfide and ferrous iron, both of which reduce oxygen and lower the redox potential of the groundwater in the backfill. A low redox potential is important for achieving low radionuclide mobility in the case of an accidental release of radionuclides. In such a scenario, the microbes would protect the environment from the products from radiolysis of water by efficiently recombining the oxygen and hydrogen produced to water, and thus, buffering the redox downwards.

5. Survival of microorganisms in compacted bentonite

Microbial processes in anaerobic environments commonly result in the formation of gas and sulfide. Gas formation may give rise to disruptive mechanical effects on the buffer, and sulfide can corrode the canister. As discussed above, sulfide production must occur close to the canister and must be vigorous for damaging corrosion to be possible. This is mainly due to the fact that sulfide has a low solubility in groundwater and that its diffusive transport capacity is therefore very low. Many bacteria consume oxygen during their degradation of organic carbon. Such a process would be beneficial anywhere in the buffer but particularly in backfill where the oxygen content in pores is expected to be significant.

Research has been ongoing regarding the effects of microbial processes in buffer and backfill. Some processes have been studied for a long time; others are now becoming available for study, with the new research facilities at the Äspö HRL. The results obtained so far from buffer and backfill research and their interpretation, as well as ongoing and planned experiments, are presented below (see also Chapter 9, this volume).

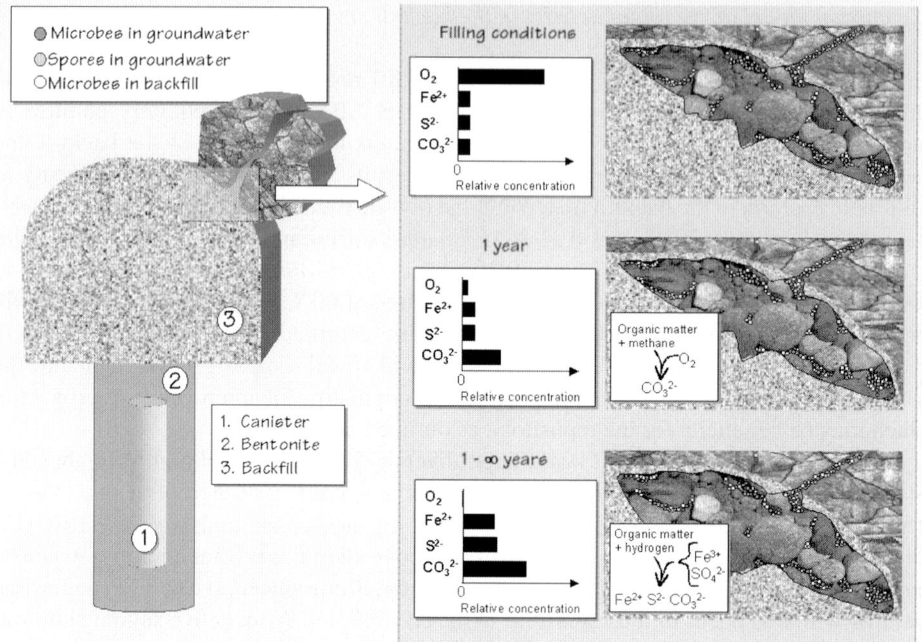

Fig. 9. A schematic model illustrating how microbial populations will inhabit and change the geochemistry of the backfill. See text for an explanation.

The buffer mass container experiment

A full-scale experiment with buffer material consisting of 50/50% bentonite/sand was performed at Atomic Energy of Canada Limited's (AECL) underground laboratory in Canada. The results showed that microbes, with a few exceptions, could only be cultured from buffer samples with a water content of 15% or more, which is approximately equivalent to a 100% bentonite density of 2000 kg m^{-3} (Stroes-Gascoyne et al., 1997). Elevated temperatures had no effect on the microbes. These results were interpreted as an effect of limited availability of water. The cell walls of most microbes (except for some fungi) are freely permeable to water so microbes cannot keep more free water inside than outside the cell. They therefore compensate for a low water content in the environment by intake of salt ions to adjust the inner osmotic pressure to the outside level. Some microbes can, alternatively, produce polyalcohols or other osmotically active organic molecules. This production requires energy and organic carbon, which in buffer and backfill is available only in limited quantities. The result of the Buffer Mass Container (BMC) experiment invoked questions about the survival of microbes, and especially SRB, in buffer materials in 100% bentonite and led to detailed laboratory experiments, as described next.

Survival of sulfate-reducing bacteria under laboratory conditions

Two species of sulfate-reducing bacteria were mixed with MX-80 bentonite at varying densities, from 1500 to 2000 kg m^{-3} (Pedersen et al., 1995; Motamedi et al., 1996). The species were *Desulfovibrio aespoeensis* and *Desulfomicrobium baculatum*, both isolated from deep groundwater at the Äspö HRL. None of the species survived 60 days at densities above 1800 kg m^{-3}. *Desulfomicrobium baculatum* survived the better of the two, remaining culturable for 60 days at 1500 kg m^{-3}. It can, however, be argued that the laboratory conditions during this experiment may have added some extra constraints to the ones found in the field situation. The laboratory experiment represents a closed situation (Fig. 2), while field conditions would be of the open system type (Fig. 3). The differences between these two situations are discussed above, in the section 'Microbial processes in closed and open systems'. A long-term field experiment was therefore initiated and the results are discussed next.

Survival under field conditions

The long-term test (LOT) of buffer performance aims to study models and hypotheses of the physical properties of a bentonite buffer (SKB AB, 1999). Processes related to microbiology, radionuclide transport, copper corrosion and gas transport under conditions similar to those found in a HLW repository (Fig. 1) are also investigated. The project is ongoing but some of the experiments have been analysed. The long-term test offers the possibility of exposing bacteria to conditions representative of those found in a repository, with the exception of high radiation.

For the LOT, several species of bacteria with different relevant characteristics were chosen (Pedersen et al., 2000a). The sulfate-reducing bacteria included were *Desulfovibrio aespoeensis* isolated from deep Äspö groundwater, the moderately halophilic bacterium *Desulfovibrio salexigens*, and the thermophilic, spore-forming *Desulfotomaculum nigrificans*. Aerobic bacteria included *Deinococcus radiophilus*, which can tolerate high doses of radiation, the chemoheterotrophic bacterium *Pseudomonas aeruginosa* that frequently occurs in soil, the chemoorganotrophic and chemolithotrophic (hydrogen-utilising) organism *Alcaligenes eutrophus*, the chemoheterotrophic, spore-forming bacterium *Bacillus subtilis*, and the thermophilic, spore-forming bacterium *Bacillus stearothermophilus*.

Suspensions of the sulfate-reducing bacteria (anaerobic) and aerobic bacteria were mixed with bentonite clay to give approximately 100 million bacteria per gram (dry weight) of clay. The clay with bacteria was subsequently formed into cylindrical plugs with a 20 mm length and diameter, and installed in bentonite blocks (Fig. 10) exposed to low (20–30°C) and high (50–70°C) temperatures. The blocks were installed in the LOT boreholes immediately after the bacteria plugs were introduced (Pedersen et al., 2000a). The experiment was terminated after 15 months. The major outcome was the effective elimination, to below detection limits, of all bacteria except the spore-forming ones (Fig. 11).

All of the three spore formers survived at the low temperature. The numbers remaining were, however, much lower than those initially introduced. The approximately 100 million spore-forming bacteria per gram (dry weight) of clay were reduced 100- to 10,000-fold. This can be interpreted as showing that the cell death rate was higher than the growth rate, which may have been zero, or close to zero. The spore-forming sulfate-reducing bacterium, *D. nigrificans*, was the only one of the three which survived at high temperature. This species was obviously best suited, possibly genetically, to the conditions in the clay. It can

Fig. 10. Block No. 29 from one of the LOT experiments was ruptured and a set of plugs that were inoculated with bacteria became visible. The plugs were observed and sampled after 15 months exposure. (Photograph: M. Motamedi.)

be concluded that the spore formers most probably survived as inactive spores, and that spores do not produce sulfide. Survival is, as pointed out previously in the section on 'The microbial enigma – death or survival', not equivalent to activity. Since the methods used in this experiment did not reveal activity, additional experiments were set up to include measurement of sulfide production in simulated repository conditions, as described below.

Microbial mixing and survival during the buffer swelling process

Bentonite blocks have a low water content (approximately 10–17%) at the time of deposition of the canisters. There is a thin gap between the bentonite and the canister (1 cm), and between the bentonite and the rock (5 cm), to enable smooth lowering of the canister and the blocks into the deposition hole. These gaps are filled with groundwater from the rock, or alternatively, with water added at the time of deposition. The bentonite begins to swell and eventually will reach the planned full compaction density (2000 kg m^{-3}) and water content (approximately 30%). It was expected that microbes could be mixed with part of the clay during the swelling process and it was also of interest to examine whether microbes could migrate into the bentonite from groundwater. A series of experiments was therefore set up to investigate the course of these events (Pedersen et al., 2000b).

A

Bacillus subtilis

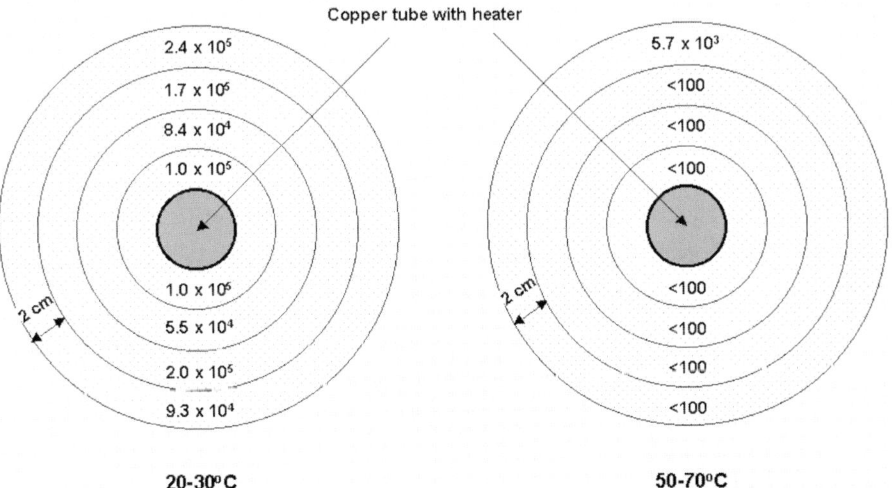

Fig. 11a.

Bacillus stearothermophilus

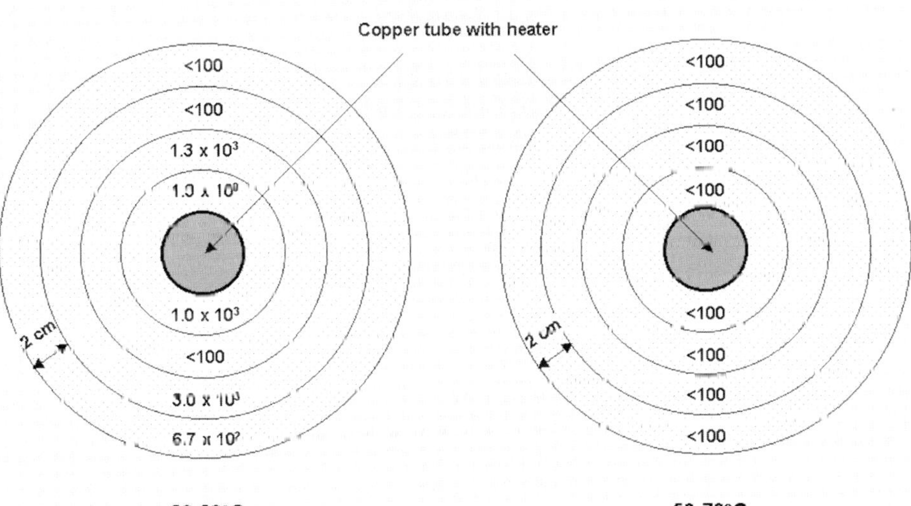

Fig. 11b.

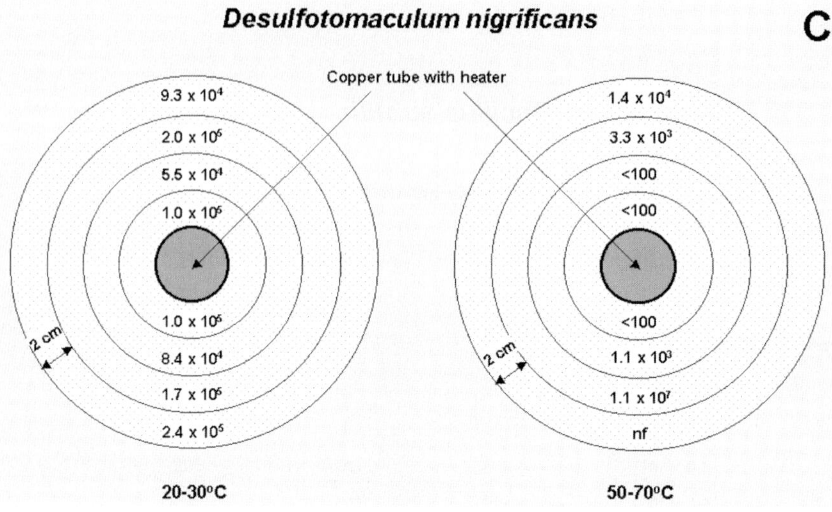

Fig. 11c.

Fig. 11. Three different spore-forming bacteria survived after 15 months exposure to different temperatures in the LOT experiment. They were exposed to gradients of the temperatures indicated, with the highest temperature being closest to the heater. The numbers given are per gram dry weight bentonite clay. (nf = plug not found.)

Swelling pressure oedometers were installed with compacted bentonite with 10% water content, and a gap was left between the bentonite and the filter lid of the oedometer, to mimic the gap in a deposition hole. Mixtures of bacteria were added to the gap and the oedometers were left for sampling at different times between 8 hours and 28 weeks. The following anaerobic bacteria were used: *Desulfomicrobium baculatum*, which has been isolated from the deep groundwater of the Äspö HRL (SKB AB 1999), the moderately halophilic *Desulfovibrio salexigens*, the thermophilic, spore-forming *Desulfotomaculum nigrificans* and the thermophilic *Thermodesulfobacterium commune*, which has an optimal temperature for growth of 70°C. Aerobic bacteria included *Deinococcus radiophilus*, which can tolerate high doses of radiation and desiccation, *Pseudomonas aeruginosa*, a chemoheterotroph which frequently occurs in soil, the chemoorganotrophic, chemolithotrophic *Alcaligenes eutrophus*, the chemoheterotrophic, spore-forming *Bacillus subtilis* and the thermophilic spore-forming *Bacillus stearothermophilus*. The incubation temperatures varied from 30 to 80°C, depending on the respective optimum temperature for the bacteria used.

The part of the bentonite that came in contact with the microbes was sliced into layers perpendicular to the gap and the different microbes were analysed. Survival varied significantly from species to species and between different depths (Table 2). *Deinococcus radiophilus* and *Bacillus subtilis* showed the best survival rates. They also could be found in the deepest layer analysed, 3–6 mm, meaning that they mixed with the clay to a depth of at least 3 mm. The other bacteria tested also survived, but only for shorter times and not at depth in the clay.

Table 2

The number of viable cells of bacteria in swelling, compacted bentonite, analysed at different times of exposure

Species introduced	Viable cells ml^{-1} suspension at time = 0	Viable cells per gram dry weight bentonite			
		8 h	14 days	84 days	196 days
Aerobic bacteria					
Deinococcus radiophilus	5.7×10^8				
A[a] (0–1 mm)		2.5×10^7	2.7×10^5	2.0×10^4	1.4×10^4
B (1–3 mm)		1.6×10^4	1.0×10^5	1.6×10^4	2.5×10^4
C (3–6 mm)		1.4×10^4	1.5×10^5	8.3×10^3	2.4×10^4
Pseudomon as aeruginosa	5.0×10^9				
A		2.8×10^8	1.7×10^5	9.0×10^3	1.4×10^4
B		7.7×10^4	4.0×10^4	–[b]	–
C		–	–	–	–
Alcaligenes eutrophus	2.4×10^9				
A		3.1×10^6	1.0×10^4	–	–
B		n.d.	3.0×10^3	–	–
C		1.3×10^4	6.3×10^3	–	–
Bacillus subtilis	n.d.				
A		4.3×10^5	1.0×10^8	4.9×10^4	5.9×10^4
B		n.d.	1.9×10^6	5.5×10^4	1.7×10^3
C		1.3×10^4	1.6×10^5	1.6×10^4	1.9×10^4
Bacillus stearothermophilus	n.d.				
A		1.2×10^5	2.3×10^3	–	–
B		–	–	–	–
C		–	–	–	–
Desulfovibrio salexigens	1.7×10^8				
A		1.7×10^5	7.3×10^3	–	–
B		2.4×10^3	4.9×10^2	–	–
C		–	–	–	–
Desulfovibrio baculatum	1.3×10^8				
A		2.8×10^7	1.3×10^5	2.4×10^2	
B		3.3×10^4	3.3×10^2	–	–
C		–	–	–	–
Desulfotomaculum nigrificans	1.7×10^7				
A		9.4×10^5	3.5×10^5	7.9×10^4	1.4×10^2
B		2.4×10^4	2.3×10^2	–	–
C		–	–	–	–
Thermodesulfobacterium commune		7.9×10^7			
A		3.3×10^2	–	–	–
B		–	–	–	–
C		–	–	–	–

[a] A, B and C represent the positions of the samples, measured from the surface.

[b] <100.

n.d. = not determined.

Microbes occurring naturally in MX-80 bentonite

The bentonite is not sterile at the outset. Inoculation of dry bentonite samples (with a water content of 10%) in our different culture media revealed several different species to be present. Typically, we found spore-forming genera and species such as *Bacillus subtilis*, *Bacillus cereus* and *Brevibacillus brevis*, and desiccation-resistant species such as *Pseudomonas stutzeri* and the actinomycete *Thermoactinomyces* (Pedersen et al., 2000b). The results of such culturing experiments should be viewed as a minimum and, as indicated by molecular investigations of Canadian buffer masses, many more species would most probably have been discovered had the experiments been more extensive (Stroes-Gascoyne et al., 1997). It has, however, become apparent that there will be two distinct sources of microbes in the buffer, specifically (1) those naturally present in the commercially available MX-80 bentonite, and (2) those introduced via the groundwater.

6. Microbial activity at the copper waste canister

The worst case scenario in copper canister corrosion would be if sulfate-reducing bacteria formed biofilms on the canisters or grew intensively in the buffer close to the canister. The corrosion process would be controlled by the transport of sulfate to the canister, provided enough hydrogen or degradable organic carbon was available for such growth. This could lead to considerably accelerated corrosion since the mass transport of sulfate is expected to be much larger than the transport of sulfide, since sulfate concentrations in the bentonite can be up to tens of mmol dm^{-3}, much higher than observed for sulfide.

During the initial phase, the temperature in the repository will be elevated, with a maximum temperature of 90°C on the copper surface. This is not an absolute constraint because, at present, the known temperature limit for life is 113°C (Stetter, 1996) so sulfate-reducing bacteria may survive. The radiation will be high at the canister surface, which will further stress the microbes. Finally, low availability of water in the buffer (i.e. the water content relative to groundwater) will also constrain long-term survival. Altogether, conditions for survival will be very difficult for microbes existing close to the canister. Some preliminary investigations have been performed.

Swelling pressure oedometers (Pedersen et al., 1995; Motamedi et al., 1996) were loaded with bentonite at different densities, corresponding to different water activity values. A copper disc was placed between the bottom lid and the bentonite was compacted. Different sulfate-reducing bacteria were added to the clay and the discs, together with $^{35}SO_4^{2-}$. The species used were laboratory cultures of *Desulfomicrobium baculatum* (optimum temperature for growth 30°C) and *Desulfotomaculum nigrificans* (optimum temperature for growth 55°C). Finally, oxidised silver foil was placed between the disc and the clay. The oedometers were reassembled and incubated for 4 weeks at the respective optimum temperatures and three different densities: 1500, 1800 and 2000 kg bentonite m^{-3}, corresponding to the water activities 0.999, 0.994 and 0.964, respectively. After incubation, $^{35}S\text{-}Ag_2S$ was localised on the silver foils and quantified by electronic autoradiographic imaging (Packard instant imager electronic autoradiography system, Meriden, USA). The amount of Ag_2S formed was used as a measure of the sulfate-reducing activity.

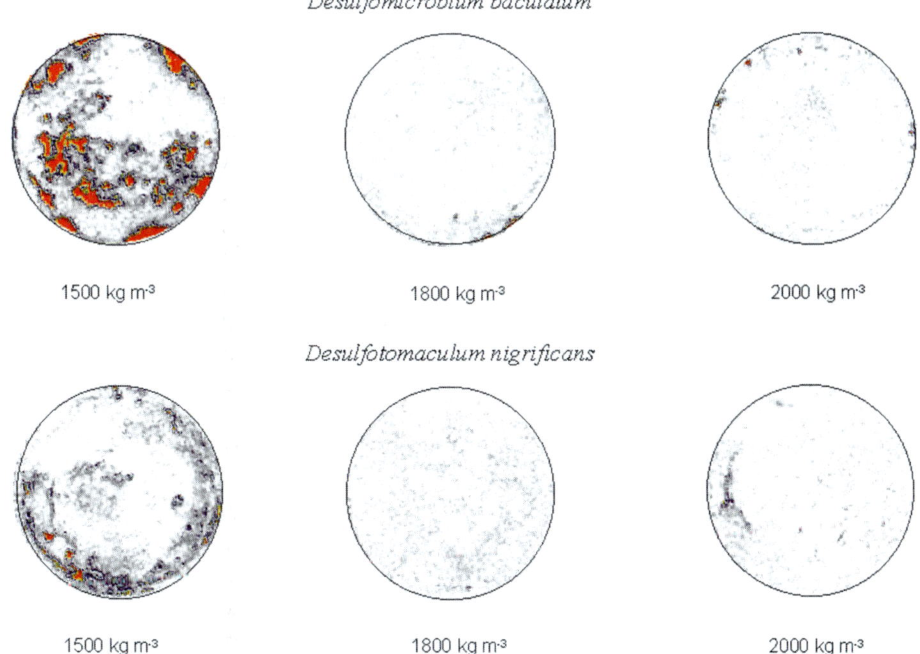

Fig. 12. Radioisotope (^{35}S-sulfide) images of copper discs are shown after incubation with sulfate-reducing bacteria and $^{35}SO_4^{2-}$. The discs were incubated with bentonite at three different densities and with two different species. Darkening indicates ^{35}S-sulfide, with intensity proportional to concentration.

Figure 12 shows that the sulfate-reducing bacteria used were active at a bentonite density of 1500 kg m^{-3}, but that sulfide production was virtually absent at higher densities. This experiment indicates that sulfate-reducing bacteria probably cannot be active at the canister surface at a repository density of 2000 kg m^{-3}. However, the experiment was run with two laboratory species and it may be that other species which were not tested could survive better. Therefore, additional experiments will be carried out using natural groundwater which commonly contains many hundreds of different microbial species, including several naturally occurring species of sulfate-reducing bacteria.

The current model of microbial survival in compacted bentonite

The results concerning survival and activity of microbes in compacted bentonite can be summarised in a conceptual model, as depicted in Fig. 13. At the time of deposition, there will be a canister, bentonite blocks and a hole in the rock. The next step will be to allow water to fill up all the void volume. This water can be groundwater from the rock or, alternatively, groundwater or water added from above at deposition. Irrespective of the source, microbes will be present in the water and these microbes will mix with the buffer, as described above.

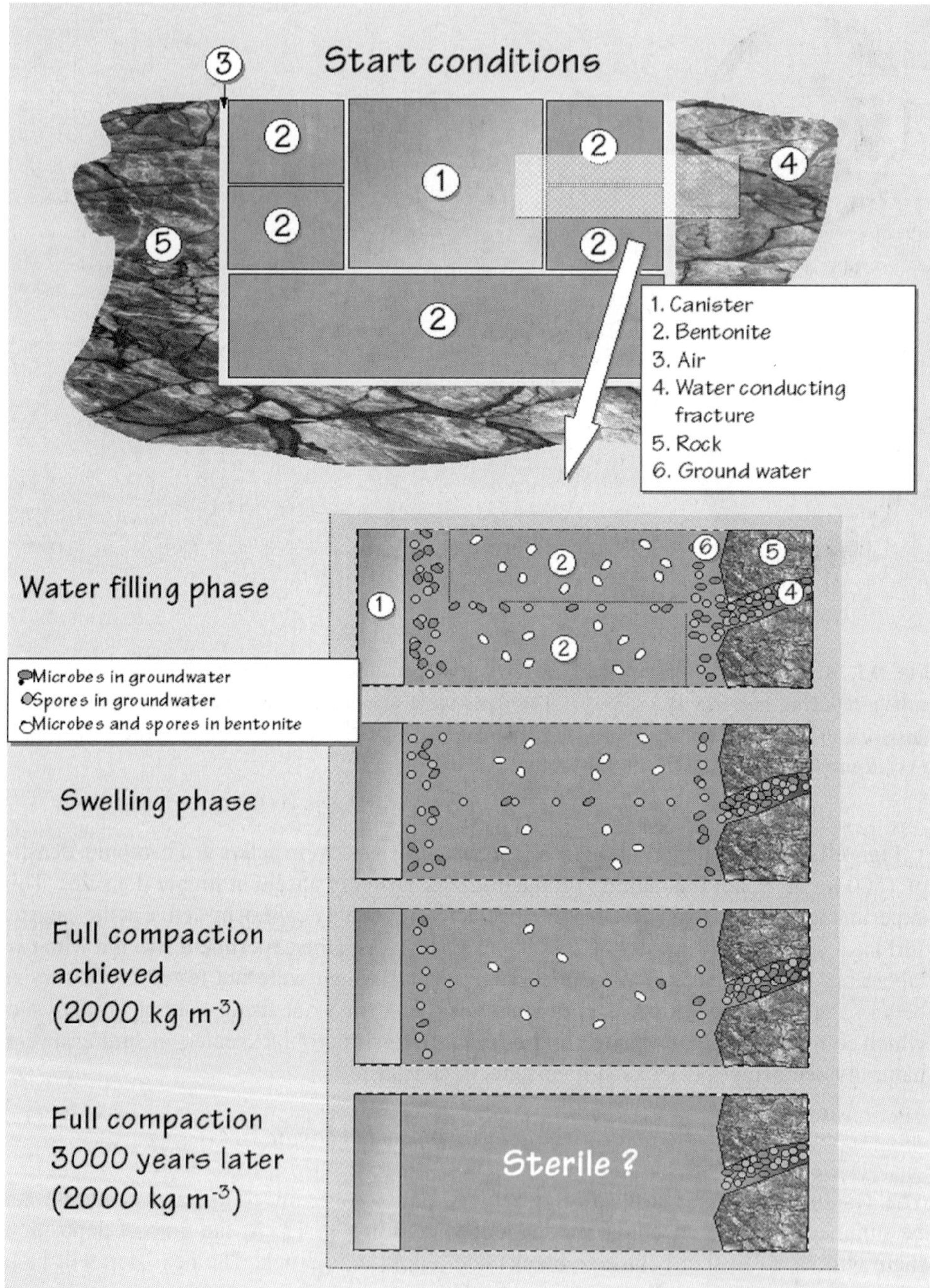

Fig. 13. A schematic model of how microbial populations will alter their presence in the buffer. See text for explanation.

The swelling of the clay will seed groundwater microbes to depth, so they may reach as close as a couple of centimetres from the canister and rock surfaces while the microbes indigenous to the bentonite will be present both inside the bentonite and in the mixing zone. The experimental data on survival and activity of microbes in bentonite suggest that the number of viable microbes will decrease rapidly during swelling and that very few viable cells will be present at full compaction. Sulfate-reducing activity will also approach zero when full compaction is achieved and the only survivors will be microbes that have formed spores. Our results indicate that viable cell activity will be impossible at full compaction, as spores are inactive. Although spores generally are very resistant to difficult environmental conditions, they do still die off. All our experiments so far indicate a decrease in the number of viable spores at full compaction so that a slow but significant death rate of spores would eventually lead to the complete eradication of life in the buffer. It has not yet been clarified whether this will occur in the lifetime of a radioactive repository. Once the bentonite becomes sterile, it will probably not be reinfected. The theoretical pore size of the clay is 100–1000 times smaller than the average-sized microbe, meaning that no new microbes can enter into the buffer.

The model presented is based on current data, obtained with laboratory cultures. It could be argued that naturally occurring microbes are more tolerant, although the working hypothesis remains total eradication of all life in the buffer. Ongoing and planned experiments will continue to test this hypothesis under increasingly relevant field conditions.

7. Retention and transport of radionuclides

Bacteria and metals

The majority of the radionuclides are metals. The transport, chemical speciation, and ultimate fate of dissolved metals in aqueous systems are largely controlled by reactions that occur at solid surfaces (Stumm & Morgan, 1996). Recognition of the importance of solid-phase reactivity in aqueous geochemistry has fostered the development of the surface complexation-precipitation theory (SCPT) as the leading model for understanding the behaviour of dissolved metals in pristine and contaminated waters (Dzombak & Morel, 1990; Stumm & Morgan, 1996). This concept embraces the principles of thermodynamics and chemical equilibria to predict when solid-phase partitioning of metal ions is likely to occur in response to sorption, and quantify subsequent surface precipitation reactions. The SCPT approach has thus far been applied almost exclusively to minerals, particularly to hydrous iron oxides (Dzombak & Morel, 1990); however, Warren & Ferris (1998) recently demonstrated that a continuum exists between ferric iron sorption and precipitation reactions on bacterial surfaces, as anticipated with SCPT. Pedersen & Albinsson (1991) report a similar process with the iron-reducing bacterium *Shewanella putrefaciens*. This is an important step forward as bacteria are at least as widely distributed in aqueous systems and probably as reactive as many inorganic solids. Moreover, if SCPT is to emerge as a true guiding paradigm for aqueous geochemistry, it must be firmly established to be applicable to both organic and inorganic solids.

The behaviour of bacteria as geochemically reactive solids can be inferred from extensive research documenting their performance as sorbents of dissolved metals, and nucleation templates for a wide range of authigenic minerals (Pedersen & Albinsson, 1991, 1992; Konhauser et al., 1994; Konhauser & Ferris, 1996). This reactivity stems directly from the presence of amphoteric surface functional groups (i.e. carboxyl, phosphoryl, and amino constituents), which are associated with structural polymers in the cell walls and external sheaths or capsules of individual cells (McLean & Beveridge, 1990). Direct interaction between these surface functional groups and dissolved metals accounts for the sorptive properties of bacteria, while surficially sorbed metals provide discrete sites for subsequent mineral nucleation and precipitation reactions (Pedersen & Albinsson, 1991).

Because of their ubiquitous distribution and reactive surface properties, hydrous iron oxides are considered to be dominant sorbents of dissolved metals in aquatic environments (Stumm & Morgan, 1996). This perception is somewhat tempered by work which shows that natural iron oxides often contain significant amounts of silica (e.g. siliceous ferrihydrite) and sulfate (e.g. jarosite and schwertmannite), as well as organic matter, including intact bacterial cells (Konhauser & Ferris, 1996). This intermixing of compositionally variable iron oxides and organic matter produces composite multiple sorbent solids with highly variable metal retention properties, so-called 'bacteriogenic iron oxides' (BIOS).

Accumulation of metals by bacteriogenic iron oxides

Bacteriogenic iron oxides and groundwater samples were collected underground at the Stråssa mine in central Sweden and from the Äspö HRL tunnel. Ferrous iron-oxidising bacteria, including stalked *Galionella ferruginea* and filamentous *Leptothrix* sp., were prominent in the BIOS samples from Stråssa, while *Galionella ferruginea* dominated in the Äspö HRL samples. The goal of these investigations was to understand the accumulation of various metals by BIOS. Strontium, caesium, lead, and uranium were studied in the Stråssa BIOS (Ferris et al., 2000), and sodium, cobalt, copper, chromium and zinc were studied in Äspö HRL BIOS (Ferris et al., 1999).

The BIOS samples were found to contain only amorphous hydrous ferric oxide, as determined by X-ray diffraction. Inductively coupled plasma mass spectroscopy revealed hydroxylamine-reducible iron and manganese oxide contents ranging from 55 to 90% on a dry weight basis. Distribution coefficients (K_d values), calculated as the ratio between BIOS and dissolved heavy metal concentrations, revealed solid-phase enrichments of 10^0 to 10^5, depending on the metal and iron oxide content of the sample (Ferris et al., 1999, 2000). At the same time, however, a strong inverse linear relationship was found between log Kd values and the corresponding mass fraction of reducible oxide in the samples, implying that metal uptake was strongly influenced by the relative proportion of bacterial organic matter in the composite solids. Based on the metal accumulation properties of the BIOS, an important role can be inferred for intermixed iron oxides and bacterial organic matter in the transport and fate of dissolved metals in groundwater systems.

References

Apps, J. A. & Van de Kamp, P. C. (1993). Energy gases of abiogenic origin in the Earth's crust. *The Future of Energy Gases US Geologic Survey Professional Paper, 1570*, 81–132.

Banwart, S. (1995). The Äspö redox investigations in block scale. Project summary and implications for repository performance assessment. SKB Technical Report, 95-26 (pp. 1–47). Stockholm: Swedish Nuclear Fuel and Waste Management Co.

Banwart, S., Tullborg, E-L., Pedersen, K., Gustafsson, E., Laaksoharju, M., Nilsson, A.-C., Wallin, B. & Wikberg, P. (1996). Organic carbon oxidation induced by largescale shallow water intrusion into a vertical fracture zone at the Äspö Hard Rock Laboratory (Sweden). *Journal of Contaminant Hydrology, 21*, 115–125.

Birch, L. & Bachofen, R. (1990). Complexing agents from microorganisms. *Experientia, 46*, 827–834.

Des Marais, D. J. (1999). Stable light isotope biogeochemistry of hydrothermal systems. In G. R. Bock & J. A. Goode J. A (Eds), *Evolution of Hydrothermal Ecosystems on Earth (and Mars?)* (pp. 83–98). Chichester, UK: John Wiley.

Dzombak, D. A. & Morel, F. M. M. (1990). *Surface Complex Modelling: Hydrous Ferric Oxide*. New York: Wiley InterScience.

Ekendahl, S. & Pedersen, K. (1994). Carbon transformations by attached bacterial populations in granitic ground water from deep crystalline bedrock of the Stripa research mine. *Microbiology, 140*, 1565–1573.

Ekendahl, S., Arlinger, J., Ståhl, F. & Pedersen, K. (1994). Characterisation of attached bacterial populations in deep granitic groundwater from the Stripa research mine with 16S-rRNA gene sequencing technique and scanning electron microscopy. *Microbiology, 140*, 1575–1583.

Ferris, F. G., Konhauser, K. O., Lyvén, B. & Pedersen, K. (1999). Accumulation of metals by bacteriogenic iron oxides in a subterranean environment. *GeoMicrobiology, 16*, 181–192.

Ferris, F. G., Hallberg, R. O., Lyvén, B. & Pedersen, K. (2000). Retention of strontium, caesium, lead and uranium by bacterial iron oxides from a subterranean environment. *Applied Geochemistry, 15*, 1035-1042.

Fisk, M. R., Giovannoni, S. J. & Thorseth, I. H. (1998). Alteration of oceanic volcanic glass: textural evidence of microbial activity. *Science, 281*, 978–980.

Fredrickson, J. K. & Onstott, T. C. (1996). Microbes deep inside the earth. *Scientific American, 275*, 42–47.

Fuhrman, J. A. & Campbell, L. (1998). Microbial microdiversity. *Nature, 393*, 410–467.

Gàal, G. & Gorbatschev, R. (1987). An outline of the Precambrian evolution of the Baltic shield. *Precambrian Research, 35*, 15–52.

Grigorova, R. & Norris, J. R. (1990). Techniques in microbial ecology. In R. Grigorova & J. R. Norris (Eds), *Methods in Microbiology* (Vol. 22) (627pp). London: Academic Press

Haveman, S. A. & Pedersen, K. (2002). Distribution of culturable microorganisms in Fennoscandian Shield groundwater. *FEMS Microbiology* (in press).

Haveman, S. A., Pedersen, K. & Routsalainen, P. (1999). Distribution and metabolic diversity of microorganisms in deep igneous rock aquifers of Finland. *GeoMicrobiology, 16*, 277–294.

Hindrichs, K-U., Hayes, J. M., Sylva, S. P., Brewer, P. G. & DeLong, E. F. (1999). Methane-consuming archaebacteria in marine sediments. Nature, 398, 802–805.

Kalyuzhnaya, M. G., Khmelenina, V. N., Kotelnikova, S., Holmqvist, L., Pedersen, K. & Trotsenko, Y. A. (1999). *Methylomonas scandinavica*, sp. nov. A new methanotrophic psychrotrophic bacterium isolated from deep igneous rock ground water of Sweden. *Systematic and Applied Microbiology, 22*, 565 572.

Konhauser, K. O. & Ferris, F. G. (1996). Diversity of iron and silica precipitation by microbial mats in hydrothermal waters, Iceland: implication for Precambrian iron formations. *Geology, 24*, 323–326.

Konhauser, K. O., Schultze-Lam, S., Ferris, F. G., Fyfe, W. S., Longstaffe, F. J. & Beveridge, T. J. (1994). Mineral precipitation by epilithic biofilms in the Speed River, Ontarion, Canada. *Applied and Environmental Microbiology, 60*, 549–553.

Kotelnikova, S. & Pedersen, K. (1998). Distribution and activity of methanogens and homoacetogens in deep granitic aquifers at Äspö Hard Rock Laboratory, Sweden. *FEMS Microbiology Ecology, 26*, 121–134.

Kotelnikova, S. & Pedersen, K. (1999). Technical Report. The microbe-REX project: microbial O_2 consumption in the Äspö tunnel (pp. 1–73). Stockholm: Swedish Nucelar Fuel and Waste Management Co.

Kotelnikova, S., Macario, A. J. L. & Pedersen, K. (1998). *Methanobacterium subterraneum*, a new species of Archaea isolated from deep groundwater at the Äspö Hard Rock Laboratory, Sweden. *International Journal of Systematic Bacteriology, 48*, 357–367.

Laaksoharju, M., Skarman, C. & Skarman, E. (1999). Multivariate mixing and mass balance (M3) calculations, a new tool for decoding hydrogeochemical information. *Applied Geochemistry, 14*, 861-871.

McLean, R. J. C. & Beveridge, T. J. (1990). Metal-binding capacity of bacterial surfaces and their ability to form mineralised aggregates. In H. L. Ehrlich & C. L. Brierley (Eds), *Environmental Biotechnology Series; Microbial Mineral Recovery* (pp. 185–222). New York, Auckland: McGraw-Hill Publishing Co.

Motamedi, M. & Pedersen, K. (1998). *Desulfovibrio aespoeensis* sp. nov. A mesophilic sulfate-reducing bacterium from deep groundwater at Äspö hard rock laboratory, Sweden. *International Journal of Systematic Bacteriology, 48*, 311–315.

Motamedi, M., Karland, O. & Pedersen, K. (1996). Survial of sulfate-reducing bacteria at different water activities in compacted bentonite. *FEMS Microbiology Letters, 141*, 83–87.

Pace, N. R. (1997). A molecular view of microbial diversity and the biosphere. *Science, 276*, 734–740.

Pedersen, K. (1993). The deep subterranean biosphere. *Earth-Science Reviews, 34*, 243–260.

Pedersen, K. (1997a). Microbial life in granitic rock. *FEMS Microbiology Reviews, 20*, 399–414.

Pedersen, K. (1997b). Investigations of subterranean microorganisms and their importance for performance assessment of radioactive waste disposal. Results and conclusions achieved during the period 1995 to 1997 (pp. 1–283). Technical Report 97-22. Stockholm: Swedish Nuclear Fuel and Waste Management Co.

Pedersen, K. (1999). Subterranean microorganisms and radioactive disposal in Sweden. *Engineering Geology, 52*, 163–176.

Pedersen, K. (2000a). Exploration of the intraterrestrial biosphere – current perspectives. *FEMS Microbiology Letters, 185*, 9-16.

Pedersen, K. (2000b). Diversity and activity of microorganisms in deep igneous rock aquifers of the Fennoscandian Shield. In M. Fletcher & J. Fredrickson (Eds), *Subsurface Microbiology, and Biogeochemistry* (pp. 97–139). New York: Wiley-Liss

Pedersen, K. & Albinsson, Y. (1991). Effect of cell number, pH and lanthanide concentration on the sorption of promethium by *Shewanella putrefaciens* . *Radiochimica Acta, 54*, 91–95.

Pedersen, K. & Albinsson, Y. (1992). Possible effects of bacteria on trace element migration in crystalline bed-rock. *Radiochimica Acta, 58/59*, 365–369.

Pedersen, K. & Ekendahl, S. (1990). Distribution and activity of bacteria in deep granitic groundwaters of southeastern Sweden. *Microbial Ecology, 20*, 37–52.

Pedersen, K. & Ekendahl, S. (1992a). Incorporation of CO2 and introduced organic compounds by bacterial populations in groundwater from the deep crystalline bedrock of the Stripa mine. *Journal of General Microbiology, 138*, 369–376.

Pedersen, K. & Ekendahl, S. (1992b). Assimilation of CO_2 and introduced organic compounds by bacterial communities in ground water from Southeastern Sweden deep crystalline bedrock. *Microbial Ecology, 23*, 1–14.

Pedersen, K. & Karlsson, F. (1995). Investigations of subterranean microorganisms – their importance for performance assessment of radioactive waste disposal (pp. 1–222). Stockholm: Swedish Nuclear Fuel and Waste Management Co.

Pedersen, K., Motamedi, M. & Karnland, O. (1995). Survival of bacteria in nuclear waste buffer materials – the influence of nutrients, temperature and water activity. SKB Technical Report 95-27. Stockholm: Swedish Nuclear Fuel and Waste Management Co.

Pedersen, K., Arlinger, J., Ekendahl, S. & Hallbeck, L. (1996a). 16S rRNA gene diversity of attached and unattached groundwater bacteria along the access tunnel to the Äspö Hard Rock Laboratory, Sweden. *FEMS Microbiology Ecology, 19*, 249–262.

Pedersen, K., Arlinger, J., Hallbeck, L. & Pettersson, C. (1996b). Diversity and distribution of subterranean bacteria in ground water at Oklo in Gabon, Africa, as determined by 16S-rRNA gene sequencing technique. *Molecular Ecology, 5*, 427–436.

Pedersen, K., Arlinger, J., Erlandson, A-C. & Hallbeck, L. (1997a). Culturability and 16S rRNA gene diversity of microorganisms in the hyperalkaline groundwater of Maqarin, Jordan. In K. Pedersen (Ed.), *Investigations of Subterranean Microorganisms and Their Importance for Performance Assessment of Radioactive Waste Disposal. Results and Conclusions Achieved During the Period 1995 to 1997* (pp. 239–262). Stockholm: Swedish Nuclear Fuel and Waste Management Co.

Pedersen, K., Ekendahl, S., Tullborg, E-L., Furnes, H., Thorseth, I-G. & Tumyr, O. (1997b). Evidence of ancient life at 207 m depth in a granitic aquifer. *Geology, 25*, 827–830.

Pedersen, K., Hallbeck, L., Arlinger, J., Erlandson, A-C. & Jahromi, N. (1997c). Investigation of the potential for microbial contamination of deep granitic aquifers during drilling using 16S rRNA gene sequencing and culturing methods. *Journal of Microbiological Methods, 30*, 179–192.

Pedersen, K., Motamedi, M., Karnland, O. & Sandén, T. (2000a). Cultivability of microorganisms introduced into a compacted bentonite clay buffer under high-level radioactive waste repository conditions. *Engineering Geology, 58(2)*, 149–161.

Pedersen, K., Motamedi, M., Karnland, O. & Sandén, T. (2000b). Mixing and sulfate-reducing activity of bacteria in swelling compacted bentonite clay under high-level radioactive waste repository conditions. *Journal of Applied Microbiology, 89*, 1038-1047.

Sherwood Lollar, B., Frape, S. K., Fritz, P., Macko, S. A., Welhan, J. A., Blomqvist, R. & Lahermo, P. W. (1993). Evidence for bacterially generated hydrocarbon gas in Canadian shield and Fennoscandian shield rocks. *Geochimica et Cosmochimica Acta, 57*, 5073–5085.

SKB AB, (1999). Äspö hard rock laboratory. Annual report (1998). Technical Report TR-99-10. Stockholm: Svensk Kärnbränslehantering AB.

Stetter, K. O. (1996). Hyperthermophilic procaryotes. *FEMS Microbiology Reviews, 18*, 145–148.

Stroes-Gascoyne, S. & Sargent, F. P. (1998). The Canadian approach to microbial studies in nuclear waste management and disposal. *Journal of Contaminant Hydrology, 35*, 175–190.

Stroes-Gascoyne, S., Pedersen, K., Haveman, S. A., Dekeyser, K., Arlinger, J., Daumas, S., Ekendahl, S., Hallbeck, L., Hamon, C. J., Jahromi, N. & Delaney, T.-L. (1997). Occurrence and identification of microorganisms in compacted clay-based buffer material designed for use in a nuclear fuel waste disposal vault. *Canadian Journal of Microbiology, 43*, 1133–1146.

Stumm, W. & Morgan, J. (1996). *Aquatic Chemistry* (3rd edn). New York: John Wiley.

Tabor, P. S. & Neihof, R. A. (1982). Improved microautoradiographic method to determine individual microorganisms active in substrate uptake in natural waters. *Applied and Environmental Microbiology, 44*, 945–953.

Torstensson, B. A (1984). A new system for ground water monitoring. *Groundwater Monitoring Review, 3*, 131–138.

Warren, L. A. & Ferris, F. G. (1998). Continuum between sorption and precipitation of Fe(III) on microbial surfaces. *Environmental Science and Technology, 32*, 2331–2337.

Wellsbury, P., Goodman, K., Barth, T., Cragg, B. A., Barnes, S. P. & Parkes, R. J. (1997). Deep marine biosphere fuelled by increasing organic matter availability during burial and heating. *Nature, 388*, 573–576.

Wersin, P., Spahiu, K. & Bruno, J. (1994). Time evolution of dissolved oxygen and redox conditions in a HLW repository. SKB Technical Report 94-02 (pp. 1–32). Stockholm: Swedish Nuclear Fuel and Waste Management Co.

Miranda J. Keith-Roach and Francis R. Livens (Editors)
© 2002 Elsevier Science Ltd. All rights reserved

Chapter 11

Biochemical basis of microbe-radionuclide interactions

Jon R. Lloyd[a], Lynne E. Macaskie[b]

[a]*Department of Earth Sciences and the Williamson Research Centre for Molecular Environment Science, University of Manchester, Manchester M13 9PL, UK*
[b]*School of Biosciences, University of Birmingham, Birmingham B15 2TT, UK*

1. Introduction

Microbial metabolism can significantly alter the mobility of radionuclides in the environment, and is increasingly being proposed as the basis of novel remediation programmes. If biotechnological approaches are to be used successfully, however, it is vital to understand the mechanisms of microbe-radionuclide interactions. Cross-disciplinary research combining molecular biology, microbial physiology, solid-state chemistry and the rapidly emerging fields of genomics and proteomics is playing a significant role in uncovering the underlying biochemistry of such processes and will be described in this chapter.

Enzymatically-catalysed redox transformations will be discussed with respect to radionuclides including uranium, neptunium, plutonium and technetium. Metabolism-dependent uptake of radionuclides such as caesium will also be discussed. Finally, the biochemical basis of indirect 'biomineralisation' processes, driven by microbial products including sulfide, phosphate and iron minerals will be described. The broad range of techniques available to help characterise such transformations will also be discussed. The microbial reduction of technetium will be used as a case study to illustrate the integrated use of: (1) advanced spectroscopic techniques to characterise end products of radionuclide transformations; (2) physiological studies to characterise biochemical pathways of importance; (3) the construction of defined mutants using molecular biology to confirm the involvement of key enzymes in radionuclide transformations; (4) the generation of high activity strains by genetic engineering; and finally (5) the use of immobilised cell bioreactors to quantify the performance of natural and engineered organisms against simulated waste under laboratory conditions.

2. Radionuclide-microbe interactions

Several recent reviews have proposed that biotechnological processes will play an increasingly important role in the treatment of radioactive waste (Macaskie, 1991; Macaskie

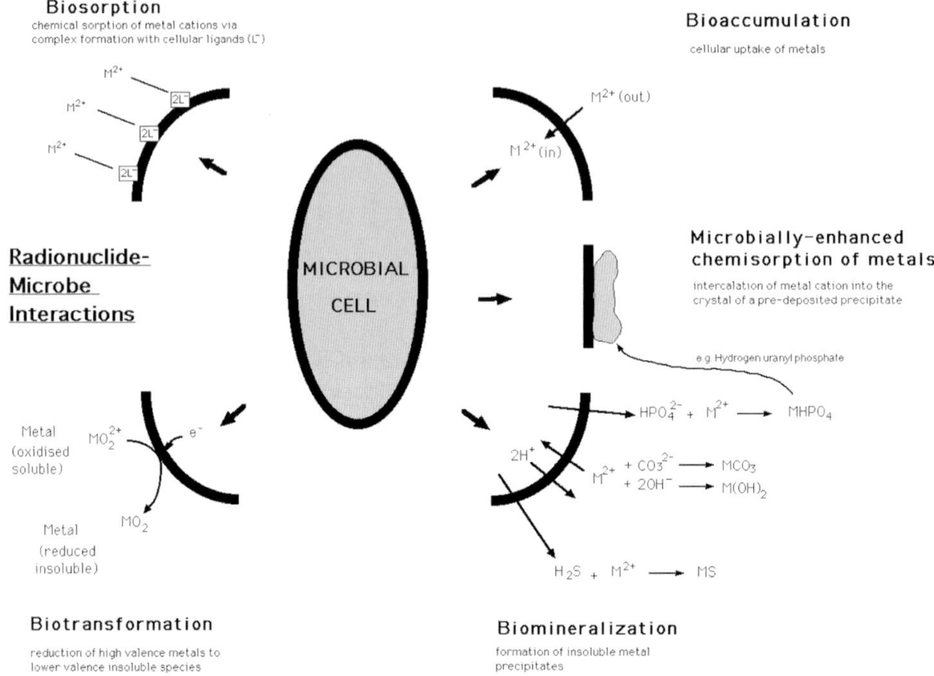

Fig. 1. Mechanisms of radionuclide-microbe interactions.

et al., 1996; Gadd, 1997; Eccles, 1999; Lloyd & Macaskie, 2000). If these processes are to be modelled effectively, it is important to understand the underlying mechanisms of radionuclide-microbe interactions. Several mechanisms are possible (Fig. 1), and can be separated into: biosorption, bioaccumulation, biotransformations, biomineralisation, and microbially-enhanced chemisorption of heavy metals (MECHM).

Biosorption

The term biosorption is used to describe the metabolism-independent sorption of heavy metals and radionuclides to biomass. It encompasses both adsorption, defined here as the accumulation of substances at a surface or interface and absorption, defined here as the almost uniform penetration of atoms or molecules of one phase forming a solution with a second phase (Gadd & White, 1989). Both living and dead biomasses are capable of biosorption and ligands involved in metal binding include carboxyl, amine, hydroxyl, phosphate and sulfhydryl groups. Biosorption of metals has been reviewed extensively (Volesky, 1990, 1994; McHale & McHale, 1994; Tobin et al., 1994; Volesky & Holan, 1995), with the recent review by Suzuki & Banfield (1999) focusing on the geomicrobiology of uranium, with a detailed review of uranium biosorption. Volesky & Holan (1995) also give an excellent overview of metal biosorption and a numerical assessment of uranium and thorium biosorption. For a more recent and extensive review of radionuclide biosorption, the reader is directed to a review by the authors (Lloyd & Macaskie, 2000).

Dead biomass often sorbs more metal than its live counterpart (Brady et al., 1994; Volesky & May-Phillips, 1995), presumably due to an increase in accessible metal-binding sites. Thus, dead biomass may be better suited to treatment of highly toxic radioactive wastes. Biosorption is generally rapid and unaffected over modest temperature ranges and, in many cases, can be described by isotherm models such as the Langmuir, Freundlich and Brunauer-Emmett-Teller (BET) isotherms (Volesky, 1990; deRome and Gadd, 1991; Volesky & Holan, 1995). Ultimately, however, the amount of residual metal remaining in solution at equilibrium is governed by the stability constant of the metal-ligand complex (Macaskie, 1991), and the only way to change the equilibrium position is to transform the metal from a poorly sorbing species to one which has a higher ligand-binding affinity, e.g. by a change of metal oxidation state or by modifying the binding ligand to one which has a greater binding affinity for the given radionuclide. It is in this latter context that the tools of molecular biology may potentially be employed to enhance the performance of biosorbents and this approach has indeed received recent attention for the biosorption of metals. Kortba et al. (1999) described delivery of histidine (Gly–His–His–Pro–His–Gly) or cysteine (Gly–Cys–Gly–Cys–Pro–Cys–Gly–Cys–Gly) containing metal-binding motifs to the surface of the bacterium *Escherichia coli*. Secretion to the outer membrane of the Gram-negative cell was achieved through fusion of the gene sequence encoding the metal-binding motif, to the gene encoding an outer membrane protein LamB. A similar approach was also used to target yeast and mammalian metallothioneins (cysteine rich metal-binding proteins) to the outer membrane of *E. coli* (Sousa et al., 1998). Uptake of divalent cations (e.g. Cd^{2+}) by the engineered strains was enhanced in both studies (Sousa et al., 1998; Valls et al., 1998). The same authors have also targeted a mouse metallothionein to the outer membrane of the soil bacterium *Ralstonia eutropha* (formerly *Alcaligenes eutrophus*) (Valls et al., 2000). The engineered strain accumulated more Cd^{2+} than its wild type counterpart, and also offered tobacco plants some protection to Cd^{2+} when inoculated into contaminated soil (Valls et al., 2000). Although these approaches have shown potential to engineer surface binding of heavy metals, they have yet to be applied to improve the biosorption of radionuclides. Other techniques that could also prove useful include the application of surface display techniques to discover proteins that bind radionuclides (e.g. the actinides) tightly. Such an approach has already been used successfully to screen for ZnO binding peptides fused to fimbriae on the surface of cells of *E. coli* (Kjaergaard et al., 2000).

Metabolism-dependent bioaccumulation

Energy-dependent metal uptake has been demonstrated for most physiologically important metal ions, and some radionuclides enter the cell as chemical ‘surrogates’ using these transport systems. Once in the cell, toxic metals (and potentially radionuclides), may be sequestered by cysteine-rich metallothioneins (Higham et al., 1984; Turner & Robinson, 1995) or, in the case of fungi, compartmentalised into the vacuole (Okorov et al., 1977; Gadd & White, 1989). In this context, it should be emphasised that the uptake of higher mass radionuclides e.g. the actinides, into microbial cells has been reported sporadically and remains poorly characterised (Strandberg et al., 1981; Marques et al., 1991; Volesky & May-Phillips, 1995; Jeong et al., 1997). Indeed, a recent review proposed that intracellular accumulation of uranium in such studies was due to increased membrane permeability caused by uranium toxicity, and was not driven by metabolism-dependent transport mechanisms (Suzuki & Banfield, 1999).

Factors that inhibit cellular energy metabolism can prevent bioaccumulation of metals and therefore limit the practical application of this mode of metal uptake to the treatment of radioisotopes with low toxicity and radioactivity. For example, monovalent cation transport, e.g. for K^+ and analogues, is linked to the plasma membrane-bound H^+-ATPase via the membrane potential, and is affected by increasing metal concentrations that de-energize the cell membrane (White & Gadd, 1987). Other factors that can reduce metal and radionuclide uptake by microorganisms include the absence of substrate, anaerobiosis, incubation at low temperatures and the presence of respiratory inhibitors such as cyanide (White & Gadd, 1987).

Despite the potential limitations of metabolism-dependent bioaccumulation for the treatment of most forms of radioactive waste, this type of mechanism is potentially useful for the treatment of at least one radionuclide: radiocaesium. Indeed, ^{137}Cs remains the best studied example of an environmentally relevant radionuclide that is actively taken up by microbial cells. In common with other alkali metals, and unlike most other radionuclides discussed in this chapter, Cs^+ is a very weak Lewis Acid with a low tendency to interact with ligands (Hughes & Poole, 1989). It forms electrostatic (ionic) rather than covalent bonds with oxygen-donor ligands, binding only weakly to organic and inorganic ligands. Current chemical methods of treatment (e.g. ion exchange using zeolites) suffer from interference by K^+ and Na^+, with a stronger effect exerted by K^+, which has an ionic radius closer to that of Cs^+. The close similarity of the K^+ and Cs^+ cations, however, dictates that both are taken up by the same metabolism-dependent transport systems, with broad-specificity alkali metal uptake transporters reported in all microbial groups (Avery, 1995). The specific mechanisms by which Cs^+ is transported into the cell have been studied in some detail and have been reviewed recently (Lloyd & Macaskie, 2000).

Enzymatically-catalysed biotransformations

Microorganisms can catalyse the direct transformation of toxic metals and metalloids to less soluble or more volatile forms via two enzymatic mechanisms. Bioreduction can result in precipitation of the metal/metalloid (Lovley, 1993), while biomethylation can yield highly volatile derivatives, for example for Se, Te and Hg (Ehrlich, 1996). Although the mechanisms are distinct, the end result is the same: a decrease in the concentration of soluble metals in contaminated water. While the microbial reduction of radionuclides including U(VI), Pu(IV), Np(V) and Tc(VII) has been demonstrated (Lovley et al., 1991; Rusin et al., 1994; Lloyd & Macaskie, 1996; Lloyd et al., 2000a), biomethylation of radionuclides has received little attention. The instability of alkylated actinides (Barnhart et al., 1980) may, in part, explain this observation. An additional enzymatic mechanism that can impact on the solubility of radionuclides is the biodegradation of radionuclide-chelate complexes, which is discussed in Chapter 12 of this volume.

Bioreduction of radionuclides

In the absence of oxygen, specialist microorganisms are able to respire through the reduction of alternative electron acceptors. The environmental relevance of nitrate, sulfate or carbon dioxide reduction has long been recognised, but more recent studies have shown that high valence metals and radionuclides such as Fe(III), Mn(IV) and U(VI) can also function as alternative electron acceptors during anaerobic respiration by specialist organisms (Lovley, 1993). Fe(III) is the most abundant electron acceptor in many sedimentary environments (see Chapter 3, this volume; Lovley, 2000), and respiration using Fe(III) has been studied in most detail (Lovley, 2000). As Fe(III) can be a surrogate for U(VI) in biological systems, outcompeting U(VI) in biosorption experiments (Suzuki & Banfield, 1999), it is not unreasonable to expect the reduction of U(VI) (and possibly other actinides) to proceed by mechanisms similar to those of Fe(III) reduction. Indeed, as far as the authors are aware, all Fe(III)-reducing bacteria that have been tested are also able to reduce U(VI). These factors, in combination with indirect effects of Fe(III) reduction on Tc(VII) and U(VI) solubility (see below), necessitate a very brief overview of the biochemical factors underlying dissimilatory Fe(III) reduction.

Most electron acceptors are soluble and can enter the microbial cell prior to reduction by an intracellular electron transport chain. Fe(III), however, is usually present as insoluble Fe(III) (hydr)oxides, presenting an unusual problem to metal-reducing bacteria; how to transfer electrons to the extracellular electron acceptor. Some analogies may also been drawn here to the reduction of high mass elements such U(VI), which may also be unable to cross biological membranes. Several recent studies have attempted to characterise the electron transport chain to Fe(III) oxides, but to date the protein responsible for the final transfer of electrons to Fe(III) has not been identified. Most studies have focused on Fe(III)-reducing bacteria belonging to the Genera *Shewanella* and *Geobacter*, and have implicated the involvement of *c*-type cytochromes in Fe(III) reduction (Myers & Myers, 1997a, b, 2000; Magnuson et al., 1999; Gordon et al., 2000; Lloyd et al., 2001b). In the cases of *Shewanella putrefaciens* MR-1 (Myers & Myers, 1992) and *Geobacter sulfurreducens* (Lloyd et al., 2001b), *c*-type cytochromes have been localised to the outer membrane and may play a direct role in transferring electrons to insoluble Fe(III) oxides, and possibly the actinides.

The biochemistry of U(VI) reduction to insoluble U(IV) is also poorly understood, but again the involvement of *c*-type cytochromes is implicated (Lovley et al., 1993c). Although the U(VI) reductase is yet to be identified in specialist metal-reducing bacteria capable of conserving energy through U(VI) reduction (e.g. *Shewanella putrefaciens* and *Geobacter metallireducens* (Lovley et al., 1991)), a small soluble *c*-type cytochrome was identified as a U(VI) reductase in pregrown cells of the sulfate-reducing bacterium *Desulfovibrio vulgaris*. It should be noted that most sulfate-reducing bacteria available for study in pure culture, including *Desulfovibrio vulgaris*, are unable to conserve energy through U(VI) reduction (Lovley, 1993), although one recent isolate is a notable exception (Tebo & Obraztsova, 1998). Cytochrome-mediated mechanisms may also be involved in the reduction of Np(V) by *Shewanella putrefaciens* (Lloyd et al., 2000a), but this remains to be confirmed. This contrasts with Tc(VII) reduction by the Fe(III)-reducing bacterium *G. sulfurreducens*, where Tc is reduced via a mechanism distinct from that used to reduce other metals and radionuclides (Lloyd et al., 2000b). Acetate supports efficient reduction of both Fe(III) and U(VI), but not Tc(VII) which is reduced only in the presence of hydrogen (Lloyd et al., 2000b). This observation implicates the involvement of a hydrogenase in Tc(VII) reduction in a mechanism analogous to that noted in *E. coli* (Lloyd et al., 1997a) and *Desulfovibrio desulfuricans* (Lloyd et al., 1999a).

Potential advantages of enzymatic transformations for bioremediation applications
Bioremediation processes employing direct enzymatic transformations of radionuclides have attracted much interest for several reasons. The biochemical processes driving such transformations are often simple and active in non-growing or even non-living biomass. Thus, growth-decoupled cultures can be used to treat radiotoxic effluents, yielding a waste material with a low organic content. The high radioresistance of several enzymes potentially useful for the bioremediation of radionuclides has also been confirmed (Strachan et al., 1991; Lloyd et al., 1997b). For biological reductions, e.g. U(VI) or Tc(VII), cheaply available electron donors such as hydrogen, lactate, acetate or formate can be supplied, negating the requirement for cofactor regeneration. The relative simplicity of the enzyme systems described, in which often only one or two enzymes are required to catalyse the target transformation, also suggests that improvements are possible using the tools of molecular biology. This is important because radionuclides are rarely the natural substrate for an enzyme (e.g. hydrogenase-mediated reduction of Tc(VII) (Lloyd & Macaskie, 1997a; Lloyd et al., 1999a) or U(VI) reduction by Fe(III)-reducing (Lovley et al., 1991; Truex et al., 1997) and sulfate-reducing bacteria (Lovley & Phillips, 1992a; Lovley et al., 1993, c; Tucker et al., 1996). Several new techniques for 'directed evolution' of proteins, such as polymerase chain reaction (PCR)-based DNA shuffling (Crameri et al., 1997), may prove very useful in altering the kinetic constants of enzymes, thus improving process efficiency. Finally, in common with some examples of bioaccumulation and biomineralisation, direct enzymatic transformations can be described by Michaelis-Menten kinetics;

$$v = V_{\max}[\mathrm{S}]/K_{\mathrm{m}} + [\mathrm{S}]$$

where v is the observed velocity of the reaction at a given substrate concentration [S], $V_{\max}$ is the maximum velocity at a saturating concentration of substrate and K_{m} is the Michaelis constant.

This allows the description of a flow-through bioreactor using an integrated form of the Michaelis-Menten equation (Macaskie, 1990; Macaskie et al., 1995, 1997; Tolley et al., 1995; Yong & Macaskie, 1995). It is possible, therefore, to predict the degree of biomass loading or the operational temperature needed to maintain metal removal at a given efficiency within the constraints set by the reactor volume available, the background solution composition and flow rates required (e.g. for bioprecipitation: (Macaskie, 1990), of U (Macaskie, 1990; Macaskie et al., 1995, 1997; Yong & Macaskie, 1995); La (Tolley et al., 1995; Yong & Macaskie, 1997) and for bioreduction: Tc (Lloyd et al., 1997b).

Biomineralisation via microbially-generated ligands

Metals and radionuclides can precipitate with enzymatically-generated ligands, e.g. phosphate, sulfide or carbonate (see below). The concentration of residual free metal at equilibrium is governed by the solubility product of the metal complex (typically 10^{-20} to 10^{-30} for the sulfides and phosphates, higher for the carbonates). Most of the metal or radionuclide should therefore be removed from solution if an excess of ligand is supplied. This is difficult to achieve using chemical precipitation methods in dilute solutions; the advantages of microbial ligand generation are that high concentrations of ligand are achieved in juxtaposition to the cell surface and the surfaces can provide nucleation foci for the rapid onset of metal precipitation. Effectively the metals are concentrated 'uphill' against a concentration gradient. In many cases, the production of the ligand can also be fine-tuned by the application of Michaelis-Menten kinetics.

Phosphates
Bioprecipitation of metal phosphates via hydrolysis of stored polyphosphate by *Acinetobacter* spp. is dependent upon alternating aerobic (polyphosphate synthesis) and anaerobic (polyphosphate hydrolysis and phosphate release) periods (Dick et al., 1995). This obligately aerobic organism is fairly restricted in the range of carbon sources utilised, but the preferred substrates (acetate and ethanol) are widely and cheaply available. The best-documented organism for metal phosphate biomineralisation, a *Citrobacter* sp., grows well on cheaply available substrates and viable cells are not required for metal uptake since this relies on hydrolytic cleavage of a supplied organic phosphate donor (Macaskie et al., 1992). The role of a phosphatase in metal accumulation was confirmed by the finding that the expression of the cloned *phoN* gene in *E. coli* conferred the ability to bioprecipitate uranyl phosphate (Basnakova et al., 1998).

Sulfides
Sulfide precipitation, catalysed by mixed cultures of sulfate-reducing bacteria, has been utilised to treat water co-contaminated by sulfate and zinc (Barnes et al., 1991) and more recently, soil leachate contaminated with sulfate alongside metal and radionuclides (Kearney et al., 1996). Ethanol was used as the electron donor for the reduction of sulfate to sulfide in both examples. A more recent study has confirmed the practicality of integrating the action of the sulfur cycling bacteria (White et al., 1998) employed by Kearney et al. (1996). In the first step of a two-stage process, sulfur-oxidising bacteria were used to leach metals from contaminated soil via the generation of sulfuric acid, and in the second step the metals were stripped from solution in an anaerobic bioreactor containing sulfate-reducing bacteria. The ubiquitous distribution of sulfate-reducing bacteria in acid, neutral and alkali environments (Postgate, 1979), suggests that they have the potential to treat a variety of effluents and contaminated sediments, while the ability of the organisms to metabolise a wide range of electron donors may also allow co-treatment of other organic contaminants. The mechanism of sulfate reduction has been studied in several mesophilic *Desulfovibrio* spp., and involves at least three cytoplasmic enzymes: (1) ATP sulfurylase is required to activate sulfate forming adenylyl sulfate; (2) adenylyl sulfate is in turn reduced to bisulfite by adenylyl sulfate reductase; and (3) bisulfite reductase catalyses the

reduction of bisulfite to sulfide (Peck, 1993). The physiology underlying sulfate reduction by extremophilic sulfate-reducing bacteria remains poorly defined, but is probably similar to that described above for mesophilic sulfate-reducing bacteria, given conservation of bisulfite reductase genes amongst archaeal and bacterial sulfate-reducing bacteria (Wagner et al., 1998).

Carbonates and hydroxides

Ralstonia eutropha (formerly *Alcaligenes eutrophus*) is able to precipitate metals (and potentially some radionuclides) via plasmid-borne resistance mechanisms. Here, proton influx countercurrent (antiport) to metal efflux results in localised alkalisation at the cell surface (Diels et al., 1995; Van Roy et al., 1997). Metal hydroxides and carbonates are formed at these high local pH values, the latter from carbon dioxide evolved in respiration. Although this system has not been tested against radionuclides, microbial respiration has been shown to play a role in the formation of extracellular strontium carbonate by *Pseudomonas fluorescens* (Anderson & Apanna, 1994). Microbial activity was also implicated in the deposition of a strontium-containing calcite phase at a groundwater discharge zone (Ferris et al., 1995). The deposition of the mineral was thought to be the result of carbonate precipitation by epilithic cyanobacteria (driven by HCO_3^-/OH^- exchange during photosynthesis), with a Sr/Ca ratio that promoted deposition of $SrCO_3$ in calcite (up to 1% Sr). Although Sr carbonates are insoluble, this is not true of uranium-carbonate complexes; in fact, carbonate has been used as a lixiviant to extract uranium from contaminated soils (Phillips et al., 1995; Francis et al., 1999). Furthermore, recent studies have shown that microbial activity (CO_2 production) can enhance the dissolution of uranyl hydroxide and uranyl hydroxophosphate species, via the formation of soluble uranyl carbonate species (Francis et al., 2000).

Microbially-enhanced chemisorption of heavy metals

Microbially-enhanced chemisorption of heavy metals (MECHM) is a generic term to describe a class of reactions whereby microbial cells first precipitate a biomineral of one metal ('priming deposit'). The priming deposit then acts as a nucleation focus, or 'host crystal' for the subsequent deposition of the metal of interest ('target' metal), acting to promote and accelerate target metal precipitation reactions. (Macaskie et al., 1994, 1996; Tolley & Macaskie, 1994). The priming deposit is made initially by the sulfide or phosphate biomineralisation routes described above and the biochemical bases of these mechanisms are, therefore, well understood. Recent studies have also shown that Fe-containing minerals can also act as foci for the precipitation of radionuclides.

Metal phosphate can be used as the priming deposit in two ways. $LaPO_4$ pre-deposited onto metal-accumulating *Citrobacter* sp. via biogenic phosphate release can be used as the 'priming' deposit for subsequent deposition of actinide phosphates (Macaskie et al., 1994). Also, by use of a priming deposit with an appropriate, well-defined cystalline lattice, the target metal can be intercalated within the 'host' lattice, effectively by a mechanism of bioinorganic ion-exchange, well established as a chemical process (Pham-Thi & Columban, 1985; Pozas-Tormo et al., 1986, 1987; Clearfield, 1988) and best characterised biologically in the case of Ni^{2+} removal into biomass-bound $HUO_2PO_4 \cdot nH_2O$ (Bonthrone et al., 1996). Hydrogen uranyl phosphate (HUP) consists of sheets of uranyl phosphate ions separated by water molecules which create a regular network of hydrogen bonds (Clearfield, 1988; Hunsberger & Ellis, 1990). The overall outcome of this highly organised crystal lattice is a high mobility of protons in the interlamellar space (Pham-Thi & Columban, 1985) which gives rise to the ion-exchange intercalative property (Hunsberger & Ellis, 1990). Other hexavalent actinides (AnO_2^{2+}) can also substitute directly for UO_2^{2+} in the backbone of the lattice, displacing the uranyl ion to give a hybrid crystal (Dorhout et al., 1989). In theory, it would be possible to produce a lattice consisting of a uranyl phosphate 'carrier' matrix with other actinides intercalated into the backbone and simple radionuclide cations intercalated within the lumen; such remediation of multiple species simultaneously has not yet been achieved, but is the subject of ongoing research in the laboratory of one of the authors (L. E. Macaskie).

Three alternative MECHM mechanisms involve the formation of iron minerals as the priming deposit. In the first mechanism, H_2S produced by sulfate-reducing bacteria reacts with iron to form FeS, with cell-bound FeS acting as a sorbent for 'target' metals, including radionuclides (Watson & Ellwood, 1988, 1994; Ellwood et al., 1992). The use of FeS is notable because this provides the mechanism for rapid biomass separation from the liquor via the magnetic properties of FeS in a high-gradient magnetic separator (Watson & Ellwood, 1988, 1994; Ellwood et al., 1992). In the second mechanism, insoluble Fe(III) oxide minerals formed by, for example, Fe(II)-oxidising bacteria (Ehrlich, 1996), also sorb considerable quantities of target radionuclide, for example U(VI) (Lloyd & Macaskie, 2000). Finally, Fe(II)-bearing minerals (e.g. magnetite) produced by iron-reducing bacteria, are also able to precipitate high valence metals and radionuclides e.g. Cr(VI) and Tc(VII) (Fendorf & Li, 1996; Lloyd et al., 2000b). Here, the biogenic Fe(II) is able to reduce the target metal resulting in precipitation of Cr(III) or Tc(IV) on the mineral surface. Use of biologically-produced magnetite as a reductant and sorbent is appealing as it could also be removed from effluents using a magnetic separating device. Tetravalent uranium formed by the action of Fe(III)-reducing bacteria is also able to reduce and precipitate Tc(VII) abiotically (Lloyd et al., 2002).

3. Methods for studying the biochemical basis of microbe-radionulide interactions

The comparatively recent realisation that microbes play important roles in mediating geologically important reactions has stimulated much interest, and led to rapid advances in the field of 'geomicrobiology'. Although some of the biochemical factors underlying processes described in this chapter have been uncovered in great detail, the study of microbial interactions with radionuclides of environment relevance is a young discipline. There is still clearly much to learn about the molecular mechanisms mediating microbe-radionuclide interactions, and a blend of several disparate disciplines is required to drive research forward in this area. The following section is intended to give an overview of some of the techniques that are available to those starting work in this area. Some commonly encountered pitfalls are also highlighted.

Bioavailability of radionuclides: choice of background matrix

Microorganisms are often cultured in media that can potentially complex metals and radionuclides (Beveridge et al., 1997). When studying radionuclide-microbe interactions, it is important that a medium is selected for laboratory studies that will maintain the bioavailability of the radionuclide of interest. If possible, a fully defined medium should be selected, free from complex organic constituents such as yeast extract, peptone and amino acids, that bind metals. If it is impossible to devise a fully defined medium that will support growth of the microorganism of interest, it is sometimes possible to use cells that have been pre-grown in a complex medium, prior to washing and resuspending the biomass in a buffered solution containing the required reactants. This approach is particularly suitable for experiments where the required activity can be uncoupled from growth (e.g. biosorption, metal reduction and some other enzymatically catalysed transformations). In experiments using both growing and non-growing (washed or 'resting') cells, it is important to choose a buffer system that offers negligible binding of the target radionuclides. Suitable buffers include the sulfonic acids MES, PIPES, MOPS, TES and HEPES (pKa values 6.15, 6.8, 7.2, 7.5 and 7.55 respectively; (Gueffroy, 1990)). Alternatively, a popular choice for U(VI) experiments using subsurface organisms is 30 mM bicarbonate under a headspace of N_2/CO_2 (80:20, pH 6.8) (Lovley et al., 1991, 1993a; Lovley & Phillips, 1992a, b). Uranium is held in solution as a soluble carbonate complex approximating to that found in some groundwater conditions. The pH of the buffer system should also be given suitable consideration, as it can affect both radionuclide speciation and protonation/deprotonation of cell constituents that bind radionuclides (Suzuki & Banfield, 1999).

Another parameter, which is often ignored but is especially relevant to the study of the actinides, is the oxidation state of the radionuclide (see Chapters 4 and 12, this volume, or Lloyd & Macaskie, 2000) for a more detailed description of the chemical characteristics of the different valence states of the actinides). As a rule, reductants that are often added to media to poise the redox potential when growing anaerobes (e.g. ascorbate, thioglycollate, cysteine or sulfide) should be excluded or removed by washing the cells prior to use, as they may potentially alter the oxidation state of the target radionuclide.

Quantification of radionuclides

It is beyond the scope of this chapter to describe the full range of techniques available to quantify radionuclide uptake by microbial cells. For a more detailed overview, the reader is directed to numerous specialist texts on the subject (e.g. L'Annunziata, 1998). Here we aim only to give a very short description of the techniques most commonly used in biochemical studies, most of which are available in non-specialist laboratories. It should also be noted that, irrespective of whichever analytical method is chosen, most techniques require the separation of cell-bound radionuclide from the bulk solution. This is generally achieved by filtering through a 0.2 μM nitrocellulose filter (Beveridge et al., 1997), or by centrifugation, sometimes through a non-aqueous phase (Garnham et al., 1992). Radionuclide uptake can then be quantified by assaying for radioactivity associated with the biomass, or by measuring loss from the filtrate or supernatant.

Although many approaches make use of the α-, β- or γ-emission of a radioisotope, colorimetric techniques are also suitable for studies with some radionuclides. For example, arsenazo III can be used to quantify comparatively low concentrations of uranium, reacting with the radionuclide to form a pink complex that can be quantified by measuring the absorbance at 652 nm. This technique can also be adapted to quantify the oxidation state of uranium using ion chromatography (Coetzee & deBeer, 1992), important when monitoring microbial reduction and precipitation of uranium. Here U(VI) and U(IV) are separated using a suitable column (e.g. Dionex HPLC-AS5, Dionex Corp., CA). Post column derivatisation precedes quantification of the radionuclide by absorption at 652 nm in a flow-through detector. Another approach commonly used to monitor uranium reduction by microorganisms is kinetic phosphorescence analysis (KPA). Here, U(VI) is quantified by measuring its fluorescence when illuminated by 337 nm laser light. This technique measures U(VI) specifically since U(IV) does not fluoresce under these conditions. Although KPA apparatus is comparatively expensive, advantages over other techniques for uranium determination include high throughput (low analysis time) and very high sensitivity. X-ray absorption spectroscopy has also been used to monitor the microbial reduction of radionuclides, including U(VI) (Francis et al., 1994, 2000) and Tc(VII) (Lloyd et al., 2000b; Wildung et al., 2000) in addition to metals such as Cr(VI) (Park et al., 2000). This powerful synchrotron-based technique can be used to identify both the oxidation state of the target radionuclide from the corresponding XANES (X-ray absorption near edge) spectrum, and also the nearest neighbour to the radionuclide and its bonding by analysis of the extended X-ray absorption fine structure (EXAFS) spectrum. Although restricted in availability, this technique is sensitive to low concentrations of radionuclide (e.g. 25 μM Tc (Lloyd et al., 2000b)) in environmental samples. Many other potentially useful spectroscopic techniques are also available for the determination of uranium and other radionuclides in specialist laboratories, but are beyond the scope of this review. The reader is referred to detailed texts on the subject (e.g. Burns & Finch (1999) for uranium).

The most commonly used technique for measurement of radionuclides via their intrinsic radioactivity is undoubtedly that of liquid scintillation counting. Most radionuclides of interest (α- or β- emitters) can be quantified using this approach, and other specialist counting devices are rarely required. One refinement which can be advantageous for studies involving mixtures of radionuclides is the ability to discriminate between α- and β-emitting isotopes. This is a prerequisite when studying, for example, microbial transformations of ^{237}Np (α-emitter) which is contaminated with the β-emitting daughter product ^{233}Pa (Lloyd et al., 2000a). Autoradiography techniques employing a kinetic phosphorimager are also useful for quantifying β- and γ-emitting isotopes (Lloyd & Macaskie, 1996; Macaskie et al., 1996; Lloyd et al., 2000a), although this technique is probably not useful for isotopes emitting weakly penetrating α-radiation. A particular advantage of the phosphorimager technique, is that it can be used to quantify spatially separated radionuclides, e.g. different oxidation states of Tc or Np (separated by paper chromatography (Lloyd & Macaskie, 1996; Lloyd et al., 2000a)) or used to image radionuclides accumulated by immobilised cells in a bioreactor (Lloyd et al., 1997b, 1999a) or biofilm.

Localisation of radionuclides by electron microscopy

Transmission electron microscopy (TEM) is a useful technique that can both localise and help identify radionuclides deposited within or around microbial cells. Identification of the site of accumulation is important as it can give clues as to the biochemical mechanism driving radionuclide accumulation, for example when correlated with cell fractionation studies aimed at characterising enzyme systems (e.g. reductases) active against target radionuclides. The TEM techniques commonly used to visualise metallic precipitates (and also radionuclides) have been reviewed in detail by Beveridge et al. (1994, 1997). A high-energy electron beam, when passed through a sample in a TEM, is deflected by high mass elements including metals and radionuclides as they pass through the electron shells or nuclei. Biological materials, which are largely composed of lighter elements, such as C, N, H, O, P and S, do not deflect the electron beam to the same degree. Thus, it is possible to visualise radionuclides (and metals) against the faint image of a bacterial cell. Indeed, it is standard procedure to stain biological material with a metallic stain (e.g. uranium or osmium) during conventional electron microscopy to visualise cellular constituents. Obviously, for the purposes of this review, such supplementary stains are often unnecessary and should be avoided when localising bioaccumulated radionuclides. Samples can be prepared by two methods when visualising radionuclides associated with microbial biomass. The simplest techniques employ the use of 'whole mounts', where cells that have been washed carefully in distilled water are layered on TEM grids coated with carbon and Formvar (Beveridge et al., 1997). For precise localisation of radionuclides upon or within the cell, biomass should be fixed, embedded in resin blocks and sectioned to approximately 60 nm thickness prior to viewing using TEM.

Two additional techniques are available which can help characterise biominerals during TEM studies. The first technique, energy dispersive X-ray analysis (EDAX) can provide elemental information via the analysis of X-ray emissions stimulated by a high energy electron beam. X-rays with energies characteristic of the sample element are given off as the remaining electrons are rearranged in the atom (Beveridge et al., 1997). A related technique, proton induced X-ray emission analysis (PIXE) uses a high-energy proton beam to generate the X-ray emissions from the sample (Johansson & Campbell, 1988). The sensitivity is much greater than EDAX but the low spatial resolution (typically tens of microns or above) makes PIXE more applicable to bulk sample analysis. The second TEM-based technique of interest is selected-area electron diffraction (SAED), which can be used to identify crystalline materials. The electron beam is passed through the sample, and diffracted electrons are imaged as reflections on the TEM screen. It is possible to identify the mineral phase from the atomic lattice characteristics of the material (Beveridge et al., 1997).

Physiological studies

Much can be learnt of the biochemical basis of radionuclide-microbe interactions through physiological investigations using whole cells. For example, microbial metabolism is regulated tightly, with expression of key pathways induced or repressed in response to the external environment. This is a prerequisite if an organism is to be competitive in its chosen niche. Thus, it may be possible to identify the biochemical pathways involved by determining the growth conditions giving optimal activity against a radionuclide. This is particularly true when redox transformations are under investigation (e.g. reduction of radionuclides such as Tc(VII)). Parameters which can effect microbial metabolism and should be considered include aeration (and anaerobiosis), growth substrates (e.g. C, N and P sources) and their relative availability, as well as the addition of radionuclides and trace metals. Another important factor that may modulate gene expression, and hence enzyme activity, is attachment to the surface of, for example, a mineral phase.

For enzymatically catalysed redox reactions, it is also important to identify the optimal electron donor using suspensions of washed cells, and also to quantify the effect of competing electron acceptors. Special emphasis should be placed on the effect of nitrate on radionuclide-microbe interactions for several reasons. First, the use of nitric acid in fuel reprocessing dictates that high concentrations of nitrate are present in low activity waste streams (300 mM) and contaminated groundwaters (Lloyd & Macaskie, 2000). High concentrations of nitrate, in turn, have an inhibitory effect on the biological treatment of uranium (Tucker et al., 1998; Yong & Macaskie, 1997) and technetium (Lloyd et al., 1997a). This may be due in part to competitive inhibition of enzyme activity by nitrate. Indeed, turning to redox transformations again, several early studies discussed the role of nitrate reductases as broad specificity metal reductases, on the basis of competitive inhibition of metal reduction by nitrate, for example Fe(III) (Lovley, 1991) and Mn(IV) (Karavaiko et al., 1988). The validity of this interpretation has since been questioned (Lovley, 1991). A more likely explanation is that nitrate is a potent modulator of gene regulation in bacteria (Stewart, 1988), switching on the synthesis of enzymes needed to reduce nitrate, and repressing expression of alternative respiratory systems that interact with radionuclides. Indeed, Tc(VII) reduction by *E. coli* is very sensitive to nitrate, although Tc(VII) is not reduced by a nitrate reductase in this organism (Lloyd et al., 1997a, 1999b). Inhibition of Tc(VII) reduction by nitrate was due to a combination of factors, including downregulation of the Tc(VII) reductase by nitrate, competition between Tc(VII) and nitrate at the active site of the Tc(VII) reductase, and diversion of reducing equivalents from the Tc(VII) reductase to the electron transfer chain to nitrate. Finally, a range of metabolic inhibitors is also available to help type respiration-linked enzyme systems responsible for radionuclide-microbe interactions (Dawson et al., 1986). Such approaches have been applied to Fe(III) and Mn(IV) reduction; (Myers & Myers, 1993; Myers & Nealson, 1990), but have not been used extensively in the context of microbial interactions with radionuclides.

Enzyme purification and analysis

Once an enzymatic activity has been characterised and optimised in whole cells, a logical extension is to study the relevant process in cell extracts, followed by purification of the enzyme(s) involved. If this approach is adopted, cells must first be broken, e.g. by sonication, passage through a French Press or by more gentle techniques involving controlled lysis of sphaeroplasts (cells that have been treated with enzymes to degrade the rigid cell wall). Fractionation by centrifugation can then be employed to generate soluble and membrane fractions from the broken cells (Chart, 1994). This step is vital to help determine the cellular localisation of enzymatic activity. Purification of the enzyme can then be attempted from the relevant cellular fraction (Ausubel et al., 1999). If the enzyme is localised at the membrane, solubilisation using a suitable detergent will generally be required, with detergent added to buffer systems used in later purification steps. A range of detergents is available for such studies (Findlay, 1990), with choice guided by factors including the solubilisation efficiency, effect on enzyme activity and compatibility with the purification technique of choice. Several purification strategies are available and have been the focus of numerous specialist texts such as Anon (1999). The most common approaches, however, utilise column chromatography with either separation by size, charge, hydrophobicity or affinity to an immobilised ligand. Once purified there are many tests that can be conducted to help further elucidate the mechanism of radionuclide-protein interactions (e.g. analysis for metal content and presence of cofactors). A particularly useful approach is to determine the N-terminal sequences of the purified protein after Edman degradation (Ausubel et al., 1999), and this information can be used to identify closely related proteins of known function from databases of reference sequences.

Although several studies have successfully adopted such classical biochemical approaches to purify enzymes and enzyme complexes that catalyse metal reduction (Yanke et al., 1995; Schroder et al., 1997; Krafft & Macy, 1998; Magnuson et al., 1999; Park et al., 2000), only one study has adopted this approach for the purification of a radionuclide-reducing enzyme system (Lovley et al., 1993c). In this elegant study, the enzyme system responsible for hydrogen-dependent reduction of U(VI) was dissected in the sulfate-reducing bacterium *Desulfovibrio vulgaris* (Lovley et al., 1993c). The majority (95%) of the enzymatic activity was localised to the soluble fraction of broken cells, but was removed when the soluble fraction was passed through a cation exchange column. The enzyme responsible for U(VI) reduction was eluted with 400 mM NaCl, and restored hydrogen-dependent U(VI) reduction activity when added back to the wash fraction that had not adhered to the column. The U(VI)-reducing protein that bound to the column was subsequently identified as cytochrome c_3, and direct reduction of U(VI) by the purified protein was demonstrated by a simple spectrophotometric assay; the purified protein was reduced in a sealed cuvette by the addition of dithionite, and was reoxidised when anaerobic stocks of U(VI) were added. A hydrogenase was also purified from the periplasmic fraction of the cell, and when combined with the purified *c*-type cytochrome, rapid reduction of U(VI) was demonstrated in vivo. Thus, it was proposed that hydrogen dependent reduction of U(VI) by *Desulfovbrio desulfuricans* is catalysed by cytochrome c_3, with electrons abstracted from hydrogen by a periplasmic hydrogenase, the normal physiological electron donor for this enzyme.

The impact of molecular biology

The tools of molecular biology have also been employed to characterise the mechanisms of radionuclide-microbe interactions at a genetic level. Using such approaches, it is possible to (1) confirm the physiological role of key proteins by the generation of mutants; (2) enhance radionuclide-uptake through overexpression of the relevant protein; and (3) further improve uptake through targeted mutagenesis of the gene encoding the protein. Finally, if the genes encoding key biochemical processes controlling radionuclide mobility are known, it is possible to use molecular techniques to monitor expression of the genes and hence microbial activity in situ. An advantage of this approach, over more conventional microbiological methodologies, is that it is often quicker and more accurate to amplify and quantify the relevant genes required for activity (e.g. by PCR (polymerase chain reaction; see Chapters 2 and 8, this volume) than it is to count the organisms with the designated metabolic capability by culturing them from environmental samples in the laboratory. It is also difficult to make direct measurement of microbial activity in situ, and this is especially true for radionuclide reductase activity. It is beyond the scope of this article to describe the full range of molecular techniques available for such studies, but the reader is referred to two of the many excellent texts on this subject (Sambrook et al., 1989; Ausubel et al., 1999).

The generation of defined mutants is required to confirm the physiological role of a given protein, and tools are available for knocking out genes in several environmentally relevant organisms that may play important roles in mediating the solubility of radionuclides (e.g. the Fe(III)-reducing bacteria *Geobacter sulfurreducens* (Coppi et al., 2001) and *Shewanella putrefaciens* (Myers & Myers, 2000), the sulfate-reducing bacterium *Desulfovibrio vulgaris* (Rousset et al., 1998), and the radiation-resistant bacterium *Deinococcus radiodurans* (Meima & Lidstrom, 2000). Mutants can be generated by disrupting the corresponding gene by two methods. First, if the gene has already been identified it may be possible to engineer a 'knock-out' mutant by insertion of a selectable marker into the gene (e.g. a kanamycin-resistance cassette; Coppi et al., 2001) through homologous recombination. If this approach is adopted, identification of the gene for disruption is aided considerably by the availability of the full genetic sequence (genome) for several environmentally relevant organisms, including species of *Deinococcus* (White et al., 1999), *Geobacter*, *Shewanella* and *Desulfovibrio* (the latter three examples are currently being completed and full genome sequences should be available by the time that this chapter is published). If sufficient biochemical and physiological information is available, the genes most likely to encode the relevant proteins can be identified by searching the genome database for sequences homologous to those previously characterised in other model organisms. Alternatively, if the protein has been purified and characterised in vitro, it is possible to identify the corresponding gene in the genome database from the N-terminal sequence of the protein. Finally, if no biochemical or genetic information is available (e.g. the physiological trait is completely novel), useful information can be generated using another approach: transposon mutagenesis. Using this approach, transposons (small mobile genetic elements) insert randomly into the microbial chromosome disrupting genes, effectively tagging them. After selection for a desired physiological trait (e.g. the inability to reduce or accumulate a metal/radionuclide), the phenotype of the mutant can be confirmed, and the gene that has been disrupted can be identified by cloning using the transposon as a marker, or by amplification by PCR.

The continuing advances in microbial genomics will also play an important role in identifying proteins mediating radionuclide-microbe interactions. With the availability of full genetic sequences for relevant organisms, future goals will be to determine which genes are expressed when these microorganisms are grown under defined conditions (e.g. in the presence of radionuclides) using 'genomic' approaches. Several techniques can be adopted to achieve this goal (Soll & Winzeler, 2000). For example, expression profiles can be obtained using miniature arrays of nucleic acids (microarrays) representative of the genome, to which labelled RNA molecules hybridise. A 'transcription profile' is obtained showing which genes are switched on under any designated growth condition. Alternatively, proteins expressed in the presence of radionuclides can be identified using a 'proteomics' approach. Here all the proteins produced by a microorganism are separated on a single gel, using two-dimensional polyacrylamide gel electrophoresis, and the extra bands produced when a radionuclide is added can be eluted and identified by mass spectrometry. Finally, mutant collections will probably be available for some model organisms, and will be useful for confirming the physiological role of proteins shown to be expressed when microbes are grown in the presence of radionuclides. Work is already under way to produce a complete set of 6000 single-gene deletion mutants covering all open-reading frames in the yeast *Saccharomyces cerivisiae* (Oliver et al., 1998), and this task will surely be repeated for other model organisms.

4. Case study: microbial reduction of Tc(VII)

Tc contamination and the need for biological treatment technologies

Technetium-99 is a long-lived β-emitter (half-life = 2.3×10^5 years) that is produced during the fission of ^{235}U. Subsequently, it is a significant pollutant of both nuclear waste streams (Macaskie, 1991) and of the subsurface where nuclear waste has escaped from storage facilities (McCullough et al., 1999). Several factors make Tc a problematic contaminant within these scenarios. First, it is normally present in the heptavalent form as the pertechnetate anion (TcO_4^-), which is both highly soluble and mobile in the environment (Wildung et al., 1979). Second, although artificial in nature, the biological activity of Tc(VII) is high since it behaves as an analogue for sulfate (Cataldo et al., 1989) with assimilation by plants facilitating entry into the food chain (Dehut et al., 1989). Indeed, these factors, when considered in combination, led Trabalka & Garten (1983) to conclude that Tc may be the critical radionuclide in determining the long-term impact of the nuclear fuel cycle.

Tc(VII) has weak ligand complexing capabilities and is difficult to remove from solution using conventional 'chemical' approaches (Macaskie, 1991), particularly against a background of competing ions (e.g. nitrate) normally present in nuclear waste. This has led to considerable interest in the development of alternative and highly selective biotechnological methods to treat Tc contaminated water. Indeed, we have been involved in the development of such a process for several years, and we will now focus our attention on a multidisciplinary study that has encompassed many of the techniques described in this chapter.

Interactions of Tc(VII) with microorganisms: preliminary studies

Several early studies suggested that microbes may interact with Tc(VII) via several mechanisms. Uptake of the pertechnetate anion by biosorption was negligible, and offered little potential for treating Tc(VII) contaminated waste (Garnham et al., 1992).

However an alternative approach utilising metal-reducing microorganisms proved more encouraging. Although Tc(VII) is highly soluble, several low valence oxides of Tc are insoluble (Kotegov et al., 1968). Early studies by Henrot (1989) and Pignolet et al. (1989) considered that microbially-derived H_2S played a major role in Tc reduction and precipitation by mixed cultures of microorganisms isolated from a marine sediment. A novel conclusion from the latter study was that Tc(VII) reduction and removal by oxygen-limited cultures of *Moraxella* sp. and *Planococcus* sp. may have been mediated by an enzymatic mechanism. Lloyd and Macaskie (1996) subsequently developed a phosphorimager-based technique to monitor the microbial reduction of Tc(VII). Using this technique, direct enzymatic reduction of Tc(VII) by resting cells of the Fe(III)-reducing bacteria *Shewanella putrefaciens* and *Geobacter metallireducens* was unequivocally demonstrated. Tc products were species-specific; only soluble, reduced species were detected in supernatants from *S. putrefaciens*, but appreciable quantities of the radionuclide were precipitated as a low valence oxide by *G. metallireducens*. The phosphorimager-based technique was also used to identify other organisms capable of reducing and precipitating the radionuclide. Thus, we have demonstrated that the ability to reduce Tc(VII) is not restricted to Fe(III)-reducing bacteria, but this activity is shared amongst laboratory cultures of *Rhodobacter sphaeroides*, *Paracoccus denitrificans*, some *Pseudomonads* (J. R. Lloyd unpublished), *E. coli* (Lloyd et al., 1997a) and a range of sulfate-reducing bacteria (Lloyd et al., 1998, 1999a, 2001a). Other workers have used this technique to show that *Thiobacillus ferrooxidans* and *T. thiooxidans* (Lyalikova & Khizhnyak, 1996) and the hyperthermophile *Pyrobaculum islandicum* (Kashefi & Lovley, 1999) are also able to reduce Tc(VII). X-ray absorption spectroscopy studies have recently identified insoluble Tc(IV) as the final oxidation state produced when Tc(VII) is reduced enzymatically by *Geobacter sulfurreducens* (Lloyd et al., 2000b), *E. coli* (Lloyd & Sole, unpublished) and *Shewanella putrefaciens* (Wildung et al., 2000).

Physiology and biochemistry of Tc reduction by Escherichia coli

Escherichia coli was subsequently selected as the model system in which to characterise the enzyme catalysing Tc(VII) reduction. This decision was guided by the wealth of biochemical and physiological information on this organism, and also by availability of well characterised mutants required to help identify the biochemical factors involved in Tc(VII) reduction.

Initial studies demonstrated that anaerobic, but not aerobic, cultures of *E. coli* reduced Tc(VII) with the reduced radionuclide precipitated within the cell (Lloyd et al., 1997a). Washed cells coupled the oxidation of formate to Tc(VII) reduction, and formate could be also be replaced as an electron donor by hydrogen, pyruvate, glucose or glycerol, but not by acetate, lactate, succinate or ethanol. Tc(VII) reductase activity was repressed during anaerobic growth in the presence of nitrate, but this effect was counteracted by the addition of formate to the growth medium. In order to identify the anaerobically induced enzyme system responsible for Tc(VII) reduction, 34 defined mutants defective in the synthesis of regulatory proteins or electron transfer proteins were tested for the ability to reduce Tc(VII) (Lloyd et al., 1997a). Mutants defective in the synthesis of the transcription factor FNR, the so-called 'anaerobic switch' protein in *E. coli* (Spiro, 1994), were unable to synthesise active Tc(VII) reductase. This phenotype was also shared by a mutant unable to synthesise molybdenum cofactor (molybdopterin guanine dinucleotide; MGD). Thus, it was concluded that a component of the Tc(VII) reductase system of *E. coli* was a molybdenum containing enzyme, the synthesis of which was activated under anaerobic conditions by FNR. Mutants unable to synthesise a range of anaerobically induced enzymes, some of which contain molybdenum cofactor, were then tested for the ability to reduce Tc(VII). Despite evidence indicating the involvement of a nitrate reductase in Mn(IV) reduction by *Acinetobacter calcoaceticus* (Karavaiko et al., 1988), mutants of *E. coli* that were unable to synthesise a membrane-bound or periplasmic nitrate reductase were still able to reduce Tc(VII). Similar results were also obtained with mutants that were unable to synthesise the two known nitrite reductases of *E. coli*, despite reports proposing the involvement of nitrite reductases in Se(IV) and U(VI) reduction by *Thauera selanitis* and *Shewanella putrefaciens* respectively (DeMoll-Decker & Macy, 1993; Wade & DiChristina, 2000). *C*-type cytochromes, fumarate reductase, dimethyl sulfoxide reductase, and trimethylamine-*N*-oxide reductase were also not required for Tc(VII) reduction by *E. coli*.

Formate dehydrogenases, three of which have been characterised in *E. coli*, require metal cofactors for activity including MGD and selenocysteine. To determine if the MGD requirement for Tc(VII) reduction was due to the involvement of one or more of the formate dehydrogenases, a strain that was unable to synthesise selenocysteine was also tested. This strain was unable to reduce and precipitate Tc(VII). Furthermore, a mutant unable to synthesise active FdhH, the formate dehydrogenase of the formate hydrogenylase complex (FHL), was also tested and was unable to reduce Tc(VII), suggesting a role for FHL in Tc(VII) reduction. This hypothesis was further supported by the observation that Tc(VII) reductase activity was enhanced by the addition of formate to the growth medium, and repressed by the addition of nitrate. Similar effects were reported on FHL activity (Birkman et al., 1987). The normal physiological role of FHL is to convert the fermentation product formic acid to hydrogen and CO_2, thus minimising acidification of the medium during anaerobic growth. Although not directly regulated by FNR, FHL contains a Ni-containing 'hydrogen-evolving' hydrogenase (hydrogenase 3), and transcription of the Ni^{2+} uptake system of *E. coli* is under the control of FNR. Furthermore, addition of elevated concentrations of Ni^{2+} to the growth medium reactivated Tc(VII) reductase activity in the FNR mutant, confirming that the requirement for FNR was indirect, and stemmed from a requirement for the Ni^{2+} uptake system. Finally, dihydrogen could replace formate as an effective electron donor for Tc(VII) reduction by the wild type strain and also in mutants defective in MGD synthesis or selenocysteine incorporation (and hence formate dehydrogenase activity). These results were used to construct a model for Tc(VII) reduction by *E. coli* (Fig. 2), proposing that the hydogenase 3 component of FHL catalyses the transfer of electrons from dihydrogen to Tc(VII). According to this model, the formate dehydrogenase component (FdhH) is required only if formate, or a precursor, is supplied as an electron donor for Tc(VII) reduction in place of hydrogen. This model has been validated by the observations that a mutant unable to synthesise hydrogenase 3 is unable to reduce Tc(VII) when either hydrogen or formate was supplied as an electron donor (Lloyd et al., 1997a). Furthermore, reduced Tc(IV) is precipitated within the cell, mostly towards the cytoplasmic membrane (see Fig. 3 and above), consistent with the proposed cellular

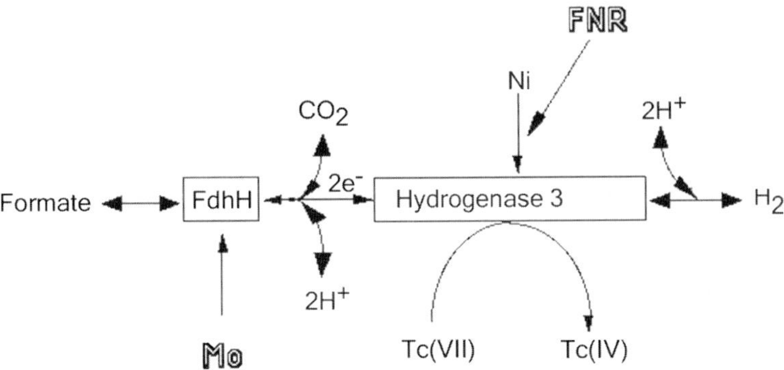

Fig. 2. The mechanism of Tc(VII) reduction by hydrogenase 3 of the formate hydrogenlyase complex of *E. coli*. Hydrogen oxidation is coupled to Tc(VII) reduction by the Ni-containing enzyme, hydrogenase 3. If formate is supplied as an electron donor for Tc(VII) reduction, a formate dehydrogenase (FdhH) is also required. Ni uptake, and therefore Tc(VII) reductase activity, is modulated by the transcription activator FNR (regulates Ni uptake). Tc(VII) reductase activity is also dependent upon processing of Mo, required as a cofactor for formate dehydrogenase activity.

location of hydrogenase 3 (Sauter et al., 1992).

Screening for bacteria with naturally enhanced activities against Tc(VII)

The identification of hydrogenase 3 of FHL as the Tc(VII) reductase of *E. coli* opened up the way for a program to screen for organisms with naturally enhanced activities against Tc(VII). Several organisms known to have naturally high FHL, or uptake hydrogenase activities, were tested, resulting in the identification of several strains of sulfate-reducing bacteria that were able to couple hydrogen oxidation to Tc(VII) reduction (Lloyd et al., 2001a). Rates of reduction in some strains were approximately 64-fold greater than those recorded in anaerobic cultures of *E. coli* (Lloyd et al., 1999b). *Desulfovibrio desulfuricans* (Lloyd et al., 1999a) and related strains (Lloyd et al., 2001a) were also able to utilise formate as an efficient electron donor for Tc(VII) reduction. This is consistent with the existence of a rudimentary FHL complex (consisting of a formate dehydrogenase coupled to a hydrogenase via a cytochrome) located in the periplasm of these strains (Peck, 1993). Accordingly, the site of reduced Tc precipitation was identified as the periplasm in *D. desulfuricans* (Lloyd et al., 1999a), and more recent studies have confirmed a role for a periplasmic Ni–Fe hydrogenase in Tc(VII) reduction by a close relative, *Desulfovibrio fructosovorans* (De Luca et al., 2000).

Although the mechanisms for Tc(VII) reduction by enteric bacteria such as *E. coli*, and sulfate-reducing bacteria such as *D. desulfuricans* are similar, there are important differences between the two biocatalysts. Hydrogen-dependent Tc(VII) reductase activity in *D. desulfuricans* is especially rapid and robust; cells remained active when sparged with air for 15 minutes or stored for several weeks under nitrogen (Lloyd et al., 1999a). The pH optima for both organisms are approximately pH 5.5, but temperature optima of 40°C (*E. coli*) and 20°C (*D. desulfuricans*) suggest that the sulfate-reducing bacterium would be more useful for treatment under realistic operating conditions (Lloyd et al., 1999b).

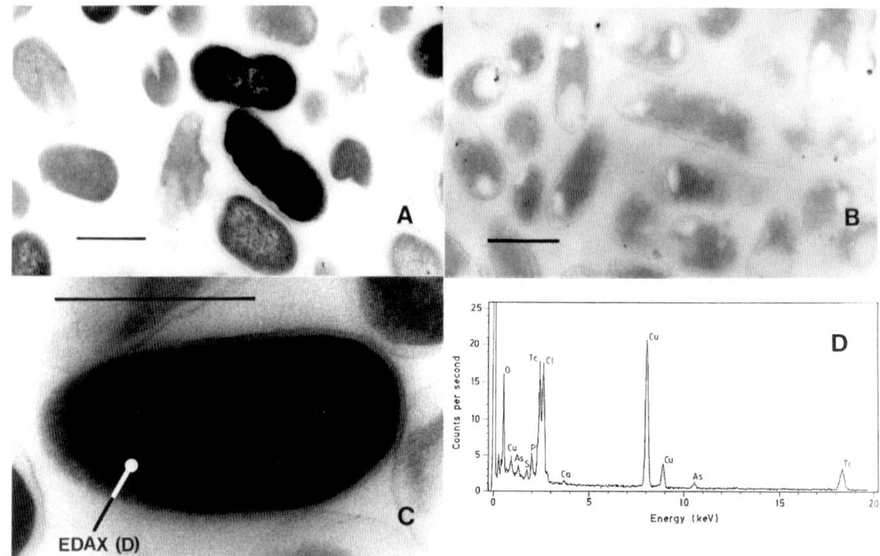

Fig. 3. Transmission electron micrographs of thin sections of *E. coli* cells grown under anaerobic conditions in the presence and absence of 1 mM Tc(VII). Cells grown in the presence of the radionuclide were electron dense (A). Cells which were not challenged with the metal were unstained (B). Solid state analysis of the sections (area arrowed; C) using EDAX (D) confirmed that Tc was accumulated within the cell. Cu was from the electron microscope grid. Bar = 1 μM. Reprinted from (Lloyd et al., 1997a) with permission.

Also, although the K_m for Tc(VII) is similar in both organisms (approximately 0.5 mM), the saturation concentration for electron donor (formate), which supports maximal rates of Tc(VII) reduction, is lower for *D. desulfuricans* (0.5 mM vs >25 mM for *E. coli*), attributable to the more accessible periplasmic localisation of the Tc(VII) reductase in the sulfate-reducing bacteria (Lloyd et al., 1999b). Thus, if formate is used in a process to treat Tc(VII), lower concentrations of electron donor would be required for efficient Tc(VII) reduction by the sulfate-reducing bacterium. Finally, hydrogen-dependent Tc(VII) reduction by *D. desulfuricans* is also insensitive to high concentrations of nitrate which may be present in nuclear waste streams (Lloyd et al., 1999b). In comparison, low concentrations of nitrate (1 mM) inhibited the reduction of 250 μM Tc(VII) by *E. coli*, possibly via the diversion of reducing equivalents to denitrifying enzymes or by affecting the regulation of the Tc-reducing formate hydrogenlyase complex (Lloyd et al., 1999b).

Development of an immobilised cell process to treat Tc(VII)

In order to treat large volumes of Tc(VII)-contaminated water, it is necessary to develop a suitable flow-through bioreactor configuration, containing high densities of active metal-reducing bacteria, and able to treat liquid effluents at high flow rates. A range of immobilisation matrices is available to retain active biomass and reduced insoluble Tc(IV) within such a system, and early studies aimed to identify the optimal support matrix for this purpose (Raihan et al., 1997). For safety reasons, a process that did not use a radioactive substrate was used in this study (the formate-dependent nitrite reductase of *E. coli*). This was chosen as model for the Tc(VII) reductase of the same organism because both enzymes are induced under anaerobic conditions and couple the oxidation of formate to the reduction of an oxyanion. The performance of column reactors containing cells entrapped in membrane, foam and gel matrices was compared to those containing biofilms grown on several support materials. Optimal performance was obtained with reactors containing cells immobilised within hollow-fibre membranes. This factor, coupled with ease of use and operation flexibility of hollow-fibre membrane reactors (Bunch, 1988), made this configuration the preferred choice for later studies employing Tc(VII).

The hollow-fibre membrane system used consisted of a 17-ml reactor containing 12 Romicon XM50 acrylic hollow-fibre membranes (Fig. 4a). *Escherichia coli* cells were grown in situ for 24 hours with growth medium supplied via the side port of the reactor (Lloyd et al., 1997b). At the end of the growth phase, the medium was replaced with phosphate-buffered saline containing 50 μM Tc(VII) and 25 mM formate. At a flow rate of 2 ml h^{-1}, removal of the Tc(VII) was sustained at 50–80% for the duration of the 112 h experiment, thus demonstrating significant radioresistence of the enzyme system. Negligible removal was noted in a control reactor containing a mutant unable to synthesise active FHL (Tc(VII) reductase) (Fig. 4b). Hydrogen also supported Tc(VII) reduction in the reactor (Lloyd et al., 1997b). Although these experiments demonstrated that sustained removal of Tc(VII) was possible using resting cells immobilised in reactor supplied with a simple electron donor (formate), the residence times necessary for efficient removal were considered too long for practical application. Therefore, the performance of *D. desulfuricans* and an *E. coli* mutant that overexpressed FHL were compared to that of the wild type strain of *E. coli* in the bench-scale hollow-fibre bioreactor system (Fig. 5). In this study, 80% removal of the radionuclide was obtained at the following residence times; *E. coli*, 24.3 hours; *E. coli* mutant overexpressing FHL, 4.25 hours; *D. desulfuricans*, 1.5 hours (Lloyd et al., 1999b). At residence times above 2.1 hours, Tc was no longer detected in the effluent exiting the reactor containing the sulfate-reducing bacterium (limit of detection 1 μM). These results demonstrate that sustained removal of Tc is possible at low residence times using a simple flow-through system containing an immobilised biocatalyst. The superiority of a wild type sulfate-reducing bacterium, compared to a genetically enhanced *E. coli* strain, is advantageous given the perceived risk of using genetically modified organisms for biotechnological applications. The direct utilisation of a simple cheap electron donor such as formate is also attractive since it negates the need for complicated cofactor regeneration steps. Finally, if hydrogen is used as the electron donor, the process is environmentally benign and leaves no additional residue in the effluent. The use of hydrogen is attractive since it can be generated electrochemically on demand and targeted to the cells via appropriate gas transfer membranes.

5. Future directions

Several recent reviews have proposed that biological processes offer the potential to help the nuclear industry meet tighter legislation regarding emissions of radionuclides into the

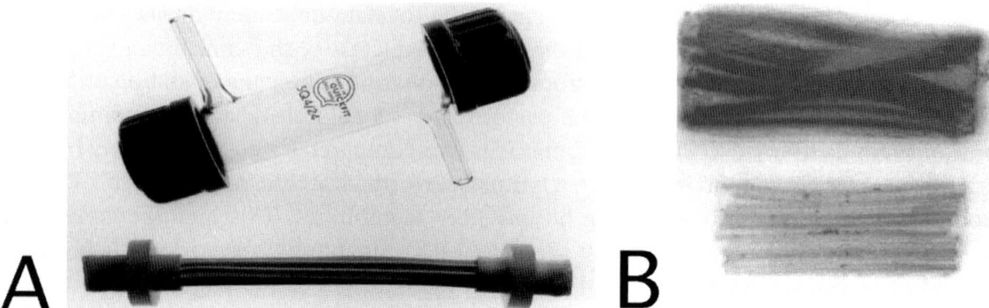

Fig. 4. (a) A hollow-fibre membrane bioreactor for treatment of Tc(VII) contaminated test solutions at laboratory scale. The 17 ml bioreactor contained 12 Romicon XM50 hollow-fibre membranes held in PTFE end plugs by araldite adhesive. (b) Tc accumulation by membrane entrapped *E. coli* cells. Hollow-fibre membrane bioreactors were operated for 112 hours in transverse mode with formate (25 mM) added to phosphate buffer as an electron donor for Tc(VII) reduction (Tc(VII) concentration = 50 μM). The top fibre bundle contained anaerobically grown cells with Tc(VII) reductase activity and was heavily stained by the radionuclide. The lower fibre bundle contained a derivative strain unable to synthesise FNR, and therefore Tc(VII) reductase, and was lightly stained with the radionuclide. Tc was visualised using a PhosphorImager (exposure time to storage phosphor screen = 6 hours). Reprinted from (Lloyd et al., 1997b) with permission.

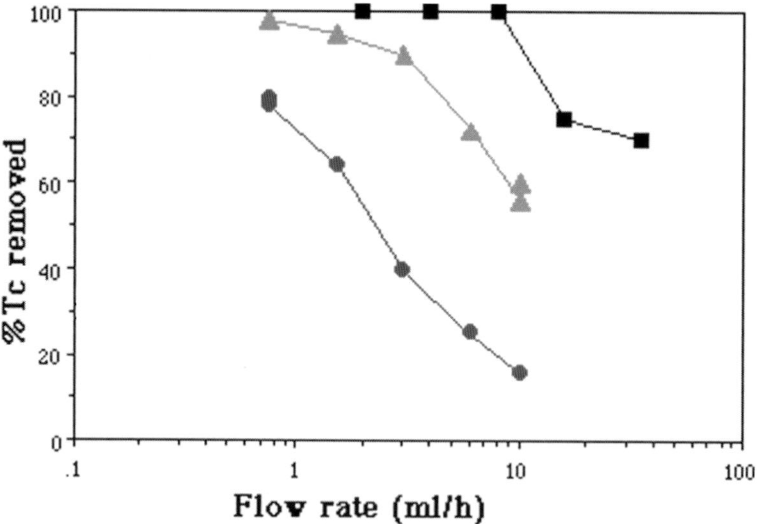

Fig. 5. Flow rate activity plot showing the performance of three Tc(VII)-reducing bacteria in a membrane bioreactor. A wild type *E. coli* strain (circles), a *hycA E. coli* mutant overexpressing FHL (Tc(VII) reductase) (triangles) and *D. desulfuricans* (squares) were immobilized in the reactor and challenged with phosphate buffered saline supplemented with 50 μM Tc(VII). Formate (25 mM) was supplied as the electron donor. Reprinted from (Lloyd et al., 1999b) with permission.

environment (Macaskie, 1991; Eccles, 1999; Lloyd & Macaskie, 2000). Biotechnological approaches may also offer the only credible solution for the remediation of subsurface environments contaminated with radioactive waste (McCullough et al., 1999; Lloyd & Macaskie, 2000). Multidisciplinary research will be required if the potential of biotechnology is to be realised within these scenarios. The involvement of specialist laboratories able to handle large quantities of radioactive waste is needed to help scale-up end of pipe approaches described in this review and in Chapter 12 of this volume. The involvement of Government agencies and laboratories is also required if in situ biotechnological techniques are to be used successfully in the field. In this context, the US Department of Energy's NABIR program is funding the development of the knowledge-base required to clean up numerous Superfund sites contaminated with radioactive waste. A significant component of this program aims to uncover the biochemical basis of radionuclide-microbe interactions. Such projects are in their infancy, but rapid advances in microbial biochemistry, genomics and proteonomics will make it possible to identify the molecular mechanisms by which microorganisms interact with radionuclides. This information is required to model biological processes in the field accurately, for the development of molecular tools to monitor microbial activities in sediments and bioreactors and for the development of biosensors to quantify bioavailable concentrations of radionuclides in the environment after cleanup.

Acknowledgements

Financial support from the Natural and Accelerated Bioremediation Research (NABIR) program, Biological and Environmental Research (BER), US Department of Energy is acknowledged.

References

Anderson, S. & Apanna, V. D. (1994). Microbial formation of crystalline strontium carbonate. *FEMS Microbiology Letters, 116*, 42–48.

Anon (1999). *Protein Purification Handbook*. Uppsala, Sweden: Amersham Pharmacia Biotechnology.

Ausubel, F. M., Brent, R., Kingston, R. E., Moore, D. D., Seidman, J. G., Smith, J. A. & Struhl, K. (1999). Short Protocols in Molecular Biology. In *A Compendium of Methods from Current Protocols in Molecular Biology* (4th edn). New York: John Wiley.

Avery, S. V. (1995). Microbial interactions with caesium – implications for biotechnology. Journal of *Chemical Technology and Biotechnology, 62*, 3–16

Barnes, L. J., Janssen, F. J., Sherren, J., Versteegh, J. H., Koch, R. O. & Scheeren, P. J. H. (1991). A new process for the microbial removal of sulphate and heavy metal from contaminated waters extracted by a geohydrological control system. *Chemical Engineering Research and Design, 69A*, 184–186.

Barnhart, B. J., Campbell, E. W., Martinez, E., Caldwell, D. E. & Hallett, R. (1980). Potential microbial impact on transuranic wastes under conditions expected in the waste isolation pilot plant (WIPP). Los Alamos National Laboratory.

Basnakova, G., Stephens, E. R., Thaller, M. C., Rossolini, G. M. & Macaskie, L. E. (1998). The use of *Escherichia coli* bearing a *phoN* gene for the removal of uranium and nickel from aqueous flow. *Applied Microbiology and Biotechnology, 50*, 266–272.

Beveridge, T. J., Popkin, T. J. & Cole, (1994). Electron microscopy. In P. Gerhardt (Ed.), *Methods for General Molecular Bacteriology* (pp. 42–71). Washington, DC: American Society for Microbiology.

Beveridge, T. J., Hughes, M. N., Lee, H., Leung, K. T., Poole, R. K., Savvaidis, I., Silver, S. & Trevors, J. T. (1997). Metal-microbe interactions: contemporary approaches. *Advances in Microbial Physiology, 38*, 177–243.

Birkman, A., Zinoni, F., Sawers, G. & Bock, A. (1987). Factors affecting transcription regulation of the formate-hydrogen-lyase pathway of *Escherichia coli. Archives of Microbiology, 148*, 129–143.

Bonthrone, K. M., Basnakova, G., Lin, F. & Macaskie, L. E. (1996). Bioaccumulation of nickel by intercalation into polycrystalline hydrogen uranyl phosphate deposited via an enzymatic mechanism. *Nature Biotechnology, 14*, 635–638.

Brady, D., Stoll, A. & Buncan, J. R. (1994). Biosorption of heavy metal cations by non-viable yeast biomass. *Environmental Technology, 15*, 429–439.

Bunch, A. W. (1988). The uses and potential of microbial hollow-fibre bioreactors. *Journal of Microbiological Methods, 8*, 103–119.

Burns, P. C. & Finch, R. (1999). *Uranium: Mineralogy, Geochemistry and the Environment*. Washington, DC: Mineralogical Society of America.

Cataldo, D. A., Garland, T. R., Wildung, R. E. & Fellows, R. J. (1989). Comparative metabolic behaviour and interrelationships of Tc and S in soybean plants. *Health Physics, 57*, 281–288.

Chart, H. (1994). Bacterial fractionation and membrane protein characterization. In H. Chart (Ed.), *Practical Laboratory Bacteriology* (pp. 1–10). Boca Raton, FL: CRC Press.

Clearfield, A. (1988). Role of ion exchange in solid-state chemistry. *Chemical Reviews, 88*, 125–148.

Coetzee, P. P. & deBeer, H. (1992). Ion chromatographic separation and spectrophotometric determination of U(IV) and U(VI). *Radiochimica Acta, 57*, 113–117.

Coppi, M. V., Leang, C., Lovley, D. R. & Sandler, S. J. (2001). Development of a genetic system for *Geobacter sulfurreducens. Applied and Environmental Microbiology, 67*, 3180–3187

Crameri, A., Dawes, G., Rodriguez, E., Silver, S. & Stemmer, W. P. C. (1997). Molecular evolution of an arsenate detoxification pathway by DNA shuffling. *Nature Biotechnology, 15*, 436–438.

Dawson, R. M. C., Elliot, D. C., Elliot, W. H. & Jones, K. M. (1986). *Data for Biochemical Research.* Oxford, UK: Oxford Science Publications.

De Luca, G., Rousset, M., Dermoun, Z., Belaich, J. P. & Vermeglio, A. (2000). Microbial reduction of technetium by *Desulfovibrio fructosovorans*: implication of the periplasmic nickel-iron hydrogenase. *The 2nd Euroconference on Bacterial-Metal/Radionuclide Interactions: Basic Research and Bioremediation*. Dresden, Germany.

Dehut, J. P., Fosny, K., Myttenaere, C., Deprins, D. & Vandecasteele, C. M. (1989). Bioavailability of Tc incorporated in plant material. *Health Physics, 57*, 263–267.

DeMoll-Decker, H. & Macy, J. M. (1993). The periplasmic nitrite reductase of *Thauera selanatis* may catalyse the reduction of selenite to elemental selenium. *Archives of Microbiology, 160*, 241–247.

deRome, L. & Gadd, G. M. (1991). Use of pelleted and immobilized yeast and fungal biomass for heavy metal and radionuclide recovery. *Journal of Industrial Microbiology, 7*, 97–104.

Dick, R. E., Boswell, C. D. & Macaskie, L. E. (1995). Uranyl phosphate accumulation by *Acinetobacter* sp. In C. A. Jerez, T. Vargas, H. Toledo & J. V. Wiertz (Eds.), *International Symposium of Biohydrometallurgy* (pp. 177–186). Vina del Mar, Chile: The University of Chile.

Diels, L., Dong, Q., Van der Lelie, D., Baeyens, W. & Mergeay, M. (1995). The *czc* operon of *Alcaligenes eutrophus* CH34: from resistance mechanism to the removal of heavy metals. *Journal of Industrial Microbiology, 14*, 142–153.

Dorhout, P. K., Kissane, R. J., Abney, K. D., Avens, L. R., Eller, P. G. & Ellis, A. B. (1989). Intercalation reactions of the neptunyl (VI) dication with hydrogen uranyl phosphate and hydrogen neptunyl host lattices. *Inorganic Chemistry, 28*, 2926–2930.

Eccles, H. (1999). Nuclear waste management: a bioremediation approach. In G. R. Choppin & M. K. Khankhasayev (Eds), *Chemical Separation Technologies and Related Methods of Nuclear Waste Management* (pp. 187–208). Dordrecht: Kluwer Academic Publishers.

Ehrlich, H. L. (1996). *Geomicrobiology* (3rd edn). New York: Marcel Dekker.

Ellwood, D. C., Hill, M. J. & Watson, J. H. P. (1992). Pollution control using microorganisms and metal separation. In J. C. Fry, G. M. Gadd, R. A. Herbert, C. W. Jones & I. A. Watson-Craik (Eds.), *Microbial Control of Pollution* (pp. 89–112). Cambridge, UK: Cambridge University Press.

Fendorf, S. & Li, G. (1996). Kinetics of chromate reduction by ferrous iron. *Environmental Science and Technology, 30*, 1614–1617.

Ferris, F. G., Fratton, C. M., Gertis, J. P., Schultzelam, S. & Lollar, B. S. (1995). Microbial precipitation of a strontium calcite phase at a groundwater discharge zone near Rock Creek, British Columbia, Canada. *Journal of Geomicrobiology, 13*, 57–67.

Findlay, J. B. C. (1990). Purification of membrane proteins. In E. L. V. Harris & S. Angal (Eds), *Protein Purification Applications: A Practical Approach* (pp. 59–82). Oxford, UK: Oxford University Press.

Francis, A. J., Dodge, C. J., Lu, F., Halada, G. P. & Clayton, C. R. (1994). XPS and XANES studies of uranium reduction by *Clostridium* sp. *Environmental Science and Technology, 28*, 636–639.

Francis, C. W., Timpson, M. E. & Wilson, J. H. (1999). Bench- and pilot-scale studies relating to the removal of uranium from uranium-contaminated soils using carbonate and citrate lixiviants. *Journal of Hazardous Materials, 66*, 67–87.

Francis, A. J., Dodge, C. J., Gillow, J. B. & Papenguth, H. W. (2000) Biotransformation of uranium compounds in high ionic strength brine by a halophilic bacterium under denitrifying conditions. *Environmental Science and Technology, 34*, 2311–2317.

Gadd, G. M. (1997). Roles of microorganisms in the environmental fate of radionuclides. *CIBA Foundation Symposia, 203*, 94–104.

Gadd, G. M. & White, C. (1989). Heavy metal and radionuclide accumulation and toxicity in fungi and yeasts. In R. K. Poole & G. M. Gadd (Eds), *Metal-Microbe Interactions* (pp. 19–38). Oxford, UK: Society for General Microbiology, IRL Press.

Garnham, G. W., Codd, G. A. & Gadd, G. M. (1992). Uptake of technetium by fresh water green microalgae. *Applied Microbiology and Biotechnology, 37*, 679–684.

Gordon, E. H. J., Pike, A. D., Hill, A. E., Cuthbertson, P. M., Chapman, S. K. & Reid, G. A. (2000). Identification and characterization of a novel cytochrome c_3 from *Shewanella frigidimarina* NCIMB400 that is involved in Fe(III) respiration. *Journal of Biochemistry, 349*, 153–158.

Gueffroy, D. E. (1990). *Buffers: A Guide for the Preparation and Use of Buffers in Biological Systems*. San Diego, CA: Calbiochem Corporation.

Henrot, J. (1989). Bioaccumulation and chemical modification of Tc by soil bacteria. *Health Physics, 57*, 239–245.

Higham, D. P., Sadler, P. J. & Scawen, M. D. (1984). Cadmium-resistant *Pseudomonas putida* synthesizes novel cadmium binding proteins. *Science, 225*, 1043–1046.

Hughes, M. N. & Poole, R. K. (1989). *Metals and Micro-organisms*. London, UK: Chapman and Hall.

Hunsberger, L. R. & Ellis, A. B. (1990). Excited-state properties of lamellar solids derived from metal complexes and hydrogen uranyl phosphate. *Coordination Chemistry Reviews, 97*, 209–224.

Jeong, B. C., Hawes, C., Bonthrone, K. M. & Macaskie, L. E. (1997). Localization of enzymically enhanced heavy metal accumulation by *Citrobacter* sp. and metal acumulation in vitro by liposomes containing entrapped enzyme. *Microbiology, 143*, 2497–2507.

Johansson, S. A. E. & Campbell, J. L. (1988). *PIXE. A Novel Technique for Elemental Analysis*. Chichester, UK: John Wiley.

Karavaiko, G. I., Yurchenko, V. A., Remizo, V. L. & Klushnikova, T. M. (1988). Reduction of manganese dioxide by free-cell *Acinetobacter calcoaceticus* extracts. *Microbiologiya, 55*, 709–710.

Kashefi, K. & Lovley, K. (1999). Reduction of Fe(III), Mn(IV), and toxic metals at 100°C by *Pyrobaculum islandicum. Applied and Environmental Microbiology, 66*, 1050–1056

Kearney, T., Eccles, H., Graves., D. & Gonzalez, A. (1996). Bench and pilot-scale demonstration of an innovative bioremediation process. In *Proceedings of the 18th Annual Conference of the National Low-Level Waste Management Program*. Salt Lake City, Utah, USA.

Kjaergaard, K., Sorenson, J. K., Schembri, M. A. & Klemm, P. (2000). Sequestration of zinc oxide by fimbrial designer chelators. *Applied and Environmental Microbiology, 66*, 10–14.

Kortba, P., Dolekova, L., De Lorenzo, V. & Ruml, T. (1999). Enhanced bioaccumulation of heavy metal ions by bacterial cells due to surface display of short metal binding peptides. *Applied and Environmental Microbiology, 65*, 1092–1098.

Kotegov, K. V., Pavlov, O. N. & Shvendov, V. P. (1968). Technetium. In H. J. Emelius & A. G. Sharpe (Eds), *Advances in Inorganic Chemistry and Radiochemistry* (pp. 1–90). New York: Academic Press.

Krafft, T. & Macy, J. M. (1998). Purification and characterization of the respiratory arsenate reductase of *Chrysiogenes arsenatis*. *European Journal of Biochemistry, 255*, 647–653.

L'Annunziata, M. F. (1998). *Handbook of Radioactivity Analysis*. London, UK: Academic Press.

Lloyd, J. R. & Macaskie, L. E. (1996). A novel phosphorImager based technique for monitoring the microbial reduction of technetium. *Applied and Environmental Microbiology, 62*, 578–582.

Lloyd, J. R. & Macaskie, L. E. (1997). Microbially-mediated reduction and removal of technetium from solution. *Research in Microbiology, 148*, 530–532.

Lloyd, J. R. & Macaskie, L. E. (2000). Bioremediation of radioactive metals. In D. R. Lovley (Ed.), *Environmental Microbe-Metal Interactions* (pp. 277–327). Washington, DC: ASM Press.

Lloyd, J. R., Cole, J. A. & Macaskie, L. E. (1997a). Reduction and removal of heptavalent technetium from solution by *Escherichia coli*. *Journal of Bacteriology, 179*, 2014–2021.

Lloyd, J. R., Harding, C. L. & Macaskie, L. E. (1997b). Tc(VII) reduction and precipitation by immobilized cells of *Escherichia coli*. *Biotechnology and Bioengineering, 55*, 505–510.

Lloyd, J. R., Nolting, H.-F., Solé, V. A., Bosecker, K. & Macaskie, L. E. (1998). Technetium reduction and precipitation by sulphate-reducing bacteria. *Geomicrobiology Journal, 15*, 43–56.

Lloyd, J. R., Ridley, J., Khizniak, T., Lyalikova, N. N. & Macaskie, L. E. (1999a). Reduction of technetium by *Desulfovibrio desulfuricans*: biocatalyst characterisation and use in a flow-through bioreactor. *Applied and Environmental Microbiology, 65*, 2691–2696.

Lloyd, J. R., Thomas, G. H., Finlay, J. A., Cole, J. A. & Macaskie, L. E. (1999b). Microbial reduction of technetium by *Escherichia coli* and *Desulfovibrio desulfuricans*: enhancement via the use of high activity strains and effect of process parameters. *Biotechnology and Bioengineering, 66*, 123–130.

Lloyd, J. R., Yong, P. & Macaskie, L. E. (2000a). Biological reduction and removal of pentavalent Np by the concerted action of two microorganisms. *Environmental Science and Technology, 34*, 1297–1301.

Lloyd, J. R., Sole, V. A., Gaw, C. & Lovley, D. R. (2000b). Direct and Fe(II)-mediated reduction of technetium by iron-reducing bacteria. *Applied and Environmental Microbiology, 66*, 3743–3749.

Lloyd, J. R., Mabbett, A., Williams, D. R. & Macaskie, L. E. (2001a). Metal reduction by sulfate-reducing bacteria: physiological diversity and metal specificity. *Biohydrometallurgy, 59*, 327–337.

Lloyd, J. R., Hodges-Myerson, A. L. & Lovley, D. R. (2001b). Localization of *c*-type cytochromes in *Geobacter sulfurreducens* and their potential role in Fe(III) oxide reduction (manuscript submitted).

Lloyd, J. R., Chesnes, J., Glasauer, S., Bunker, D. J., Livens, F. R. & Lovley, D. R. (2002). Reduction of actinides and fission products by Fe(III)-reducing bacteria. *Geomicrobiology Journal* (in press).

Lovley, D. R. (1991). Dissimilatory Fe(III) and Mn(IV) reduction. *Microbiology Review, 55*, 259–287.

Lovley, D. R. (1993). Dissimilatory metal reduction. *Annual Review of Microbiology, 47*, 263–290.

Lovley, D. R. (2000). Fe(III) and Mn(IV) reduction. In D. R. Lovley (Ed.), *Environmental Microbe-Metal Interactions* (pp. 3–30). Washington, DC: ASM Press.

Lovley, D. R., Phillips, E. J. P., Gorby, Y. A. & Landa, E. (1991). Microbial reduction of uranium. *Nature, 350*, 413–416.

Lovley, D. & Phillips, E. J. (1992a). Reduction of uranium by *Desulfovibrio desulfuricans*. *Applied and Environmental Microbiology, 58*, 850–856.

Lovley, D. R. & Phillips, E. J. P. (1992b). Bioremediation of uranium contamination with enzymatic uranium reduction. *Environmental Science and Technology, 26*, 2228–2234.

Lovely, D. R., Roden, E. E., Phillips, E. J. P. & Woodward, J. C. (1993a). Enzymatic iron and uranium reduction by sulfate-reducing bacteria. *Marine Geology, 113*, 41–53.

Lovley, D. R., Giovannoni, S. J., White, D. C., Champine, J. E., Phillips, E. J. P., Gorby, Y. A. & Goodwin, S. (1993b). *Geobacter metallireducens* gen. nov. sp. nov., a microorganism capable of coupling the complete oxidation of organic compounds to the reduction of iron and other metals. *Archives in Microbiology, 159*, 336–344.

Lovley, D. R., Widman, P. K., Woodward, J. C. & Phillips, E. J. P. (1993c). Reduction of uranium by cytochrome c_3 of *Desulfovibrio vulgaris*. *Applied and Environmental Microbiology, 59*, 3572–3576.

Macaskie, L. E. (1990). An immobilized cell bioprocess for the removal of heavy metals from aqueous flows. *Journal of Chemical Technology and Biotechnology, 49*, 357–379.

Macaskie, L. E. (1991). The application of biotechnology to the treatment of wastes produced from nuclear fuel cycle: biodegradation and bioaccumulation as a means of treating radionuclide-containing streams. *Critical Reviews in Biotechnology, 11*, 41–112.

Macaskie, L. E., Empson, R. M., Cheetham, A. K., Grey, C. P. & Skarnulis, A. J. (1992). Uranium bioaccumulation by a *Citrobacter* sp. as a result of enzymically-mediated growth of polycrystalline HUO_2PO_4. *Science, 257*, 782–784.

Macaskie, L. E., Jeong, B. C. & Tolley, M. R. (1994). Enzymically-accelerated biomineralization of heavy metals: application to the removal of americium and plutonium from aqueous flows. *FEMS Microbiology Reviews, 14*, 351–368.

Macaskie, L. E., Empson, R. M., Lin, F. & Tolley, M. R. (1995). Enzymatically-mediated uranium accumulation and uranum recovery using a *Citrobacter* sp. immobilised as a biofilm within a plug-flow reactor. *Journal of Chemical Technology and Biotechnology, 63*, 1–16.

Macaskie, L. E., Lloyd, J. R., Thomas, R. A. P. & Tolley, M. R. (1996). The use of microoorganisms for the remediation of solutions contaminated with actinide elements, other radionuclides and organic contaminants generated by nuclear fuel cycle activities. *Nuclear Energy, 35*, 257–271.

Macaskie, L. E., Yong, P., Doyle, T. C., Roig, M. G., Diaz, M. & Manzano, T. (1997). Bioremediation of uranium-bearing wastewater: biochemical and chemical factors influencing bioprocess application. *Biotechnology and Bioengineering, 53*, 100–109.

Magnuson, T. S., Hodges-Myerson, A. L. & Lovley, D. R. (1999). Characterization of a membrane-bound NADH-dependent Fe^{3+} reductase from the dissimilatory Fe^{3+}-reducing bacterium *Geobacter sulfurreducens*. *FEMS Microbiology Letters, 185*, 205–211.

Marques, A. M., Roca, X., Simon-Pujol, M. D., Fuste, M. C. & Francisco, C. (1991). Uranium acumulation by *Pseudomonas* sp. EPS-5028. *Applied Microbiology and Biotechnology, 35*, 406–410.

McCullough, J., Hazen, T. C., Benson, S. M., Metting, F. B. & Palmisano, A. C. (1999). *Bioremediation of metals and radionuclides . . . what is it and how it works*. Berkeley, CA: Lawrence Berkeley National Laboratory.

McHale, A. P. & McHale, S. (1994). Microbial biosorption of metals: potential in the treatment of metal pollution. *Advances in Biotechnology, 12*, 647–652.

Meima, R. & Lidstrom, M. E. (2000). Characterization of the minimal replicon of a cryptic *Deinococcus radiodurans* SARK plasmid and development of versatile *Escherichia coli-D-radiodurans* shuttle vectors. *Applied and Environmental Microbiology, 66*, 3856–3867.

Myers, C. R. & Myers, J. M. (1992). Localization of cytochromes to the outer membrane of anaerobically grown *Shewanella putrefaciens* MR-1. *Journal of Bacteriology, 174*, 3429–3438.

Myers, C. R. & Myers, J. M. (1993). Ferric reductase is associated with the membranes of ananerobically grown *Shewanella putrefaciens* MR-1. *FEMS Microbiology Letters, 108*, 15–21.

Myers, C. R. & Myers, J. M. (1997a). Cloning and sequence of cymA, a gene encoding a tetraheme cytochrome *c* required for reduction of iron(III), fumarate and nitrate by *Shewanella putrefaciens* MR-1. *Journal of Bacteriology, 179*, 1143–1152.

Myers, C. R. & Myers, J. M. (1997b). Outer membrane cytochromes of *Shewanella putrefaciens* MR-1: spectral analysis, and purification of the 83-kDa *c*-type cytochrome. *Biochimica and Biophysica Acta, 1326*, 307–318.

Myers, C. R. & Myers, J. M. (2000). Role of tetraheme cytochrome CymA in anaerobic electron transport in cells of *Shewanella putrefaciens* MR-1 with normal levels of menaquinone. *Journal of Bacteriology, 182*, 67–75.

Myers, C. R. & Nealson, K. H. (1990). Respiration-linked proton translocation coupled to anaerobic reduction of manganese(IV) and iron(III) in *Shewanella putrefaciens* MR-1. *Journal of Bacteriology, 172*, 6232–6238.

Okorov, L. A., Lichko, L. P., Kodomtseva, V. M., Kholodenko, V. P., Titovsky, V. T. & Kulaev, I. S. (1977). Energy-dependent transport of manganese into yeast cells and distribution of accumulated ions. *European Journal of Biochemistry, 75*, 373–377.

Oliver, S. G., Winson, M. K., Kell, D. B. & Baganz, F. (1998). Systematic functional analysis of the yeast genome. *Trends in Biotechnology, 16*, 373–378.

Park, C. H., Keyhan, M., Wielinga, B., Fendorf, S. & Matin, A. (2000). Purification to homogeneity and characterization of a novel *Pseudomonas putida* chromate reductase. *Applied and Environmental Microbiology, 66*, 1788–1795.

Peck, H. D. (1993). Bioenergetic strategies of the sulfate-reducing bacteria. In J. M. Odom & R. Singleton (Eds), *Sulfate-Reducing Bacteria: Contempary Perspectives*. New York: Springer-Verlag.

Pham-Thi, M. & Columban, P. (1985). Cationic conductivity, water species motions and phase transitions in $H_3OUO_2PO_4{\cdot}3H_2O$ (HUP) and HUP related compounds ($M^+ = Na^+, K^+, Ag^+, Li^+, NH_4^+$). *Solid State Ionics, 17*, 295–306.

Phillips, E. J. P., Landa, E. R. & Lovley, D. R. (1995). Remediation of uranium contaminated soils with bicarbonate extraction and microbial U(VI) reduction. *Journal of Industrial Microbiology, 14*, 203–207.

Pignolet, L., Fonsny, K., Capot, F. & Moureau, Z. (1989). Role of various microorganisms on Tc behaviour in sediments. *Health Physics, 57*, 791–800.

Postgate, J. R. (1979). *The Sulphate-Reducing Bacteria*. Cambridge. UK: Cambridge University Press.

Pozas-Tormo, R., Moreno-Real, L., Martinez-Lara, M. & Bruque-Gamez, S. (1986). Layered metal uranyl phosphates. Retention of divalent ions by amine intercalates of uranyl phosphates. *Canadian Journal of Chemistry, 64*, 30–34.

Pozas-Tormo, R., Moreno-Real, L., Martinez-Lara, M. & Bruque-Gamez, S. (1987). Intercalation of lanthanides into $H_3OUO_2PO_4{\cdot}3H_2O$ and $CH_4H_9NH_3UO_2PO_4{\cdot}3H_2O$. *Inorganic Chemistry, 26*, 1442–1445.

Raihan, S., Ahmed, N., Macaskie, L. E. & Lloyd, J. R. (1997). Immobilisation of whole bacterial cells for anaerobic biotransformations. *Applied Microbiology and Biotechnology, 47*, 352–357.

Rousset, M., Casalot, L., Rapp-Giles, B. J., Dermoun, Z., De Philip, P., Belaich, J. P. & Wall, J. D. (1998). New shuttle vectors for the introduction of cloned DNA in *Desulfovibrio. Plasmid, 39*, 114–122.

Rusin, P. A., L., Q., Brainard, J. R., Strietelmeier, B. A., Tait, C. D., Ekberg, S. A., Palmer, P. D., Newton, T. W. & Clark, D. L. (1994). Solubilization of plutonium hydrous oxide by iron-reducing bacteria. *Environmental Science and Technology, 28*, 1686–1690.

Sambrook, J., Fritsch, E. F. & Maniatis, T. (1989). *Molecular Cloning: A Laboratory Manual* (2nd edn). New York: Cold Spring Harbor Press.

Sauter, M., Bohm, R. & Bock, A. (1992). Mutational analysis of the operon (*hyc*) determining hydrogenase 3 formation in *Escherichia coli. Molecular Microbiology, 6*, 1523–1532.

Schroder, I., Rech, S., Krafft, T. & Macy, J. M. (1997). Purification and characterization of the selenate reductase from *Thauera selenatis. Journal of Biological Chemistry, 272*, 23765–23768.

Soll, D. R. & Winzeler, E. A. (2000). Genome-wide approaches to cell function. *Current Opinion in Microbiology, 3*, 283–284.

Sousa, C., Kotrba, P., Ruml, T., Cebolla, A. & De Lorenzo, V. (1998). Metalloadsorption by *Escherichia coli* cells displaying yeast and mammalian metallothioneins anchored to the outer membrane protein LamB. *Journal of Bacteriology, 180*, 2280–2284.

Spiro, S. (1994). The FNR family of transcription factors. *Antonie Leeuwenhoeek, 66*, 23–36.

Stewart, V. (1988). Nitrate respiration in relation to faculatative metabolism in Enterobacteria. *Microbiology Reviews, 52*, 190–232.

Strachan, L. F., Tolley, M. R. & Macaskie, L. E. (1991). Radiotolerance of phosphatases of a *Citrobacter* sp.; potential for the use of this organism in the treatment of wastes containing radiotoxic actinides. In Proceedings of the 201st Meeting of the American Chemical Society. *Symposium: Biotechnology for Wastewater Treatment* (pp. 128–131). American Chemical Society, Atlanta, GA.

Strandberg, G. W., Shumate II, S. E. & Parrott, J. R. (1981). Microbial cells as biosorbents for heavy metals; accumulation of uranium by *Saccharomyces cerevisiae* and *Pseudomonas aeruginosa. Applied and Environmental Microbiology, 41*, 237–245.

Suzuki, Y. & Banfield, J. F. (1999). Geomicrobiology of uranium. In P. C. Burns & R. Finch (Eds), *Uranium: Mineralogy, Geochemistry and the Environment* (pp. 393–432). Washington, DC: Mineralogical Society of America.

Tebo, B. M. & Obraztsova, A. Y. (1998). Sulfate-reducing bacterium grows with Cr(VI), U(VI), Mn(IV), and Fe(III) as electron acceptors. *FEMS Microbiology Letters, 162*, 193–198.

Tobin, J. M., White, C. & Gadd, G. M. (1994). Metal accumulation by fungi: applications in environmental biotechnology. *Journal of Industrial Microbiology, 13*, 126–130.

Tolley, M. R. & Macaskie, L. E. (1994). Metal removal from aqueous solutions. PCT Patent Application No. GB94/00626.

Tolley, M. R., Strachan, L. F. & Macaskie, L. E. (1995). Lanthanum accumulation from acidic solutions using *Citrobacter* sp. immobilized in a flow-through bioreactor. *Journal of Industrial Microbiology, 14*, 271–280.

Trabalka, J. R. & Garten, C. T., Jr. (1983). Behaviour of the long-lived synthetic elements and their natural analogues in food chains. In J. T. Lett, U. K. Ehman & A. B. Cox (Eds), *Advances in Radiation Biology* (pp. 68–73). London, UK: Academic Press.

Truex, M. J., Peyton, B. M., Valentine, N. B. & Gorby, Y. A. (1997). Kinetics of U(VI) reduction by a dissimilatory Fe(III)-reducing bacterium under non-growth conditions. *Biotechnology and Bioengineering, 55*, 490–496.

Tucker, M. D., Barton, L. L. & Thomson, B. M. (1996). Kinetic coefficients for simultaneous reduction of sulfate and uranium by *Desulfovibrio desulfuricans. Applied Microbiology and Biotechnology, 46*, 74–77.

Tucker, M. D., Barton, L. L. & Thompson, B. M. (1998). Removal of U and Mo from water by immobilized *Desulfovibrio desulfuricans* in column reactors. *Biotechnology and Bioengineering, 60*, 90–96.

Turner, J. S. & Robinson, N. J. (1995). Cyanobacterial metallothioneins: biochemistry and molecular genetics. *Journal of Industrial Microbiology, 14*, 119–125.

Valls, M., Atrian, S., De Lorenzo, V. & Fernandez, L. A. (2000). Engineering a mouse metallothionein on the cell surface of *Ralstonia eutropha* CH34 for immobilization of heavy metals in soil. Nature Biotechnology, 18, 661–665.

Valls, M., Gonzalez-Duarte, R., Atrian, S. & De Lorenzo, V. (1998). Bioaccumulation of heavy metals with protein fusions of metallothionein to bacterial OMPs. *Biochimie, 10*, 855–61.

Van Roy, S., Peys, K., Dresselaers, T. & Diels, L. (1997). The use of an *Alcaligenes eutrophus* biofilm in a membrane bioreactor for heavy metal recovery. *Research in Microbiology, 148*, 526–528.

Volesky, B. (1990). *Biosorption of Heavy Metals*. Boca Raton, FL: CRC Press.

Volesky, B. (1994). Advances in biosorption of metals: selection of biomass types. *FEMS Microbiology Reviews, 14*, 291–302.

Volesky, B. & Holan, Z. R. (1995). Biosorption of heavy metals. *Biotechnology Progress, 11*, 235–250.

Volesky, B. & May-Phillips, H. A. (1995). Biosorption of heavy metals by *Saccharomyces cerevisiae. Applied Microbiology and Biotechnology, 42*, 797–806.

Wade Jr., R. & DiChristina, T. J. (2000). Isolation of U(VI) reduction-deficient mutants of *Shewanella putrefaciens. FEMS Microbiology Letters, 184*, 143–148.

Wagner, M., Roger, A. J., Flax, J. L., Brusseau, G. A. & Stahl, D. (1998). Phylogeny of dissimilatory sulfite reductases supports an early origin of sulfate respiration. *Journal of Bacteriology, 180*, 2975–2982.

Watson, J. H. P. & Ellwood, D. C. (1988). A biomagnetic separation process for the removal of heavy metal ions from solution. In *Proceedings of the International Conference on Control of Environmental Problems from Metal Mines*. Federation of Norwegian Industries and State Pollution Control Authority. Roros, Norway.

Watson, J. H. P. & Ellwood, D. C. (1994). Biomagnetic separation and extraction process for heavy metals from solution. *Minerals Engineering, 7*, 1017–1028.

White, C. & Gadd, G. M. (1987). Inhibition of H^+ efflux and K^+ uptake and induction of K^+ efflux in yeast by heavy metals. *Toxicological Assessment, 2*, 437–447.

White, C., Sharman, A. K. & Gadd, G. M. (1998). An integrated microbial process for the bioremediation of soil contaminated with toxic metals. *Nature Biotechnology, 16*, 572–575.

White, O., Eisen, J. A., Heidelberg, J. F., Hickey, E. K., Peterson, J. D., Dodson, R. J., Haft, D. H., Gwinn, M. L., Nelson, W. C., Richardson, D. L., Moffat, K. S., Qin, H. Y., Jiang, L. X., Pamphile, W., Crosby, M., Shen, M., Vamathevan, J. J., Lam, P., McDonald, L., Utterback, T., Zalewski, C., Makarova, K. S., Aravind, L., Daly, M. J., Minton, K. W., Fleischmann, R. D., Ketchum, K. A., Nelson, K. E., Salzberg, S., Smith, H. O., Venter, J. C. & Fraser, C. M. (1999). Genome sequence of the radioresistant bacterium *Deinococcus radiodurans* R1. *Science, 286*, 1571–1577.

Wildung, R. E., Gorby, Y. A., Krupka, K. M., Hess, N. J., Li, S. W., Plymale, A. E., McKinley, J. P. & Fredrickson, J. K. (2000). Effect of electron donor and solution chemistry on products of dissimilatory reduction of technetium by *Shewanella putrefaciens. Applied and Environmental Microbiology, 66*, 2451–2460.

Wildung, R. E., McFadden, K. M. & Garland, T. R. (1979). Technetium sources and behaviour in the environment. *Journal of Environmental Quality, 8*, 156–161.

Yanke, L. J., Bryant, R. D. & Laishley, E. J. (1995). Hydrogenase I of *Clostridium pasteurianum* functions as a novel selenite reductase. *Anaerobe, 1*, 61–67.

Yong, P. & Macaskie, L. E. (1995). Enhancement of uranium bioaccumulation by a *Citrobacter* sp. via enzymatically-mediated growth of polycrystalline $NH_4UO_2PO_4$. *Journal of Chemical Technology and Biotechnology, 63*, 101–108.

Yong, P. & Macaskie, L. E. (1997). Effect of substrate concentration and nitrate inhibition on product release and heavy metal removal by a *Citrobacter* sp. *Biotechnology and Bioengineering, 55*, 821–830.

Miranda J. Keith-Roach and Francis R. Livens (Editors)
© 2002 Elsevier Science Ltd. All rights reserved

Chapter 12

Microbial interactions with radioactive wastes and potential applications

Lynne E. Macaskie[a], Jon R. Lloyd[b]

[a]*School of Biosciences, University of Birmingham, Birmingham B15 2TT, UK*
[b] *Department of Earth Sciences, University of Manchester, Manchester M13 9PL, UK*

1. The nuclear fuel cycle: waste sources and target areas

The nuclear fuel cycle centres around the civil use of uranium for energy production. The nuclear industry is in decline in many parts of the world, partly due to a number of well-publicised accidents, which, together with problems of safe waste disposal, have undermined public confidence. However, in many countries the nuclear industry continues to be a major supplier of electricity and, even if phased out immediately, large problems would persist with run-off of mine waters from disused uranium mines, safe disposal of the stockpiles of solid and liquid waste and the decommissioning of existing plants. Decontamination, for example of redundant plants by the use of chemical agents, converts the radioactivity into a liquid form which ideally then requires volume reduction and solidification for final disposal. Historically, some wastes have been dumped in shallow trenches or buried but it is now appreciated that radioactive species can migrate substantial distances underground (Means et al., 1978; Cleveland & Rees, 1981; Means & Alexander, 1981; Penrose et al., 1990) to reappear in local rivers, groundwaters and other sources of potable water. Technologies for soil remediation lag behind those for remediation of liquid wastes and the strategy of 'dig and dump' only shifts the problem. Excavated soil can be treated off-site (ex situ) or treated in situ by the use of agents which mobilise the metal species, prior to the use of organisms that can remove the metals from solution ex situ, concentrating metals into a small volume for easier disposal (see later). Alternatively, it may be possible to immobilise the metals in situ, assuming that this offers a satisfactory endpoint for bioremediation. Many metal species bind tightly to humic acids and other reactive components in the soil (see Chapters 3, 4 and 5 of this volume; Allen, 1984; Choppin & Allard, 1985; Bulman & Baker, 1987), or in marine or freshwater sediments, reducing their bioavailability, and it could be considered acceptable to rely on this approach of natural attenuation. However the radioactive species could remobilise over time (Burton et al., 1986; Assinder et al., 1990; Bryan et al., 1992) and, moreover, radioactive decay results in the production of daughter elements with different chemical properties, which may not include strong binding to soil or sediment components. An extreme example of this is the production of radon gas from uranium and thorium decay in nature, and another, involving anthropogenic elements, is the decay of ^{241}Pu (half-life of 14.4 years: a significant component of fresh waste (see later)) which decays via ^{241}Am (half-life of 432 years) to ^{237}Np, a long-lived α-emitter which has a half-life of 2.13 million years (Weigel et al., 1986). Thus, on a geological timescale, ^{237}Np will eventually predominate. This has major implications since, unlike most of the other actinide elements, Np is relatively mobile in the environment (Chapters 3 and 4, this volume; Cleveland et al., 1985). The environmental chemistry of radioactive waste disposal and fate of ^{237}Np in the environment are outside the scope of this article but have been reviewed in depth elsewhere (Duffield & Williams, 1986; Lieser & Muhlenweg, 1988; Hursthouse et al., 1991). These factors, and the ease of remediation by current practices, are typically taken into account when identifying the radionuclides which ultimately limit the acceptability of the waste (Oversby, 1987). Human activities are not the only means by which radionuclides enter the environment and, in a few relatively rare instances, nuclear chain reactions have occurred as a result of natural fission of uranium in the environment. These 'natural nuclear reactors' have provided an opportunity to study the generation and environmental fate of radionuclides resulting from geological sources (Nagy et al., 1991). ^{210}Po, the last unstable member of the ^{238}U decay series, has also attracted recent attention, accounting for all α-emitting activity in some samples of groundwater from the Central Florida Phosphate District (Burnett et al., 1987). Levels of over 16 Bq l^{-1} have been recorded, originating from either the naturally occurring phosphate rock of the area (high in uranium and daughter elements) or from phosphogypsum, a byproduct of the wet process manufacture of phosphoric acid, which is high in ^{226}Ra, ^{210}Pb and ^{210}Po. ^{210}Po has also been identified as a troublesome isotope, contaminating and fouling oil drilling equipment in areas rich in uranium ores (A. Clerkin, personal communication). Sulfate-reducing bacteria are able to precipitate Po when biogenic sulfate concentrations reach approximately 10 μM and above, and this mechanism has been proposed as a biotechnological approach to remediating Po-contaminated waters.

Wastes are produced from uranium mining and nuclear fuel fabrication and reprocessing and a review of waste treatment processes is given by Boegley & Alexander (1986). A recent survey (Lloyd & Macaskie, 2000) covered newer developments in the bioremediation of the fission products and minor elements. This chapter will focus specifically on the remediation of minewaters and on the removal of major problematic species from reprocessing wastes. The principal elements of interest are U, Th and Ra, which tend to be present in minewaters at relatively low specific activities but in large volumes, and U and the transuranic elements, which are significant components of reprocessing wastes. In addition, the fission product ^{99}Tc is of interest since the heptavalent Tc(VII) species exists as the pertechnetate ion (TcO_4^-) which is poorly removed by current practices and has a high biological significance due to its ability to mimic the sulfate ion (SO_4^{2-}) and thus enter food chains. This has resulted in the 'elevation' of the status of ^{99}Tc to a nuclide which may ultimately be limiting in waste discharge. The development of a bioprocess to treat Tc(VII) is discussed in detail in Chapter 11, alongside the biochemical factors underlying several other bioprocesses developed to treat radioactive waste.

It is now appreciated that wastes containing metals and chelating agents in combination can give rise to special problems (e.g. Delegard et al., 1984). For example, enhanced actinide migration in the environment has been attributed to formation of mobile complexes with natural or anthropogenic chelating agents (Nelson et al., 1989; Choppin, 1992; Baik & Lee, 1994; Silva & Nitsche, 1995). This chapter will also briefly cover salient aspects of the bioremediation of metal complexing agents since in many cases it is necessary to degrade the ligand in order to release the metal into a state suitable for physicochemical or bioremediation.

Mining and milling wastes

Wastewaters arise naturally from mine drainage waters and run-off. The composition of a typical wastewater is shown in Table 1a. Wastes arising from the extraction of uranium from ores ('mill tailings'; see Chapter 8) are a particular class of low-level waste in which the radioactivity is naturally occurring and long-lived, comprising mainly ^{238}U with small amounts of ^{235}U. Thorium is an additional component of uranium bearing rock strata which is less of a problem per se but, as noted earlier, its decay products can be a serious issue. The products of U decay comprise isotopes of Th, Ra, Po, Bi and Pb, with widely varying half-lives (Flowers et al., 1986), although the mass of the radioactive components is usually small in comparison with that of the non-radioactive species. The composition of a typical Ra-containing tailings stream is shown in Table 1b.

A major problem occurs with disused mines which continue to discharge drainage waters into the environment. The primary consideration is prevention of release of uranium but the minor radioactive species must also be considered. Table 1a shows that the major anion present is sulfate, produced mainly via the activity of acidophilic bacteria such as *Thiobacillus ferrooxidans*. The pH of such waters is typically outside the physiological pH range of most microorganisms and they also usually contain substantial concentrations of species which may influence microbial activity or metal speciation. In addition to the species shown in Table 1a, mine wastes may contain other metals such as selenium and molybdenum which are not removed by the ion exchange treatment processes which are often used to reduce the uranium content of the water (Kauffman et al., 1986). Table 1b also illustrates that the low mass concentration of the radioactive species may be problematic. $RaSO_4$, like $BaSO_4$, is highly insoluble but, even in the presence of the high concentration of SO_4^{2-}, significant Ra^{2+} remains in solution. One large-scale method for treating mining and milling wastewaters consists of adding lime to raise the pH into the 7–9 range (Tsezos & Keller, 1983) This chemical treatment precipitates most contaminants except ^{226}Ra, which persists. Another current treatment for ^{226}Ra is coprecipitation as barium radium sulfate by addition of $BaCl_2$ to sulfate-rich effluents. However, as with most coprecipitation methods, settling ponds are required, with residence times of several weeks. Effluent from these ponds often results in substantial releases of ^{226}Ra to surface water courses (Tsezos & Keller, 1983). Possible approaches to the bioremediation of ^{226}Ra have been described previously (Lloyd & Macaskie, 2000).

In addition to drainage waters and tailings wastes, some liquors are produced from the leaching of low grade uranium deposits with the aim of improved uranium extraction from the orebody. For example, an economic in situ bioleaching process which involves

Table 1a
Composition of a typical uranium mine wastewater

Analyte	Concentration (mg l^{-1})	Concentration (mM)
SO_4^{2-}	3339	34.8
NH_4^+	10.6	0.59
Mg	639	26.3
Ca	499	12.4
Mn	79.3	1.44
Na	78.7	3.42
UO_2^{2+}	38.6	0.14
Al	29.9	1.11
Fe	12.9	0.23
Zn	4.37	0.066
Ni	2.90	0.049
Cu	0.17	0.0027

The minewater was obtained from ENUSA (Spain) in October 1993 and was analysed by Dr. M. G. Roig (University of Salamanca, Spain) (Macaskie et al., 1997). The pH was 3.5.

Table 1b
Composition of a typical radium-loaded wastewater

Analyte	Concentration (mg l^{-1})
SO_4^{2-}	1393
NO_3^-	76
Fe	0.3
Zn	0.03
UO_2^{2+}	0.25
Cu	0.02
Pb	0.06
Mn	0.32
Ra	8×10^{-7}

Data from Tsezos et al. (1987).

the bacterial oxidation of pyrite and generation of sulfuric acid (here pH 1–2) to solubilise the uranium in the ore was developed at Denison Mines in Canada (McCready & Lakshmanan, 1986). In this case, the uranium solution concentration was 200–500 mg l^{-1} but the solution was held underground; the large volume and associated pumping costs make concentration in situ desirable. In this last example, uranium recovery, and not environmental protection, is the main focus but these examples illustrate that a biological process should be sufficiently versatile to remove uranium over a wide range of target concentrations, often from aggressive solution matrices, and should also preferably remove co-contaminating trace nuclides.

Wastes from nuclear fuel reprocessing

The composition of a typical unreprocessed fuel is shown in Table 2a, while the reduction factors required for final elemental releases to the environment are shown in Table 2b. Spent nuclear fuel is often reprocessed and most nuclear fuel reprocessing plants employ variants of the PUREX process (Swanson, 1990). Spent fuel rods are dissolved in nitric acid to solubilise the metals. Uranium and plutonium are coextracted into a solution of tributyl phosphate in odourless kerosene, leaving fission products and other impurities in the aqueous phase. Neptunium is poorly extracted and its persistence in aqueous waste streams can be problematic. Pu is selectively reduced chemically or electrochemically; reduced Pu, and recovered U, are back-extracted separately into aqueous streams. The U is still enriched in ^{235}U with respect to natural U and is reused in fuel fabrication, while recovered Pu can be used in mixed oxide fuel for some types of nuclear reactor. Effluent streams consist of byproducts from these extractions, with the highly active waste fission products remaining in the acidic aqueous waste from the first partition. Further wastes are produced during cleanup and recycling of the TBP solvent; for example, actinide concentrations of 0.5– 5.0 kg per tonne of fuel processed may be present in the Na_2CO_3 solutions used to clean up TBP (Horwitz et al., 1980). Although solvent extraction of actinides adds greatly to the total waste volume, it is still viewed as preferable to alternative technologies (McKay et al., 1990). Organic ion exchange resins are unsuitable for high levels of radioactivity; radiologically degraded resin cannot be regenerated and spent resin would form a substantial waste. Inorganic ion exchangers, while radiostable, can react explosively with nitric acid. Natural zeolites are radiostable, but tend to be selective for caesium and strontium and so are used specifically to remove these elements. High-level wastes are stored to await vitrification and final disposal. Examples of typical low-level liquid wastes are given in Table 3, and those which are currently discharged into environmental waters are therefore among the most problematic wastes at present.

Wastes from nuclear decontamination operations: the problem of chelated wastes

The discussion above shows that some wastes will contain carbonate ions, which readily form complexes with several actinide elements. In some cases organic chelating agents are used deliberately in order to extract metals. They are also used in nuclear reactor decontamination (Means et al., 1978; Riley et al., 1992; Jardine et al., 1993), and have been discharged along with metal species. There is a wealth of information available from US Department of Energy (DOE) sites regarding the nature of and current problems with nuclear wastes (Banaszak et al., 1999). There are many examples where chelating agents used in nuclear fuel processing were co-disposed with the actinide contaminants. The most common chelates found on US DOE lands include NTA, EDTA, citric and oxalic acids (Riley et al., 1992), while liquid wastes (Reed et al., 1998) and some high level wastes (Delegard et al., 1984) also contain complexing ligands. Enhanced actinide

Table 2a

Typical activity at discharge for pressurised water reactor unreprocessed fuel

Nuclide	Half-life (years)	Type of decay	Activity at discharge[a] (TBq/GW[e]y)	% of total activity[b]
^{90}Sr	28.5	β	9.6×10^4	24
^{93}Zr	1.5×10^6	β	3.87	
^{99}Tc	2.1×10^5	β	18.3	
^{129}I	1.6×10^7	β	0.0338	
^{139}Cs	2.0×10^6	β	0.42	
^{137}Cs	30.1	β	1.32×10^5	33
^{234}U	2.4×10^5	α	0.0129	
^{235}U	7.0×10^8	α	0.0245	
^{238}U	4.5×10^9	α	0.412	
^{237}Np	2.1×10^6	α	0.304	0.00014
^{239}Pu	2.4×10^4	α	438	0.1
^{240}Pu	6540	α	663	0.2
^{241}Pu	14.4	β	1.7×10^5	33
^{242}Pu	3.8×10^5	α	2.65	
^{241}Am	433	α	170	0.7
^{243}Am	7400	α	22.5	0.006

[a] Inventory data adapted from Flowers (1986).
[b] Calculated from data presented in Oversby (1987).

Table 2b

Reduction factors for elemental releases[a]

Element	Reduction factor
Americium	18,300
Plutonium	12,300
Thorium	457
Uranium	52
Neptunium	46
Radium	38
Nickel	13
Zirconium	5
Technetium	3

[a] From Oversby (1987). The data are reduction factors for maximum US Regulatory Commission (NRC)-allowed release. This is the amount by which their release at the value given for the Engineered Barrier System Performance Objective on Release Rate exceeds the amount that could be released into the accessible environment (Oversby, 1987).

Table 3a

Composition of a typical low active waste stream (volume 1000 m^3 day^{-1}; acidity approx. 300 mM as HNO_3)

Non-metal, inactive and trace active species	Concentration ($mg\ l^{-1}$)[a]	Total discharge ($kg\ d^{-1}$)[b]
NO_3^-	17,000	17,000
SO_4^{2-}	500	500
Tributyl phosphate	30	30
Kerosene	30	30
Fe	150	150
Na	400	400
K	30	30
Ca	10	10
Mg	10	10
U	10	10

Active species	Concentration ($mg\ l^{-1}$)[a]	Total discharge ($mg\ d^{-1}$)[b]	Concentration ($kBq\ l^{-1}$)[a]	Total discharge ($GBq\ d^{-1}$)[b]
^{106}Ru	3.2×10^{-6}	3.2	410	407
^{144}Ce	3.1×10^{-7}	310	37	37
^{90}Sr	8.3×10^{-6}	8.3	44	44
^{95}Zr	1.0×10^{-7}	0.1	2.2	2.2
^{95}Nb	2.3×10^{-7}	0.23	81	81
^{137}Cs	4.1×10^{-5}	41	133	133
^{99}Tc	2.9×10^{-2}	29,000	19	19
Pu-α	1.4×10^{-2}	14,000	63	63
^{241}Pu	6.4×10^{-4}	640	2370	2370
^{241}Am	4.8×10^{-5}	48	6.3	6.3
^{237}Np	8.5×10^{-2}	85,000	2.2	2.2
^{60}Co	3.9×10^{-7}	0.39	16	16
^{134}Cs	6.3×10^{-7}	0.63	31	31

[a] Data adapted from Ashley et al. (1987).
[b] Authors' calculations.

migration is associated with organic complexants (Nelson et al., 1989; Choppin, 1992; Dozol et al., 1993; Baik & Lee, 1994; Silav & Nitsche, 1995). For example, in at least two radioactive waste burial sites (Oak Ridge National Laboratory, Tennessee and Maxey Flats, Kentucky), radionuclides were found up to several hundred metres away from the original burial trenches. In each case, EDTA-enhanced nuclide migration was implicated (Means et al., 1978; Cleveland & Rees, 1981; Means and Alexander, 1981). These studies

Table 3b

Composition of a typical fuel rod storage pond stream (volume 1000 m^3 day^{-1}; acidity: pH approx. 12)

Non-metal and inactive metal species	Concentration ($mg\ l^{-1}$)[a]	Total discharge ($kg\ day^{-1}$)[b]
NaOH	1400	1400
$NaNO_3$	120000	120000
Na_2CO_3	150	150
Tributyl phosphate	80	80
Kerosene	130	130

Trace metals (all isotopes)	Concentration ($mg\ l^{-1}$)[a]	Total discharge ($mass\ day^{-1}$)[b]
Sr	1.4×10^{-8}	14 μg
Tc	1.8×10^{-4}	180 mg
Ru	4.6×10^{-5}	46 mg
Cs	4.3×10^{-8}	43 μg
Ce	5.5×10^{-7}	550 μg
U	0.34	340 g
Pu	1.4×10^{-5}	14 mg
Np	4.3×10^{-4}	430 mg

Active species	Concentration ($mg\ l^{-1}$)[a]	Total discharge ($mg\ d^{-1}$)[b]	Concentration ($kBq\ l^{-1}$)[a]	Total discharge ($GBq\ d^{-1}$)[b]
^{90}Sr	7.5×10^{-9}	0.007	0.04	41
^{99}Tc	1.8×10^{-4}	180	0.11	110
^{106}Ru	1.1×10^{-7}	0.11	13.3	13300
^{129}I	3.9×10^{-13}	3.9×10^{-7}	0.29	290
^{134}Cs	5.3×10^{-10}	5.3×10^{-4}	0.02	20
^{137}Cs	1.8×10^{-8}	0.018	0.06	60
^{144}Ce	9.7×10^{-10}	9.7×10^{-4}	0.11	110
^{235}U	0.10	100000	0.008	8
^{241}Pu	1.2×10^{-6}	1.2	5.2	5200

[a] Data adapted from Ashley et al (1987).
[b] Authors' calculations.

are of more than just academic importance; the Savannah River Plant (South Carolina) commercial burial site bears many hydrological and climatic similarities to the Maxey Flats Site; in particular, the depth to the nearest aquifer is only 10–20 m (Means and Alexander, 1981).

Clearly, wastes containing chelating agents should be treated prior to disposal, but they are stable in water at temperatures up to about 220-260°C. EDTA (but not NTA) is sensitive to photolysis and radiolysis, and the existing literature suggests that EDTA and its homologues can be decomposed either thermally or chemically (Means & Alexander 1981). For large volumes of waste this may not be practical, and there is clearly scope for biological processes involving biological degradation of these chelating agents. An early review (Macaskie, 1991) highlighted the paucity of information regarding the biodegradation of chelating agents. During the last decade this problem has been addressed. The biodegradation of NTA is now well established and the processes of EDTA biodegradation are becoming better understood (see below). In contrast to the man-made chelating agents, citrate has also received attention because, although it binds metals less avidly, it is a naturally produced chelating agent which is degraded by many microorganisms. However, the presence of the bound metal ion may make the complex more recalcitrant due to reduced recognition, not increased toxicity (Brynhildsen & Allard, 1994; JoshiTope & Francis, 1995). The current status of research into biodegradation of chelating agents is reviewed in a later section.

2. The scale of industrial problems

Liquid wastes are generally presented in one of two ways. The simplest is the production of discrete batches with on-site storage facilities. Here simple precipitation methods are applicable and settling out techniques can be used. This may be less feasible if the metals are held as soluble complexes or chelates (see above) where the concentration of free metal can be very low. However, many operations utilise continuous processes with little storage facility, resulting in the continuous production of high volume, often low activity, wastes which require continuous treatment. For example, wastewater discharges from tailings in the Elliot Lake area of Canada are 2.5×10^4 m^3 day^{-1} (Tsezos & Keller, 1983). For mine run-off, although the radioactivity is low, the total metal requiring treatment may be several hundred milligrams per litre (Tsezos et al., 1989). With in situ leaching operations (McCready and Gould, 1990), a mine was flooded over 3 days; the solution was allowed to react in the mine for 3 weeks before repeating the drain-flood cycle. The drain flow to be treated was approximately 6500 m^3; the drain time was not stated, but the need to concentrate the uranium underground in situ (Tsezos et al., 1989) suggests that, in addition to pumping costs, the speed of uranium recovery may be an economic consideration.

For uranium recovery operations, large total quantities of metal may be handled but the recovered metal is of low specific radioactivity. Nuclear fuel reprocessing streams are another matter. The quantities may be large, up to about 1000 m^3 day^{-1} (Tables 3a and b), but here the waste is a complex 'cocktail' of fission and activation products and long-lived actinide isotopes, together with inactive metals and other species. With isotopes such as ^{241}Pu, the specific activity is high and although the solutions are dilute, substantial quantities of radioactivity may be discharged daily. In terms of mass concentration, stable metals and those of low specific activity are present in large excess over the high specific activity radioactive species and an efficient process would be required to remove short-lived actinides against the background of low active uranium and stable elements such as Fe. The chemical behaviour of Fe and actinides is similar in some respects and a selective accumulation technique (or a high total capacity one) would be required. Even without competing Fe, the problem of a 100-fold excess of U remains, and the need for a highly selective transuranic element accumulation technique is therefore evident.

Another problem is the presence of large quantities of anionic species, which may form complexes with metals and retain these as mobile ions. Anion removal (nitrate in the case of fuel reprocessing wastes, sulfate in the case of minewaters) may be needed prior to a final biological 'polishing' step. Also, the pH may be low. While a minor problem in chemical engineering, this is an important consideration for processes designed to operate at physiologically permissive pH values, and neutralisation (or the use of acidophilic microorganisms) may be required. In some cases, microorganisms are surrounded by polysaccharidic materials which contain ionisable functional groups such as –COOH and can act as localised buffers in the same way as organic acids in the bulk solution. Uranium minewater remediation is a particular example where, although the pH of the water may be as low as 3.5, uranium was deposited on a film of biomass which was surrounded by a polymeric matrix synthesised by the organism. By contrast, the mediating biochemical reaction was non-functional at this low pH in free, non-enmeshed cells (Macaskie et al., 1997).

Prediction of the likelihood and ease of bioremediation is difficult, requiring understanding of the behaviour of both the biomass and the target metal species under conditions which may not be homogeneous within the reactor system, or which may involve unstable metal species or short-lived isotopes. However, modelling is a useful aid in optimising contactor systems and in obtaining an estimate of process costs and longevity. It is essential to consider the chemistry of the target transuranic elements carefully, since the chemical properties of any one will depend on the pH, redox potential, presence of other metals or transuranic elements (and their concentrations), salts, chelating agents, light (and its wavelength), and also the intrinsic radioactivity of the solution, the type of radioactivity, and radiolytic effects on the other species present (Banaszak et al., 1999). A particularly good example of harnessing photochemistry to facilitate remediation was given by Dodge & Francis (1994) where photodegradation of uranium-citrate complexes permitted uranium recovery, circumventing the need for citrate biodegredation to release the metal. Such potential synergism of chemical and biotechnology has been neglected in the literature.

3. Toxicity of radioactive waste

The acidic nature of some mining wastes may pose a problem for existing biotreatment processes, many of which utilise neutrophilic bacteria. Wastes from fuel processing and reprocessing can also be presented at extremes of pH, yielding high salt content wastes after neutralisation. Dilution of wastewater to a physiologically compatible composition is possible, but increases the volume of the waste, perhaps unacceptably. In this context, it may be useful to investigate the application of extremophilic microorganisms. The radiotoxicity of effluents containing high activity radionuclides may also adversely affect the microbial component of a bioprocess. However, the large microbial gene pool may provide novel radioresistant organisms with very efficient DNA repair mechanisms.

For example, bacteria of the genus *Deinococcus* are able to withstand the normally lethal mutagenic effects of DNA-damaging agents, particularly ionising radiation. The biochemical basis for this phenomenon, and the potential use of these organisms to treat radiotoxic waste, are areas of intense research activity (see later).

Most wastes from mining and fuel reprocessing contain a large background of non-fissile residual ^{238}U. The acquisition of heavy deposits of uranium around cells accumulating U(VI) as the phosphate or as U(IV) oxide (see below) seems not to interfere with continued removal. The non-penetrating α-emissions of the transuranic elements contrast with the moderately penetrating β-emissions of ^{99}Tc and ^{241}Pu and highly penetrating γ-rays from nuclides such as ^{60}Co and ^{241}Am. Radioactivity lethal to cell viability may not inhibit bioremediation where only one or a few enzymatic step(s) are required; for example, the phosphatase activity of *Citrobacter* sp. accumulating uranyl phosphate was highly radioresistant after complete loss of cell viability (Strachan et al., 1991), while hydrogenase-mediated ^{99}Tc removal by *Escherichia coli* or *Desulfovibrio desulfuricans* (see Chapter 11) continued up to high metal loads and very high local β-activity. Finally, other organic contaminants, such as TBP or odourless kerosene, may add to the chemical toxicity of the effluent but could be utilised as energy sources by bacteria and, indeed, may have to be removed in order to convert the target metals into available forms. In general, there have been few attempts to develop versatile technologies for the treatment of aggressive wastes although several possible systems in the analogous area of metal bioremediation of geothermal wastes have been described (Premuzic et al., 1997).

4. The problematical actinide species: salient features of actinide chemistry

In contrast to mine waters (high volume, relatively low metal concentrations), nuclear fuel reprocessing waste streams (lower volume, higher concentrations of Fe and U; Macaskie, 1991) contain long-lived isotopes of transuranic elements such as americium, neptunium and plutonium, as well as various fission products. Although the transuranic elements can exist in the hexavalent state, in practice the major environmental contaminants are commonly tetravalent (plutonium), trivalent (americium) or pentavalent (neptunium) and studies on uranium bioremediation, while a useful model, cannot be extrapolated directly. The metal oxidation state, and hence metal speciation and amenability to physicochemical or bioremediation, are dependent on the physicochemical and radiochemical properties of the carrier solution. For example, the presence of chelating ligands can stabilise otherwise unstable species such as Pu(V) and promote reduction of hexavalent Pu and Np species (Reed et al., 1998), while the plutonyl (VI) ion (PuO_2^{2+}) readily undergoes photochemical (Toth et al., 1980; Bell et al., 1983) or chemical (Allard & Rydberg, 1983) reduction to Pu(IV). At low pH and under reducing conditions, Pu(III) forms (Allard & Rydberg, 1983); this can also occur photochemically. Conversely, radiolytic oxidation of Am(III) and Pu(IV) has been suggested to occur in saline solution (Buppelman et al., 1986). It is clear that study of transuranic element remediation is far from straightforward and the first steps should employ a reasonably well-defined model system to establish basic parameters. Detailed consideration of actinide chemistry is outside the scope of this chapter, but has been discussed thoroughly by Banaszak et al. (1999).

5. Microbiological approaches to heavy metal removal

New microbially based technologies can find applications at the point where traditional physicochemical treatments fail. For example, chemical precipitation of heavy metals is convenient and economic but is not applicable to the very low residual nuclide concentrations of typical low level wastes. The resulting wet sludge is difficult and hazardous to handle and transport and also requires compaction for final disposal. Methods such as reverse osmosis are effective but expensive. Various separation and ion exchange technologies are available but are are usually costly. Natural zeolite ion exchangers are a cost-effective option but a ready supply of the zeolite must be secured for the projected lifetime of the plant. Extraction using zeolites, although cheap and effective, is subject to the same constraints as other ion exchange processes: a generally low metal selectivity, sensitivity to low pH and the presence of co-contaminating components, often to excess, in the solution. Biologically based processes are attractive if they can overcome one or more of these constraints and can be operated economically, effectively and reproducibly. Many potentially useful biomaterials have been identified but very few have been evaluated with respect to the above criteria. The radiotoxicity and longevity of residual nuclides require a highly effective means of removal but most proposed bioprocesses have focused on uranium, with only scant attention being paid to the removal of the transuranic elements whose chemistry and relatively low concentrations often make them more difficult to remove. In addition, the presence of fission products, such as ^{99}Tc, poses additional problems due to the formation of anionic species, such as the pertechnetate ion (TcO_4^-), which are not removed by cation-specific extraction mechanisms. Electrochemical treatments are often feasible, e.g. reduction of Tc(VII) to insoluble Tc(IV), but fouling of the electrode and the need for metal removal to regenerate the electrode are likely to be problematic. Large-scale electrochemical reactors of this type are not well developed and the size of the electrode required may ultimately be limiting. The main approaches to biological metal removal are biosorption, bioaccumulation and biomineralisation. The biochemical bases of these processes are described in Chapter 11, with additional factors pertinent to the bioremediation of radioactive waste discussed below.

Biosorption

For cation removal, dead biomass has potential as a metal sorbent. Biosorption occurs via interactions with various ligand groups associated with the surfaces of microbial cells. These essentially chemical reactions, subject to the same constraints as ion exchange technology, have formed the subject of many investigations and are summarised in numerous reviews (e.g. Volesky, 1994; Volesky & Holan, 1995; Gadd, 1996, 1997). A primary consideration is the economics of biosorbent production; ideally this would be as a waste from another process, e.g. spent brewer's yeast has been used for biosorption of uranyl ion (Omar et al., 1996, Riordan et al., 1997), while metal-accumulating fungal biomass such as *Aspergillus* is available in large quantities from commercial fermentations (Yakubu & Dudeney, 1986). Some algal biomasses (e.g. *Sargassum natans*: Kuyucak & Volesky, 1989) have commercial potential but require harvesting from the source. With some notable exceptions, the biosorption capacity of biomass is often low and, in general, such processes have not yet been developed to the point of general commercial application although examples do exist where a particular waste has been treated by biosorption (e.g. McCready and & Lakshmanan, 1986; Tsezos et al., 1987a, b, c; 1989). Biosorption of uranium has been reviewed recently (Lloyd & Macaskie, 2000) and, since then, further studies have added to the portfolio of potential uranyl biosorbents: *Myxococcus xanthus* (Gonzalez-Munoz et al., 1997), and *Sargassum* (Yang & Volesky, 1999a). In the case of *M. xanthus*, uranyl uptake was mediated by the cells and also by the extracellular polymeric matrix, giving an approximately 3-fold greater capacity than previously reported U-sorbing biomass such as *Rhizopus arrhizus* (Gonzalez-Munoz et al., 1997). A similar uptake (~ 50% of the biomass dry weight) was reported in the case of *Sargassum fluitans* (Yang & Volesky, 1999a). Uranyl uptake (a proton-exchange mechanism: the hydrolysed ion exchange model) was modelled in terms of the speciation of uranium and its hydrolysed species and the pH of the solution matrix (Yang & Volesky, 1999b). These two biomasses look particularly promising for future process development and it will be interesting to see how they function under realistic waste conditions, and whether the removal of other hexavalent actinide species can be achieved in a similar manner to uranyl ion.

Bioaccumulation and radioresistant microorganisms

Living microbial cells can accumulate substantial quantities of heavy metals but the metals (or, indeed, co-contaminants) can be toxic. Furthermore, living biomass usually requires physiologically compatible carrier solutions and biomass growth may be unavoidable, adding to the organic load of the final solid waste and hence the cost of its transportation to the site of reprocessing or final burial. In the case of radionuclides, the radiotoxicity is an important consideration. Experiments using different isotopes of the same actinide (^{238}Pu and ^{239}Pu: i.e. with the same chemical properties) clearly demonstrate radiotoxic effects (Reed et al., 1999). A large amount of work has established the effects of radiation upon living cells (e.g. Coggle, 1983). Radiation damage occurs via the effect of oxidising species from the radiolysis of water (the hydroxyl free radical (OH·), the hydroperoxyl free radical ($HO_2\cdot$), hydrogen peroxide (H_2O_2) and $\cdot O_2^-$). Oxidative damage results in cell death and mutations. The way in which the cells are grown has profound effects on their radioresistance. For example, oxygen sensitises bacteria to radiation damage by increasing the net yield of oxidising radicals (Coggle, 1983). This is thought to occur via a fixation process which involves the combination of oxygen with free radicals produced in the target molecule to produce peroxy-radicals; this 'fixation' of radiation damage occurs within 10^{-2} to 10^{-3} μs (Coggle, 1983). By contrast, the presence of chloride ion (a free radical scavenger: Spinks & Woods, 1976) enhances radiation tolerance so halophilic bacteria (those which prefer or require a high salt content in the growth medium) such as *Halobacterium halobium* are relatively radioresistant (Banaszak et al., 1999). Thiol groups (e.g. the sulfhydryl-containing cysteine and glutathione) also have a free radical sequestering function. Here, radioprotection is thought to be via a transfer of H atoms from the sulfhydryl compound (M-SH group) to the biological free radical ($R^\bullet$) which occurs in the target molecule as a result of attack by the chemical free radical species (above). The $R^\bullet$ is thus repaired by conversion to RH. In the case of the radionuclides found in typical wastes (Tables 1–3), although the mechanism of radiation damage is ultimately the same, the 'quality' of the radiation is responsible for the actual quantity of damage sustained. The 'relative biological effectiveness' (RBE) of a type of radiation is expressed relative to a dose of a standard type of radiation (Coggle, 1983). The factor having the greatest influence on the RBE of a given type of radiation is the distribution of ionisations and excitations described by the term linear energy transfer (LET), which is expressed in terms of the mean energy release (keV per micron of the tissue traversed). The LET is a function of the nature, mass, velocity and charge of the ionising radiation. Alpha particles, neutrons and protons are high LET radiations while X-rays, γ-rays and fast electrons are low LET radiations. To obtain the biologically effective dose for different types of radiation the concept of 'dose equivalence' takes into acccount the differences in the relative effectiveness of different radiations. The Sievert (Sv) is the unit of dose equivalence and is numerically equal to the dose in Grays (Gy: defined as 1 joule per kg, = 100 rad: Coggle, 1983) multiplied by the 'quality factor' of the radiation (Q). Table 4 shows the quality factors that have been specified by the International Commission on Radiological Protection and the associated LET values. What this means in terms of the specimen waste shown in, for example, Table 3a, is that the α-emitting species of, for example, Pu and Np present more of a hazard in terms of the 'quality' of the dose than the gamma emitters such as ^{60}Co. Gamma rays are highly penetrating and so deposit energy along their entire path, whereas alpha particles have little penetrating power and so cause a very high amount of local damage, typically restricted to a few tens of μm in tissues. These factors are of critical importance in consideration of the radiotoxic effects. For example, an assessment of radiotoxicity, made using a ^{60}Co γ-source, will be difficult to extrapolate to irradation by the α-emitting transuranic elements. Cobalt, as an essential metal in metabolism, is likely to have specific mechanisms for cellular uptake and sequestration, while the ability of microbial cells to take up actinide elements is not clear (Strandberg et al., 1981; Marques et al., 1991). It was proposed recently that intracellular accumulation of uranium was attributable to increased membrane permeability caused by uranium toxicity and was transport-independent (Suzuki & Banfield, 1999). Thus, assuming cell integrity is not compromised, little cellular accumulation of transuranic elements should occur. The radiotoxicity would be expected to be lower when a dissolved plutonium complex is the source of ionising radiation than when the Pu atom is located in/on the cell, where there is a higher probability of direct interaction between the ionised particle track and the cell components (Banaszak et al., 1999). In practice, this means that, when the concentration of an actinide is sufficiently high for the cellular permeability barrier to be compromised, there will be a greater radiotoxic effect than in the case of intact cells. In a real situation, even though the concentration of, for example, plutonium is low, the presence of chemically toxic concentrations of uranium may enhance access of Pu to the cell. Even if the cells remain intact, the various biosorptive mechanisms will tend to associate the actinide element with the cell surface. This concept was confirmed by Reed et al. (1999) who showed that with the NTA-degrading bacterium *Chelatobacter heintzii*, the observed loss of cell viability was much greater when α-emitting plutonium was the source of ionising radiation, rather than gamma radiation; this was attributed to bioassociation of Pu in the system.

Metal accumulating microorganisms, on the whole, have no great radiotolerance. On the other hand, radiation resistant bacteria have been isolated. For example strains of *Deino-*

Table 4
The dependence of quality factor (Q) on LET for some radiations

LET in water (keV μm^{-1})	Quality factor	Radiation
≤3.5	1	X-rays, γ-ray or electrons
7	2	X-rays, γ-ray or electrons
23	5	Protons, neutrons
53	10	Protons, neutrons
≥175	20	α-particles

From Coggle (1983).

coccus radiodurans can withstand doses of 5000–30,000 Gray without loss of viability (Binks, 1996), attributed to highly effective DNA repair mechanisms (Banaszak et al., 1999) although nutritional factors play an important role (Venkateswaran et al., 2000). Under nutrient-limiting conditions, DNA repair was limited by the organisms' metabolic capabilities. Radiation resistance was highly dependent on an abundant exogenous amino acid source. While these amino acids are not specifically required for growth, some exogenous amino acids are necessary as transamination substrates since the organism cannot assimilate inorganic nitrogen (Venkateswaran et al., 2000). Recent work has focused on the potential for *D. radiodurans* as a host for genes encoding useful bioremediation properties. Thus, Lange et al. (1998) successfully cloned and expressed genes imparting the ability to oxidise toluene, chlorobenzene, 3,4-dichloro-1-butene and indole, with synthesis of toluene dioxygenase in an environment of 60 Gy/hour using ^{137}Cs. Application to heavy metal remediation was pioneered by Brim et al. (2000) who cloned the Hg(II) resistance gene (*merA*) from an *E. coli* strain. The mercury resistance system is well understood, with mercuric reductase-mediated bioreduction of Hg^{2+} to elemental Hg^0 which is volatile. This reductive route to mercury removal has not been exploited extensively in bioremediation because of the need for a trapping mechanism for Hg^0, which has been achieved recently using a two-phase bioreactor system for bioremediation of an industrial wastewater.

There is good potential for the approach of cloning of metal resistance genes into radiation-resistant organisms. Metal resistance genes conferring other functions leading to metal deposition are well known, e.g. the plasmid-encoded metal efflux genes of *Ralstonia eutropha* (formerly *Alcaligenes eutrophus*: Diels et al., 1995), but as yet no specific genetically-encoded actinide resistance mechanisms have been identified. A recent study has also shown that *D. radiodurans* catalyses the reduction of Fe(III) NTA and the humic analogue anthraquinone-2,6-disulfonate (AQDS), while the reduced form of AQDS can reduce U(VI) and Tc(VII) abiotically. *D. radiodurans* was, however, unable to conserve energy for growth via these processes (Fredrickson et al., 2000). Thus, although this organism shows considerable promise for ex situ treatment applications, it will probably be of limited use for in situ treatment of radionuclides in contaminated environments, where it is unlikely to compete effectively with specialist dissimilatory metal-reducing bacteria.

With respect to biodegradation of organic species, studies have concentrated on organic solvents often found in radioactive mixed wastes, largely because these systems are well understood and appropriate genes are identified. The understanding of biodegradative pathways for the problematic chelating agents is still poor (see below) but the ability to clone these into *D. radiodurans* would greatly facilitate remediation of radioactive wastes. While metals can be removed via biosorption or biomineralisation reactions using dead or resting cells (see below), biodegradation of organics requires complete metabolic pathways and hence necessitates a live cell background for expression. Progress on our understanding of the breakdown of chelating agents is reviewed in a later section.

Biomineralisation

Potentially useful hybrid technologies that could overcome many of the above problems are based on metabolically mediated generation of ligands (or reducing power) to promote the formation of insoluble metal deposits, which remain attached to the microbial cells via a 'scaffold' function of cell surface materials. This is termed biomineralisation and it is common for the deposited metal mineral to exceed the biomass dry weight by several fold. Concepts of biomineralisation are discussed by Mann (1988, 1993, 1997) who says that: 'the traditional view of inorganic solids as 'condensed matter' is currently being reshaped to accommodate the notion of organised matter chemistry. This transformation represents a shift in emphasis away from solid phases as thermodynamic states of consolidated matter, towards solids as organisational states determined by local (rather than global) energy minima and constructional mechanisms' (Mann, 1997). Possible mechanisms leading to biomineralisation have been discussed elsewhere both generally (White et al., 1995; Schultze-Lam et al., 1996) and with specific emphasis on the possible treatment of radionuclides (Francis, 1994, 1995; Barton et al., 1994; Macaskie et al., 1994, 1995, 1996; White et al., 1995; McClean et al., 1996).

Precipitation of solids from solution can occur once the solubility product has been exceeded. This critical value is dependent on the product of the concentrations of the metal and ligand species, regardless of whether the two reactant species are present at equal or different concentrations. Examples of precipitant ligands are the sulfide, phosphate and carbonate ions. Bioproduction of these for metal remediation is discussed later, and a description of the biochemical factors underlying formation of these ligands is given in Chapter 11. As the concentration of ligand increases, the concentration of metal required for precipitation decreases, so continuous production of ligand will drive metal precipitation at low concentrations, an important consideration where the target metal is in a very dilute solution. However, in practice, precipitation may not occur. The initial precipitation event requires the overcoming of an initial 'energy barrier', which can lead to the phenomenon of supersaturation. In this case, the product of the metal and ligand species gives a value which exceeds the solubility product but precipitation only occurs after the solution has become supersaturated to a critical point where the energy released during precipitation is greater than that needed for the initial nucleation of the solid (Morel et al., 1993). The presence of pre-formed nuclei can decrease the amount of supersaturation required to initiate precipitation by lowering the energy barrier. The nuclei can be preformed deposits of the same substance or can be an entirely different type of solid if it has a structure compatible with that of the precipitating species. In biological systems, the nucleating surfaces can be organic templates (Mann, 1997). The degree to which the nucleation free energy barrier is reduced is related to the structural similarity between the nuclei and the precipitate; the more similar they are, the lower the energy barrier to precipitation. Thus, pre-deposition of a small amount of the desired precipitate can facilitate continuous removal as long as the precipitant ligand is applied, until the residual free metal concentration reaches a value consistent with the solubility product.

Microorganisms can facilitate chemical precipitation in several ways. Cell surface components can be used as heterogeneous nucleation sites (McLean et al., 1996; Schultze-Lam et al., 1996). In this case the cell surface acts to reduce the free energy barrier and provides a convenient surface for initiation of precipitation. Once nucleated, the metal precipitate forms a focus where the energy barrier is low with respect to both the bulk solution and also the rest of the cell surface so deposition of more precipitate will occur at this site. This has the effect of locally depleting the concentration of metal, so more metal will diffuse to the site down the ensuing concentration gradient and the bulk solution will gradually become depleted.

Mann (1988) points out that the form of the metal deposit (its spatial and temporal morphological features) is contingent on the spatial and chemical modification of growth processes (see below), but is also driven by preorganised biological structures such as proteins and membrane vesicles (e.g. Ferris & Beveridge, 1986a). Early studies using deuterium and ^{31}P nuclear magnetic resonance examined the interaction of metal ions with phosphatidylcholine bilayer membranes, showing metal binding to the polar head groups, specifically to the polar phosphate head groups (Akutsu & Seelig, 1981). Mann et al. (1986) extended this concept to the use of phosphatidylcholine-based phospholipid vesicles as a model for biomineralisation. Here, Fe(II) and Fe(III) ions were trapped within the space bounded by the vesicles. The encapsulation of the crystallochemical reactions allowed control of chemical regulation (ion transport into the vesicles), the localised supersaturation levels, the stereochemical requirements for ion binding, redox and nucleation events at the organic matrix interface and the spatial organisation of crystal growth and morphology. These early in vitro tests illustrated the important role of membrane phospholipids in biomineralisation. Later work used membrane vesicles (liposomes) with entrapped phosphatase to drive the biomineralisation of metal phosphates using glycerol 2-phosphate as the substrate and source of phosphate for biomineralisation of metal phosphate which remained attached to the liposome (Jeong et al., 1997). The formation of inorganic phosphate was followed using ^{31}P NMR, with electron opaque material identified as uranyl phosphate on the surface of the liposome. It is not clear how the bulky glycerol 2-phosphate substrate and uranyl cation were able to enter the liposome and gain access to the entrapped enzyme but some radial discontinuities in the liposome were visible which may have provided access channels to the entrapped enzyme (Jeong et al., 1997). This model system suggested that the phospholipid membrane surfaces of whole cells may provide a nucleating surface for formation of metal phosphates, and uranyl ion deposition on to the surfaces of the membranes of Gram negative bacteria was observed (Strandberg et al., 1981; Marques et al., 1991; Jeong et al., 1997).

Where enzymatic activity generates the precipitant ligand, the enzymatic step(s) leading to sustained metal biomineralisation can often be decoupled from microbial growth, allowing metal removal from hostile solutions; the process constraint is that of the metal accumulating step(s) and 'resting' (growth-decoupled) or even non-viable biomass can be used. One example of biomineralisation occurs via a metal detoxification reaction in *Ralstonia eutropha* (formerly *Alcaligenes eutrophus*). Here, a plasmid-encoded metal efflux from the cell occurs in antiport to proton uptake and the resulting localised exocellular alkalinisation causes the precipitation of metal hydroxides and carbonates (Diels et al., 1995; Taghavi et al., 1997). When immobilised in a mixed-species biofilm in a moving bed sandfilter, *R. eutropha* was recovered as a dominant organism after more than a year of continuous operation in metals removal from an on line reactor at an industrial plant (Pümpel et al., 2001). An alternative method of biomineralisation exploits the sulfate dissimilatory pathway of sulfate-reducing bacteria (SRB) and precipitation of metal cations as the corresponding insoluble metal sulfides. This has been used commercially (Barnes et al., 1991) and more recently has been considered within an integrated process where metal sulfide bioprecipitation is applied following the solubilisation of bound metals via the activity of oxidising/leaching bacteria (White and Gadd, 1996; White et al., 1998; Eccles, 1999). Where the primary metal deposit is FeS, the magnetism of the loaded cells can be used for their easy recovery (Ellwood et al., 1992; Watson & Ellwood, 1994) and these studies were further extended to the remediation of trace amounts of radionuclides by the use of the bioaccumulated FeS as a chemical sorbent for radionuclides (Ellwood et al., 1992; Watson & Ellwood, 1994). As an alternative to sulfide, biogenic phosphate has also been used as a precipitant ligand produced by whole cells. In the case of *Citrobacter* sp. N14, hydrolysis of an applied phosphomonoester substrate results in precipitation of cell bound $MHPO_4$ (where M = heavy metal cation: Macaskie et al., 1992a, 1994, 1996) and the role of the *PhoN* acid phosphatase (common to several enterobacterial species) was confirmed by cloning of *phoN* into *Escherichia coli*, which conferred the ability to accumulate uranyl phosphate (Basnakova & Macaskie, 1998). Addition of phosphate to a waste stream could be undesirable industrially and an alternative approach utilised the ability of *Acinetobacter* sp. to accumulate polyphosphate aerobically (Boswell et al., 1999). A shift to anaerobiosis triggered polyphosphate hydrolysis and efflux of inorganic phosphate, which was harnessed to metal bioprecipitation using an aerobic/anaerobic bioreactor system. Other approaches to heavy metal biomineralisation have utilised microbial redox reactions. Biooxidation of metals is relatively uncommon in bioremediation processes but metal bioreduction, for example Fe(III) to Fe(II), is well known (Chapter 7; Lovley, 1993), e.g. by *Geobacter metallireducens* (Lovley et al., 1993a); this organism (Lovley et al., 1991; Gorby & Lovley, 1992) and sulfate-reducers such as *Desulfovibrio desulfuricans* (Lovley and Phillips, 1992, Lovley et al., 1993b), can reduce U(VI) to U(IV) forming insoluble UO_2 (uraninite: Gorby & Lovley, 1992).

An interesting development is the coupling of the bioreduction of Fe(III) to the reduction of Tc(VII) with nascent Fe(II) used as the Tc(VII) reductant (Lloyd et al., 2001), while FeS has been shown to promote the reduction of U and Tc (Moyes et al., 2000; Wharton et al., 2000). Uranium reduction by *D. vulgaris* occurred via cytochrome c_3 (Lovley et al., 1993c), while uranium was reduced by *Shewanella putrefaciens* with hydrogen as the sole electron donor (Lovley et al., 1991), suggesting the involvement of a hydrogenase. In addition to remediation of metal cations it is also possible to generate base metals via hydrogenase-mediated metals reduction, e.g. for the production of Pd(0) from Pd(II) using *D. desulfuricans* (Lloyd et al., 1998a) or Se(0) from selenite using the hydrogenase activity of *Clostridium pasteurianum* (Yanke et al., 1995). The remainder of this chapter will illustrate the applicability of biomineralisation processes to remediation of radionuclides, highlighting some of the factors likely to contribute to their industrial application.

6. Bioremediation of heavy elements

Uranium, thorium and radium are naturally occurring nuclides of concern in mines, tailings and ground waters. Their bioremediation has been the subject of recent reviews (Volesky & Holan, 1995; Lloyd & Macaskie, 2000) to which the reader is referred. Radium can be removed via biosorption (e.g. Tsezos et al., 1987a, b, c) or by coprecipitation into e.g. $BaSO_4$ (see Lloyd & Macaskie, 2000). The major method for Th removal is biosorption (see Lloyd & Macaskie, 2000) while the uranyl ion (UO_2^{2+}) can be removed via biosorption, or bioreduction to U(IV) (see above). The removal of uranium via biogenic H_2S release is more controversial. The geochemical literature provides evidence for sorption of U(VI) on to sulfide minerals with concomitant partial reduction of U(VI) to U(IV) (Wersin et al., 1994). Mohagheghi et al. (1994) suggested that aqueous sulfide species can reduce U(VI) to U(IV) with precipitation of uraninite but this is disputed by Lovley and Coates (1997) who state that uranium is resistant to reduction by sulfide. Indeed, the presence of sulfate did not affect the rate of U(VI) reduction by *D. desulfuricans* (Tucker et al., 1998). Enzymatically-mediated UO_2^{2+} reductions appear to predominate over chemical reduction, either directly or via microbial generation of reductive intermediates such as Fe(II) anoxically. A more detailed appraisal of the mechanisms of U(VI) reduction is given in Chapter 7.

Bioremediation of the transuranic elements has been little studied. Pu(IV) was removed by *Pseudomonas aeruginosa* on a plasma treated polypropylene web but it is likely that PuO_2 entrapment, rather than true biosorption, was responsible (Tengerdy et al., 1981). Natural sediments removed ^{239}Pu after 4 months with little removal by heat-killed controls, but the mechanism was not determined (Peretrukhin et al., 1996). Evidence has been obtained for microbially-mediated reduction of Pu(IV) to Pu(III), but the Pu(III) reoxidised spontaneously (Rusin et al., 1994). In contrast to Pu(III), Am(III) is stable (Allard et al., 1980). ^{237}Np is a major, long-lived α-emitting contaminant of some wastes (Ashley & Roach, 1990; Macaskie, 1991) that is not removed well by current practices, largely due to its chemistry. Np exists in solution predominantly as the pentavalent species (NpO_2^{+}) and is not strongly complexed by most ligands or readily precipitated from solution (Ahrland et al., 1973), precluding the use of simple biosorption or bioprecipitation techniques. Accordingly, little Np removal was seen using various microbial biomasses (Strandberg & Arnold, 1988). There is little evidence in the literature for the biologically mediated removal of Pu and Np from solution and hence these formed a focus of recent investigations

7. Case history: development of a biomineralisation process for removing uranium and transuranic elements from liquid waste

Phosphate-based biomineralisation

Phosphate-based biomineralisation processes rely on the liberation of high concentrations of inorganic phosphate at the cell surface via the enzymatic cleavage of phosphate donor molecules e.g. glycerol 2-phosphate (see above) or tributyl phosphate (Thomas & Macaskie, 1996, 1998), or efflux of phosphate via hydrolysis of stored polyphosphate (Boswell et al., 1999). The high local concentration of phosphate allows the solubility product of the metal phosphate to be exceeded locally and thus metals can be scavenged from very dilute concentrations in the bulk solution. The most studied system is the acid phosphatase (*PhoN*) mediated metal uptake of a *Citrobacter* sp. that was originally isolated from a lead polluted site in the UK. Early studies showed stoichiometric metal and phosphate deposition, e.g. as HUO_2PO_4 (Macaskie et al., 1992a, 1994, 1996). The role of acid phosphatase was confirmed since a phosphatase deficient mutant did not accumulate UO_2^{2+} (Jeong et al., 1997) and cloning and expression of *phoN* in *Escherichia coli* promoted uranyl phosphate accumulation (Basnakova et al., 1998b). A model was developed whereby the enzyme liberates inorganic phosphate to promote metal deposition at the cell surface with the outer membrane (presumably phospholipids) suggested as nucleation sites for biomineral formation (Ferris & Beveridge 1986a; Mann et al., 1986; Jeong et al., 1997). However, several inconsistencies suggested that this model was over simplistic. The pH optimum of the enzyme is from 5–7 (Jeong et al., 1998) with little activity below pH 4.5, while the K_m for glycerol 2-phosphate increases several-fold as the pH falls, presumably attributable to increased protonation of the substrate (Tolley et al., 1995). Preliminary attempts to remediate uranium-acid mine drainage waters (adjusted to pH 4) using immobilised cells were unsuccessful although phosphate was detected in the column outflow (Macaskie et al., 1997). Either the liberated phosphate or carboxyl groups within the extracellular polymeric matrix may have acted as localised pH buffers in the cellular micro environment. After 2 days of column operation, uranyl ions were removed by the immobilised cells at an efficiency of 70% (input concentration 0.2 mM; flow rate 1 column volume h^{-1}) and this continued at steady state (Macaskie et al., 1997). This was interpreted in terms of the need for a period of time for initiation of nucleation foci prior to steady state crystal growth of $HUO_2PO_4{\cdot}4H_2O$, which was identified by X-ray powder diffraction analysis (Yong & Macaskie, 1995a; Macaskie, et al., 2000a). The need for nucleation was also suggested by the observation that metal removal was greatly enhanced if the columns were allowed to stand overnight in the test solution before initiation of flow-through conditions (Macaskie et al., 1992b). These tests suggested that the simple enzymatic model alone could not explain metal biomineralisation. For successful removal of transuranic elements, which are present at much lower concentrations, often in solutions of low pH and heavily loaded with nitrate (Macaskie, 1991), it was necessary to refine the model further.

A conceptual model for uranyl phosphate biomineralisation

The experiments discussed above suggested that the energetic barrier to nucleation is an important factor in the initiation of sustained metal removal; low phosphatase activity columns could remove metal as effectively as normal columns if a quiescent period was incorporated (Macaskie et al., 1992b). The possibility of additional nucleation sites was considered. Experiments using extracted extracellular polymeric material (EPM) showed uranyl ion uptake to 25% of the polymer dry weight (Bonthrone et al., 2000). Analysis of the extracted material showed a substantial amount of phosphate, confirmed by ^{31}P nuclear magnetic resonance, which gave spectra attributable to the phosphate groups of the lipid A component of lipopolysaccharide (LPS) (Bonthrone et al., 2000; Macaskie et al., 2000a) as described by Strain et al. (1983a, b). Addition of Cd^{2+} promoted a chemical shift downfield with production of a single peak from two original major peaks, suggesting Cd-mediated crosslinking of the EPM strands that was confirmed by a change in the morphology of the EPM (by electron microscopy) from amorphous and diffuse to structured and electron opaque upon metal addition (Bonthrone et al., 2000). This suggested the main nucleation site was the lipid A component of the LPS, in accordance with earlier studies on metal binding to the LPS of *E. coli* (Ferris & Beveridge, 1986b). Upon addition of UO_2^{2+} to the EPM, the solution ^{31}P NMR signal disappeared and a yellow precipitate, confirmed as uranyl phosphate by X-ray powder diffraction analysis, was observed (Macaskie et al., 2000a). In a similar experiment, whole cells were examined under the atomic force microscope and the formation of crystalline material around the cells was clearly seen (Macaskie et al., 2000a).

Coordination and precipitation of metal within the EPM provides the rationale for the nucleation process, with the nascent biomineral further consolidated at the expense of enzymatically-fed phosphate. Since cell loadings of 9 g of uranium per gram of dry biomass have been observed, it is surprising that access to the cell surface is not blocked by the metal phosphate coating. Further examination by electron microscopy in conjunction with immunogold labelling (a specific electron opaque 'tag' for phosphatase localisation) showed a phosphatase pool entrapped within the extracellular matrix (Macaskie et al., 2000a) and in some cases the enzyme was visible as small beads on fibrils of extruded material. Two very similar but distinct phosphatase isoenzymes were seen by enzymological (Jeong et al., 1998) and physiological (Jeong and Macaskie, 1999) studies. Export of enzymes in Gram negative bacteria by secretion pathways (e.g. alkaline phosphatase via a precursive form) is now well established (Nesmeyanova et al., 1991; Kadurugamuwa & Beveridge, 1995). Early studies suggested that phosphatase export occurs in association with LPS (Ingram et al., 1973) and that LPS vesicle production is a secretion process per se. The native enzyme was recovered as a high molecular mass complex (Jeong et al., 1998), in accordance with its occurrence in association with exocellular materials. The idea that EPM may act as an immobilising medium for extracellular enzymes was proposed by Frolund et al. (1995) in the context of the activated sludge floc matrix. Kadurugamuwa and Beveridge (1995) further suggested that phosphatase scavenges orthophosphate from membrane phospholipids as it is released, providing an additional set of nucleation sites (free phosphate) as well as free phosphate which could act in pH stasis.

A conceptual model can be developed in which incoming metal coordinates with the phosphate groups of the lipid A component of the lipopolysaccharide close to the LPS-bound enzyme. Metal incorporation into the nascent crystal creates a localised depletion, facilitating entry of more metal into the extracellular matrix from the bulk solution down the concentration gradient generated. The incoming metal encounters a high local concentration of phosphate liberated via phosphatase activity and the kinetics of metal phosphate precipitation are accelerated by the presence of a pre-formed nucleating surface. Such a mechanism would permit the accumulation of transuranic elements from very low concentration bulk solutions provided that the initial nucleation events and steady state are established initially. Pre-deposition of a 'surrogate' metal phosphate species (La(III) for Am(III) and Th(IV) for Pu(IV)) promoted metal removal (see Macaskie et al., 1994, and below).

Once the barrier to nucleation is overcome, metal uptake proceeds at steady state in a flow through column. Phosphate and metal incorporation are stoichiometric, although an excess of phosphate is required in the solution for maximum removal efficiency in the case of UO_2^{2+}. This intrinsic inefficiency is related to the kinetics of metal phosphate crystallisation and depends on the metal species, the composition of the carrier solution (e.g. the pH and presence of anions) and, in the case of the transuranic elements, the metal valence; trivalent species are precipitated easily, hexavalent species moderately so and the tetravalent species very poorly, attributable to their high tendency to hydrolyse in solution and to form strong chelate complexes (Yong & Macaskie, 1997a). For any given set of conditions, this 'inefficiency factor' will be constant assuming that the column remains at steady state. Modelling trials used entrapped, chemically coupled and biofilm immobilised cells (Macaskie et al., 1995; Finlay et al., 1999; Yong & Macaskie 1997b, 1999). The activity of flow-through columns (in terms of product liberation and metal removal) can be expressed in terms of the available biomass and phosphatase specific activity, the substrate concentration and the flow rate by an integrated form of the Michaelis-Menten equation (see Chapter 11) with the additional incorporation of the inefficiency factor (Macaskie et al., 1995) to allow for the metal crystallisation characteristics. This approach enables predictions to be made of the activity of the column with varying parameters, permitting extrapolation of benchscale models to actual process situations. The predictive model was tested initially to quantify and overcome the interfering effect of nitrate (Yong & Macaskie, 1997b) which is often present to a large excess in nuclear waste solutions (Macaskie, 1991). A similar approach was used to predict the effect of sulfate (against which the system was more robust) and the model was used to define the bioreactor activity against acid mine drainage water which contained approximately 35 mM $SO_4{}^{2-}$ (Yong & Macaskie, 1999) in addition to 0.2 mM uranyl ion and a variety of other metallic co-contaminants (Macaskie et al., 1997).

The model assumes that the biomass in contact with the perfusing flow is constant but, in reality, biofilms display a typical (Costerton et al., 1994) 'stack and channel' format (shown for *Citrobacter* sp. in Finlay et al., 1999). The additional surface area offered by the convoluted structure introduces another level of complexity to the bioreactor model which is currently being quantified in terms of structure, flow and diffusional processes (Paterson-Beedle et al., 2000; Allan et al., 2002).

Application to removal of transuranic elements

The elements considered in this study were ^{241}Am, ^{239}Pu and ^{237}Np (in oxidation states (III), (IV) and (V) respectively). It was anticipated that Am(III) would crystallise easily since it behaves chemically like La(III) which was removed effectively as $LaPO_4$ (Tolley et al., 1995). Problems were anticipated in the removal of Pu(IV) because in vitro precipitation tests had shown that the analogous Th(IV) did not precipitate readily as the phosphate in the presence of citrate which was incorporated to suppress formation of colloidal hydroxide species (Tolley et al., 1995); the strong complexation of actinide(IV) by citrate (substantially greater than than for the hexavalent actinide-citrate complexes; see Macaskie, 1991) resulted in a very low free metal concentration and correspondingly slow phosphate precipitation. Also, the accumulated thorium phosphate was amorphous by X-ray powder diffraction, should read 'the accumulated thorum phosphate was anorphous by X-ray powder diffraction, even when the solubility of thorium phosphate was reduced by the incorporation of ammonium ions (Yong & Macaskie, 1995b). Both HUO_2PO_4 and $LaPO_4$ formed highly crystalline material on the cells but only cell-bound $LaPO_4$ could function as a nucleating surface for $Th(HPO_4)_2$ (Yong & Macaskie, 1998). The XRD pattern of $LaPO_4$ acquired additional peaks in the presence of Th(IV), probably due to the additional $Th(HPO_4)_2$ component (Yong & Macaskie, 1998) but the relationship between the $Th(HPO_4)_2$ and $LaPO_4$ is not clear.

Neptunium was not removed at all, consistent with its ready formation of soluble complexes (see above). It was concluded that removal of ^{241}Am from wastes would not be problematic but that a more efficient system would be needed for Pu and Np removal. Since pre-nucleation with $LaPO_4$ promoted Th(IV) removal (see above), the use of $LaPO_4$ as the nucleating surface for transuranic element removal was evaluated; a non-homologous crystal 'priming layer' could facilitate the deposition of the actinide species, with formation of a hybrid crystal (Yong & Macaskie, 1998). 'Biological' crystals of metal deposit can have advantages over chemically prepared material. For example, in the case of microbially generated FeS, the 'biocrystal' was reported to be a superior sorbent for nuclide species as compared to 'geochemical' FeS (Watson & Ellwood, 1994); here, the discrete intertwined fibrils of FeS (Ellwood et al., 1992) gave a very high surface area which produced a metal sorbent 10–100 times better than the 'geochemical' FeS counterpart (Watson & Ellwood, 1994). In another study, sulfate-reducing bacteria reduced Pd(II) to cell-bound Pd(0) (Lloyd et al., 1998a); the 'biological' Pd(0) had a smaller crystal size, reflected in a greater chemical catalytic activity of the biological Pd(0) deposits as compared to their chemically-prepared counterparts (Macaskie et al., unpublished data). In the present case, the organic template of the cell surface housing the nucleation sites could provide a finely divided 'scaffold' of $LaPO_4$ for further metal precipitation.

These concepts were applied to the phosphatase mediated removal of Pu(IV) and Np(V). Using cells of *Citrobacter* sp. precoated with $LaPO_4$; effective removal of both nuclides was observed whereas none was removed in the absence of the 'priming' deposit (Basnakova & Macaskie, 1998). The solubility product of tetravalent actinide phosphates is low (see Yong and Macaskie, 1998); Np(V) phosphate is thought to be relatively soluble yet Np was removed with a comparable efficiency to Pu(IV) in these tests. It is possible that the species removed was Np(IV) (arising from the disproportionation of Np(V) into Np(IV) and Np(VI) and stabilisation of the tetravalent species by formation of the citrate complex, Reed et al., 1998); the actual species accumulated was not identified since this analysis requires relatively large amounts of material and a specialised laboratory for handling large quantities of α-emitting material.

The role of the 'priming layer' was suggested to be as follows. It is assumed that nascent metal phosphate is not easily crystallised from solution unless a nucleating surface (local energy minimum) is present (Mann, 1988). During the 'priming' step in columns at a slow flow rate (100% removal of La^{3+} from solution as $LaPO_4$), it is likely that substrate (glycerol 2-phosphate) is depleted at the front end of the column and metal deposition is confined to that region; effectively a large amount of $LaPO_4$ may be present but the surface area to volume ratio of the highly localised deposit may be small and, with this, the ability to capture nascent actinide phosphate is limited. In contrast, a column that is prechallenged with La^{3+} and substrate at a rapid flow rate (or a column of lower phosphatase activity) will not deplete substrate until later in the column and hence will have nucleating surfaces distributed throughout the length of the column. This greater surface area will thus be potentially more effective at actinide phosphate removal. This was confirmed using both Pu and Np, where the extent of actinide removal during challenge was greater in columns that had been previously primed with $LaPO_4$ at a greater flow rate (Basnakova & Macaskie, 1998).

Towards in situ biomineralisation processes

A potential problem with the system described above is that, although it is applicable to the removal of transuranic elements by phosphate precipitation, the requirement for an organic phosphate donor molecule is likely to limit the economic attractiveness of the technique and, indeed, when applied to acidic mine wastewater, the cost of the phosphate addition was the limiting factor. The ability of some organisms to utilise tributyl phosphate as the phosphate donor for bioprecipitative uranyl phosphate deposition was demonstrated (Thomas & Macaskie, 1996; Thomas et al., 1998) but this property was unstable and often irretrievably lost; this instability and the undesirability of addition of TBP to a natural water are likely to limit the application of the technology. However, a recent study has pointed the way towards a possible hybrid technology with potential application in situ. Nash et al. (1998) suggested that there may be a role for the hydrolytically unstable organophosphorus complexant, calcium phytate (a natural product from beans and leafy vegetables) as a combined nutrient source for microorganisms and precipitating agent. They suggested addition of this agent to contaminated soils would promote metal complexation, and that subsequent natural hydrolysis would produce inorganic phosphate precipitates. They also emphasised that thermodynamic calculations suggest phosphate is an optimum mineral phase for actinide sequestration. The normal half-life of phytate in synthetic groundwater solutions at 25°C is 100–150 years (Jensen et al., 1996) but microorganisms can accelerate hydrolysis by orders of magnitude (Suzumura & Kamatani, 1995). Thus, there is clear scope for the bioaugmentation of soils and waters with appropriate microorganisms and phytic acid to promote a similar biomineralisation process; this is currently under evaluation in the authors' laboratory.

8. Problems associated with the study of ^{237}Np, and Np removal via bioreduction

Study of Np removal is not trivial. The most common isotope, ^{237}Np (α-emitting species) is inevitably contaminated with the daughter nuclide ^{233}Pa (β-emitter). Tests based on removal of radioactivity to indicate removal of ^{237}Np, therefore, require a method for nuclide separation or counting with α, β discrimination. The γ-emitting ^{239}Np can be used as a tracer, but its short half-life (2.35 days) restricts routine use. Methods for detecting ^{237}Np in the presence of ^{233}Pa include scintillation counting with α, β discrimination and chromatographic separation followed by quantification of radioactive spots on paper using a phosphorImager (Lloyd & Macaskie, 1996) or by scintillation counting of paper sections. Both methods have been used to establish biological removal of Np. Np and Pa are both commonly pentavalent but Pa(V), uncommonly among the pentavalent actinide species, hydrolyses extensively to give a colloidal suspension which is immobile in paper chromatography (see Lloyd et al., 2000a). Thus Np(V) (and reduced Np species) and Pa(V) can be separated using paper chromatography in conjunction with visualisation and quantification of the radioactive spots using a phosphorImager (Lloyd et al., 2000a). Chromatograms are usually wrapped in clingfilm to prevent contamination of the equipment. The technique blocks non-penetrating α-radiation (^{237}Np), requiring the use of a 'spike' of ^{239}Np for visualisation of the Np (Lloyd et al., 2000a).

Since Np(V) does not form insoluble species readily, methods for its removal would aim to promote its oxidation to Np(VI) with removal by methods well documented for U(VI) (see above, and see Chapter 11 of this volume which describes intercalation of Np(VI) within the lattice of HUO_2PO_4), or its reduction to Np(IV), where the reduced product would be amenable to removal as for Th(IV) and Pu(IV). *Citrobacter* sp. removed Pa(V) since this, unlike Np(V), forms insoluble phosphates (Lloyd et al., 2000a), but did not remove Np per se (see earlier). The organism *Shewanella putrefaciens* is well known to reduce species such as Tc(VII) (Lloyd & Macaskie, 1996) and U(VI) (Lovley et al., 1991; Lovley, 1993b) and was used to reduce Np(V) to generate a soluble species of Np(IV), demonstrated by a shift in the chromatographic spot to a position similar to that of a Th(IV) standard (Lloyd et al., 2000a). *S. putrefaciens* reduced, but did not remove, Np. Removal of the Np(IV) species was achieved using *Citrobacter* sp. in the presence of glycerol 2-phosphate, with NH_4^+ to enhance precipitation of Np(IV) phosphate (Lloyd et al., 2000a); 100% removal of Np was achieved using the concerted action of the two organisms, but their co-immobilisation and long-term Np removal studies, with identification of the accumulated solid species, will require a specialist facility for handling the larger quantities of long-lived α-emitters required for solid state analysis.

As a possible alternative future route to the reduction of Np(V), it was observed (Garnov et al., 1998) that Pd(0) is a chemical catalyst for the reduction of Np(V) in the presence of formate (80% reduced within 1 hour at 80°C). Biomass of *Desulfovibrio desulfuricans* reduces Pd(II) to Pd(0) at the expense of hydrogen in an electrobioreactor (Macaskie et al., unpublished data). This 'palladised biomass' rapidly reduced Cr(VI) to Cr(III) with the coupled oxidation of formate or hydrogen, whereas neither native biomass nor chemically-prepared Pd(0) did so (Macaskie et al., unpublished data), and evaluation of this 'bioinorganic catalyst' against Np(V) would be a logical extension to these studies. In particular, it would be interesting to evaluate the possibility of formation of Np(III) and

Pu(III) since neither of these species (which, like Am(III), would be much more readily removed as the phosphates) is usually stable except under highly reducing conditions (Rusin et al., 1994).

9. A new approach: bioinorganic catalysis

The above suggests that a coupling of reductive and precipitative mechanisms can promote removal of metals even in oxidation states that are recalcitrant. However, under some circumstances there may be no need to invoke a bioreductive step. It is known from the literature that zero valent iron can promote reduction and precipitation of uranium from groundwater (Cantrell et al., 1995; Farrell et al., 1999) and iron nodules (iron oxides) can have a similar effect (Sato et al., 1997); U(VI) is reduced during Fe corrosion and sorbs to the Fe hydroxide/oxide corrosion products (Farrell et al., 1999). This is currently under evaluation as a method of groundwater decontamination at several US Department of Energy facilities (Farrell et al., 1999). The mechanism of reductive U deposition was investigated by X-ray photoelectron spectroscopy (Fiedor et al., 1998). Under aerobic conditions, U(VI) was found to sorb strongly and rapidly to hydrous ferric oxide particulates but under anaerobic conditions it was only slowly and incompletely reduced to U(IV). Uranium removal rates were increased in sodium chloride solution but reduced in sodium nitrate solutions; this could have implications in the treatment of nitrate, rich waters. In this context, bioreduction could prove superior, since studies using sulfate-reducing bacteria to reduce Tc(VII) suggest the mechanism to be nitrate insensitive (Lloyd et al., 1999a, b). Lloyd et al. (2001) have also shown that bioreduction of Fe(III) by Fe-reducing bacteria reduces Tc(VII) by an indirect abiotic mechanism at the expense of Fe(II). In addition to Fe, other metals have been shown to catalyse actinide reduction. For example, reduction of Np(V) and precipitation of Np(IV) hydroxide were observed using a matrix of anion exchanger coated with palladium or platinum metal with formate as the electron donor (Garnov et al., 1998). In an interesting juxtaposition of bio- and chemical technology, Mabbett et al. (unpublished data) harnessed the hydrogenase activity of *Desulfovibrio desulfuricans* initially to reduce Pd(II) to Pd(0) at the expense of H_2; the base metal formed a nanocrystalline layer with a high chemical catalytic activity attributable to the formation of Pd(0) nanocrystals on the biomass (Yong et al., unpublished data). Neither biomass alone, nor chemically reduced Pd(0), could reduce Cr(VI) but palladised biomass successfully reduced 1 mM Cr(VI) over several weeks in a flow-through reactor, with quantitative recovery of Cr(III). Although this process was not tested on the actinide elements, these results suggest that biocatalytic activity could be harnessed to the production of an improved catalyst for actinide remediation; this approach is particularly attractive since the Pd(II) was recovered from scrap automotive catalysts as a low cost source of Pd(0) (Yong et al., unpublished data).

10. Biodegradation of chelating agents

Most published bioremediation studies have focused on the treatment of liquid wastes and the technology for soil remediation, particularly in situ, has lagged behind. At the same time the problems of large-scale decontamination of nuclear plants from power stations and submarines are becoming apparent. Both of these problems can be tackled by the application of chelating agents which effectively solubilise the radionuclides but generate a secondary waste containing chelated metals which are difficult to treat using established physicochemical or biotechnologies. Citrate, for example, is a readily biodegraded natural chelating agent which becomes recalcitrant when bound to many metals (Francis et al., 1992; JoshiTope & Francis, 1995). The recalcitrance is not attributable to the toxicity of the metal but to resistance of the metal complex to biodegradation (Brynhildsen & Allard, 1994; JoshiTope & Francis, 1995). In the case of uranyl citrate, the complex was not degraded by whole cells but was degraded by cell free extracts (JoshiTope & Francis, 1995), suggesting that the metal complex was not recognised by the appropriate uptake system. The recalcitrance varies depending on the exact nature of the complex; some metal citrate complexes, such as Ca, Fe(III) and Zn, are readily biodegradable (Brynhildsen & Rosswall, 1989; Francis et al., 1992; Brynhildsen & Allard, 1994), whereas those of Cu, Cd and U are not (Francis et al., 1992). In general, bidentate citrate-metal complexes are biodegradable whereas tridentate ones are recalcitrant, e.g. in the case of iron, the tridentate [Fe(II)-citrate] complex was not degraded but after oxidation and hydrolysis this was converted to the bidentate [Fe(III)$(OH)_2$-citrate] complex and biodegradation occurred (Francis & Dodge, 1993). Recently *Pseudomonas putida* was reported to grow at the expense of a variety of metal citrate complexes (including those of Cu and Cd: see above) and was applied to the treatment of Ni-citrate secondary waste, with biodegradation of the citrate ligand to release Ni(II) for subsequent precipitation (Thomas et al., 2000). Francis et al. (1992) showed that the Ni complex was only ca 70% degraded and the pH of the culture increased to 7.6 during growth (Francis et al., 1992). At acid to neutral pH, the complex is bidentate but, at alkaline pH, the citrate dissociates further to form a tridentate complex (Glusker, 1980), which is resistant to further degradation. An increase in pH can also have implications for the bioremediation of actinide elements (see below).

For the most effective decontamination, a stronger chelating agent is usually required. For example, the stability constants (log β values for the 1:1 Am(III) complexes with citrate, nitrilotriacetate (NTA) and ethylenediaminetetraacetate (EDTA) are 7.7, 11.5 and 18.2, respectively; see Macaskie, 1991), so EDTA is the preferred decontaminant, even though it is the least biodegradable (see below). In the case of Pu/EDTA complexes, the stability constants for the 1:1 complexes with Pu(III), (IV), (V) and (VI) are 18.1, 25.6, 12.9 and 16.4 (see Macaskie, 1991) so EDTA is most effective when removing tetravalent species. Importantly, Banaszak et al. (1998) point out that biodegradation of the organic ligand can have profound effects on the actinide speciation. Hence, biodegradation of a Np(IV) citrate complex would suggest the precipitation of $Np(OH)_4$, but the production of CO_2 as a metabolic product would increase the (pH-dependent) concentration of soluble bicarbonate and carbonate species, increasing the solubility of the Np(IV) species as the (bi)carbonate complex and effectively negating the effect of biodegradation. In the case of U(VI), the log β values for the citrate and carbonate complexes are 7.4 and 16.3, respectively. This very important factor means that the application of microorganisms which can degrade chelating agents must be considered very carefully in relation to the pathway of metabolism and the final metabolic products (Banaszak et al., 1998). However, Np(V) is an interesting example since the toxicity of the metal to *Chelatobacter heintzii* was moderated by its complexation to NTA. Biodegradation of the latter released Np, but rather than forming soluble carbonate complexes, as proposed above, Np was desolubilised, probably as a phosphate (Banaszak et al., 1998). Solubility data are not available for the Np(V)-phosphate system, but other studies have suggested that Np(V) is not desolubilised by phosphate liberating *Citrobacter* sp. (Lloyd et al., 2000a), and Nash et al. (1998) state that phosphate alone has little effect on the solubility of $NpO_2{}^+$ except at pH above 7 and a PO_4^{3-} concentration above 10 μM. The actual oxidation state of the Np in the cellular microenvironments could not be determined in these two studies but when the applied Np(V) was bioreduced it was readily desolubilised (Lloyd et al., 2000a), probably as $Np(HPO_4)_2$. Clearly the benefits of biodegradation of chelating agents need to be considered with respect to the target environment and its pH, redox potential and the presence of phosphate ligands. Thermodynamic calculations suggest phosphate as the optimum mineral phase for actinide sequestration (Nash et al., 1998).

For decontamination operations, citrate complexes may not be strong enough to achieve the required decontamination levels but EDTA, which would be more suitable, has until recently, been regarded as non-biodegradable. As a compromise, NTA received early attention and its biodegradation is well established (e.g. Egli, 1994; Xun et al., 1996; White & Knowles, 2000) via nitrilotriacetate monooxygenase activity (Xun et al., 1996). Like citrate, the formation of NTA-metal complexes can prevent biodegradation (White and Knowles, 2000). It is interesting to note that the NTA-complexes of Fe(III), Mn(II), Zn(II) and Co(II) were degraded by *Chelatobacter heintzii* ATCC 29600, but those of Cu(II) or Ni(II) were not (White & Knowles, 2000).

Recently, the biodegradation of EDTA has become well established (e.g. Thomas et al., 1998; Henneken et al., 1998; Nortemann, 1999; Witschel et al., 1999) and, as with citrate and NTA, the biodegradability is dependent on the metal cation in the complex. Nortemann (1999) has reviewed the current state of the art, while Henneken et al. (1998) and Witschel et al. (1999) give kinetic and uptake data, respectively. In particular, strain DSM9103 was effective in degrading the Ca-EDTA complex. This organism was identified using 16S rRNA gene sequence analysis as being located in the *Rhizobium-Agrobacterium* branch within the α-subclass of the Proteobacteria (see Witschel et al., 1999). In this organism, a monooxygenase was shown to catalyse the first attack on EDTA, with oxidative removal of two acetyl groups from the molecule (Witschel et al., 1997). In his recent review, Nortemann (1999) points out that the biochemical basis for EDTA degradation appears to be similar to that of the analogous NTA, with similar catabolic enzymes involved. The apparent recalcitrance of EDTA in the environment is probably due to its occurrence as metal complexes (Henneken et al., 1998); indeed, Palumbo et al. (1994) reported that although Co-EDTA was not degraded by *Agrobacterium* sp., replacement of the Co by Fe(III) permitted biodegradation. EDTA biodegradation can be promoted by the use of immobilised cells. Thomas et al. (1998) immobilised a mixed culture biofilm on shale particles and obtained removal of 20% of the EDTA (1mM) with a flow residence time of 2.5 hours, but tests at slower flow rates were not reported. Henneken et al. (1998) immobilised a mixed culture on various carriers, achieving 95–99% removal of EDTA (up to 12.8 kg m^{-3} day^{-1}) at an EDTA concentration of 450 mg l^{-1} and with a hydraulic retention time of less than 1 hour.

11. Towards in situ remediation

The above discussion suggests that our understanding of soil and subsurface processes is insufficient at the present time to be able to predict the efficacy of bioremediative systems in the soil and subsurface environment. It is possible that actinides might be remobilised in the form of (bi)carbonate complexes, but it may be that sequestration by phosphates will prevail over the carbonate system (see above). This has led to the proposal of addition of calcium phytate, a hydrolytically unstable, non-toxic organophosphorus complexant, to contaminated soils in order to promote mineralisation of the actinide species (Nash et al., 1998), effectively creating an in situ phosphate mineralisation system analogous to the laboratory system described earlier. Since bioreduction of U(VI), other actinide species and fission products is well documented, it is likely that soluble high valence actinide species would be reduced in the subsurface environment. This is also particularly likely since the reducing properties of Fe(II), generated via the use of Fe(III) as the terminal electron acceptor in anaerobic respiration, has been demonstrated (e.g. for Tc(VII) reduction: Lloyd et al., 2001). Formation of insoluble species in the subsurface would immobilise the elements but it may be more desirable to remove them altogether since over a geological timescale they will transform through radioactive decay into a series of other elements, which may have different solubility behaviour. To this end, it may be possible to resolubilise precipitated radionuclides in situ using certain metal-reducing bacteria, by virtue of their ability to couple radionuclide reoxidation to nitrate reduction (D. Lovley, personal communication). Thus, radionuclides could be precipitated in situ using a 'biobarrier' engineered downstream from a contaminated aquifer plume, prior to recovery from the barrier in a low volume of nitrate-rich liquor. Alternatively, washing soils with chelating agents would mobilise the metals but could lead to additional problems in addition to the relatively high cost of the process and need to protect the environment from transfer into subsurface waters. It may be more realistic to use this approach ex situ. An alternative approach has harnessed the ability of sulfur-oxidising bacteria to leach metals into a soluble form and coupled this to the ability of sulfate-reducing bacteria to bioprecipitate metal sulfides (White et al., 1998). However, this would still require the metals to be pumped from the soil and results in the need to treat a large volume. Earlier studies suggested that it is possible to move metallic contaminants through soil in an electric field, concentrating them for recovery without the need for soil invasion (Ho et al., 1995). This process, electrokinetics, generates hydroxide ions at the cathode and hydrogen ions at the anode. These migrate through the soil, with H^+ displacing soil bound metal ions into the liquid phase. It is possible to maintain an acidic pH throughout the soil and the solubilised metal ions are transported by electromigration through the soil for recovery at the cathode (Maini et al., 2000). The process has been effective in both model and real systems and has been applied to sites in the USA and in Europe (Lageman et al., 1989)). Electrokinetics may mobilise metal species but not all metal precipitates (e.g. metal sulfides) can be mobilised in this way. However, this problem can be overcome by using sulfur-oxidising bacteria to lower the soil pH (by formation of sulfuric acid) and solubilise metal precipitates, followed by electrokinetic decontamination. An additional benefit is that the electrokinetic treatment appears to stimulate the activity of sulfur-oxidising bacteria (Jackman et al., 1999; Maini et al., 2000). Until now, this approach has been applied to common metals like Cu which frequently occur as the sulfide, particularly as a result of the activity of sulfate-reducing bacteria. Thus, this approach should be particularly effective for targeting reduced U and Tc sulfides, in addition to enzymatically-produced UO_2 and TcO_2. The different speciation of the solids may dictate the preferred electrokinetic route (with or without the participation of sulfur-oxidising bacteria); an ideal strategy would aim to solubilise and recover the Tc and actinide species via electrokinesis while leaving the Fe behind. However, given that FeS can sorb actinide species (see above), and that the exact nature of the hybrid solid phase formed is not known, this may prove very difficult. As far as is known to the authors, this problem is not being addressed and the application of electrokinetic approaches to radionuclide-contaminated soils is an obvious step towards environmental remediation.

References

Ahrland, S., Liljenzin, J. O. & Rydberg, J. (1973). In J. C. Bailar, H. J. Emelius, R. Nyholm and A. F. Trotman-Dickenson (Eds), *Comprehensive Inorganic Chemistry*, Vol. 5: *The Actinides* (pp. 465–635). Oxford, UK: Pergamon Press.

Akutsu, H. & Seelig, J (1981). Interaction of metal ions with phosphatidylcholine bilayer membranes. *Biochemistry, 20*, 7366–7373.

Allan, V. J. M., Paterson-Beedle, M., Nott, K., Hall, L. D. & Macaskie, L. E. (2002). The effect of nutrient limitation on biofilm formation and phosphatase activity of a *Citrobacter* sp. *Microbiology* (in press).

Allard, B. & Rydberg, J. (1983). Behaviour of plutonium in natural waters. In W. T. Carnall and G. R. Choppin (Eds), *Plutonium Chemistry*. American Chemical Society Symposium Series, 216 (pp. 275–295). Washington, DC: American Chemical Society.

Allard, B., Kipatsi, H. & Liljenzin, J. O. (1980). Expected species of uranium, neptunium and plutonium in neutral aqueous solutions. *Journal of Inorganic and Nuclear Chemistry, 42*, 1015–1027.

Allen, S. E. (1984). Radionuclides in natural terrestrial ecosystems. *Science of the Total Environment, 35*, 285–300.

Ashley, N. V., Pope, M. R. & Roach, D. J. W. (1987). Feasibility study of the application of biotechnology to nuclear waste treatment. DOE/RW/88.008, Department of the Environment, UK.

Ashley, N. V. & Roach, D. W. (1990). Review of biotechnology applications to nuclear waste treatment. *Journal of Chemical Technology and Biotechnology, 49*, 381–394.

Assinder, D. J., Hamilton-Taylor, J., Kelly, M., Mudge, S. & Bradshaw, K (1990). Field and laboratory measurements of the rapid remobilisation of plutonium from estuarine sediments. *Journal of Radioanalytical and Nuclear Chemistry, 138*, 417–424.

Baik, M. H. & Lee, K. J. (1994). Transport of radioactive solutes in the presence of chelating agents. *Annals of Nucear Energy, 21*, 81–96.

Banaszak, J. E., Reed, D. T. & Rittman, B. E. (1998). Speciation-dependent toxicity of neptunium (V) toward *Chelatobacter heintzii. Environmental Science and Technology, 32*, 1065–1091.

Banaszak, L. J., Rittman, B. E. & Reed, D. T. (1999). Subsurface interactions of actinide species and microorganisms: implications for the bioremediation of actinide-organic mixtures. *Journal of Radioanalytical and Nuclear Chemistry, 241*, 385–435.

Barnes, L. J., Janssen, F. J., Sherren, J., Versteegh, J. H., Koch, R. O. & Scheeren, P. J. H. (1991). A new process for the microbial removal of sulfate and heavy metals from contaminated waters extracted by a geohydrological control system. *Chemical Engineering Research and Design, 69A*, 184–186.

Barton, L. L., Fekete, F. A., Huybrechts, M. M. E., Sillerud, L. O., Blake II, R. C. & Pigg, C. J. (1994). Application of biotechnology in management of industrial wastes containing toxic metals. *Radioactive Waste Management and Environmental Restoration, 18*, 13–25.

Basnakova, G. & Macaskie, L. E. (1998). Microbially-enhanced chemisorption of heavy metals: a novel method for the bioremediation of solutions containing long-lived isotopes of neptunium and plutonium. *Environmental Science and Technology, 32*, 184–187.

Basnakova, G, Stephens, E., Thaller, M. C., Rossolini, G. M. & Macaskie, L. E. (1998). The use of *Escherichia coli* bearing a *phoN* gene for the removal of uranium and nickel from aqueous flows. *Applied Microbiology and Biotechnology, 50*, 266–272.

Bell, J. T., Toth, L. M. & Friedman, H. A. (1983). Photochemistry of aqueous plutonium solutions. In W. T Carnall and G. R Choppin (Eds), *Plutonium Chemistry*. American Chemical Society Symposium Series, 216 (pp. 263–274). Washington DC: American Chemical Society.

Binks, P. R. (1996). Radioresistant bacteria: have they got industrial uses? *Journal of Chemical Technology and Biotechnology, 67*, 319–322.

Boegley, W. J. Jr. & Alexander, H. J. (1986). Radioactive wastes. *Journal of the Water Pollution Control Federation, 58*, 594–600.

Bonthrone, K. M., Quarmby, J., Hewitt, C. J., Allan, V. J. M., Paterson-Beedle, M., Kennedy, J. F. & Macaskie, L. E. (2000). The effect of the growth medium on the composition and metal binding behaviour of the extracellular polymeric material of a metal-accumulating *Citrobacter* sp. *Environmental Technology, 21*, 123–124.

Boswell, C. D., Dick, R. E. & Macaskie, L. E. (1999). The effect of heavy metals and other environmental conditions on the anaerobic phosphate metabolism of *Acinetobacter johnsonii. Microbiology, 145*, 1711–1720.

Brim, H., McFarlan, S. C., Fredrickson, J. K., Minton, K. W., Zhai, M., Wackett, L. P. & Daly, M. J. (2000). Engineering *Deinococcus radiodurans* for metal remediation in radioactive mixed waste environments. *Nature Biotechnology, 18*, 85–90.

Brynhildsen, L. & Roswall, T. (1989). Effects of copper, magnesium and zinc on the decomposition of citrate by a *Klebsiella sp. Applied and Environmental Microbiology, 55*, 1375–1379.

Brynhildsen, L. & Allard, B. (1994). Influence of metal complexation on the metabolism of citrate by *Klebsiella oxytoca. Biometals, 7*, 163–169.

Bryan, G. W. & Langston, W. J. (1992). Bioavailability, accumulation and effects of heavy metals in sediments with special reference to United Kingdom estuaries: a review. *Environmental Pollution, 76*, 89–131.

Bulman, R. A. & Baker, S. T. (1987). Investigation of interactions of transuranics and cerium with humates. *Science of the Total Environment, 62*, 213–218.

Buppelman, K., Magirius, S., Lierse, C. & Kim, J. I. (1986). Radiolytic oxidation of americium (III) to americium (IV) and plutonium (IV) to plutonium (VI) in saline solution. *Journal of Less Common Metals, 122*, 329–336.

Burnett, W. C., Cowart, J. B. & Chin, P. A. (1987). Polonium in the superficial aquifer of West Central Florida. In B. Graves (Ed.), *Radon, Radium and Other Radioactivity in Groundwater: Hydrogeologic Impact and Application to Indoor Airborne Contamination* (pp. 251–269). Boca Raton FL. Lewis Publishers.

Burton, P. J. (1986). Laboratory studies on the remobilisation of actinides from Ravenglass estuary sediment. *Science of the Total Environment, 52*, 123–145.

Cantrell, K. J., Kaplan, D. I. & Weitsma, T. W. (1995). Zero-valent iron for the in situ remediation of selected metals in ground water. *Journal of Hazardous Materials, 42*, 201–212.

Choppin, G. R. (1992). The role of natural organics in radionuclide migration in natural aquifer systems., *Radiochimica Acta, 58/59*, 113–120.

Choppin, G. R. & Allard, B. (1985). Complexes of actinides with naturally occurring organic compounds. In A. J. Freeman and C. Keller (Eds), *Handbook on the Physics and Chemistry of the Actinides* (pp. 407–429). Amsterdam: Elsevier Science.

Cleveland, J. M. & Rees, T. F. (1981). Characterisation of plutonium in Maxey Flats radioactive trench leachates. *Science, 212*, 1506–1509.

Cleveland, J. M., Rees, T. F. & Nash, G. L. (1985). Plutonium, americium and neptunium speciation in selected groundwaters. *Nuclear Technology, 69*, 380–387.

Coggle, J. E. (1983). *Biological Effects of Radiation*. London: Taylor and Francis Ltd.

Costerton, J. W., Lewandowski, Z., Caldwell, D. E., Korber, D. R., de Beer, D. & James, G. (1994). Biofilms: the customised microniche. *Journal of Bacteriology, 176*, 2137–2142.

Delegard, C. H., Barney, G. S. & Gallagher, S. A. (1984). Effects of Hanford high-level waste components on the solubility and sorption of cobalt, strontium, neptunium, plutonium and americium. In G. S. Barney, J. D. Navratil & W. W. Schulz (Eds), *Geochemical Behavior of Disposed Radioactive Waste*. American Chemical Society Symposium Series, 246, (pp. 95–112). Washington DC: American Chemical Society.

Depoorter, G. L. & Rofer-Depoorter, C. K. (1980) Application of photochemical techniques to actinide separation processes. In J. D. Navratil & W. W Schulz (Eds), *Actinide Separations*. American Chemical Society Symposium Series, 117, (pp. 267–276). Washington DC: American Chemical Society.

Diels, L., Dong, Q., Van der Lelie, D., Baeyens, W. & Mergeay, M. (1995). The *czc* operon of *Alcaligenes eutrophus* CH34: from resistance mechanisms to the removal of heavy metals. *Journal of Industrial Microbiology, 14*, 142–153.

Dodge, C. J. & Francis, A. J. (1994). Photodegradation of uranium-citrate complex with uranium recovery. *Environmental Science and Technology, 28*, 1300–1306.

Dozol, M. & Hagemann, R. (1993). Radionuclide migration in groundwaters: review of the behaviour of actinides. *Pure and Applied Chemistry, 65*, 1081–1102.

Duffield, J. R. & Williams, D. R. (1986). The environmental chemistry of radioactive waste disposal. *Chemical Society Reviews, 15*, 2291–307.

Eccles, H. (1999). Nuclear waste management: a bioremediation approach. In G. R. Choppin & M. K. Khankhasayev (Eds), *Chemical Separation Technologies and Related Methods of Nuclear Waste Management* (pp. 187–208). Dordrecht, The Netherlands: Kluwer Academic Publishers.

Egli, T. (1994). Biochemistry and physiology of the degradation of nitrilotriacetic acid and other metal complexing agents. In: C. Ratledge (Ed.), *Biochemistry of Microbial Degradation* (pp. 79–195). Dordrecht, The Netherlands: Kluwer Academic Publishers.

Ellwood, D. C., Hill, M. J. & Watson, J. H. P. (1992). Pollution control using microorganisms and metal separation. In J. C. Fry, G. M. Gadd, R. A. Herbert, C. W. Jones & I. A. Watson-Craik (Eds), *Microbial Control of Pollution*. SGM Symposium No. 48, (pp. 89–112). Cambridge, UK: Cambridge University Press.

Farrell, J., Bostick, W.D., Jarabek, R.J. & Fiedor, J. N. (1999). Uranium removal from groundwater using zero valent iron media. Groundwater, 37, 618–624.

Ferris, F. G. & Beveridge, T. J. (1986a). Binding of a paramagnetic metal cation to *Escherichia coli* K-12 outer membrane vesicles. *FEMS Microbiology Letters, 24*, 23–46.

Ferris, F. G. & Beveridge, T. J. (1986b). Site specificity of metallic ion binding in *Escherichia coli* K-12 lipopolysaccharide. *Canadian Journal of Microbiology, 32*, 52–55.

Fiedor, J. N., Bostick, W. D., Jarabek, R. J. & Farrell, J. (1998). Understanding the mechanism of uranium removal from groundwater by zero-valent iron using X-ray photoelectron spectroscopy. *Environmental Science and Technology, 32*, 1466–1473.

Finlay, J. A., Allan, V. J. M., Conner, A., Callow, M. E., Basnakova, G. & Macaskie, L. E. (1999). Phosphate release and heavy metal accumulation by biofilm-immobilised and chemically coupled cells of a *Citrobacter* sp. pre-grown in continuous culture. *Biotechnology and Bioengineering, 63*, 87–97.

Flowers, R. H., Roberts, L. E. J. & Tymons, B. J. (1986). Characteristics and quantities of radioactive wastes. *Philisophical Transactions of the Royal Society of London, A319*, 5–16.

Francis, A. J. (1994). Microbial transformations of radioactive wastes and environmental restoration through bioremediation. *Journal of Alloys and Compounds, 213/214*, 226–231.

Francis, A. J. & Dodge, C. J. (1993). Influence of complex structure on the biodegradation of iron citrate complexes. *Applied and Environmental Microbiology, 59*, 109–113.

Francis, A. J., Dodge, C. J. & Gillow, J. B. (1992). Biodegradation of metal citrate complexes and implications for toxic metal mobility. *Nature, 356*, 140–142.

Fredrickson, J. K., Kostandarithes, H. M., Li, S. W., Plymale, A. E. & Daly, M. J. (2000). Reduction of Fe(III), Cr(VI), U(VI) and Tc(VII) by *Deinococcus radiodurans* R1. *Applied and Environmental Microbiology, 66*, 2006–2011.

Frolund, B., Greibe, T. & Nelson, P. H. (1995). Enzymatic activity in the activated sludge floc matrix. *Applied Microbiology and Biotechnology, 43*, 755–761.

Gadd, G. M. (1996). Influence of microorganisms on the environmental fate of radionuclides. *Endeavour, 20*, 150–156.

Gadd, G. M. (1997). Roles of microorganisms in the environmental fate of radionuclides. *CIBA Foundation Symposia, 203*, 94–104.

Garnov, A. Y., Gelis, A. V., Bessonov, A. A. & Shilov, V. P. (1998). Reduction of neptunium (V) in alkali solutions with formate ion in the presence of palladium and platinum. *Radiochemistry, 40*, 319–320. Translated from *Radiokhimiya, 40*, 309–310.

Glusker, J. P. (1980). Citrate conformation and chelation: enzymatic implications. *Accounts of Chemical Research, 13*, 3445–3452.

Gonzalez-Munoz, M. T., Merroun, M. L., Omar, B. & Arias, J. M. (1997). Biosorption of uranium by *Myxococcus xanthus. International Biodeterioration and Biodegradation*, 40, 107–114.

Gorby, Y. A. & Lovley, D. R. (1992). Enzymatic uranium precipitation. *Environmental Science and Technology, 26*, 205–207.

Henneken, L., Nortemann, B. & Hempel, D. C. (1998). Biological degradation of EDTA: reaction kinetics and technical approach. *Journal of Chemical Technology and Biotechnology, 73*, 144–152.

Ho, S. V., Sheridan, P. W., Athmer, C. J., Heitkamp, M. A., Brackin, J. M., Weber, D. & Brodsky, P. H. (1995). Integrated in situ soil remediation technology: the Lasagna process. *Environmental Science and Technology, 29*, 2528–2534.

Horwitz, E. P., Mason, G. W., Bloomquist, C. A. A., Leonard, R. A. & Bernstein, G. J. (1980). The extraction of DBP and MBP from actinides; Application to the recovery of actinides from TBP-sodium carbonate scrub solutions In J. D. Navratil & W. W. Schulz (Eds), *Actinide Separations. American Chemical Society Symposium Series, 33*, (pp. 475–483). Washington D C: American Chemical Society

Hursthouse, A. S., Baxter, M. S., Livens, F. R. & Duncan, H. J. (1991). Transfer of Sellafield-derived ^{237}Np to and within the terrestrial environment. *Journal of Environmental Radioactivity, 14*, 147–174.

Ingram, J. M. K., Cheng, K. J. & Costerton, J. W. (1973). Alkaline phosphatase of *Pseudomonas aeruginosa*: the mechanism of secretion and release of the enzyme from whole cells. *Canadian Journal of Microbiology, 19*, 1407–1415.

Jackman, S. A., Maini, G., Sharman, A. K., Knowles, C. J. (1999) The effects of direct electric current on the viability and metabolism of acidophilic bacteria. *Enzyme and Microbiology Technology, 24*, 316–324.

Jardine, P. M., Jacobs, G. K. & O'Dell, J. D. (1993). Unsaturated transport processes in disturbed heterogeneous porous media. 2. Cocontaminants *Soil Science Society of America Journal, 57*, 945–953.

Jensen, M. P., Nash, K. L., Morss, L. R., Appelman, E. H. & Schmidt, M. A. (1996). Immobilisation of actinides in geomedia by phosphate precipitation. In J. S. Gaffney, N. A. Marley & S. B. Clark (Eds), *Humic and Fulvic Acids: Isolation, Structure and Environmental Role*. American Chemical Society Symposium Series, 651, (pp. 272–285). Washington DC: American Chemical Society

Jeong, B. C. & Macaskie, L. E. (1999). Production of two phosphatases by a *Citrobacter* sp. grown in batch and continuous culture. *Enzyme and Microbial Technology, 24*, 218–224.

Jeong, B. C., Hawes, C., Bonthrone, K. M. & Macaskie, L. E. (1997). Localisation of enzymically-enhanced heavy metal accumulation by *Citrobacter* sp. and metal accumulation in vitro by liposomes containing entrapped enzyme. *Microbiology, 143*, 2497–2507.

Jeong, B. C., Poole, P. S., Willis, A. J. & Macaskie, L. E. (1998). Purification and characterisation of acid-type phosphatases from a heavy metal-accumulating *Citrobacter* sp. *Archives of Microbiology, 169*, 166 173.

JoshiTope, G. & Francis, A. J. (1995). Mechanisms of biodegradation of metal citrate complexes by *Pseudomonas fluorescens. Journal of Bacteriology, 177*, 1989–1993.

Kadurugamuwa, J. L. & Beveridge, T. J. (1995). Virulence factors are released from *Pseudomonas aeruginosa* in association with membrane-vesicles during normal growth and exposure to gentamicin – a novel mechanism of enzyme-secretion. *Journal of Bacteriology, 177*, 3998–4008.

Kauffman, J. W., Laughlin, W. C. & Baldwin, R. A. (1986). Microbiological treatment of uranium mine wastes. *Environmental Science and Technology, 20*, 243–248.

Kuyucak, N. & Volesky, B. (1989). Accumulation of cobalt by marine algae. *Biotechnology and Bioengineering, 33*, 809–814.

Lageman, R., Pool, W. & Seffinga, G. (1989). Electro-reclamation- theory and practice. *Chemistry and Industry, 18*, 585–590.

Lange, C. C., Wackett, L. P., Minton, K. W. & Daly, M. J. (1998). Engineering a recombinant *Deinococcus radiodurans* for organopollutant degradation in radioactive mixed waste environments. *Nature Biotechnology, 16*, 929–934.

Lieser, K. H. & Muhlenweg, U. (1988) Neptunium in the hydrosphere and the geosphere. *Radiochimica Acta, 43*, 27–35.

Lloyd, J. R. & Macaskie, L. E. (1996). A novel phosphorImager-based technique for monitoring the microbial reduction of technetium. *Applied and Environmental Microbiology, 62*, 578–582.

Lloyd, J. R. & Macaskie, L. E. (2000). Bioremediation of radionuclide-containing wastewaters. In D. R. Lovley (Ed.), *Environmental Microbe-Metal Interactions* (pp. 277–327). American Society for Microbiology.

Lloyd, J. R., Yong, P. & Macaskie, L. E. (1998a). Enzymatic recovery of palladium by using sulfate-reducing bacteria. *Applied and Environmental Microbiology, 64*, 4607–4609.

Lloyd, J. R., Nolting, H-F., Sole, V. A., Bosecker, K and Macaskie, L. E. (1998b). Technetium reduction and precipitation by sulfate-reducing bacteria. *Geomicrobiology Journal, 15*, 43–56.

Lloyd, J. R., Ridley, J., Khizniak, T., Lyalikova, N. N. & Macaskie, L. E (1999a). Reduction of technetium by *Desulfovibrio desulfuricans*: biocatalyst characterisation and use in a flow-through bioreactor. *Applied and Environmental Microbiology, 65*, 2691–2696.

Lloyd, J. R., Thomas, G. H., Finlay, J. A., Cole, J. A. & Macaskie, L. E. (1999b). Microbial reduction of technetium by *Escherichia coli* and *Desulfovibrio desulfuricans*: enhancement via the use of high activity strains and effect of process parameters. *Biotechnology and Bioengineering, 66*, 122–130.

Lloyd, J. R., Yong, P. & Macaskie, L. E. (2000a). Biological reduction and removal of Np(V) by two microorganisms. *Environmental Science and Technology, 34*, 1297–1301.

Lloyd, J. R., Sole, V. A., Praagh, C. V. G. & Lovley, D. R. (2000b). Direct and Fe(II)-mediated reduction of technetium by Fe(III)-reducing bacteria. *Applied and Environmental Microbiology, 66*, 3743–3749.

Lloyd, J. R., Mabbett, A. N., Williams, D. R. & Macaskie, L. E. (2001). Metal reduction by sulfate-reducing bacteria: physiological diversity and metal specificity. *Hydrometallurgy, 59*, 327–337.

Lovley, D. R. (1993). Dissimilatory metal reduction. *Annual Review of Microbiology, 47*, 263–290.

Lovley, D. R., Phillips, E. J. P., Gorby, Y. A. & Landa, E. (1991). Microbial reduction of uranium. *Nature, 350*, 413–416.

Lovley, D. R. & Phillips, E. J. (1992). Reduction of uranium by *Desulfovibrio desulfuricans. Applied and Environmental Microbiology, 58*, 850–856.

Lovley, D. R. & Coates, J. D. (1997). Bioremediation of metal contamination. *Current Opinion in Biotechnology, 8*, 285–289.

Lovley, D. R., Giovannoni, S. J., White, D. C., Champine, J. E., Phillips, E. J. P., Gorby, Y. A. & Goodwin, S. (1993a). *Geobacter metallireducens* gen. nov. sp. nov., a microorganism capable of coupling the complete oxidation of organic compounds to the reduction of iron and other metals. *Archives of Microbiology, 159*, 336–344.

Lovley, D. R., Roden, E. E., Phillips, E. J. P. & Woodward, J. C. (1993b). Enzymatic iron and uranium reduction by sulfate-reducing bacteria. *Marine Geology, 113*, 41–53.

Lovley, D. R., Widman, P. K., Woodward, J. C. & Phillips, E. J. P. (1993c). Reduction of uranium by cytochrome c_3 of *Desulfovibrio vulgaris. Applied and Environmental Microbiology, 59*, 3572–3576.

Macaskie, L. E. (1991). The application of biotechnology to the treatment of wastes produced from the nuclear fuel cycle: biodegradation and bioaccumulation as a means of treating radionuclide-contaminated streams. *Critical Reviews in Biotechnology, 11*, 41–112.

Macaskie, L. E., Empson, R. M., Cheetham, A. K., Grey, C. P. & Skarnulis, A. J. (1992a). Uranium bioaccumulation by a *Citrobacter* sp. as a result of enzymically-mediated growth of polycrystalline HUO_2PO_4. *Science, 257*, 782–784.

Macaskie, L. E., Gilbert, J and Tolley, M. R. (1992b). The effects of ageing on the accumulation of uranyl ion by a biofilm bioreactor and promotion of uranium deposition in stored biofilms. *Biotechnology Letters, 14*, 525–530.

Macaskie, L. E., Jeong, B. C. & Tolley, M. R. (1994). Enzymically-accelerated biomineralisation of heavy metals: application to the removal of americium and plutonium from aqueous flows. *FEMS Microbiology Reviews, 14*, 351–368.

Macaskie, L. E., Empson, R. M., Lin, F. & Tolley, M. R. (1995). Enzymatically-mediated uranium accumulation and uranium recovery using a *Citrobacter* sp. immobilised as a biofilm within a plug-flow reactor. *Journal of Chemical Technology and Biotechnology, 63*, 1–16.

Macaskie, L. E., Lloyd, J. R., Thomas, R. A. P. & Tolley, M. R. (1996). The use of microorganisms for the remediation of solutions contaminated with actinide elements, other radionuclides and organic contaminants generated by nuclear fuel cycle activities. *Nuclear Energy, 35*, 257–271.

Macaskie, L. E., Yong, P., Doyle, T. C., Roig, M. G., Diaz, M. & Manzano, T. (1997). Bioremediation of uranium-bearing wastewater: biochemical and chemical factors influencing bioprocess application. *Biotechnology and Bioengineering, 53*, 100–109.

Macaskie, L. E., Bonthrone, K. M., Yong, P. & Goddard, D. T. (2000a). Enzymically-mediated bioprecipitation of uranium by a *Citrobacter* sp.: a concerted role for exocellular lipopolysaccharide and associated phosphatase in biomineral formation. *Microbiology, 146*, 1855–1867.

Maini, G., Sharman, A. K., Sunderland, G., Knowles, C. J., Jackman, S. A. (2000). An integrated method incorporating sulfur-oxidising bacteria and electrokinetics to enhance removal of copper from contaminated soil. *Environmental Science and Technology, 34*, 1081–1087.

Mann, S. (1988). Molecular recognition in biomineralisation. *Nature, 332*, 119–124.

Mann, S. (1993). Molecular tectonics in biomineralisation and biomimetic materials chemistry. *Nature, 365*, 499–505.

Mann, S. (1997). Biomineralisation: the form(id)able part of bioinorganic chemistry. *Journal of the Chemical Society – Dalton Transactions, 1997*, 3953–3961.

Mann, S., Hannington, J. P. & Williams, R. J. P. (1986). Phospholipid vesicles as a model system for biomineralisation. *Nature, 324*, 565–567.

Marques, A. M., Roca, X., Simon-Pujol, M. D., Fuste, M. C. & Francisco, C (1991). Uranium accumulation by *Pseudomonas* sp. EPS-5028. *Applied Microbiology and Biotechnology, 35*, 406–410.

McCready, R. G. L. & Lakshmanan, V. I. (1986). Review of bioadsorption research to recover uranium from leach solutions in Canada. In H. Eccles & S. Hunt (Eds), *Immobilisation of Ions by Bio-sorption* (pp. 219–226). Chichester, UK: Ellis Horwood.

McCready, R. G. L. & Gould, W. D. (1990). Bioleaching of uranium at Denison Mines. In J. Salley, R. G. L. McCready & P. L. Wichlacz (Eds), *Biohydrometallurgy* (pp. 477–489). 1999. Proceedings of an International Symposium on Biohydrometallurgy. Ottawa, Canada. Canmet.

McKay, H. A. C., Miles, J. H & Swanson, J. L. (1990). The Purex Process. In W. W. Schulz, L. L. Burger & J. D Navratil (Eds), *Science and Technology of Tributyl Phosphate*, Vol. III. *Applications of Tributyl Phosphate in Nuclear Fuel Reprocessing* (pp. 8–37). Boca Raton, FL: CRC Press.

McLean, R. J. C., Fortin, D. & Brown, D. A. (1996). Microbial metal-binding mechanisms and their relation to nuclear waste disposal. *Canadian Journal of Microbiology, 42*, 392–400.

Means, J. L. & Alexander, C. A. (1981). The environmental biogeochemistry of chelating agents and recommendations for the disposal of chelated radioactive wastes. *Nuclear and Chemical Waste Management, 2*, 183–196.

Means, J. L., Crerar, D. A. & Duguid, J. O. (1978). Migration of radioactive wastes: radionuclide mobilisation by complexing agents. *Science, 200*, 1477–1479.

Milodowski, A. E., West, J. M., Pearce, J. M., Hyslop. E. K., Basham, I. R. & Hooker, P. J. (1990). Uranium mineralised microorganisms associated with uraniferous hydrocarbons in southwest Scotland. *Nature, 347*, 465–467.

Miyake, C., Kondo, T. & Imoto, S. (1986). Direct evidence of uranium (V) intermediates by electron spin resonance in photo- and electrolytic reduction processes of uranyl complexes in organic solutions. *Journal of Less Common Metals, 122*, 313–317.

Mohegheghi, A., Updegraff, D. M. & Goldhaber, M. B. (1994). The role of sulfate-reducing bacteria in the deposition of sedimentary uranium ores. *Geomicrobiology Journal, 4*, 153–173.

Morel, F. M. M. & Herring, J. G. (1993). *Principles and Applications of Aquatic Chemistry.* New York: John Wiley.

Moyes, L.N., Parkman, R. H., Charnock, J. M., Vaughan, D. J., Livens, F. R., Hughes, C. R. & Braithwaite, A. (2000). Uranium uptake from aqueous solution by interaction with goethite, lepidocrocite, muscovite and mackinawite: an X-ray absorption spectroscopy study. *Envronmental Science and Technology, 34*, 1062–1068.

Nagy, B., Gauthier-Lafaye, F., Holliger, P., Davis, D. W., Mossman, D. J., Leventhal, J. S., Rigali, M. J. & Parnall, J. (1991). Organic matter and containment of uranium and fissionogenic isotopes at the Oklo natural reactors. *Nature, 345*, 472–475.

Nash, K. L., Cleveland, J. M. & Rees, T. F. (1988). Speciation patterns of actinides in natural waters: a laboratory investigation. *Journal of Environmental Radioactivity, 7*, 131–157.

Nash, K. L., Jensen, M. P. & Schmidt, M. A. (1998). Actinide immobilisation in the subsurface environment by in situ treatment with a hydrolytically unstable organophosphorus complexant: uranyl uptake by calcium phytate. *Journal of Alloys and Compounds, 271-273*, 257–261.

Nelson, D. M., Penrose, K. A. & Penrose, W. R. (1989). Oxidation states of plutonium in carbonate-rich natural waters. *Journal of Environmental Radioactivity, 9*, 189–198.

Nesmeyanova, M. A., Tsfasman, I. M., Karamyshev, A. L. & Suzina, N. E. (1991). Secretion of the overproduced periplasmic *phoA* protein into the medium and accumulation of its precursor in *phoA*-transformed *Escherichia coli* strains: involvement of outer membrane vesicles. *World Journal of Microbiology and Biotechnology, 7*, 394–406.

Nortemann, B. (1999). Biodegradation of EDTA. *Applied Microbiology and Biotechnology, 51*, 751–759.

Omar, N. B., Merroun, M.T., Gonzalez-Munoz, M.T. & Arias, J. M. (1996). Brewery yeast as a biosorbent for uranium. *Journal of Applied Bacteriology, 81*, 283–287.

Oversby, V. M. (1987). Important radionuclides in high level waste disposal: determination using a comparison of the U.S. EPA and NRC regulations. *Nuclear and Chemical Waste Management, 7*, 149–161.

Palumbo, A. V., Lee, S. Y. & Boerman, P (1994). Effect of media composition on EDTA degradation by *Agrobacterium* sp. *Applied Biochemistry and Biotechnology, 45*, 811–822.

Paterson-Beedle, M., Nott, K. P., Macaskie, L. E. & Hall, L. D. (2001). Study of biofilm in a packed bed reactor by three dimensional magnetic resonance imaging. *Methods in Enzymology, 337*, 285–305.

Penrose, W. R., Polzer, W. L., Essington, E. H., Nelson, D. M. & Orlandini, K. A. (1990). Mobility of americium and plutonium through a shallow aquifer in a semiarid region. *Environmental Science and Technology, 24*, 228–234.

Peretrukhin, V. F., Khizniak, T., Lyalikova, N. N. & German, K. E. (1996). Biosorption of technetium-99 and some actinides by bottom sediments of Lake Belsso Kosino of the Moscow region. *Radiochemistry, 38*, 440–443.

Pümpel, T., Ebner, C., Pernfuss, B., Schinner, F., Diels, L., Keszthelyi, A., Macaskie, L. E., Tsezos, M. & Wouters, H. (2001). Removal of nickel from plating rinsing water by a moving-bed sandfilter inoculated with metal sorbing and precipitating bacteria. *Hydrometallurgy, 59*, 383–393.

Premuzic, E. T., Lin, M. S., Jin, J. Z. & Hamilton, K. (1997). Geothermal waste treatment biotechnology. *Energy Sources, 19*, 9–17.

Reed, D. T., Wygmans, D. G., Aase, S. B. & Banaszak, J. E. (1998). Reduction of Np(VI) and Pu(VI) by organic chelating agents. *Radiochimica Acta, 82*, 109–114.

Reed, D. T., Vojta, Y., Quinn, J. W., Richmann, M. K. (1999). Radiotoxicity of plutonium in NTA-degrading *Chelatobacter heintzii* cell suspensions. *Biodegradation, 10*, 251–260.

Riley, R. G., Zachara, J. M. & Wobber, F. J. (1992). Chemical contaminants on DOE lands and selection of contaminant mixtures for subsurface science research. DOE/ER-0547T, Office of Energy Research, US Department of Energy, Washington DC.

Riordan, C., Bustard, M., Putt, R. & McHale, A. P. (1997). Removal of uranium from solution using residual brewery yeast: combined biosorption and bioprecipitation. *Biotechnology Letters, 19*, 385–387.

Rusin, P. A., Brainard, J. R., Strietelmeier, B. A., Tait, C. D., Ekberg, S. A., Palmer, P. D., Newton, T. W. & Clark, D. L. (1994). Solubilisation of plutonium hydrous oxide by iron-reducing bacteria. *Environmental Science and Technology, 28*, 1686–1690.

Sato, T., Murakami, T., Yanase, N., Isobe, H., Payne, T. E., Airey, P. L. (1997) Iron nodules scavenging uranium from groundwater. *Environmental Science and Technology, 31*, 2854–2858.

Schultze-Lam, S., Fortin, D., Davis, B. S. & Beveridge, T. J. (1996). Mineralization of bacterial surfaces. *Chemical Geology, 132*, 171–181.

Silva, R. J. & Nitsche, H. (1995). Actinide Environmental Chemistry. *Radiochimica Acta, 70/71*, 377–396.

Spinks, J. W. T. & Woods, R. J. (1976). *An Introduction to Radiation Chemistry*. New York: John Wiley.

Strachan, L., Jeong, B. C. & Macaskie, L. E. (1991). Radiotolerance of phosphatases of a *Citrobacter* sp.: potential for the use of this organism in the treatment of wastes containing radiotoxic actinides. *Proceedings of the 201st Meeting of the American Chemistry Society, 31*, (pp. 128–131). American Chemical Society, Washington DC.

Strain, S. M., Fesick, S. W. & Armitage, I. M. (1983a). Structure and metal binding properties of lipopolysaccharides from heptoseless mutants of *Escherichia coli* studied by ^{13}C and ^{31}P nuclear magnetic resonance. *Journal of Biological Chemistry, 258*, 13466–13477

Strain, S. M., Fesick, S. W. & Armitage, I. M. (1983b). Characterisation of lipopolysaccharide from a heptoseless mutant of *Escherichia coli* by carbon 13 nuclear magnetic resonance. *Journal of Biological Chemistry, 258*, 2906–2910.

Strandberg, G. W. & Arnold, W. D. (1988). Microbial accumulation of neptunium. *Journal of Industrial Microbiology, 3*, 329–331.

Strandberg, G. W., Shumate II, S. E. & Parrott, J. R. (1981). Microbial cells as biosorbents for heavy metals; accumulation of uranium by *Saccharomyces cerevisiae* and *Pseudomonas aeruginosa. Applied and Environmental Microbiology, 41*, 237–245.

Suzuki, Y., Banfield, J. F. (1999). Geomicrobiology of uranium. In: P. C. Burns & R. Finch (Eds), *Uranium: Mineralogy, Geochemistry and the Environment* (pp. 393–432). *Reviews in Mineralogy, 38*. Washington DC: Mineralogical Society of America

Suzumura, M. & Kamatani, A. (1995). Mineralisation of inositol hexaphosphate in aerobic and anaerobic marine sediments – implications for the phosphorus cycle. *Geochimica et Cosmochimica Acta, 59*, 1021–1026.

Swanson, J. L. (1990). Purex process flowsheets. In W. W. Schulz, L. L. Burger & J. D. Navratil (Eds), *Science and Technology of Tributyl Phosphate*, Vol. III: *Applications of Tributyl Phosphate in Nuclear Fuel Reprocessing* (pp. 55–72). Boca Raton, FA: CRC Press

Taghavi, S., Mergeay, M., Nies, D. & Van der Lelie, D (1997). *Alcaligenes eutrophus* as a model system for bacterial interactions with heavy metals in the environment. *Research in Microbiology, 6*, 536–555.

Tengerdy, R. P., Hollo, J. E. & Toth, J. (1981) Denitrification and removal of heavy metals from waste water by immobilised microorganisms. *Applied Biochemistry and Bioetchnology, 6*, 3–13.

Thomas, R. A. P. & Macaskie, L. E. (1996). Biodegradation of tributyl phosphate by naturally occurring microbial isolates and coupling to the removal of uranium from aqueous solution. *Environmental Science and Technology, 30*, 2371–2375.

Thomas, R. A. P. & Macaskie, L. E. (1998). The effect of growth conditions on the biodegradation of tributyl phosphate and potential for the remediation of acid mine drainage waters by a naturally occurring mixed microbial culture. *Applied Microbiology and Biotechnology, 49*, 202–209.

Thomas, R. A. P., Lawlor, K., Bailey, M., Macaskie, L. E. (1998) Biodegradation of metal-EDTA complexes by an enriched microbial population. *Applied Environmental Microbiology, 64*, 1319–1322.

Thomas, R. A. P., Beswick, A. J., Basnakova, G., Moller, R. & Macaskie, L. E. (2000). Growth of naturally occurring microbial isolates in metal-citrate medium and bioremediation of metal-citrate wastes. *Journal of Chemical Technology and Biotechnology, 75*, 187–195.

Tolley, M. R., Strachan, L. F. & Macaskie, L. E. (1995). Lanthanum accumulation from acidic solutions using *Citrobacter* sp. immobilised in a flow-through bioreactor. *Journal of Industrial Microbiology, 14*, 271–280.

Toth, L. M., Bell, J. T. & Friedman, H. A. (1980). Photochemistry of the actinides. In: J. D Navratil & W. W. Schulz (Eds), *Actinide separations. American Chemical Society Symposium Series, 117*, (pp. 253–266). Washington DC: American Chemical Society.

Trabalka, J. R. & Garten, C. T. Jr. (1983). Behaviour of the long-lived synthetic elements and their natural analogues in food chains. In J. T. Lett, U. K. Ehman & A. B. Cox (Eds), *Advances in Radiation Biology* (pp. 68–73). (Vol. 10). London, UK: Academic Press.

Tsezos, M. & Keller, D. M. (1983). Adsorption of radium-226 by biological origin adsorbents. *Biotechnology and Bioengineering, 25*, 201–215.

Tsezos, M., Baird, M. H. I. & Shemilt, L. W. (1987a). The use of immobilised biomass to remove and recover radium from Elliott Lake uranium tailing streams. *Hydrometallurgy, 17*, 357–368.

Tsezos, M., Baird, M. H. I. & Shemilt, L. W. (1987b). The elution of radium adsorbed by microbial biomass. *Chemical Engineering Journal, B34*, 57–64.

Tsezos, M., Baird, M. H. I. & Shemilt, L. W. (1987c). The kinetics of radium biosorption. *Chemical Engineering Journal, 33*, B35–B41.

Tsezos, M., McCready, R. G. L. & Bell, J .P. (1989). The continuous recovery of uranium from biologically leached solutions using immobilised biomass. *Biotechnology and Bioengineering, 34*, 10–17.

Tucker, M. D., Barton, L. L & Thompson, B. M. (1998). Removal of U and Mo from water by immobilised *Desulfovibrio desulfuricans* in column reactors. *Biotechnology and Bioengeering, 60*, 90–96.

Venkateswaran, A., McFarlan, S. C., Ghosal, D., Minton, K. W., Vasilenko, A. & Makarova, K., (2000). Physiologic determinants of radiation resistance in *Deinococcus radiodurans. Applied and Environmental Microbiology, 66*, 2620–2626.

Volesky, B. (1994). Advances in biosorption of metals: selection of biomass types. *FEMS Microbiology Reviews, 14*, 281–302.

Volesky, B. & Holan, Z. R. (1995). Biosorption of heavy metals. *Biotechnology Progress, 11*, 235–250.

Wackett, L. P. & Daly, M. J. (2000). Physiologic determinants of radiation resistance in *Deinococcus radiodurans. Applied and Environmental Microbiology, 66*, 2620–2626.

Watson, J. H. P. & Ellwood, D. C. (1994). Biomagnetic separation and extraction process for heavy metals from solution. *Minerals Engineering, 7*, 1017–1028.

Weigel., F., Katz, J. J. & Seaborg, G. T. (1986). Plutonium. In J. J. Katz, G. T. Seabourg & L. R. Morss (Eds), *The Chemistry of The Actinide Elements* (2nd ed), (pp. 499–886). London UK: Chapman and Hall.

Wersin, P., Hochella Jr., M. F., Persson, G., Redden, G., Leckie, J. O. & Harris, D. W. (1994). Interaction between aqueous uranium (VI) and sulfide minerals: spectroscopic evidence for sorption and reduction. *Geochimica et Cosmochimica. Acta, 58*, 2829–2843.

Wharton, M. J., Atkins, B., Charnock, J. M, Livens, F. R., Pattrick, R. A. D. & Collison, D. (2000). An X-ray absorption study of the coprecipitation of Tc and Re with mackinawite (FeS). *Applied Geochemistry, 15*, 347–354.

Witschel, M., Nagel, S. & Egli, T. (1997). Identification and characterisation of the two enzyme system catalysing the oxidation of EDTA in the EDTA-degrading bacterial strain DSM 9103. *Journal of Bacteriology, 179*, 6937–6943.

Witschel, M., Egli, T., Zehnder, A. J. B., Wehrli, E. & Spycher, M. (1999). Transport of EDTA into cells of the EDTA-degrading bacterial strain DSM 9103. *Microbiology, 145*, 973–983.

White, C. & Gadd, G. M. (1996). Mixed sulfate-reducing bacterial cultures for bioprecipitation of toxic metals: factorial and response-surface analysis of the effects of dilution rate, sulfate and substrate concentration. *Microbiology, 142*, 2197–2205.

White, C., Wilkinson, S. C. & Gadd, G. M. (1995). The role of microorganisms in biosorption of toxic metals and radionuclides. *International Biodeterioration and Biodegradation, 35*, 17–40.

White, C., Sharman, A. K. & Gadd, G. M. (1998). An integrated microbial process for the bioremediation of soil contaminated with toxic metals. *Nature Biotechnology, 16*, 572–575.

White, V. E. & Knowles, C. J. (2000). Effect of metal complexation on the bioavailability of nitrilotriacetic acid to *Chelatobacter heintzii* ATCC 29600. *Archives of Microbiology, 173*, 373–382.

Xun, L. Y., Reeder, R. B., Plymale, A. E., Girvin, D. C. & Bolton, H. Jr (1996). Degradation of metal-nitrilotriacetate complexes by nitrilotriacetate monooxygenase. *Environmental Science and Technology, 30*, 1752–1755.

Yakubu, N. A. & Dudeney, A. W. L. (1986). Biosorption of uranium with *Aspergillus niger*. In H. Eccles & S. Hunt (Eds), *Immobilisation of Ions by Bio-sorption* (pp. 183–200). Chichester, UK: Ellis Horwod.

Yang, J. & Volesky, B. (1999a). Biosorption of uranium on *Sargassum* biomass. *Water Research, 33*, 3357–3363.

Yang, J. & Volesky, B (1999b). Modelling uranium-proton ion exchange in biosorption. *Environmental Science and Technology, 33*, 4079–4085.

Yanke, L. J., Bryant, R. D. & Laishley, E. J. (1995). Hydrogenase I of *Clostridium pasteurianum* functions as a novel selenite reductase. *Anaerobe, 1*, 61–67.

Yong, P. & Macaskie, L. E. (1995a). Enhancement of uranium bioaccumulation by a *Citrobacter* sp. via enzymically-mediated growth of polycrystalline $NH_4UO_2PO_4$. *Journal of Chemical Technology and Biotechnology, 64*, 89–95.

Yong, P. & Macaskie, L. E. (1995b). Removal of the tetravalent actinide thorium from solution by a biocatalytic system. *Journal of Chemical Technology and Biotechnology, 64*, 101–18.

Yong, P. & Macaskie, L. E. (1997a). Removal of lanthanum, uranium and thorium from the citrate complexes by immobilised cells of *Citrobacter* sp. in a flow through reactor: implications for the decontamination of solutions containing plutonium. *Biotechnology Letters, 19*, 251–255.

Yong, P. & Macaskie, L. E. (1997b). Effect of substrate concentration and nitrate inhibition on product release and heavy metal removal by a *Citrobacter* sp. *Biotechnology and Bioengineering, 55*, 821–830.

Yong, P. & Macaskie, L. E. (1998). Bioaccumulation of lanthanum, uranium and thorium and use of a model system to develop a method for the biologically mediated removal of plutonium from solution. *Journal of Chemical Technology and Biotechnology, 71*, 15–26.

Yong, P. & Macaskie, L. E. (1999). The role of sulfate as a competitive inhibitor of enzymatically-mediated heavy metal uptake by *Citrobacter* sp.: implications in the bioremediation of acid mine drainage water using biogenic phosphate precipitant. *Journal of Chemical Technology and Biotechnology, 74*, 1149–1156.

Miranda J. Keith-Roach and Francis R. Livens (Editors)
© 2002 Elsevier Science Ltd. All rights reserved

Chapter 13

Microbial interactions with radionuclides – summary and future perspectives

Miranda J. Keith-Roach[a], Francis R. Livens[b]

[a] *Department of Nuclear Safety Research, Risø National Laboratory, P.O. Box 49, 4000 Roskilde, Denmark*

[b] *Centre for Radiochemistry Research, Department of Chemistry, University of Manchester, Oxford Road, Manchester M13 9PL, UK*

1. Introduction

In this volume, we have attempted to describe the current understanding of microbial interactions with radionuclides, the techniques which represent the state of the art in the study of these interactions in complex, heterogeneous systems, and provide diverse examples of current, topical research. The breadth and depth of this are quite remarkable, given that the subject is relatively new, and was first addressed specifically in 1998 (1st International Conference on Bacterial:Metal/Radionuclide Interactions, FZR Rossendorf, Dresden, Germany). Indeed, one of the major strengths of the field as a whole is its cross-disciplinary nature, which has prompted research using different approaches and techniques, and on different levels of detail. This has created a body of information with wide applicability, including, for example, the identification of biotechnologically exploitable species from environmental processes, and identification of specific biochemical processes which may help explain environmental observations. Understanding microorganisms' adaptability and ability to survive, especially in hostile environments such as highly polluted sites and waste repositories, is also of fundamental importance. Overall, this diverse field has already achieved a great deal, generating information and technologies that are of real practical benefit in dealing with nuclear wastes, from removal at source to the disposal of low-, intermediate- and high-level wastes, remediation of contaminated land and understanding the behaviour, transport and fate of environmental contamination.

However, while scientific understanding in this area is strong and provides an excellent basis for further work, the range of research described within this book also illustrates that gaining a fundamental understanding of microbial interactions with radionuclides is not a simple task. This is particularly true if accurate predictions are to be made of the extent to which key radionuclides will be affected within a variety of natural and engineered environments. There are many different biochemical and physiological processes which can allow microorganisms to interact with radionuclides, both directly and indirectly, and these are dependent both on the properties of the chemical species and the microbial consortia involved. Since the physicochemical environment affects the chemistry of the radionuclides, the microbial community and, particularly, the interplay between them, it is essential to adopt an integrated approach. Studying geochemical controls on the behaviour of radionuclides in the natural environment will always be difficult because of the generally very low contaminant concentrations, and including the specific effects of microorganisms in and alongside the geochemical processes presents a very significant additional challenge. Meeting this is exacerbated by the sensitivity of microbes to change, with community structure being rapidly and irreversibly affected by physical perturbation as well as chemical change, and the predominance of unculturable organisms in the environment, so it is difficult to characterise and work with environmentally realistic communities in the laboratory. We have read about attempts to gain an understanding of the interrelationships between microbiological activity, radioactivity and the broader environment. While very significant progress has been made, important issues remain unresolved and problems unsolved, so here we synthesise what has been achieved and identify future aims and priorities.

2. Discussion

Natural microbial communities: diversity and stability

Before the effects of any microbial community or culture can be assessed or predicted, its fundamental diversity and robustness must be understood, as discussed in Chapter 1 of this volume. The relatively recent development of techniques such as PLFA, 16S RNA and DGGE analysis has provided immensely useful probes into community structure, and 16S RNA analysis has allowed an unprecedented examination of genetic diversity (see Chapters 2 and 8, this volume). The remarkable diversity of microbes, with an estimated 10^9 prokaryotic species (Dykhuizen, 1998), indicates their adaptability and, as described in Chapters 8, 9 and 10 of this volume, there are abundant communities even within highly contaminated or extreme environments, as well as dormant organisms waiting for the onset of life-supporting conditions in environments such as deep hard rock or salt crystals (Vreeland et al., 2000). As microbial life adapts to polluted conditions, mechanisms evolve to protect the species from radioactive or heavy metal contaminants. For example, the microdiverse strains of *Acidithiobacillus ferrooxidans* found in uranium waste heaps demonstrate different tolerances to uranium (see Chapter 8, this volume). The strains found in the more polluted wastes limit U accumulation to levels below the lethal limit, whereas the strains found in less polluted wastes have not needed to adapt to the same stresses and thus have a lower tolerance. As described in Chapter 10 of this volume, new species are regularly identified in extreme environments, and their potential influences on pollutant behaviour are unclear.

The interactions of microbes with their environment, the degree to which they affect the chemistry of the system as whole, and the specific ways they interact with major and trace metals have been discussed in detail throughout this book. However, it is not clear whether the geochemical effects of a microbial community represent the sum of the activities of all component species, weighted according to the activity and abundance of each species, or the effects of certain key species in the community. This distinction is important since it relates to the idea of redundancy, with redundancy giving the community a certain robustness so that minor perturbations will not affect the principal microbially-driven processes. Only if a change is maintained for a longer time, or is extreme, will the functional diversity of the community start to decrease and the community's geochemical role consequently begin to change. In contrast, if the geochemical effects of a natural microbial consortium arise from the presence of a few key species, then removal of one or more of these may well have a significant and relatively prompt effect on geochemical behaviour.

Sedimentary diagenetic processes are an excellent example of changes in community structure in response to stress, in this case specifically from changes in terminal electron acceptor availability, which in turn cause distinct geochemical change. Oxygen depletion in certain systems during times of high biological activity also leads to seasonal changes in microbial community structure, and associated geochemical change. It is therefore important to understand community functional stability both spatially and temporally.

While the function of a natural microbial community, defined in terms of effects on pH, redox status and concentrations of metabolites, may be measured relatively easily, it is more difficult to isolate the effects of an individual component species or group of species. This problem is accentuated by the difficulty of isolating and growing in culture most of the species which make up the natural community. However, it is possible to exploit individual species and mixed cultures which can be grown in the laboratory, or on a larger scale in bioreactors, in order to determine their effects. Using radiolabelling in conjunction with these techniques may give extra precision in identifying the species or functional groups involved in key metabolic processes (see Chapter 2, this volume). While PLFA, 16S RNA and DGGE analysis have reduced the problems of non-cultivability in terms of examining natural communities, they cannot directly help ascertain the role of different species within the overall function of the community. However, comparing the DNA of non-cultivable species with that of cultivable species, especially the functional genes, should help determine their role. Information on genetic function and diversity is increasing at an extremely rapid rate, and with this should come the ability to deduce the properties of non-cultivable species. Also, these techniques can yield information in the field on the effects of these microorganisms in changing or extreme environments, where the changes in the community can be related to the stresses they receive, or the resultant geochemical change.

Environmental radiochemistry and the effects of microbial processes

The behaviour of radionuclides in the environment has been studied more or less since the onset of the nuclear programme, largely within a geochemical framework. This means that while microbes have only occasionally been the main focus in these studies, their effects have, to some extent, been taken into account implicitly. More recently, field studies have started considering microbial effects explicitly, and this should develop further, aided by the new techniques for characterising microbial communities. However, the research undertaken so far in this area gives us a good insight into the behaviour of key radionuclides within different systems, especially those that are currently contaminated with radioisotopes. It is not our wish to repeat the reviews of transuranium element biogeochemistry in Chapters 3 and 4 of this volume, or the discussions of of the specific interactions between radionuclides and microbial products and processes in Chapters 3, 6 and 11 here. Instead we shall focus on the main themes which emerge. First, it is important to note that only in special circumstances will there be a high enough radionuclide concentration (e.g. in effluent streams, waste heaps) or external radiation field (e.g. in a HLW repository) to have significant direct effects on the microbial community. The presence of most radionuclides at the low (below micromolar) mass concentrations in which they often occur in the environment is unlikely to have much effect on microbial viability.

Research efforts have generally focused on the most radiologically important radionuclides which, because of the timescales for which they will persist, are frequently long-lived. Uranium and its decay products have received the most attention because it is primordial and, in terms of mass and volume if not activity, wastes from uranium mining and extraction represent a significant problem. We have read quite a lot about uranium here, and it is certainly the best understood radioelement in terms of biogeochemistry and biotechnology. Plutonium is also well studied, being of environmental and political significance due to its importance in both civil and military nuclear programmes. Technetium has more recently become a focal radionuclide because of increased releases from Sellafield in the 1990s and groundwater contamination at sites in the USA, and because of its high solubility in environmental waters. Neptunium and americium have probably received the least attention of these radionuclides, although neptunium has received greater attention more recently, as interest in very long-lived radionuclides grows and low-level measurement has become easier with the development of heavy element mass spectrometries.

In some of this book, the basic physicochemical conditions of a site are viewed almost as a 'baseline' on which biogeochemical change is superimposed. However, it is clear that the nature of the radioelement is fundamental to its fate within any given conditions, since biogeochemical influence on an element's mobility can only be exerted within the context of the element's chemistry. So, radionuclides such as Tc and U, which are soluble in their oxidised forms and particle-reactive in their reduced forms, are potentially affected by biogeochemical change to a much greater degree than those such as Pu, which have a relatively high affinity for particulate matter in all their environmentally-accessible oxidation states. Still, variations in the oxidation state of Pu result in observed distribution coeffecients varying by 100- to 1000-fold, and so biogeochemical change can lead to the leaching and transport of significant quantities of even relatively particle-reactive species. This has particular significance in more highly contaminated environments and waste repositories.

The physicochemical factors that are generally of most importance in determining whether radionuclides are mobile or sorbed to particulate matter are the redox conditions, the content and nature of organic matter (see Chapter 5, this volume), the solid/solution ratio, ionic strength and inorganic ligand concentrations. All of these factors interact; for example, a soluble, organic-bound radionuclide will be removed from solution as river water mixes with seawater, since organic molecules flocculate in high ionic strength waters. Processes of this type are encountered in the global cycling of uranium and contribute to the importance of estuarine and near-shore areas as temporary sinks for uranium, along with the deposition of inorganic particles and the reduction of soluble U(VI) in anoxic sediments (see Chapter 4, this volume). These mechanisms are all, at least qualitatively, directly relevant to the other actinide elements. The very low concentrations present make coherent study of the global cycling of transuranics almost impossible, but the readier reduction of Pu and Am and their general environmental behaviour suggest that they would be retained in the particulate phase to a greater extent than U. However, some localised, contaminated areas are representative of systems which may be important sinks, so that this part of the overall cycle can be examined in more detail. For example, the near-shore and estuarine environments in the Irish Sea contain 'globally significant' quantities of radionuclides, and the intertidal salt marshes fringing the coast are notably contaminated. Here, seasonal cycling of Pu and Am has been observed, with microorganisms implicated in this cycle, and it is generally believed that the site now represents a net source of transuranic elements, whereas once it was a sink.

Moreover, the broader environment in which microorganisms and radionuclides interact (temperature, aeration, moisture content, amount and lability of organic matter, nature and reactivity of inorganic components) controls the development of the microbial community and influences the speciation of the radionuclides present. The behaviour of a radionuclide observed in practice will then be the resultant of the interactions of the different chemical species formed by the radionuclides with the microbial community and its products. Microbial metabolism can potentially affect radionuclide speciation and behaviour in the environment in several ways, both direct and indirect. Indirect effects include redox reactions induced by microbially driven changes in Eh, solubilisation or hydrolysis caused by changes in pH, complexation with microbial metabolites and products. Direct effects include the use of radionuclides as terminal electron acceptors, their oxidation as an energy resource, their entrainment in detoxification mechanisms, and their adsorption on to cell walls and extracellular polymeric material. Again, these processes do not occur in isolation, and the observable effect may be some combination of these. For example, as discussed in Chapter 5 of this volume, the mobility of reactive species depends on the ligands or solid phases to which they are bound. So, the affinity of humic substances for binding actinides and Tc in their reduced forms means that, in reducing systems of high organic content, the solubility and migration behaviour of the nuclide are very dependent on those of the humic substances. This also applies to other microbially produced species such as siderophores and complexing acids, which enhance the solubility of the reduced, low solubility species to a much greater degree than the oxidised species (see Chapter 6, this volume).

It is generally much easier to observe indirect than direct effects in situ, and so much more is known about these effects in the environment, as shown in Chapter 4 of this volume. It is very important that these processes are studied within natural and engineered systems, as well as in single organism or consortium microbiological and biochemical experiments, to assess the extent of their importance in more complex surroundings. Indeed, one of the most obvious indirect effects of microbial activity on radionuclide behaviour is the potential for accelerated release resulting from degradation of the waste containers or the wasteform. Work on uranium has reached the stage where the knowledge obtained within the laboratory can be used to manipulate natural microbial communities, giving workable remediation strategies. Moreover, as microbial interactions with uranium have become better understood, the whole basis of uranium geochemistry has been re-evaluated, with microbes implicated in processes which were formerly believed to be abiotic. These ideas are discussed further in the following section.

Remediation and biotechnology

In the 'real' environment, there seems to be a much better understanding of the effects of bacteria (or at least some groups) than of fungi (or viruses) and this translates to study of their interactions in the laboratory and to their potential biotechnological applications, where bacteria predominate. However, as described in Chapter 6 of this volume, fungi have several potential advantages in biotechnology, such as their tolerance of high pH conditions, high metal concentrations and the ability to produce high concentrations of complexing acids. Mechanisms by which organism(s) interact with radionuclides have been defined at the cellular level, for example, the biochemical mechanisms by which *Escherichia coli* and *Desulfovibrio desulfuricans* enzymatically reduce Tc(VII) were described in detail in Chapter 11 of this volume. Therefore, the origins of these microbial effects on radioelement chemistry are also understood at the molecular level. With this level of detail, it is possible to exploit microbial metabolic processes to achieve a predictable and potentially useful outcome in, for example, effluent treatment (see Chapters 11 and 12, this volume) or cleanup of contaminated land (see Chapters 6, 7 and 12, this volume). These processes are particularly attractive alternatives where chemical procedures are difficult, expensive, ineffective or hazardous, but in the case of most radionuclides, are not currently developed to a workable level. Desirable properties in one microorganism can, in principle, be transferred to another species through genetic engineering. Such properties may include radiation resistance, as found in bacteria of the genus *Deinococcus*, or the ability to function in high ionic strength solutions, which often arise from chemical processing and neutralisation.

However, genetic manipulation may not be necessary or even competitive with natural species, a particular attraction given public misgivings about genetic engineering. For example, we have seen that enzymatic Tc redution was carried out more efficiently and robustly in a bioreactor by the naturally occurring soil sulfate-reducing bacteria *Desulfovibrio desulfuricans* than mutant strains of *E. coli* (see Chapter 11, this volume). Uranium can be reduced enzymatically by a variety of species, processes in which some Fe(III) and SO_4^{2-} reducers, including *Geobacter* sp., gain energy for growth. *Geobacter* species are the dominant species in the Fe(III)-reducing community of the subsurface environment and are found in natural aquifer sediments. They couple U(VI) reduction to growth using acetate, one of the most common degradation intermediates in aquifer sediments. This was discussed specifically in Chapter 7 of this volume, and the application of acetate for stimulating *Geobacter* in situ was suggested as a potential remediation technique for polluted aquifers. Indeed, Chapter 7 neatly illustrates the many ways in which natural bacterial populations affect uranium, and the potential applications in bioremediation and biotechnology. The oxidation of U(IV) causes it to be leached from the solid phase and Fe(II)-oxidising species are implicated in and used for these purposes. One of these species, *Acidithiobacillus ferrooxidans*, is believed to be ubiquitous in uranium waste heaps and, as mentioned earlier, microdiversity has allowed more resistant strains to develop in the more toxic wastes (see Chapter 8, this volume). More generally, the bacterial groups present in waste piles are often those known to biotransform metals, highlighting the importance of knowing the structure of the microbial community in attempting to understand the effects the microorganisms will have on the pollutants. The research described in Chapters 7 and 8 really shows how laboratory and field studies can be complementary. Studying natural communities in contaminated areas not only shows the species that can survive, but also aids the search for genetic sequences which promote tolerance to a heavy metal or radionuclide.

Biochemical research has provided a wealth of information on the specific interactions of bacteria and radionuclides and the specific pathways have been identified in some cases. This aspect of the subject has been discussed in Chapters 11 and 12 of this volume, and has provided potential biotechnologies for the nuclear fuel cycle, as well as a fundamental understanding of some physiological processes. There are many ways in which microbes can be used to remove radionuclides from wastes, from simple biosorption to complex interactions involving the participation of more then one species, for example the use of *Shewanella putrefaciens*, to reduce Np, in conjunction with a *Citrobacter* sp. to precipitate the Np(IV) as a phosphate, as described in Chapter 12 of this volume.

In the course of laboratory studies, additional information has been gained which has possible implications for environmental processes. For example, minerals may form more readily in the presence of a microbial surface, which can act as a template, providing preformed nuclei to stimulate mineralisation. Biominerals are also often much more effective as sorbents for radionuclides than the abiotic equivalents because of the structure in which the mineral grows when attached to a microbial surface (see Chapter 12, this volume). In addition, while enzymatic reduction of a radionuclide may only occur in the absence of competitive chemical species, different processes resulting from the same organism's activity can affect the radionuclide indirectly. For example, Tc(VII) is reduced abiotically by the Fe(II) generated by Fe(III)-reducing bacteria (see Chapter 12, this volume), and Tc is precipitated as a sulfide during sulfate-reducing activity (see Chapter 6, this volume). These phenomena demonstrate the close and sometimes inseparable coexistence of microbial and geochemical processes.

Despite significant progress to date, there is still an almost limitless number of combinations of microbes which may be tested for their ability to take up radionuclides. This means that there is still much knowledge to gain. Industrially, these processes have not yet actually been used on a large scale, and so the challenge of actually bringing these processes into routine use also remains to be met.

Modelling

In the context of environmental radioactivity, an important aim is to reach the level of understanding where accurate predictions can be made, whether it be of the interactions of a single microbial species with a radionuclide on the basis of the organism's genetics and the radionuclide's chemistry, or the influence of a natural microbial community on the fate of a radionuclide in the field or in a repository. Given the range of potential interactions between microorganisms, their metabolic processes and products, the environment and radionuclides, it is very difficult to incorporate these effects reliably into the thermodynamically-based speciation models which underpin much work in the areas of radiological impact assessment or repository performance assessment. Since most radionuclide migration in the environment will take place in non-sterile conditions, it is likely that this will in practice often be non-thermodynamic, and, moreover, it is difficult to know whether the net effect of microbial activity will be to enhance or retard transport. The observation that a mixed culture isolated from a uranium waste pile could reduce U(VI) more robustly and 5 times as effectively as any of the individual species in the isolate (see Chapter 8, this volume) illustrates the challenge of understanding the interplay of species within communities, and these factors certainly increase the complexity of obtaining an accurate predictive understanding.

However, predictive modelling is a very important part of assessing the long-term stability and continuing function of waste repositories, and models have been developed for predicting microbial survival as well as their potential role in breakdown of the repository structure (see Chapters 9 and 10, this volume). Careful HLW repository engineering can prevent sustained microbial growth in close proximity to the primary waste container itself, and models indicate that life will not be supported in these systems over time. However, microbial effects on both the materials commonly used in radioactive waste disposal and the radionuclides themselves are certainly important in less extreme repository environments (see Chapter 9, this volume). The breakdown of physical barriers may aid migration, but depends on the physicochemical conditions of the site and the specific properties of the microbial community. The effects of humic substances, in whose production microorganisms are involved, on the migration of radionuclides have been modelled, with the model showing excellent agreement with laboratory experiments (see Chapter 5, this volume). Field predictions indicate that humic substances will enhance migration, with the extent depending on the kinetics of metal desorption from the complex and, with slow kinetics, the effect could be quite extreme. However, we do not generally have this level of understanding and remain unable to include the overall effects of the microbial populations in predictive, mechanistic models with confidence.

Since microbes have been demonstrated to affect the speciation and behaviour of radionuclides as well as the physical environment in which they are held, proceeding towards a better predictive understanding is a serious issue in the disposal of nuclear waste. Moreover, possession of this knowledge may yield safe technologies and remediation strategies as well as a much more accurate understanding of the biogeochemical behaviour of radionuclides in the environment. It is therefore important that this research field continues to advance towards this goal, aided by the techniques available and good cross-disciplinary collaboration and interest.

References

Dykhuizen, D. E. (1998) Santa Rosalia revisited: why are there so many species of bacteria? *Antonie van Leeuwenhoek International Journal of General and Molecular Microbiology, 73*, 25–33.

Vreeland, R H, Rosenzweig, W. D. & Powers, D. W. (2000). Isolation of a 250 million year old halotolerant bacterium from a primary salt crystal. *Nature, 407*, 897–900.

Glossary*

authigenic *Geology*. of mineral and rock constituents, formed in place at the same time as, or after, the formation of the rock of which they are a part.

biotin *Biochemistry*. $C_{10}H_{16}N_2O_3S$, a B vitamin that functions as a coenzyme and is ubiquitous in nature.

biotinylation *Molecular Biology*. the labelling of a probe with conjugated biotin, whose high affinity for antibiotin antibodies is exploited to mark the place at which the probe binds by indirect immunoassay.

bole the trunk of a tree, especially the highly marketable lower portion of a tree trunk.

chemoautotroph *Microbiology*. **1**. a microorganism that can derive the energy required for growth from oxidation of inorganic compounds such as hydrogen sulfide or ammonia. **2**. any autotrophic bacteria or protozoan that is not involved in the process of photosynthesis.

chemotroph *Biology*. any organism that creates its principal energy source by oxidising organic or inorganic compounds.

chimera *Biology*. an organism consisting of two or more genetically distinct cell types, produced as a result of mutation, transplantation, or grafting, the fusing of different embryos, or other similar processes. (From *Chimera*, a mythological Greek monster made up of the front of a lion, the middle of a goat, and the tail of a snake.)

chromosome *Cell Biology*. in a cell nucleus, a structure containing a molecule of DNA that transmits genetic information and is associated with RNA, histones, and nonhistone proteins. Each organism of a species normally has a characteristic number of chromosomes in its somatic cells; the normal number for humans is 46.

conservative elements *Oceanography*. those elements in seawater, such as chlorine, sodium, and magnesium, whose ratios to other such elements remain constant, regardless of variations in salinity.

denaturation of DNA *Biochemistry*. a process in which excessive heat causes the hydrogen bonds between the paired bases in the double helix to break, causing it to unravel into two separate strands.

diagenesis *Geology*. in the lithification of a sediment, the chemical, physical, and biological changes that occur after its deposition but before metamorphism and consolidation.

DNA *Biochemistry*. a nucleic acid that constitutes the genetic material of all cellular organisms and the DNA viruses; DNA replicates and controls through messenger RNA

* Harcourt's on-line science dictionary was used in compiling this glossary. The URL is http://www.harcourt.com/dictionary

the inheritable characteristics of all organisms. A molecule of DNA is made up of two parallel twisted chains of alternating units of phosphoric acid and deoxyribose, linked by crosspieces of the purine bases and the pyrimidine bases, resulting in a right-handed helical structure, that carries genetic information encoded in the sequence of the bases.

DNA denaturation *Biochemistry*. a process in which excessive heat causes the hydrogen bonds between the paired bases in the double helix to break, causing it to unravel into two separate strands.

DNA hybridization *Molecular Biology*. an analytical technique using radioisotope labelling to determine the degree of similarity of DNA base sequences in two species.

endolithic *Ecology*. of or related to a plant that grows within a stone, such as some lichens, algae, or fungi.

endonuclease *Enzymology*. an enzyme of the hydrolase class that catalyses the hydrolysis of interior bonds of ribonucleotide or deoxyribnucleotide chains, producing oligonucleotides or polynucleotides.

enrichment culture *Microbiology*. a selective nutrient medium designed to specifically promote the growth of one particular microorganism in a mixed population.

ephemeral plant *Botany*. a plant that germinates and grows rapidly, with its short life cycle often repeating several times in a growing season, such as desert plants that respond quickly following a brief rain.

epicellular *Cell Biology*. located at or on the surface of a cell.

euxinic *Hydrology*. **1**. describing an environment having restricted circulation and stagnant or anaerobic conditions. **2**. of or relating to the material deposited in such environments and the process of its deposition.

fugacious Falling or withering away very early.

fugacity *Thermodynamics*. a function f that is introduced as an effective substitute for pressure, to allow a real gas system to be considered by the same equations that apply to an ideal gas; the function is related to the molar Gibbs free energy by the equation $dG/d(\ln f) = RT$, where R is the gas constant and T is the temperature.

gene probe *Molecular Biology*. a biochemical that is labelled with radioactive isotopes or tagged in another way for ease of identification; used to identify or isolate a gene.

genome *Genetics*. the complete gene complement of an organism, contained in a set of chromosomes (in eukaryotes), in a single chromosome (in bacteria), or in a DNA or RNA molecule (in viruses).

genotype *Genetics*. the sum total of the genetic information contained in an organism; the genetic constitution of a cell or organism.

homologous *Genetics*. of two or more chromosomes or chromosome segments, having the same genetic loci and appearance.

hybridization of DNA *Molecular Biology*. an analytical technique using radioisotope labelling to determine the degree of similarity of DNA base sequences in two species.

hypha, *plural* **hyphae**. *Mycology*. a nonreproductive filament in fungi.

microaerobic *Microbiology*. **1**. of or related to an environment containing oxygen at a concentration lower than that present in the atmosphere.

microaerophilic *Microbiology*. of or related to an organism requiring oxygen for growth but at a concentration lower than that present in the atmosphere.

mRNA messenger RNA.

mycelium *Biology*. a matted mass of fungal filaments (hyphae) that forms the vegetative body of a fungus.

mycorrhiza *Ecology*. the symbiotic relationship between certain nonpathogenic or weakly pathogenic fungi and the living cells of roots of certain higher plants.

nitrification *Microbiology*. the process by which nitrogen in ammonia and organic compounds is oxidised to nitrites and nitrates by soil bacteria of the family Nitrobacteraceae.

oligonucleotide *Biochemistry*. the partial hydrolysis products of nucleic acids, consisting of a few nucleoside phosphate residues joined together.

oligotrophic *Hydrology*. deficient in plant nutrients and characterised by an abundance of dissolved oxygen in its lower layer.

PCR *Biotechnology*. polymerase chain reaction, a process for amplifying a DNA molecule by up to 10^6- to 10^9-fold; extremely important in biotechnology and in research.

pelagic *Oceanography*. of or relating to the open ocean, near the surface or in the middle depths, beyond the littoral zone and above the abyssal zone. *Hydrology*. relating to the deeper regions of a lake that are characterised by deposits of mud or ooze and by the absence of aquatic vegetation. *Ecology*. of or relating to aquatic organisms that live in the ocean, without direct dependence on the shore or bottom or on deep-sea sediment. (Going back to a Greek word meaning 'the sea'.)

phenotype *Genetics*. the appearance or other characteristics of an organism, resulting from the interaction of its genetic constitution with the environment, as opposed to its underlying hereditary determinants, or genotype.

phylogenetic *Evolution*. of or relating to the evolutionary relationships within and between groups.

primer DNA *Molecular Biology*. a short single-stranded DNA fragment that is required to initiate polymerisation of new DNA nucleotides.

propagule *Ecology*. the minimum number of individuals of a species required to colonise a habitable island. *Botany*. **1**. a seed, bud, or other offshoot capable of developing into a new and fully independent plant. **2**. a reproductive structure of brown algae.

reannealing *Molecular Biology*. the process of renaturing complementary single-stranded DNA molecules to yield duplex molecules; usually refers to strands that were originally separated prior to melting.

reassociation kinetics *Molecular Biology*. a measurement of the rate at which complementary DNA strands reassociate to form duplexes.

restriction endonuclease *Enzymology*. any of several enzymes that destroy foreign DNA molecules by cleavage at specific sites.

reverse transcriptase *Enzymology*. an RNA-dependent DNA polymerase, found in viruses, that catalyses the synthesis of DNA from deoxyribonucleoside 5′ triphosphates, using RNA as a template.

ribosome *Cell Biology*. the functional unit of protein synthesis, composed of a cellular complex of RNA and protein molecules and organised into two subunits.

rhizomorph *Mycology*. a discrete, root-like bundle of somatic fungal hyphae that functions as a single structure, carrying the fungus across non-favourable substrates.

RNA *Biochemistry*. a linear, usually single-stranded polymer of ribonucleotides, each containing the sugar ribose in association with a phosphate group and one of four nitrogenous bases: adenine, guanine, cytosine, or uracil. RNA is found in all living cells; in prokaryotic and eukaryotic cells, it encodes the information needed to synthesise proteins (i.e., it copies 'instructions' that it receives from DNA); in certain viruses, it serves as the genome. (An abbreviation for ribonucleic acid.)

streptavidin *Biochemistry*. a protein that is derived from *Streptomyces avidinii* and can be used in place of avidin, i.e. acts as a vitamin antagonist to biotin, binding tightly to it and thereby rendering it unavailable to the body (producing the syndrome known as biotin deficiency).

taxonomy *Systematics*. the theories and techniques of describing, naming, and classifying organisms.

transposon *Genetics*. specifically, a transposable element that, in addition to those genes necessary for transposition, carries genes with other functions, such as resistance to antibiotics.

type III reaction (immune complex mediated) *Immunology*. a hypersensitive reaction that is caused by the deposit of antigen-antibody complexes in tissue and blood vessels.

vital staining *Microbiology*. a method of colouring viable cells with certain biological dyes, such as methylene blue.

xeric *Ecology*. **1**. describing a location or habitat with very little moisture. **2**. describing an organism that lives in such an environment.

Index of Authors

Subject Index